Analytical Ultracentrifugation

Susumu Uchiyama • Fumio Arisaka •
Walter F. Stafford • Tom Laue
Editors

Analytical Ultracentrifugation

Instrumentation, Software, and Applications

Springer

Editors
Susumu Uchiyama
Graduate School of Engineering
Osaka University
Suita
Osaka, Japan

Fumio Arisaka
Life Science Research Center
Nihon University
Fujisawa
Kanagawa, Japan

Walter F. Stafford
Department of Neurology
Harvard Medical School
Boston
Massachusetts, USA

Tom Laue
BITC
University of New Hampshire
Durham
New Hampshire, USA

ISBN 978-4-431-55983-2 ISBN 978-4-431-55985-6 (eBook)
DOI 10.1007/978-4-431-55985-6

Library of Congress Control Number: 2016938115

Printed on acid-free paper

This Springer imprint is published by Springer Nature
The registered company is Springer Japan KK

Preface

Analytical ultracentrifugation (AUC) is one of most powerful and reliable techniques to study the biophysical behavior of solutes in solution. Since 1950, there have been more than 16,000 references to analytical ultracentrifugation, including 17 books on the topic. In the last few years, there have been steady advances made in hardware, software, and applications for AUC. In this book, therefore, we focus on (1) providing chapters that cover everything essential for beginners to the most advanced users; (2) updating the field on advances in hardware, software, and applications; and (3) encompassing AUC applications for nonbiological questions.

The first two chapters (Chaps. 1 and 2) in Part 1 provide minimum but essential theoretical and experimental explanations for studies using AUC. In Part 2, novel analytical ultracentrifuge instrumentation (Chap. 3) and detection systems (Chaps. 4 and 5) that give added dimensions to AUC are introduced, followed by three chapters describing computer programs for sedimentation data analysis (Chaps. 6, 7 and8). The final four chapters (Chaps. 9, 10, 11 and 12) in Part 2 introduce the theoretical and practical considerations for calculating hydrodynamic parameters, such as the sedimentation and diffusion coefficients, from three-dimensional molecular structure. The next three parts describe applications of AUC to specific research fields including material science (Part 3, Chaps. 13 and 14), membrane proteins (Chap. 15), protein–ligand interactions (Chap. 16), and polysaccharides (Chap. 18). Sedimentation analysis of high-salt-concentration solutions and the concomitant changes in chemical activity of biological macromolecules are introduced in Chap. 17. AUC for biopharmaceutical formulations, typically composed of proteins, buffers, salts, and additives such as sugars and surfactants, is introduced in three chapters of Part 5. First, the importance of AUC in the quantitation of oligomers and aggregates is introduced in Chap. 19. Next, Chap. 20 introduces practical approaches for the analysis of protein size distributions in biopharmaceuticals. Finally, methods for evaluating the intermolecular interactions that underlie protein behavior and govern protein aggregation by sedimentation equilibrium are presented in Chap. 21. Part 6 focuses on the AUC of high-concentration and complex systems, where hydrodynamic nonideality may influence sedimentation of macromolecules strongly. Presented in this part are recent advances in AUC detectors and developments in programs for AUC data analysis that will enable us to use AUC under these highly nonideal conditions. Theory (Chap. 22), data acquisition, and data analysis (Chaps. 23, 24 and 25) are comprehensively described. Finally, Chap. 26 in Part 7 provides an example of new application of AUC for studies of biological macromolecules.

We are convinced that AUC will remain one of the most powerful biophysical methods, and, with the advances described here, will continue to grow in terms of theory, instrumentation, and data analysis.

Osaka, Japan — Susumu Uchiyama
Fujisawa, Japan — Fumio Arisaka
Cambridge, MA, USA — Walter F. Stafford
Durham, NH, USA — Tom Laue

Contents

Part I
Introduction

Chapter 1
Important and Essential Theoretical Aspects of AUC

Susumu Uchiyama and Fumio Arisaka

Abstract Analytical ultracentrifugation (AUC) is a powerful method to reveal biophysical behavior of solute in solution. AUC has a long history and is based on well-established and concrete thermodynamic and hydrodynamic theory. AUC provides valuable parameters such as the sedimentation and diffusion coefficients, from which molar mass and information on hydrodynamic shape and solvation of the solute can be derived. Here, important and essential theoretical aspects of AUC are described.

Keywords Analytical ultracentrifugation • Sedimentation equilibrium • Sedimentation velocity • Theory

1.1 Introduction

This chapter is devoted to the basics of analytical ultracentrifugation. Efforts have been made to cover the minimum essentials and to make it qualitatively but precisely understandable. Two fundamentally important concepts in analytical ultracentrifugation are sedimentation and diffusion. Detailed theories can be found in the published literature (Fujita 1962; Cantor and Schimmel 1980; van Holde et al. 2005).

There are two modes of experiments, namely, sedimentation velocity (SV) and sedimentation equilibrium (SE). In SV experiments, both sedimentation and diffusion take place simultaneously. In SE experiments, sedimentation and diffusion are balanced and reach equilibrium.

S. Uchiyama (✉)
Department of Biotechnology, Graduate School of Engineering, Osaka University, 2-1 Yamadaoka, Suita, Osaka 565-0871, Japan

Department of Bioorganization Research, Okazaki Institute for Integrative Bioscience, 5-1 Higashiyama, Myodaiji-cho, Okazaki, Aichi 444-8787, Japan
e-mail: suchi@bio.eng.osaka-u.ac.jp

F. Arisaka
Life Science Center, College of Bioresource Science, Nihon University, 1866 Kameino, Fujisawa, Kanagawa 252-0880, Japan

S. Uchiyama et al. (eds.), *Analytical Ultracentrifugation*,
DOI 10.1007/978-4-431-55985-6_1

1.2 Principle for Sedimentation and Diffusion

1.2.1 Sedimentation

Suppose we suspend some homogeneous fine grains of sand in water and stir and leave it still (Fig. 1.1). The boundary between the air and the solution is called "meniscus." The sand will leave the meniscus and slowly sediment and the moving boundary will appear. Above the moving boundary, sand grains have sedimented down and are already gone. Below the boundary, the same concentration or the same number of grains per unit volume of sand is still there sedimenting with the same velocity. The speed of the moving boundary, v, is proportional to the acceleration of gravity, and the proportionality constant, s, will define the sedimentation coefficient of the sand particle.

$$v = sg \tag{1.1}$$

If all the sand particles have the same s-value, they sediment with the same velocity toward the bottom and the concentration at plateau region is kept constant. The particles that have reached at the bottom accumulate there.

Now, s is related to the mass, m, and specific volume, $\overline{v}$, of the particle and the density of water, ρ:

$$s = \frac{m\left(1 - \overline{v}\rho\right)}{f}, \tag{1.2}$$

where f is the frictional coefficient of the particle. The term $\overline{v}\rho$ is the excluded volume of the particle and $m\overline{v}\rho$ is the mass of the excluded water, leading to the

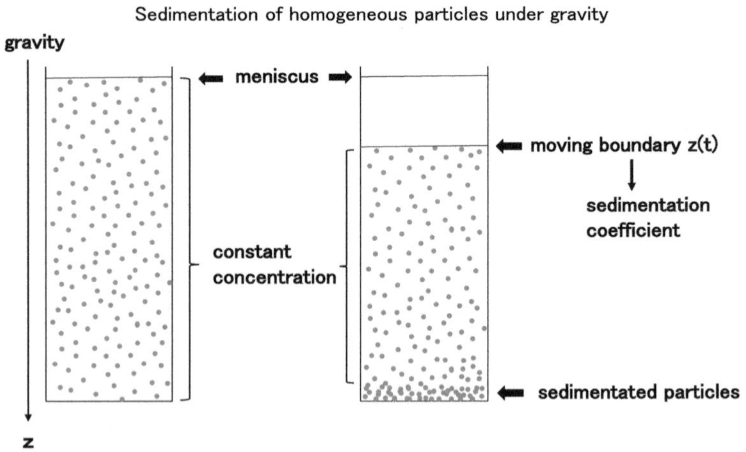

Fig. 1.1 Sedimentation of homogeneous grains in a cylinder under gravity

buoyancy. The term $m(1 - \overline{v}\rho)$ is called the buoyant mass. If the sand particles are not homogeneous, each size of the sand grain sediments with its own sedimentation coefficient, and we may observe the distribution of sedimentation coefficients.

Before invention of the analytical ultracentrifuge, Thé Svedberg measured the size distribution of colloidal gold which he was studying by the method as described above. In order to measure the position of the moving boundary precisely, he observed the boundary using a microscope. Eventually, Svedberg got interested in measuring the size of proteins which had been recognized to be very important in biological organisms. However, s-values of proteins are too small and do not sediment at all under the normal gravity due to the overwhelming diffusion. In order to let them sediment in spite of diffusion, the gravity has to be much increased, and he decided to utilize the centrifugal force of a centrifugation. Special devices, then, had to be developed in order to observe the moving boundary or the concentration gradient in the cell in a revolving rotor. Thus, Svedberg designed and constructed a prototype of analytical ultracentrifuge.

The velocity, v, of sedimentation of the moving boundary, r_b, is proportional to the centrifugal force, and the sedimentation coefficient is defined as the proportionality constant:

$$v = s r_b \omega^2, \tag{1.3}$$

where r_b is the position of the moving boundary from the center of revolution and ω is the angular velocity. It is noted that g in Eq. (1.1) is now replaced by $r\omega^2$, the acceleration of centrifugal force. The sedimentation coefficient is related to the mass of the particle, m, and the frictional coefficient, f:

$$s = \frac{m(1 - \overline{v}\rho)}{f} = \frac{M(1 - \overline{v}\rho)}{N_A f}, \tag{1.2a}$$

where N_A is the Avogadro's number, ρ is the density of the solvent, and M is the molar mass. s has the dimension of time and commonly used with the unit of S, where 1 S is 10^{-13}s. Simple classical mechanics treatment shows $\overline{v}$ to be a specific volume, but more rigorous treatment of the transport process by nonequilibrium thermodynamics shows that it is a partial specific volume or

$$\overline{v} = \left(\frac{\partial V}{\partial m}\right)_{T,p}, \tag{1.4}$$

where $\overline{v}$ is a volume increase of a solution of a large volume when 1 g of lyophilized protein or other solute molecules had been dissolved. If the concentration of the solute is low enough, it is close to the specific volume. From Eq. (1.3) or $(dr_b/dt) = s r_b \omega^2$,

$$\ln(r_b) = \ln(r_m) + s\omega^2 t \tag{1.5}$$

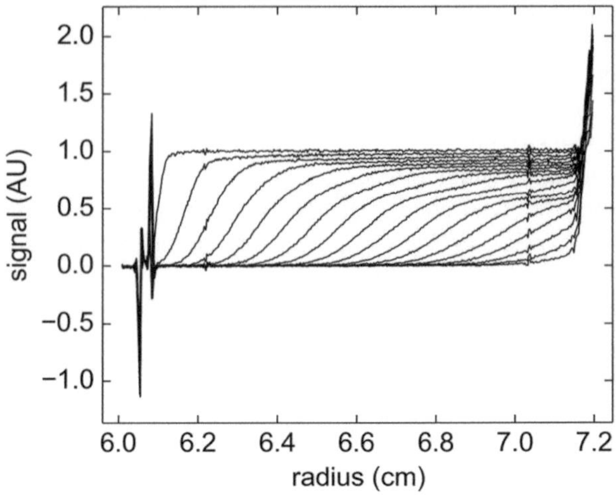

Fig. 1.2 Typical data of sedimenting boundary in sedimentation velocity experiment

so that

$$s = \ln\left(r_b/r_m\right) / \left(\omega^2 t\right),$$

where r_m denotes the position of the meniscus.

Svedberg noticed at the early stage of the development of his analytical ultracentrifuge (AUC) that the cell has to be sector shaped instead of rectangular, because the particle sediments in the direction of radius and if the cell is rectangular some particles will collide with the wall and the moving boundary will be distorted. The consequence of the sector-shaped cell is that the concentration of the solute at the plateau region, c_p, will gradually decrease due to the fact that the cross section of the flow of the solute will expand proportionally to the distance, r, from the center of revolution (Fig. 1.2):

$$C_p = C_0 e^{-2s\omega^2 t} \tag{1.6}$$

The analysis of moving boundary will be described more in detail when we introduce the Lamm equation which describes the time course of the concentration gradient in sedimentation velocity. r_b is then defined as the position of r, where the concentration is half that of the plateau region.

The value of sedimentation coefficient thus determined depends on the buffer conditions and temperature. The buffer condition changes the density and viscosity and the temperature mainly affecting the viscosity of water. In order to obtain the intrinsic physical parameter of the solute, the density and viscosity have to be corrected to reflect the experimental conditions.

It may be noted that in the above discussion, we are not looking at the behavior of each sedimenting molecule in solution but looking at the moving boundary or the concentration gradient at the boundary. All the information we extract concerning the characteristics of the solute molecules are contained in the shape and its change during time course of the moving boundary. In fact, the moving boundary contains abundant information concerning homogeneity/heterogeneity, sedimentation coefficients and their distribution, diffusion coefficient and interactions between solute molecule, non-ideality, etc. Measurement of moving boundary is similar to the frontal analysis of the size exclusion chromatography.

1.2.2 Diffusion

Sedimentation coefficient and frictional coefficient are related through Eq. (1.2a). Now, there is a simple relation between diffusion coefficient and frictional coefficient for ideal solution which is called the Einstein-Sutherland equation:

$$D = \frac{RT}{N_A f} \tag{1.7}$$

Replacing f in Eq. (1.2a) with f in Eq. (1.7) will give the Svedberg equation:

$$\frac{s}{D} = \frac{M(1 - \overline{v}\rho)}{RT} \tag{1.8}$$

This equation indicates that if we have the values for s and D, we can determine the molar mass, M, with the advance knowledge of $\overline{v}$ and ρ. The molar mass thus determined does not depend on the shape of the molecule as the equation implies.

In the case of non-ideality, D can be expressed as

$$D = \frac{RT}{N_A f}\left\{1 + C\frac{\partial \ln \gamma}{\partial C}\right\} \tag{1.9}$$

and the corresponding Svedberg equation is

$$\frac{s}{D} = \frac{M(1 - \overline{v}\rho)}{RT\left\{1 + C\frac{\partial \ln \gamma}{\partial C}\right\}}, \tag{1.10}$$

where γ is the activity coefficient of the solute.

Now, assume a cell of uniform cross section with infinite length and that a sharp concentration gradient or boundary is present at $x = 0$ at time $t = 0$ (Fig. 1.3). The time course of the change of concentration gradient can be predicted by the Fick's second law (diffusion equation):

$$\left(\frac{\partial C(x,t)}{\partial t}\right)_x = D\left(\frac{\partial^2 C(x,t)}{\partial x^2}\right)_t, \tag{1.11}$$

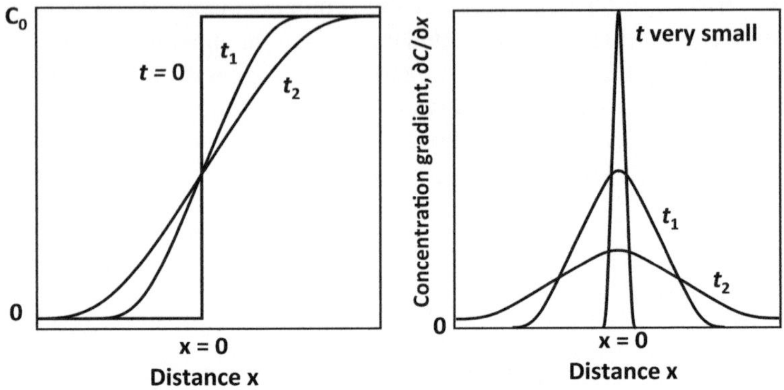

Fig. 1.3 Change of concentration gradient of solute due to diffusion

where D is the translational diffusion coefficient. Equation (1.6) can be solved with the initial condition and boundary condition. Under these conditions of free diffusion, the solution of Eq. (1.6) is

$$C(x,t) = \frac{C_0}{2}\left\{1 - \frac{2}{\sqrt{\pi}}\int_0^{x/2\sqrt{Dt}} e^{-y^2 dy}\right\} \tag{1.12}$$

and the derivative of $C(x,t)$ with respect to r is simply a Gaussian "error" curve:

$$\left(\frac{\partial C(x,t)}{\partial x}\right)_x = \frac{C_0}{2\sqrt{\pi Dt}} e^{-x^2/4Dt}, \tag{1.13}$$

where c_0 is the concentration of the solute at $t = 0$. We can get the gradient either by calculating from the experimental concentration gradient data or measure the gradient by traditional so-called schlieren optical system. The height of the bell-shaped curve, H, is given by

$$\left(\frac{\partial C(x,t)}{\partial x}\right)_{x=0} = \frac{C_0}{2\sqrt{\pi Dt}} = H \tag{1.14}$$

and

$$\left(\frac{C_0}{H}\right)^2 = 4\pi Dt \tag{1.15}$$

We can thus determine D by plotting $(C_0/H)^2$ with respect to t. There is a special cell, synthetic boundary cell, which can be used to form a sharp boundary to determine the diffusion coefficient by the method as described above. In current practice, both sedimentation coefficient and diffusion coefficient are determined by direct curve fitting to the Lamm equation solution to the raw data of sedimentation velocity.

1.3 Sedimentation Velocity

1.3.1 Lamm Equation

As soon as the solute molecules leave the meniscus and start to sediment, a concentration gradient will appear. In the concentration gradient (moving boundary), sedimentation and diffusion take place simultaneously. It was Ole Lamm, a Ph.D. student of Thé Svedberg, who reported a partial differential equation which describes the time course of the concentration gradient of the solute in sedimentation velocity (Lamm 1929):

$$\left(\frac{\partial C}{\partial t}\right)_r = -\frac{1}{r}\left\{\frac{\partial}{\partial r}\left[s\omega^2 r^2 C - Dr\left(\frac{\partial C}{\partial r}\right)_t\right]\right\} \tag{1.16}$$

This equation, called the Lamm equation after Ole Lamm, precisely describes the time course of the sedimentation together with diffusion. Detailed derivation of the Lamm equation is described elsewhere. Due to the consideration of the sector-shaped cell, the right-hand side of the equation is somewhat complicated, but, basically, it consists of two terms in the brackets [ ]. The first term has the coefficient s, and the second term contains D. The former describes sedimentation and the latter diffusion. In current methods of SV analysis, s and D are determined by nonlinear least squares curve fitting to the raw data.

Equation (1.16) assumes that s and D are constants (i.e., there is no hydrodynamic non-ideality) under the same conditions. The measured s- and D-value are usually corrected to a standard condition which is traditionally in water at 20 °C, $s_{20,w}$, $D_{20,w}$. If the actual measurements were made in a buffer solution at temperature T, $s_{T,b}$, (1.2a) and (1.7) contain the frictional coefficient, which is related to the viscosity of the solvent by the Stokes law:

$$f = 6\pi\eta R_s, \tag{1.17}$$

where R_s is the Stokes radius and has been worked out for particles of various shapes. In any event, it is directly proportional to the viscosity of the solvent. The measured $s_{T,b}$ and $D_{T,b}$ are, therefore, corrected for viscosity and the density of the solvent:

$$s_{20,w} = \frac{(1-\overline{v}\rho)_{20,w}}{(1-\overline{v}\rho)_{T,b}}\frac{\eta_{T,b}}{\eta_{20,w}}s_{T,b} \tag{1.18}$$

$$D_{20,w} = \frac{293.1}{T}\frac{\eta_{T,b}}{\eta_{20,w}}D_{T,b} \tag{1.19}$$

Correction for viscosity is mainly due to the temperature dependence of that of water. For example, the viscosity of water at 4°C is about 1.3 times larger than that of water at 20°C. The *s*- and *D*-values thus corrected are known to depend on the concentration of the solute and need to be extrapolated to zero concentration in order to obtain the real intrinsic physical parameter of the solute $s^0_{20,w}$. For common spherical soluble proteins, extrapolation to zero concentration may not be necessary if the concentration is below 1 mg/mL or so. However, care has to be taken for extremely elongated proteins such as triple-helical collagen or highly negatively or positively charged molecules such as nucleic acids. In fact, the s-value of nucleic acids has a much higher concentration dependence than proteins and has to be measured at very low concentration. Fortunately, nucleic acids, DNA and RNA, have much higher extinction coefficients than proteins, and, as a result, they can be measured at much lower concentration. Extinction coefficient of nucleic acids is about 20 times larger than those of proteins. Much less frequently, sedimentation coefficients may decrease as the concentration decreases. In such a case, subunit dissociation may be anticipated.

1.3.2 Relationship Between s and M

The molar mass, *M*, of the sedimenting particle can be estimated by Svedberg Eq. (1.18), which requires the *s*- and *D*-values. The *s*-value can be determined rather accurately with the error of one or two percent. On the contrary, *D*-values are more difficult to evaluate precisely, and they have been frequently obtained from DLS measurement and combined with the *s*-value from SV experiments. However, recent software, such as *c*(*s*) analysis in SEDANAL, SEDFIT, or ULTRASCAN, utilizes the relationship of *D* and f/f_0, called the scaling law:

$$D(s) = \frac{\sqrt{2}}{18\pi} kTs^{-1/2} \left(\eta (f/f_0)_w\right)^{-3/2} \left((1 - \overline{\upsilon}\rho)/\overline{\upsilon}\right)^{1/2} \tag{1.20}$$

The rationale of using this equation for analysis is discussed by Peter Schuck. In *c*(*s*) analysis, a common value for f/f_0 value is assumed, but there is another mode of analysis $c(s, f/f_0)$ in SEDFIT or 2DSA in ULTRASCAN, where *s* and f/f_0 are independently fitted for each molecular species. SEDANAL fits directly for *s* and *D* for each species with or without constraints relating the frictional coefficients of each component.

1.3.3 Molecular Shape and f/f_0

Molecular shape may be discussed based on the frictional ratio, f/f_0. The frictional ratio, obtained from the SV analysis, may be related to the molecular shape of the

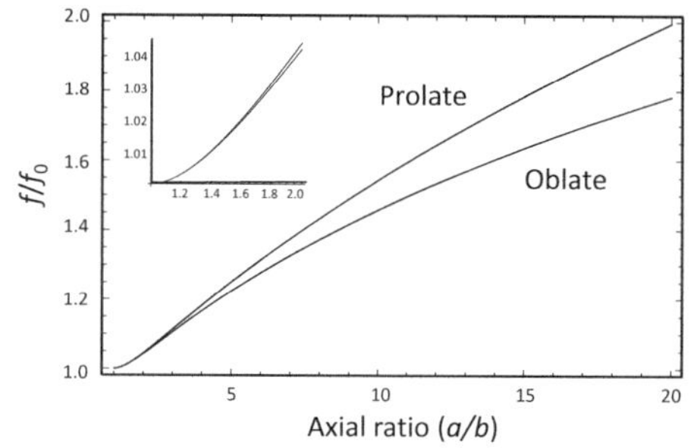

Prolate

$$f/f_0 = \frac{P^{-1/3}(P^2 - 1)^{1/2}}{\ln\,[P + (P^2 - 1)^{1/2}]}$$

Oblate

$$f/f_0 = \frac{(P^2 - 1)^{1/2}}{P^{2/3}\tan^{-1}[(P^2 - 1)^{1/2}]}$$

$$P = a/b \qquad (1.21)$$

Fig. 1.4 Relationship between axial ratio and frictional ratio

assumed prolate, oblate, or rod, through Perrin's equations (Fig. 1.4). As shown in Fig. 1.4, a small increase of f/f_0 increases the axial ratio quite a bit, but a large change in the axial ratio does not affect the f/f_0 very much, which is the rationale of using Eq.(1.21) independent of the molecular species. Note that spherical protein will give the f/f_0 of about 1.2 instead of 1.0, which is because f_0 is estimated for a protein without solvation, whereas the experimentally determined f includes solvation. We cannot discuss the molecular shape in detail but can decide if the molecule is close to sphere or elongated (or flattened). For example, native triple-helical collagen has a large f/f_0 value, larger than 2.

1.3.4 Sedimentation Coefficients Estimated from the X-ray Structure of the Proteins

If the atomic structure of a protein is known, one can estimate the hydrodynamic values including the sedimentation coefficient and diffusion constant. Although it is not so simple to predict the structure from the sedimentation coefficient, if a number of possible model structures are available, one could assess each structure by estimating each *s*-value and judge which model would fit the measured *s*-value (Rocco and Byron 2015).

1.4 Sedimentation Equilibrium

When solution is centrifuged at a relatively low speed as compared with that of the SV experiments, both sedimentation and diffusion contribute significantly to the concentration distribution. As a result, the moving boundary which we see in SV experiments will not be seen. Instead the concentration of the solute at the meniscus will decrease and that at the bottom will increase and eventually reach the equilibrium, when sedimentation and diffusion are balanced. The resultant concentration gradient will be used to estimate the molar mass:

$$\frac{d\ln C}{d\left(r^2\right)} = \frac{M\left(1-\overline{v}\rho\right)\omega^2}{2RT} \tag{1.22}$$

Equation (1.22) can be derived by assuming that the total chemical potential including centrifugal potential is common everywhere in the cell. It can be also derived from Lamm Eq. (1.16) by setting $\partial C/\partial t = 0$.

Integration of Eq. (1.22) gives

$$C(r) = C\left(r_a\right)\exp\left(M_{w,app}\left(1-\overline{v}\rho\right)\left(r^2-{r_a}^2\right)\right) + \text{baseline} \tag{1.23}$$

$$c(r) = \sum_{i=1}^{n} c_i\left(r_a\right)\exp\left(M_i\left(1-\overline{v}_i\rho\right)\left(r^2-r_a^2\right)/2RT\right) \tag{1.24}$$

In signal units:

$$S(r) = S_{\text{baseline}}(r) + \sum_{i=1}^{n} \epsilon_i c_i\left(r_a\right)\exp\left(M_i\left(1-\overline{v}_i\rho\right)\left(r^2-r_a^2\right)/2RT\right) \tag{1.25}$$

Currently, most programs utilize nonlinear fitting of the raw data to Eq. (1.25) to obtain the molar masses, M_i. Here, it should be noted that the fitting data from a mixture of molecules using the model of a single ideal component results in an "average" M that is closer to Mz (z-average molecular weight) than Mw but falls between these and is not equal to either. As in the case of SV experiment to obtain the molar mass, which utilizes Svedberg Eq. (1.8), the values of $\overline{v}$ and ρ are required. The partial specific volume $\overline{v}$ can be measured by precision densitometry using a densitometer, e.g., Anton Paar DMA5000, but in many cases the requirement of the amount of proteins, on the order of 10 mg, may not be realistic and the values are commonly calculated based on the weighted average of the values for amino acid residues as listed in Cohn and Edsall (1943). The protocol for calculation is implemented in SEDNTERP.

At early times, sedimentation equilibrium was thought to be unrealistic due to the prolonged time required for equilibration. It was van Holde and Baldwin (1958) who showed that the required time to reach equilibrium is proportional to the square of the solution column length and is inversely proportional to the diffusion constant:

$$t \approx 0.7\frac{\left(r_b-r_a\right)^2}{D}, \tag{1.26}$$

where r_b and r_a are, respectively, the distance of meniscus and the bottom from the center of rotation and D is the diffusion coefficient of the solvent. Since then, shorter columns have been used for SE experiments, and the time of measurement became more realistic. Experiment at lower temperature requires longer time to reach equilibrium. Traditionally, judgment of equilibration was made if three or four scans are superimposed. There are some programs with which one could decide if the system reached equilibrium, like HeteroAnalysis, which includes the program Match that uses a least squares fitting procedure to determine whether or not equilibrium has been achieved, available from Jim Cole. Also a calculator is available to predict the time to attain equilibration in ULTRASCAN and SEDFIT. It should be noted that the time to equilibrium may be seriously underestimated for reversibly associating systems.

It is the usual case that the mass average molar mass, $M_{w,app}$, has concentration dependence due to nonideal behavior of the solute and thus can be expressed using second virial coefficients, B_2, and third virial coefficients, D_2, as

$$\frac{1}{M_{w,app}} = \frac{1}{M_w}\left(1 + B_2C + D_2C^2 + \cdots\right) \tag{1.19}$$

At concentrations below 10 mg/mL, usually $M_{w,app}$ is linearly dependent on the concentration; thus molar mass at infinite dilution, M_w and B_2, can be estimated by fitting of $M_{w,app}$ at different concentrations.

When the solute is not homogeneous, it is shown that the obtained molar mass is that of the weight average:

$$M_w = \sum_i^n C_i M_i \Big/ \sum_i^n C_i \tag{1.20}$$

In the case of interacting systems, the association-dissociation equilibrium between or among species is achieved at any position, r; nonlinear fitting of the $C_{\text{total}}(r)$ gives equilibrium dissociation constants between or among species.

1.5 Conclusions

In this chapter, the basics of analytical ultracentrifugation including theoretical background and essential mathematical expression used for understanding SV-AUC and SE-AUC were introduced. More details will be described in the following chapters dealing different topics.

References

Cantor CR, Schimmel PR (1980) Biophysical chemistry, Part 2: Techniques for the study of biological structure and function. W. H. Freeman and Company, San Francisco

Cohn EJ, Edsall JT (1943) Proteins, amino acids, and peptides. Rheinhold, New York

Fujita H (1962) Mathematical theory of sedimentation analysis. Physical chemistry: a series of monographs. Academic Press, New York

Lamm O (1929) Die Differentialgleichung der Ultrazentrifugierung. Arkiv för matematik, astronomi och fysik 21B(2):1–4

Rocco M, Byron O (2015) Computing translational diffusion and sedimentation coefficients: an evaluation of experimental data and programs. Eur Biophys J 44:417–431

van Holde KE, Boldwin RL (1958) Rapid attainment of sedimentation equilibrium. J Phys Chem 62:734–743

van Holde KE, Curtis J, Pui Shing H (2005) Methods for the separation and characterization of macromolecules. In: Principles of physical biochemistry. Prentice Hall, New Jersey, pp 213–248

Chapter 2
Experimental Design and Practical Aspects

Fumio Arisaka and Susumu Uchiyama

Abstract This chapter summarizes basic experimental setup and precautions to be made for successful measurements in analytical ultracentrifugation. There are two types of ultracentrifuge, XL-A and XL-I from Beckman Coulter Co. Ltd. A new ultracentrifuge, CFA analytical ultracentrifuge, is described in Chap. 3.

Keywords Analytical ultracentrifugation • Sedimentation velocity • Sedimentation equilibrium • Rotor • Cell

2.1 Optical Systems in XL-A and XL-I

There are three currently available optical systems. Beckman Coulter XL-A has an absorbance optical system. XL-I has the same absorbance optical system as well as an interference optical system. Fluorescence detection system is also available as an extra attachment. In addition, two more optical systems in use in Germany, namely, multiwavelength absorbance systems, have emerged, which enable us to conduct a new analysis method using up to 500 or 2048 wavelengths to deconvolute the concentration profiles of multiple components (see Chap. 6). The ability to analyze these big data sets will enhance our ability to study heterogeneous interacting and noninteracting systems.

Usually, sample solution involving solute to be studied and solvent is applied to sample cell and to reference cell, respectively, in a centerpiece (described in detail later). Ideally, the solution is dialyzed against a solvent, and the dialysate is used

---

F. Arisaka (✉)
Life Science Research Center, College of Bioresource Science, Nihon University, 1866 Kameino, Fujisawa, Kanagawa 252-0880, Japan
e-mail: fumio.arisaka@gmail.com

S. Uchiyama
Department of Biotechnology, Graduate School of Engineering, Osaka University, 2-1 Yamadaoka, Suita, Osaka 565-0871, Japan

Department of Bioorganization Research, Okazaki Institute for Integrative Bioscience, 5-1 Higashiyama, Myodaiji-cho, Okazaki, Aichi 444-8787, Japan
e-mail: suchi@bio.eng.osaka-u.ac.jp

S. Uchiyama et al. (eds.), *Analytical Ultracentrifugation*,
DOI 10.1007/978-4-431-55985-6_2

as a reference solution. This is essential for interference optics, because the density match is critical for the interference optics, but for the absorbance optics, if the last step of purification is gel filtration, equilibration solvent for the chromatography may be used. Absorbance is measured alternately between the solution cell and reference cell, and the difference is recorded as the output. Low absorption in the reference cell would not interfere much. If a solvent for reference has no absorption at the wavelength employed for the monitoring of solute, water can be used as the reference instead of the solvent in which the solute is solubilized. Nevertheless, solvent with absorbance especially with large absorbance has to be avoided as the reference, because the amount of light reached to the detector will be reduced, the voltage of the photomultiplier will increase, and the linearity of the absorbance vs. concentration would drop.

Absorbance has advantage for its simplicity and selectivity. Typically, proteins will be measured at 280 nm where the spectrum has its local maximum. The absorbance needs to be less than 1.2 where the value will quantitatively reflect its concentration according to the Lambert-Beer law. The path length of 1.2 cm for the centerpiece has to be taken into consideration when the solution to be measured is prepared. The minimum absorbance of about 0.1 can be measured. One can choose the wavelength between 190 and 800 nm, but data acquisition at a wavelength between 220 and 700 nm is generally performed. If the absorbance at 280 nm is too low, one could choose wavelength at around 230 nm to enhance the sensitivity. The lamp intensity has its peak at 230 nm. The lamp intensity spectrum has to be checked from time to time whether there is any decrease of the intensity, because the dirt due to the oil vapor from the diffusion pump gradually accumulates on the xenon flash lamp and/or optical mirror, and when the intensity gets low as shown in the manual, one needs to clean the optical system or call a Beckman Coulter service person to clean the optical system.

When the protein has its typical absorbance in the visible region, one can use that wavelength as well. For example, hemoglobin has an intense band at about 450 nm, which is a typical absorption band for the heme, called the Soret band. If the Soret band is used for detection, as there is rarely another protein with that absorbance, one could selectively measure hemoglobin even if there are some contaminant proteins.

When the sample has a large particle size, scattering is not negligible, but as far as the molecular weight is not changed due to association or dissociation, scattering may be regarded as part of the absorption.

Rayleigh interference optics which utilizes refractive index of the solute for detection has its own advantage. When the buffer absorbs, it is the preferred optical system. Also, for samples, including polysaccharide, lipids, or some nanoparticles, which do not absorb light, interference can be used for detection. As interference optics has no selectivity, stoichiometry of hetero-association in which the two interacting proteins have different extinction coefficients, interference optics is simpler for interpretation than absorbance optics. It is more precise than absorption. As mentioned earlier, sample dialysis is essential which guarantees the matching of the density of the solvent in sample and reference cells. The menisci for the sample and reference cells have to be precisely matched.

In XL-I, the optical detection attachment has a laser ($\lambda = 670$ nm) for interference detector. The output is the fringe displacement, Y_T, which is proportional to the weight concentration of the solute, c:

$$Y_T = \delta + lc\frac{dn}{dc}\frac{1}{\lambda}$$

δ is the offset which can be taken $c = 0$, where the meniscus is depleted of the solute [Laue 1996]. $(dn/dc)/\lambda$ is analogous to the extinction coefficient in absorbance measurement. In the case of proteins, dn/dc is not significantly dependent of the amino acid composition of proteins.

Comparing the two optical systems, absorbance system has better resolution at lower concentrations, but interference optics has wider range of linearity even at higher concentrations.

By using both absorbance and interference optics, one could determine the extinction coefficient based on the fact that (dn/dc) value for proteins does not significantly depend on the amino acid sequence but is approximately 0.185.

Measuring the concentration of the solute by fluorescence is possible if one installs the fluorescence detection attachment which was originally developed by Dr. Laue and now is manufactured and distributed by Aviv Inc. It has an extremely high sensitivity and as the fluorescent spectrophotometer detects only close to the surface of the cell, much less sample volume is needed for measurement. It requires fluorescent label of a solute, but one can be detected even in a highly complex background. Fluorescence detection system is described in Chap. 4.

It is possible to scan the same cell with three different wavelengths. As the UV (and sometimes visible) spectra of individual proteins are different, it is possible to utilize the difference to distinguish the proteins in the solution which is occasionally useful [Brautigam et al. 2013]. However, due to the limited performance of the stepping motor for the wavelength control equipped on the instrument, wavelength often does not precisely return to the original wavelength once changed and occasionally causes some inconvenience. It appears to depend on the particular instrument. This problem has been solved in the newer multiwavelength instruments by Helmut Coelfen and Kristian Schilling in Germany and in the CFA developed by Spin Analytical in New Hampshire.

2.2 Rotors and Centerpieces

Two types of rotors are available, namely, An60Ti and An50Ti. The former can accommodate four cell assemblies and the latter can insert eight cell assemblies. 60 and 50 in the name of the rotors stand for the speed limit of 60,000 rpm and 50,000 rpm, respectively.

The cells for loading sample solution and solvent are called as centerpieces. The sedimentation velocity (SV) run commonly utilizes double-sector cells which are made of charcoal-embedded Epon, of which the guaranteed maximum speed limit is 42,000 rpm, above which one needs to use aluminum cells. The optical path length is 1.2 cm. Chemical resistance of the Epon cells is listed in the table and available at the website of Beckman Coulter Co. Ltd. and has to be considered when any organic solvent is used. Aluminum cells are resistant to most of the organic solvent, but acidic or basic pH has to be avoided.

For absorption optics, a counterbalance is essential and has to be inserted at cell no.4 or no.8 for An60Ti or An50Ti, respectively. It is not just a counterbalance but is used for system calibration. For interference optics, the counterbalance is not necessary once the radial calibration has been performed.

It has to be kept in mind that the more cells are used, the more time it takes before next measurement is made. One, therefore, needs to consider the s-values of the solutes, the necessary time for each scan, and the number of scans. At least 20–30 scans are necessary for satisfactory analysis.

For absorption optics, quartz windows are usually utilized. For interference optics, sapphire windows which have higher resistance to mechanical stress have to be used, because small strain would distort the correct fringe pattern.

As for centerpieces, the double-sector cells may be used for sedimentation equilibrium (SE) as well, but six-channel cells are available for SE, where three samples and the corresponding reference solutions can be applied. There are two types of six-channel centerpieces. One is the classical style, developed by Dr. Yaphantis, in which the cell is first half assembled, and then sample solutions and the reference buffer are applied, and finally the other window and a gasket are set and the screw is tightened by the torque stand. The other type of the six-channel cell is a flow-through-type assembly assembled just like the double-sector cell, and the samples are applied after assembly. For double-sector cells, 100–150 μL of sample solution is applied for the sample cell for absorbance system, while 10 μL larger volume (110–160 μL) of reference is applied for the reference cell in the case of absorption optics.

For six-channel cells, typically, 100 μL of sample solution and 110 μl of the reference solutions are applied. For interference system, not only the density match for the buffer but also, preferably, the meniscus positions have to be matched precisely.

2.3 Determination of the Appropriate Temperature and Rotor Speed for Measurements

When the cells are ready for measurement, or even prior to starting AUC experiment including sample preparation, one needs to decide to set up the conditions for measurement.

Let us first consider SV measurements. Concerning the temperature, if the sample is stable enough at room temperature, 20 °C would be the choice, because the sedimentation coefficients need to be ultimately corrected for 20° in water or $s_{20,w}$. Some of the sample solution may prefer a lower temperature for stability. In that case, the rotor has to be precooled in the analytical ultracentrifuge or a cold room, because otherwise it may take too much time to equilibrate the temperature of the rotor due to its large heat capacity. Most of the current software for the analysis will return the $s_{20,w}$ value if one inputs the density and viscosity of the solvent under experimental conditions. The speed is generally chosen as the highest in order to obtain the s-value as precise as possible, but there must be enough number of scans, at least 30 to 40, preferably 100 scans, for the accurate analysis, whereas our experience is that somewhat lower speed would give more accurate molecular weight, because it gives enough time for the molecule to diffuse for obtaining more precise diffusion coefficient, which would result in better molecular weight. It should be noted that SE is generally more reliable method for the determination of molecular weight of solute than SV.

For SE, the appropriate rotor speed may be estimated based on the molecular weight or sedimentation coefficient referring to the diagram given in the manual of XL-A or XL-I. Alternatively, programs such as SEDFIT or ULTRASCAN provide appropriate rotor speed and the time required to reach equilibrium for SE experiment of a solute. As global fitting is now possible which is more reliable than otherwise, SE is currently conducted at a number of conditions such as three rotor speeds and three loading concentrations. One may first determine the seemingly best rotor speed and choose higher and lower concentrations as, say, 20 % higher and 20 % lower. Loading concentration may be 0.15, 0.3, and 0.5 absorbance at measuring wavelength.

Scans are usually carried out every 2–3 h. It will take 20–24 h to reach equilibrium for ordinary size proteins, and it has been traditionally checked if at least three concentration gradients of a solution in a cell are superimposed. It depends on the s-value and the diffusion coefficient of the solute. A number of software are available to judge if the equilibrium is attained. One of them is WINMATCH which calculates the difference of the concentration gradient measured every 2 h (as the user sets) starting when the speed of rotation has been changed.

2.4 Data Analysis of Sedimentation Velocity and Sedimentation Equilibrium

Raw AUC data are stored in the folder of xlidata and can be retrieved to carry out the analysis.

For sedimentation velocity data, SEDFIT, UltraScan, or some other software may be used to obtain the distribution of the sedimentation coefficient, c(s). These programs have useful multiple functions, and readers are encouraged to go to the website of these programs to read the instructions to take advantage of these programs. For the interacting systems, SEDPHAT which is linked to SEDFIT, SEDANAL, and UltraScan can be used.

The Nonlin program which is provided by the manufacturer can be used to analyze SE data for single-component system and self-association model. For hetero-association, SEDPHAT, UltraScan, and SEDANAL (See Chap. 7) can be used for the analysis.

Sednterp is a versatile program with which one can calculate the partial specific volume based on the amino acid sequence or composition, density and viscosity of the buffer solutions, etc., which can be operated on the website or downloaded from the website.

2.5 Important and Useful Websites Related to Analytical Ultracentrifugation

Besides previously published books (Harding et al. 1992; Schuster and Laue 1994; Scott et al. 2005), one can access important and useful knowledge for analytical ultracentrifugation from the websites and mailing lists shown below.

Software	
1. SEDFIT	http://www.analyticalultracentrifugation.com/
2. UltraScan	http://www.ultrascan.uthscsa.edu/
3. SEDANAL	http://www.sedanal.org/
4. Sednterp	http://sednterp.unh.edu

Instrumentations, parts, and services	
1. Beckman Coulter	https://www.beckmancoulter.com/
2. Spin Analytical	http://www.spinanalytical.com/
3. Nanolytics	http://www.nanolytics.de/
	http://www.nanolytics-instruments.com
4. Alliance Protein Laboratories	http://www.ap-lab.com
5. U-medico	http://u-medico.co.jp

Mailing lists	
1. RASMB	Registration from http://www.rasmb.org
2. SEDFIT	Registration from Sedfit website
3. UltraScan	Registration from UltraScan website
4. SEDPHAT	Registration from SEDPHAT website
5. BMIA forum	Registration from http://www.biomoleculesinteractions.com

References

Brautigam CA, Padrick SB, Schuck P (2013) Multi-signal sedimentation velocity analysis with mass conservation for determining the stoichiometry of protein complexes. PLoS ONE 8(5):e62694

Harding SE, Rowe AJ, Horton JC (eds) (1992) Analytical ultracentrifugation in biochemistry and polymer science. Royal Society of Chemistry, Cambridge

Laue TM (1996) Beckman application information A-1821A, pp 1–18

Schuster TM, Laue TM (eds) (1994) Modern analytical ultracentrifugation: acquisition and interpretation of data for biological and synthetic polymer systems. Birkhäuser, Boston

Scott D, Harding SE, Rowe A (eds) (2005) Analytical ultracentrifugation: techniques and methods. Royal Society of Chemistry, Cambridge

Part II
AUC Instrumentation and Analysis

Chapter 3
The CFA Analytical Ultracentrifuge Architecture

Thomas M. Laue and J. Brett Austin

Abstract As part of the Open AUC Project, the CFA is the next generation of analytical ultracentrifuge. The design philosophy of the CFA, which is to encourage hardware and software innovation, has been published. However, the hardware and software that allow and encourage future development have not been described previously. Presented here is the CFA hardware and software architecture and the rationale for how this architecture was developed. Overall, both the hardware and software architecture is modular, allowing for updates and additions over time without the need for wholesale redevelopment. The common features needed by optical systems are contained in "stacks" of electronics for synchronizing the sources and detectors to the spinning rotor. Separate computer programs operate the stacks, the motion control, the centrifuge hardware, the experiment protocol, and each optical system. These programs communicate with one another to execute functions and to provide data and status information. Each program has a database associated with it to provide nonvolatile storage and inter-process communications. While the current implementation of the CFA uses one central computer to coordinate and operate all of the systems, the modular design includes provisions for using multiple computers should that be needed for a particular optical system.

Keywords Analytical ultracentrifuge • Ultracentrifuge instrumentation • Open AUC Project • Ultracentrifuge hardware description • Ultracentrifuge operating software description

3.1 Introduction

The Model E analytical ultracentrifuge (AUC) was built for 24 years, from 1948 to 1972. More recently, the XLA/I was introduced in 1990 and has been the only commercially available AUC for the past 25 years. This chronology suggests that it is time for the introduction of next generation of AUC. Recognizing that

T.M. Laue (✉) • J.B. Austin
Spin Analytical, Inc, 468 Portland Street, Berwick, ME 03901, USA
e-mail: Tom.Laue@Spinanalytical.com; Brett.Austin@Spinanalytical.com

S. Uchiyama et al. (eds.), *Analytical Ultracentrifugation*,
DOI 10.1007/978-4-431-55985-6_3

the development of an AUC is costly and time consuming, and that AUC has a limited commercial market, the Open AUC Project began, in part, to encourage new generations of AUC hardware and software, with the Centrifugal Fluid Analyzer (CFA, Spin Analytical, Inc.) as the base platform (Cölfen et al. 2010).

Presented here is the rationale behind the CFA hardware and software design, along with the designs' architecture and implementation.

3.2 Hardware Description

3.2.1 Hardware Design Rationale

Serving as the foundation of the new AUC, the CFA hardware will allow new or improved optical systems to be added with minimum interference. At the same time, the hardware provides the common functions needed by any optical system, such as synchronizing data acquisition with the spinning rotor, motion control, and centrifuge operations. The CFA divides the data acquisition load across several interfaces (Table 3.1). Any given optical system may need to access all of the interfaces, and the interfaces must be able to handle requests from all of the optical systems.

In order to acquire data, each optical system has sets of electronic boards, referred to as stacks. These stacks allow the CFA to identify, control, and acquire data from the device. Some devices require more than one stack. A common example is when an optical system's source and detector are located in different places in the CFA. A typical stack contains separate, bussed boards for the following functions: power supply, master clock, synchronizer, digitizer, memory, and hardware control.

Table 3.1 CFA subsystems and services

System	Function	Notes
Parallel digital	Operate data acquisition stacks, high-speed acquisition	48-bit bus, address, command, and data transfer to stacks
Low-speed serial	Board identification, low-speed operations	
Motion control	Servo, stepper, linear motor, encoder operations	Each device runs in its own thread
Centrifuge hardware	Temperature, vacuum, motor control	Operate the centrifuge hardware
Experiment	Directs the CFA operations	Common to all optical systems, rpm, temperature, etc.
Sednterp2	Sample/solvent descriptions	Calculations available that are useful to AUC analysis
Master daemon	External communications	TCP/IP socket

3.2.1.1 Data Acquisition Stacks

Each optical system will require hardware for synchronizing the spinning rotor to data acquisition. Synchronizing may be needed for the source or the detector, and the nature of the synchronizing may differ for different detector types. There are synchronizing needs that are common to all optical systems, as well as needs that are specific to a particular optical system. The common needs are a jitter-free rotor timing pulse and variable-speed clock to determine the period of one rotation (as counts of the clock), a means to determine the period of a fraction of a rotation, a means to adjust the latter count for propagation delays, and a way to produce a properly timed window when data acquisition should occur (Yphantis et al. 1984; Laue et al. 1984).

Different data rates and different signal stream types are needed to accommodate data acquisition from different optical systems, and the CFA provides two useful data buses, one a high-speed parallel digital bus and the other a lower-speed serial bus.

3.2.2 Parallel Digital Bus

One inescapable feature of AUC is that the spinning rotor provides the "heartbeat" for any optical detection. The rotor timing pulse is asynchronous with respect to any computer operations and unforgiving with respect to the timing of signal acquisition from any sample (Laue et al. 1983). Consequently, intervening data acquisition hardware must be used to: (1) synchronize acquisition with the spinning rotor, (2) digitize and buffer data as it comes from a detector, and (3) allow the computer to configure the acquisition hardware and download data. A parallel digital bus is used to operate the data acquisition hardware. This bus provides for addressing, commands, and data paths to and from the stacks.

3.2.3 Serial Bus

Many operations do not require a high-speed bus. The CFA includes a serial bus that is used to identify each board, including the board type, which optical system it is part of, and other information needed to allow optical systems to be swapped in and out of the CFA. The bus also is used to perform other low-speed functions, such as selecting optical filters.

3.2.4 Motion Control Bus

Optical systems often require motion control, and monitoring motion control systems can consume substantial computer resources. Consequently, a separate computer network operates the motion control bus. The CFA uses a commercial system capable of high-speed, high-precision, high-accuracy positioning and is compatible with a variety of motion control devices (position encoders, stepping motors, linear motors, servo motors). The motion control system operates asynchronously from the main computer, while having the capability to interrupt the main computer with information about any of several events (limit reached, move done, error condition). Calibration information for converting encoder units to instrument units (e.g., radial position or wavelength) is stored in a database (below). Depending on the servo, encoder, and leadscrew, micron-level positioning may be achieved. Motion may be initiated on several axes simultaneously and motion status read for each axis during movement. The completion of movement generates a software interrupt.

3.2.5 Power Distribution

Power for motion control and device operation is available. For each stack, voltages for digital (5 and 3.3 V) and for analog (±12 V) circuits are generated locally to minimize electrical noise and cross talk.

3.2.6 Stack Electronics

The synchronizing functions have been divided up and the electronics for each function put on separate, stackable boards. This architecture has the advantage of isolating the different optical systems from one another, thus reducing the chance that noise from one system will affect the others. The architecture also allows for new or upgraded boards to be incorporated in the stacks without requiring changes to existing boards. In addition to the power supply, each stack may contain the following boards.

3.2.6.1 Master Clock

Each stack may contain a master clock board. This board provides a high-frequency signal used to synchronize the rotor timing pulse (RTP) with other operations (Fig. 3.1). This signal is divided to produce two lower-frequency synchronizing clocks (Laue et al. 1984). Which of the two synchronizing clocks is used depends

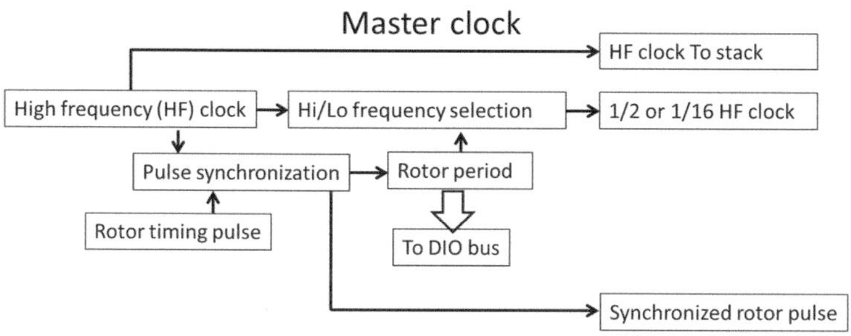

Fig. 3.1 Master clock block diagram. The master clock provides a high-frequency clock (HF) that is divided either by one-half or one-sixteenth for synchronizing purposes. The HF appears on the internal stack bus and is used to de-glitch asynchronous external signals for use by the stack. Which of the two lower-frequency clocks is used by the synchronizing circuitry depends on the rotor speed. Clock selection is made automatically to guarantee a minimum of 4000 counts per revolution while not exceeding 65,000 counts per revolution. There hysteresis built in to the clock selection circuit to prevent the clock speed from toggling rapidly between the higher and lower rates during acceleration or deceleration

on the rotor speed; above ~5000 rpm the higher-frequency clock is used, and below 5000 rpm the lower-frequency clock is used. The switch to higher clock rates occurs at a higher rotor speed than the switch back to the lower clock rate. This hysteresis prevents rapid switching between the two clock rates during acceleration or deceleration. The clock rates and rotor speed switch points were selected to guarantee that at least 4000 clock pulses will be provided per revolution while not exceeding 64,000 clock pulses per revolution. In this manner, 16-bit counters may be used for all synchronizing operations (Yphantis et al. 1984). In order to prevent "glitches" between the rotor timing pulse (RTP) and the clock, the undivided clock signal is used to synchronize the rotor timing pulse (SRTP) with the master clock (Laue et al. 1984). In so doing, there is an inherent uncertainty of up to 125 ns between the edges of the RTP and the SRTP. This uncertainty is less than the jitter specification (±0.1°) for the synchronizing circuit. The leading edge of the SRTP is used for all subsequent timing functions.

This board makes accessible the rotor period, which may be used to calculate the rotor speed to a fraction of an rpm. The SRTP operates a divide by two circuits to produce an ODD/EVEN signal. The ODD/EVEN signal is used to toggle between two banks of timers on the synchronizer board (below) and may be stored along with digitized data as a surrogate for the RTP.

3.2.6.2 Synchronizer

The synchronizer board produces the timing signal needed to time light source or detector operations with the spinning rotor (Fig. 3.2). Two sets of timers are used: for one revolution, timer 1 provides the synchronized signal while timer 2 is

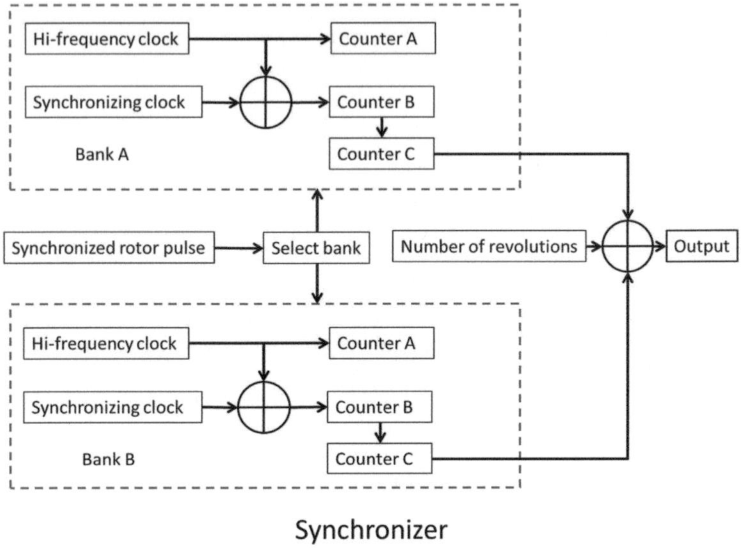

Fig. 3.2 Synchronizer block diagram. Two banks of three counters each are used. While bank one is in a "preparation" mode where the counts are loaded and adjusted for propagation delays, the other bank is generating a delay time and duration time to operate the source or detector. The circuitry can be configured to provide a single pulse per rotor revolution or to provide an equally spaced series of pulses each revolution. The synchronizer may be configured to operate for a fixed number of revolutions, or it may be allowed to run continuously

prepared for operation. On the next revolution, timer 2 provides the synchronized signal while timer 1 is prepared. This arrangement provides a correctly timed pulse each revolution, even when the timers are being updated by software.

Each timer consists of three separate 16-bit counters, A, B, and C. The counters initially are loaded with the desired count and then decrement until they reach zero counts. The A counter provides an "offset" count that compensates for propagation delays (Yphantis et al. 1984; Laue et al. 1984, and below). Counter B contains the count corresponding to the desired fraction of the rotor period (e.g., if the rotor period is 6000 counts, and you want to trigger an event at ¼ of the rotation, the B counter would start with 1500 counts). Counter C contains the period of the event in terms of a fraction of a revolution. While both the B and C counters count a fraction of the rotor period using the synchronizing clock, the A is active for a time equivalent to the propagation delays in an optical system and is used to adjust the count in the B counter as described previously (Laue et al. 1984; Yphantis et al. 1984). The operation of the A counter and consequent adjustment of the B counter occurs during the preparation phase of a timer (Yphantis et al. 1984).

The synchronizer may be operated in three modes. First, the synchronizer may be enabled a given number of rotor revolutions. This mode is used when data are collected for a fixed number of revolutions (NREV), and the signals from the samples are sorted out in software (Laue et al. 2006). Second, counter B provides the delay (in a fraction of a revolution) needed to start an event, and counter C provides the duration of the event (in a fraction of a revolution). This mode is used to operate the pulsed laser light source for interference optical system and the pulsed Xe lamp for the absorbance optics. Third, counter B provides the time between events during one revolution, and counter C provides the duration of each event. In this mode, several pulses are provided by counter C each rotor revolution. This mode may be used to collect data from several samples at a time, for example, when using a turbidity or light-scattering detector (Mächtle 1999). The latter two modes may be used in conjunction with the NREV counter so that data collection occurs only for a fixed number of revolutions.

3.2.6.3 Analog to Digital Converter (ADC)

An optical system stack may include an analog to digital converter with up to 16-bit resolution to convert analog signals to digital values. The ADC may be operated by the signal coming from the synchronizer board, or an external signal may be supplied. Data from the ADC board is put on the stack's internal data bus, along with a pulse indicating that a valid digitized value (VDV) is available. The leading edge of the VDV is used to store the data in memory, and the trailing edge is used to increment the memory address.

3.2.6.4 Memory

Each memory card in a stack holds 1 Mbyte of 16-bit data. The data may be accessed randomly. Up to four memory cards may be connected in series, so that when one is full, the next board is activated. Data are stored sequentially from the internal data bus. The memory address may be set or read from software, and the data at an address may be read from software. A typical data acquisition sequence would be: (1) set the initial memory address to zero; (2) initiate data acquisition; (3) when acquisition is done, re-zero the address; and (4) read data and increment the address (one software command will do this). The software should not change the memory address during data acquisition.

3.2.6.5 Source/Detector Control

A control board unique to each stack handles the specific signals needed to operate a light source or detector and provides a way for the source/detector pair to coordinate their operations. For example, array detectors need to "free run" in order to minimize any dark current background signal. In operation, the source must wait to fire until the array detector has completed a scan and halted. Thus, a typical operating sequence for this type of detector is: (1) the software issues an "acquire" command; (2) the array detector completes a scan and halts further scanning, and the detector stack issues a "ready" command to the light source stack; (3) the light source stack then provides properly synchronized bursts of light (the number of bursts depends on the source intensity and detector sensitivity, which must be determined separately); (4) when the light bursts are completed, the source stack signals the detector stack to resume scanning and to save the data from the next scan; and (5) the detector stack signals that data are available.

3.3 Software Description

3.3.1 Software Design Rationale

The CFA software is modular. Separate computer programs run each of the subsystems listed in Table 3.1, and each program is considered a service. The details needed to operate a particular subsystem are contained within its service. By using separate programs to operate the services, it will be possible to update the software for a particular service in response to hardware changes without requiring the complete replacement of the other services.

Likewise, separate computer programs operate each optical system (OptSys). Multiple optical systems may be running simultaneously (OptSys1, OptSys2, etc.), with some sharing a single optical track (e.g., schlieren and interference systems). Using this design, the details of operation, user interface, and data handling for each optical system are contained in its OptSys program and may be updated independently of the other optical systems. All of the programs broadcast a "heartbeat" once a second to let the other programs know they are still functioning. Likewise, every second each system broadcasts a status structure that provides information (e.g., busy, queue lengths, etc.) to the other programs.

3.3.1.1 Inter-process Communications

The OptSys programs send requests and commands to the services via inter-process communications (IPC). Similarly, the services send status updates and data to the OptSys programs by IPC. Because processing some commands may take a relatively long time, all services and OptSys programs have their own internal message queues. When a command is received, it is immediately placed on the internal queue, unblocking the calling program. Messages are processed from the internal queue in a first-in-first-out manner. The internal queue system does provide for the possibility of checking priorities on messages, however thus far this has not been needed. After a command has been processed, a reply, containing both the original command and any requested data, is sent to the calling program. The reply message and data are placed on the internal queue of the calling program for processing. The messaging system, then, operates in full-duplex mode.

Every message contains an address structure containing the sender ID, receiver ID, the time when sent, the command, a unique message number, whether the sender expects a receipt acknowledgment message prior to command execution, and, in a reply, whether the command was executed successfully. Each of the services has a unique set of commands associated with it. Some of these commands are described below.

3.3.2 Parallel Digital

This system operates the high-speed digital input/output system. The structure of a command to this system consists of a stack number, board number, command, and value. The first two determine which board will receive the command. The "value" is an optional 16-bit number that will appear on the data output lines when the command is executed. In addition to the stack, board, command, and value, an optional "delay" value may be provided. When used, the delay value causes the service to suspend operations for the number of milliseconds prior to handling another command. Ordinarily, the delay is not used, but is provided if some source/detector board commands need extra processing time.

In addition to sending single commands, a calling program may send a sequence of up to 10 commands in a single message. Because DIO operations are rapid, this capability is convenient and efficient. For example, the synchronizer board may be set to the angles needed for a particular sample and an "acquire data" command issued. Once the acquire data command has been issued, an acknowledgment is sent to the calling program indicating that the acquisition process is ongoing. The calling program may then issue status requests to the stacks to determine whether the acquisition process is complete and a command to retrieve the data after acquisition.

3.3.3 Motion Control

The commands to the motion control subsystem include ones to initialize the system and to command the axis to move. Each axis has an entry in a database maintained by the motion control system that identifies each axis and has the parameters needed to convert an encoder reading into a meaningful value (e.g., to cm from the center of the rotor or nm for a monochromator).

3.3.4 Centrifuge Hardware

The centrifuge hardware includes the vacuum system, the rotor speed control (rpm, acceleration, and deceleration), and the chamber temperature control. Because its status is important to all of the other service and OptSys programs, the centrifuge hardware service broadcasts a status structure every second and records the information in a file kept with the experimental data. The status structure includes the rotor speed, temperature, vacuum, time, $\omega^2 t$, and status flags.

3.3.5 Experiment

The experiment subsystem provides information that is common to all experimental protocols including the user, the method, the rotor setup, and the samples. The user information includes a log-in name, a password, the last data path storage path, the last method and rotor setup used, and whether the user is an administrator. An administrator can set up or modify user accounts and determine whether or not logging in is required. At the start of an experiment, the user is provided the option to save the data in a location other than the default location.

The method includes the information for operating the machine. Method information includes a description, whether to wait for temperature, how long to wait after the temperature is reached before beginning the protocol, and whether to shut the machine off after the protocol is done. Each method also has one or more steps that include the rotor speed, temperature, acceleration and deceleration, duration, and flag indicating the completion of the step. The duration may be a time or it may be the number of scans required of each optical system. The number of scans is used as the default by the optical systems and may be overridden locally by them. Completion of a step is signaled when all of the optical systems indicate that they have completed all of the scans at that rotor speed and temperature.

The rotor setup includes information on the rotor and the type of cell in each rotor hole. The cell information includes a description of the cell type, the radial positions of the top and bottom of each channel, as well as the angular offset and width of each channel. This information is needed by the optical systems for data acquisition. Tables are included for keeping an inventory of rotors and cells. This inventory allows their use to be logged and may be used to associate data with a specific cell.

The samples include solvent and solute information for each channel. Solvent and solute information are saved in Sednterp, making it easier to perform data interpretation. The concentration of each solute is kept for each channel.

3.3.6 Hardware Inventory

The CFA uses a low-speed serial bus and serial number firmware to create an inventory of what optical systems are present and how they are configured. The hardware inventory subsystem allows optical systems to be swapped in and out of the CFA.

3.3.7 Master Daemon

The master daemon is a service used to: (1) monitor the operating status of the services, (2) monitor the operating status of the optical systems, and (3) provide external communications. This service has access to all of the databases (below) and provides status updates to any program connected to its server socket. Updates are provided every second on the status of all of the other services. These status updates include whether a service exists and is: (1) initialized, (2) busy, (3) ready to receive commands, (4) writing to its database, (5) operating in simulation mode, and (6) has encountered a critical error. In addition, the message count for the input, output, and pending queues is provided. Also, the machine status (above) is broadcast every second. Data to or from tables are handled as SQL statements. Data from tables are returned as either SQL or XML strings.

3.4 Databases

3.4.1 Database Rationale

Databases provide a structured means to store data that may be accessed from several programs concurrently. The CFA operating system uses SQLite databases (www.sqlite.org). SQLite is open source and widely used. There is no requirement for a separate database server application, simplifying program distribution and operations. There are SQLite bindings for over 30 programming languages and dozens of operating systems. While SQLite provides thread-safe read access (i.e., several programs may access a database at once), it is not thread-safe for writing. To prevent potential conflicts when more than one service wishes to access a database during a write, a simple write-lock system is used that consists of creating a file for the lifetime of the write process. Any blocked process must wait for the disappearance of this file before accessing the database. Because each service and each optical system has its own database, and since writing to that database is made only by that service, conflicts do not occur. The write-lock is provided in cases where an optical system is saving data to its database and an analysis program wishes to read from that database. The databases used by the CFA operating system are presented in Table 3.2.

In addition to the CFA operating system databases, each optical system will have two types of databases associated with it. The first database will hold the information needed to operate the hardware for that optical system. The second database will hold the data from a particular experiment for that optical system. For example, the multiwavelength absorbance (MWA) optical system uses a hardware database that contains the information presented in Table 3.3. A new stand-alone data database is created for each experiment. In this way, a data database will provide an analysis programs with a complete record of an experiment.

Table 3.2 CFA operating system databases

Database	User	Tables	Notes
Processes	All services and optical systems	Paths	Folder locations
		Processes	Identifying information for inter-process communications
Hardware	Machine operations and all optical systems	Axes	Motion control
		Instrument	Service information
		Inventory	Stack and board information, offset, and speed dependence parameters
		Optical tracks	
Experiment	All optical systems	User	Information used by all optical systems
		Method and steps	
		Setup	
		Samples	
		Status	
Error	All services and optical systems	Errors	Log of all errors encountered, which service or system and when
Utility	All services and optical systems	Units	Various units, length, concentration, etc.
		Min and max	

Table 3.3 MWA hardware database

Table	Contents
Channels	Detailed angle information for each channel
Finetune	Rotor speed-dependent offset values
GainMeta	Statistics for gain scans by wavelength
GainScans	Intensity scan at each wavelength taken during gain setting
Gains	Gain and averaging used at each wavelength
Instrument	Identification and service interval information
Monochromator	Information about the monochromator
NolightScans	Intensity scans acquired using the gain settings, but without light
OpticalTrack	Angles and speed dependence parameters

Acknowledgments The authors wish to acknowledge the support of the AUC community during the development of the CFA. Particular appreciation is expressed to Bo Demeler and Walter Stafford, who have contributed to this work. The tireless work of Anthony Cipolla and Paul Vachon at Spin Analytical must be recognized. They not only have helped contribute to the design process, but they have built, assembled, and integrated components. We wish to dedicate this work to our predecessors: Ed Pickels, Howard Schachman, David Yphantis, Ken Van Holde, David Teller, and so many others who have spent time developing and improving AUC hardware over the years.

References

Cölfen H, Laue TM, Wohlleben W et al (2010) The open AUC project. Eur Biophys J 39(3):347–359

Laue TM, Domanik RA, Yphantis DA (1983) Rapid precision interferometry for the analytical ultra-centrifuge 1. A laser controller based on a phase-lock-loop circuit. Anal Biochem 131(1):220–231

Laue TM, Yphantis DA, Rhodes DG (1984) Rapid precision interferometry for the analytical ultracentrifuge. III. Determination of period of rotation, frequency of rotation and elapsed time. Anal Biochem 143(1):103–112

Laue TM, Austin JB, Rau DA (2006) A light intensity measurement system for the analytical ultracentrifuge. Progr Colloid Polym Sci 131:1–8

Mächtle W (1999) High-resolution, submicron particle size distribution analysis using gravitational-sweep sedimentation. Biophys J 76:1080–1091

Yphantis DA, Laue TM, Anderson IA (1984) Rapid precision interferometry for the analytical ultracentrifuge. II. A laser controller based on a rate-multiplying circuit. Anal Biochem 143(1):95–102

Chapter 4
Fluorescence Detection System

Tao G. Nelson, Glen D. Ramsay, and Matthew A. Perugini

Abstract One of the most exciting advances in contemporary analytical ultracentrifugation has been the development of the analytical ultracentrifuge equipped with a fluorescence detection system (AU-FDS). The AU-FDS provides both selectivity and sensitivity thereby enabling the measurement of specific fluorescently labeled macromolecules in complex solutions and/or the detection of dilute samples at low (sub-nanomolar) concentrations. This has had tremendous impact to the study of low-abundance proteins, the quantification of high-affinity interactions (K_D < nM), and the characterization of biomolecules in crowded biological backgrounds. This chapter will describe the (i) development, (ii) optical setup, (iii) advantages and disadvantages, (iv) pre-experimental requirements, (v) experimental operation procedures, and (vi) contemporary applications of the AU-FDS focused on enzyme self-association, antibody-antigen interactions, amyloid protein aggregation, and the composition of enzyme complexes.

Keywords Analytical ultracentrifugation • Analytical ultracentrifuge • Fluorescence detection system • Fluorescence optics • Green fluorescent protein • High affinity • Interaction • Sedimentation • Sensitive • Selective

4.1 Development of the Fluorescence Detection System

The basic principle of analytical ultracentrifugation (AUC) requires the detection of a sample or analyte as a function of radial position and time in a gravitational field established by the force of centrifugation (Cole et al. 2008; Howlett et al. 2006; Kroe and Laue 2009). Historically, this was first achieved by Svedberg in the 1920s using colloidal particles and proteins measured by schlieren optics (Svedberg and Pedersen 1940), which earned him the 1926 Nobel Prize in Chemistry. Rayleigh

---

T.G. Nelson • M.A. Perugini (✉)
Department of Biochemistry and Genetics, La Trobe Institute for Molecular Science, La Trobe University, Melbourne, VIC 3086, Australia
e-mail: M.Perugini@latrobe.edu.au

G.D. Ramsay
Aviv Biomedical, Inc, Lakewood, NJ 08701, USA

S. Uchiyama et al. (eds.), *Analytical Ultracentrifugation*,
DOI 10.1007/978-4-431-55985-6_4

interference optics were then developed by Schachman and colleagues in the 1950s for the Spinco Model E instrument (Schachman 1959; van Holde and Hansen 1998) and later advanced to the contemporary interference optical system of the Beckman XL-I instrument through the developmental work of Yphantis, Laue, and colleagues (Yphantis et al. 1994). Toward the end of the twentieth century, an absorbance detector was developed for the Beckman XL-A instrument (Giebeler 1992; Laue 1996), which enabled the analysis of nucleic acids, proteins, and biomolecular complexes through the detection of native chromophores. At the beginning of the twenty-first century, it was also Laue and his colleagues at the University of New Hampshire that developed the fluorescence detection system (FDS) for the Beckman XL-A/XL-I instruments (Laue et al. 1997; Laue and Stafford 1999; Laue 2004; MacGregor et al. 2004), although a fluorescence detector for the Model E was earlier described by Riesner and co-workers (Schmidt et al. 1990). The development of the FDS provided several technical challenges, including the engineering of complex electronics, optics, temperature stabilization, and a high-flux laser source. However, Laue and colleagues were able to resolve these issues through precision engineering and the utilization of a solid-state laser and black body radiation (Laue 2004; MacGregor et al. 2004). The FDS is now commercially available from Aviv Biomedical, who provide the hardware, Advanced Operating System (AOS) data acquisition software, and technical assistance.

4.2 FDS Optical Setup

The Aviv Biomedical FDS is purchased separately and fitted into the vacuum chamber of the Beckman XL-A/XL-I (Fig. 4.1a), where it is positioned above the rotor mounted with sample cell(s) (Fig. 4.1b). The basic assembly of the fluorescence optics is configured similar to that of a confocal microscope with coaxial excitation and emission (Fig. 4.1c). Light is emitted from a laser (originally a solid-state laser, now a diode laser) tuned to 488 nm, which is reflected at right angles by a mirror and expanded into a cone to a collimating lens (Fig. 4.1c). The collimated light is then directed by a dichroic mirror to the sample cell where the excitation beam is focused by a condensing lens that also functions as the objective lens for the emission beam (Fig. 4.1c). The excited fluorescently labeled sample will then emit light at longer wavelengths (>488 nm), which passes through the dichroic mirror to a cutoff filter that restricts transmission to wavelengths between 505 and 565 nm. Finally, the filtered emission beam is refocused by a lens through a 50 μm pin hole and then to the photomultiplier tube (PMT) (Fig. 4.1c). The resulting emission signal is measured as a function of radial position across the sample cell using a stepper motor capable of progressing in 20 μm steps (Cole et al. 2008; MacGregor et al. 2004). Similar to interference and absorbance measurements, the signal (in this case fluorescence intensity) versus radial position is then measured at different time points enabling the monitoring and subsequent analysis of the sedimenting (or floating) fluorescently labeled analyte.

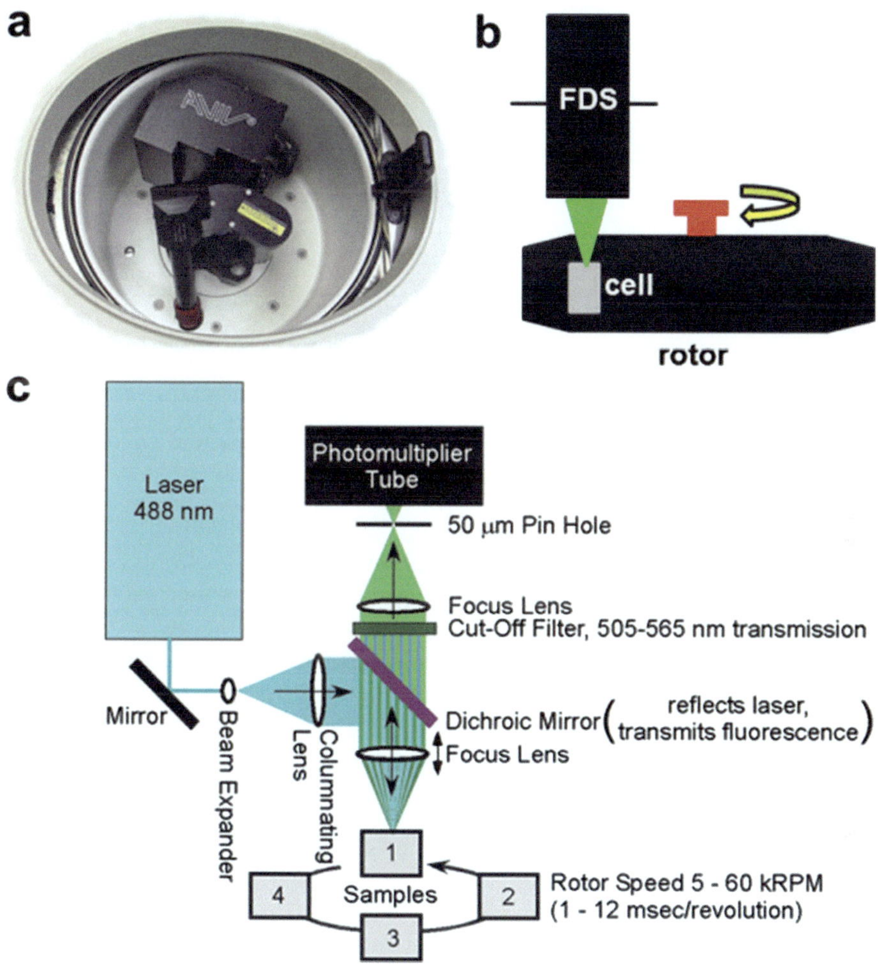

Fig. 4.1 The AU-FDS hardware. (**a**) The inside of a Beckman XL-I analytical ultracentrifuge fitted with the FDS optics box and Beckman Coulter absorbance – interference detector. (**b**) Diagrammatical representation showing the side view of the FDS optics positioned above the rotor and sample cell. (**c**) The optical path of the FDS from laser to photomultiplier tube (PMT)

4.3 Advantages and Limitations

The FDS provides significant advantages over the more conventional interference and absorbance optics. Firstly, the fluorescence detector offers 2–3 orders of magnitude higher sensitivity, which equates to the detection of picomolar concentrations of fluorescently labeled sample (Table 4.1). This capability is referred to as *normal use tracer sedimentation* (NUTS) (Kroe and Laue 2009). By contrast, absorbance or interference optics can only detect, at best, low nanomolar sample concentrations

Table 4.1 Comparison of absorbance, interference, and fluorescence optics for the analytical ultracentrifuge

Detector	Absorbance	Interference	Fluorescence
Sensitivity[a]	0.1 OD	0.1 mg/ml	100 pM
Dynamic range[b]	2–3 logs	3–4 logs	6–8 logs
K_D lower limit[c]	nM	nM	pM
Selectivity[d]	No	No	Yes
Sample throughput[e]	1–7	1–7	1–14
Perturbation of analyte required for detection	No	No	Yes[f]

[a]Minimum amount of sample required to obtain a good signal
[b]Concentration range of analyte
[c]Lowest magnitude for accurate measurement of binding affinity
[d]Specific detection of an analyte in complex or crowded solutions
[e]Number of samples per experiment (upper range using an eight-hole An-50Ti rotor)
[f]Requires labeling of analyte with a suitable fluorophore (e.g., Alexa Fluor 488 or green fluorescent protein)

(Table 4.1). Practically, the increased sensitivity of the AU-FDS affords the quantification of tight biomolecular interactions with sub-nanomolar dissociation constants (K_D) (Table 4.1) (Burgess et al. 2008; Kroe and Laue 2009; Zhao et al. 2013, 2014b). Another advantage of the FDS is the selective detection of trace amounts of analyte in complex or crowded biological solutions, such as plasma, urine, cerebrospinal fluid, or cell lysates (Table 4.1). This capability has been coined *biological online tracer sedimentation* (BOLTS) (Kroe and Laue 2009). However, the AU-FDS requires conjugation of a fluorophore in order to afford sample detection. This represents a potential limitation, since the attachment of a fluorescent partner can perturb the native properties of the analyte of interest (Romanini and Cornish 2012) (Table 4.1). By comparison, absorbance and interference optics enable the measurement of native, non-modified biomolecules (Table 4.1). Nevertheless, the AU-FDS offers greater sample throughput given that no reference solution is required (Kroe and Laue 2009). This means that both compartments of a standard double-sector cell can be occupied by sample. Accordingly, a maximum of 6 samples (when using a four-hole rotor) and 14 samples (when using an eight-hole rotor) can be accommodated in the AU-FDS per experiment, compared to only 3 and 7 samples, respectively, when using absorbance or interference optics (Table 4.1).

4.4 Pre-experimental Requirements

4.4.1 Sample Labeling

As discussed in Sect. 4.3, one of the inherent requirements of the AU-FDS is the attachment of an appropriate fluorescent probe to the sample of interest. Given that the laser is tuned to 488 nm, the selection of a fluorescent partner is limited

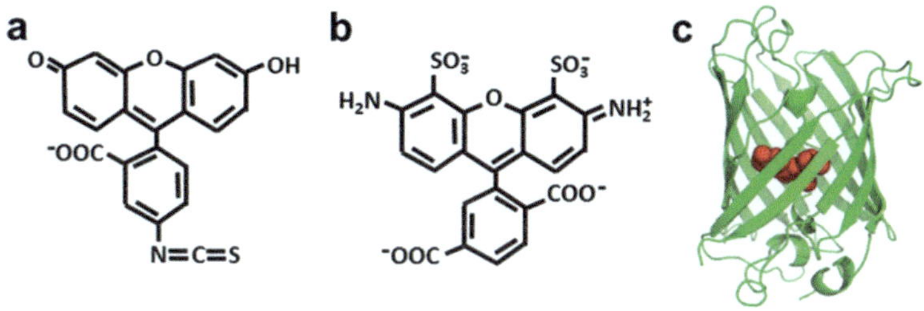

Fig. 4.2 Structures of common fluorophores used in the AU-FDS. (**a**) Fluorescein isothiocyanate (FITC), (**b**) Alexa Fluor 488, and (**c**) *green* fluorescent protein (GFP) (PDB ID: 1EMA)

to fluorophores with an excitation spectrum encompassing this wavelength. It is therefore common to use fluorescein derivatives, such as fluorescein isothiocyanate (Fig. 4.2a) and Alexa Fluor 488 (Fig. 4.2b). Alexa Fluor 488 is an excellent choice due to its high quantum yield, resistance to photobleaching, and stability over a broad pH range (Cole et al. 2008). Furthermore, Alexa Fluor 488 is commercially available in succinimidyl ester or maleimide coupling chemistries to afford efficient conjugation to native amine or thiol groups of biomolecules, respectively (Bailey et al. 2009; Burgess et al. 2008). As an example, a three-step protocol for producing 1:1 labeled protein for the AU-FDS has been developed by Bailey and colleagues (Bailey et al. 2009). Firstly, the protein is incubated with amine-reactive Alexa Fluor 488 succinimidyl ester. Free dye is then removed using a desalting column, which also buffer exchanges, and finally, a hydrophobic column is employed to separate the labeled and unlabeled biomolecules. This method can easily be adapted to label cysteine (i.e., thiol)-containing proteins using the Alexa Fluor 488 maleimide preparation.

Sample labeling can also be achieved via attachment of CyX dyes to hexa-histidine (His) tags of recombinantly expressed proteins (Hellman et al. 2011; Zhao et al. 2010). This offers another advantage of the hexa-histidine tag in addition to its traditional use in protein purification using immobilized metal-affinity chromatography (Hochuli et al. 1988; Zhao et al. 2010). Alternatively, labeling can be achieved via coupling of the sample to green fluorescent protein (GFP) (Fig. 4.2c) (Kroe and Laue 2009), which is a native fluorescent gene product originally isolated from the jellyfish *Aequorea victoria* (Tsien 1998). Commercial vectors producing GFP fusion proteins are now widely available for all types of expression systems enabling AU-FDS users to prepare recombinant proteins as GFP fusion constructs (Polling et al. 2013). However, native GFPs possess the propensity to dimerize in solution (Campbell et al. 2002), which can bias studies aimed at characterizing the quaternary structure of GFP-tagged constructs. To circumvent this, several monomeric analogs of GFP have recently been engineered, including red and blue fluorescent varieties (Campbell et al. 2002; Pettikiriarachchi et al. 2012; Wilmann et al. 2005). The use of photoswitchable versions of GFP commonly employed in super-resolution microscopy is also feasible in the AU-FDS (Zhao et al. 2014a)

4.4.2 Meniscus Detection

In order to calculate accurate sedimentation coefficients in the analytical ultracentrifuge, one needs to know the radial position of the meniscus (i.e., the air/water interface) in the sample sector (Bailey et al. 2009; Zhao et al. 2014b). This represents the origin of sedimentation. With absorbance or interference optics, the meniscus can easily be determined since an obvious spike in the signal is observed at the air/water interface. However, given that the AU-FDS measures fluorescence just below the surface of the sample, there is no opportunity for light to scatter. Accordingly, the position of the sample meniscus is not easily recognizable in radial scans generated in the AU-FDS. One solution developed by Bailey and co-workers is to overlay the sample with light mineral oil containing an uncharged fluorescent dye such as BODIPY 493/503 (Bailey et al. 2009). This produces a clear signal at the oil/water interface that correlates to the origin of sedimentation. More recently, Schuck and colleagues describe an alternative method for determining the meniscus position in the AU-FDS using the Raman scattering profile of water (Zhao et al. 2014b). This method works well at high PMT voltages, since Raman scattering of water results in a baseline signal shift at the radius corresponding to the air/water interface (Zhao et al. 2014b).

4.4.3 Sample Loading

As for absorbance- or interference-detected measurements, samples to be analyzed in the AU-FDS are loaded into double-sector cells fitted with either quartz or sapphire windows pre-torqued to 120 psi. As documented in Sect. 4.3, a reference solution is not required, which enables the user to load sample into both compartments when using conventional double-sector cells. The cells are sealed from the external environment using Teflon plugs and brass screws in the same manner employed for absorbance or interference measurements. Each cell is then loaded into either a Beckman An-60 Ti four-hole rotor or Beckman An-50 Ti eight-hole rotor. However, instead of using a counterbalance as employed for absorbance or interference measurements, a special purpose-built calibration centerpiece is required. This centerpiece has the same weight as a two-channel epon or charcoal centerpiece but contains a fluorescent strip with known dimensions. Given that it has the same weight as a two-channel centerpiece, no counterweight is required. The rotor containing sample cells and calibration centerpiece is then mounted in the AU-FDS ready for experiment initiation.

4.5 Experimental Operation Procedures

The following subsections will present a basic overview of experimental operation of the AU-FDS. It is not the intention here to describe a complete standard operation procedure for the instrument, since this is provided by Aviv Biomedical upon purchase of the FDS optics unit and AOS software. Instead, the proceeding subsections will focus on some of the unique operational features, common problems, and strategies the user can adopt to manage or circumvent these problems.

4.5.1 Experiment Initiation

Both the FDS detector and XL-A/XL-I centrifuge are controlled by the AOS software. When the software is initiated, the *rotor setup* panel appears (Fig. 4.3). Additional panels are available to configure the experiment and select cells for data collection. The AU-FDS enjoys a large dynamic range of up to five magnitudes in fluorescence intensity. This is made possible given the powerful laser source, high-sensitivity detector, high-voltage PMT, and the addition of an adjustable amplifier (Fig. 4.1c). Both the PMT voltage and the amplifier's gain can be adjusted to levels that are optimal for each sample. When samples have more than a magnitude

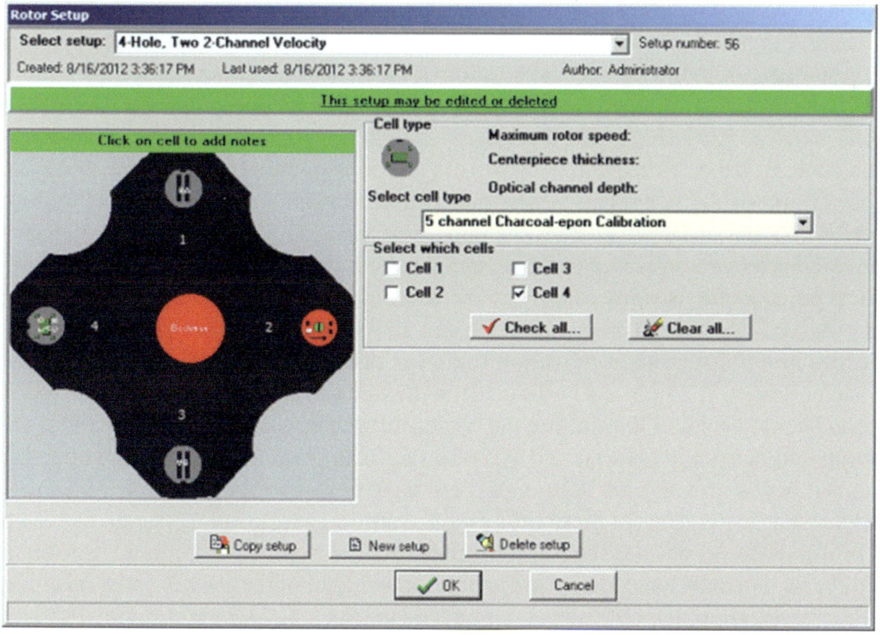

Fig. 4.3 Rotor setup screen of the AOS software

difference in fluorescent intensity, experiments can be performed with more than one set of sensitivities. However, each gain setting requires additional radial scans, which attenuates the rate of data collection. Accordingly, most experiments in the AU-FDS, particularly sedimentation velocity studies, are conducted with a single gain setting to afford maximum scan rates of ~90 s per sample.

4.5.2 Position Calibration

To measure concentration gradients in the analytical ultracentrifuge, the optical system must be able to accurately determine the position of the signal. This is achieved in the AU-FDS using a cylindrical coordinate system described by an "angle," "radius," and "height."

The "angle" is dependent on the rotor velocity and is adjusted automatically by AOS. When the rotor speed changes, AOS applies an algorithm that searches for the fluorescent strip in the calibration centerpiece. The AOS software then applies an angle offset so that the center of the strip falls in the expected angular range. This process is referred to as "locking the magnet angle."

The "radius" can be adjusted automatically or manually. This is achieved by performing a radial calibration via the *fluorescence focusing and calibration tool* panel of AOS (Fig. 4.4). The high-radius edge of the fluorescent strip serves as the reference point. A radial calibration should be performed at least monthly, when the rotor is changed or when maintenance is performed.

The "height" refers to the position of the focal point of the excitation beam, which is adjusted by moving the vertical position of the focus lens located above the sample cell (Fig. 4.1c). This is also adjusted in the *fluorescence focusing and calibration tool* panel of AOS (Fig. 4.4). The "focus height" is critical in the AU-FDS given the conical shape of the beam (Fig. 4.1c). The adjustment range is broad, since the focus point of the beam can be positioned from just above the sample solution to well below the surface, therefore encompassing the 12 mm path length of a standard double-sector centerpiece. For dilute samples (less than 0.1 optical density/cm), the focal point profile is sigmoidal (Fig. 4.5, 13 pM IgG). Moreover, little fluorescence is observed when the focus point is located above the sample; but as this is shifted deeper into the sample, a transition region is observed followed then by a signal plateau (Fig. 4.5, 13 pM IgG). To achieve maximum signal and stability, the focus point should be set to the starting portion of the plateau. It should be noted that the 4 mm height sample cells are too short for the focal point to be fully encompassed, since it begins to leave the bottom of the sample before it has fully entered the top. As a result, focus scans of these cells produce only a peak, never a plateau; therefore 4 mm cells do not provide the full sensitivity of 12 mm cells. Furthermore, Beckman Coulter 4 mm cells have a top surface below the height of the 12 mm cells, resulting in increased clipping at the high-radius edge (see Sect. 4.5.3.2 for clipping details).

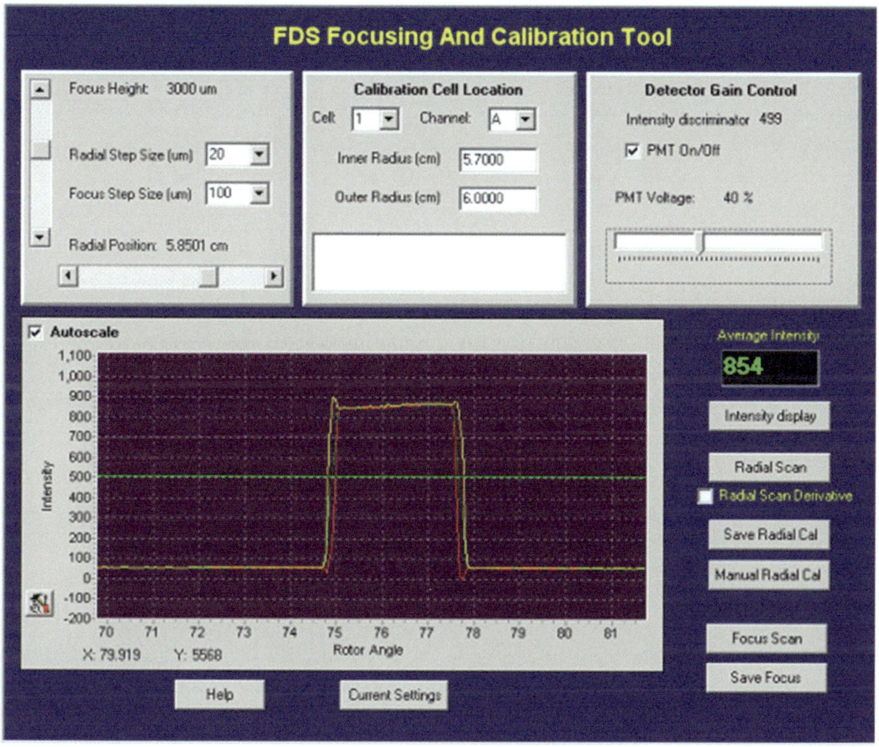

Fig. 4.4 FDS focusing and calibration tool screen of AOS

4.5.3 Common Signal Attenuation Issues

The optical nature of the FDS can give rise to common phenomena that result in attenuation of the signal, namely, (i) the "inner filter effect," (ii) "light beam clipping," and (iii) "focal height drift." The strategies employed to circumvent these potential problems will be discussed below.

4.5.3.1 Inner Filter Effect

At high concentrations of fluorescently labeled sample (>0.1 optical density/cm), significant absorption of incident and emitted light results. This is known as the "inner filter effect" (Lyons et al. 2013). A practical rule of thumb is that if the sample has a color seen by the eye, then it is at risk of having an inner filter effect. In the AU-FDS, inner filtering occurs when the focal depth is moved deeper into a sample solution that absorbs, whether it be due to the fluorophore, sample, or other buffer components. What results is a characteristic sigmoidal profile followed then by an

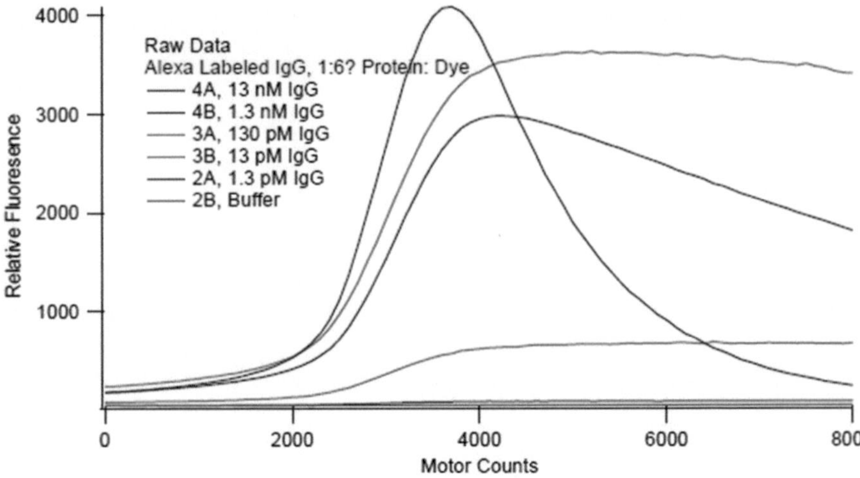

Fig. 4.5 Focus scans of Alexa Fluor 488-labeled IgG at different loading concentrations. The PMT high voltages were adjusted to keep the signal strengths in range of the electronics. As a result, the signal amplitude is not necessarily proportionate to the concentration between samples

exponential decrease in light intensity as the focus height is moved deeper into the sample (Fig. 4.5, 13 nM IgG). Intuitively, one would predict that inner filtering could be prevented in the AU-FDS by raising the focal height of the beam closer to the sample surface. However, this is avoided in practice, since the change in sample concentration during sedimentation (or flotation) results in deeper penetration of the laser light causing a change in the focus profile. This introduces nonlinearity of the signal as a function of sample concentration. Accordingly, the best method to avoid inner filtering in the AU-FDS is to reduce the label concentration of the fluorescent sample at higher sample concentrations. Moreover, concentration-dependence studies should simply employ a fixed concentration of labeled sample but an increase in the proportion of unlabeled sample (Wowor et al. 2011). As such, all samples in the concentration series will have similar fluorescence levels. Importantly, this approach will also afford the use of a single gain setting thus maximizing the rate of data collection (Sect. 4.5.1). In summary, the best way to manage the inner filter effect in the AU-FDS is to prevent it from occurring.

4.5.3.2 Light Beam Clipping

Given that the excitation beam in the AU-FDS is conical in shape, light emerging from the first focus lens to the top of the sample can be "clipped" by the edge of the cell (Fig. 4.1c). This results in attenuation of the fluorescence signal. By comparison, the light beam emerging from the absorbance and interference systems is columnar in shape and therefore is unaffected by movements in the light beam. Clipping can occur in two physical locations. First, the edges of the sample channels can clip the light if the focus is set too deep in the sample. This situation reduces the period of time the sample is clearly viewed during a rotation. To circumvent this problem, the focal point of the beam is positioned at the beginning of the focus scan plateau (Fig. 4.5). This also prevents focal height drift (Sect. 4.5.3.3). The second source of clipping is even more prominent at the high-radius (i.e., wider) end of the sample sector, due to the cell's retaining ring. However, this clipping can be mitigated by using a dense oil (such as FC-43) to raise the sample off the bottom of the cell (Bailey et al. 2009). Alternatively, a special FDS two-channel velocity cell with a slightly shorter column height is available to purchase from Spin Analytical (www.spinanalytical.com).

4.5.3.3 Focal Height Drift

Since the optical path in the FDS is not perfectly parallel to the sample surface, a slight drift of the focus height occurs during radial scans. Hence, if the focus height is placed on a sloping portion of the focus scan (Fig. 4.5), the slight vertical drift of the optics results in a signal change. Nevertheless, setting the focal point to the plateau region of the focus scan prevents this drift from occurring.

4.5.4 Data Acquisition and Analyses

Once the user has optimized the fluorescence signal for each sample and programmed the experiment method to be followed, the experiment is commenced by hitting START in the AOS main experiment window (Fig. 4.6). Temperature equilibration can be handled manually or can be included as part of the experiment method. The rotor will then accelerate until the input speed is attained at which point data collection will commence. Each scan is then measured as a function of time and plotted by the AOS software to allow the user to view all time points (Fig. 4.6). The experiment is then stopped after a predetermined time/number of scans or manually by the user when all samples have sedimented to the bottom (or floated to the top) of the cell. The AOS software packages the data in time-stamped files that are available for analysis using contemporary software suites, such as SEDANAL (Stafford and Sherwood 2004), SEDFIT (Schuck 2000; Schuck et al. 2002), SEDPHAT (Vistica et al. 2004), and ULTRASCAN (Demeler 2005). Both sedimentation velocity and equilibrium types of experiments are possible.

4.5.5 Post-experimental Considerations

The post-experiment procedures of the AU-FDS do not differ substantially from those employing absorbance or interference optics. However, the detergent used for

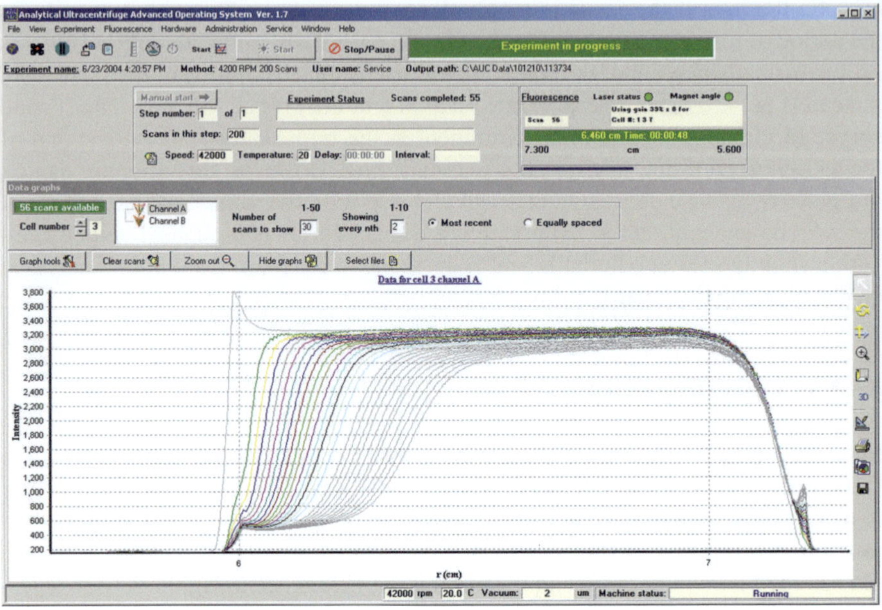

Fig. 4.6 AOS main experiment screen showing fluorescence intensity versus radial position profile of a sample at different time points

cleaning cells is critical, since many household detergents contain high amounts of fluorescein. As a consequence, we have found that trace amounts of these detergents remaining in cleaned cells can give rise to a non-sedimenting baseline in the AU-FDS. However, this problem can be avoided by using a natural detergent, which is colorless or pale yellow in appearance, compared to fluorescein-containing products that have a bright fluorescent green color.

4.6 Applications of AU-FDS

4.6.1 High-Affinity Interactions

As documented in Sect. 4.3, one of the major advantages of the AU-FDS platform is the ability to quantify high-affinity or tight interactions (Table 4.1). To illustrate this in the context of protein self-association, Burgess and colleagues employed the AU-FDS to study and quantify the self-association of a tight dimeric enzyme from the gram-positive pathogen *Staphylococcus aureus* (also refer to Chap. 16). The enzyme, dihydrodipicolinate synthase (DHDPS) (Fig. 4.7a), is of significant interest to biomedicine given that it is the product of an essential bacterial gene that is absent in humans (Dogovski et al. 2013; Kobayashi et al. 2003). Accordingly,

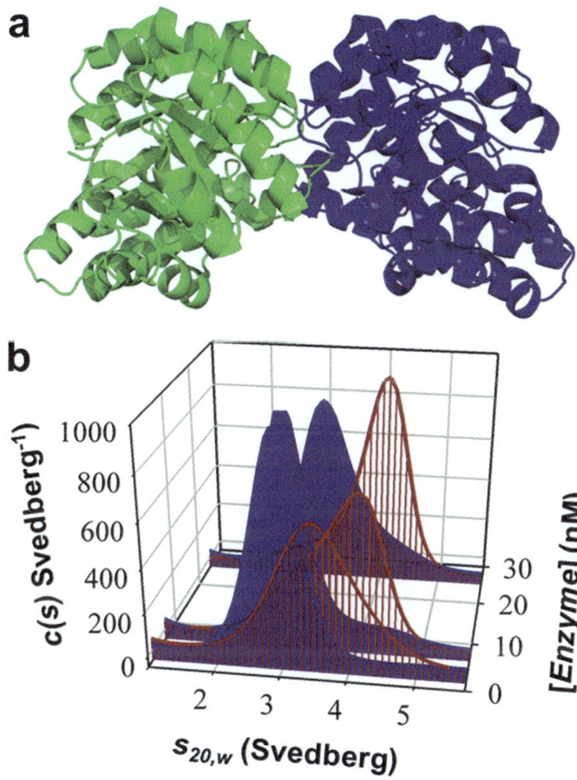

Fig. 4.7 AU-FDS studies of a high-affinity dimeric enzyme. (**a**) Structure of *S. aureus* DHDPS dimer (PDB ID: 3DAQ). (**b**) Continuous size-distribution [*c(s)*] analyses of *S. aureus* DHDPS plotted as a function of standardized sedimentation coefficient and enzyme concentration. The distributions for the apoenzyme are shown in *blue*, whereas the profiles in the presence of pyruvate are displayed in *red*

DHDPS represents a valid but as yet unexploited antibiotic target (Dogovski et al. 2009, 2012; Gerrard et al. 2007; Hutton et al. 2007). The study by Burgess et al. (2008) commenced with absorbance-detected AUC measurements at low micromolar enzyme concentrations. The resulting sedimentation velocity data, which was analyzed using enhanced van Holde-Weischet (Demeler and van Holde 2004) and continuous size-distribution (Schuck 2000; Schuck et al. 2002) algorithms, demonstrated that *S. aureus* DHDPS exists as a 4.2 S dimeric species. The dimer was shown to be highly stable at enzyme concentrations ranging from mid- to low micromolar in either the apo (i.e., unliganded) or substrate-bound forms. However, sedimentation velocity experiments conducted in the AU-FDS using Alexa Fluor 488-labeled enzyme at high picomolar to low nanomolar concentrations showed that *S. aureus* DHDPS actually exists in a monomer-dimer equilibrium (Fig. 4.7b). Interestingly, the addition of the substrate, pyruvate, shifted the equilibrium in favor of the dimer (Fig. 4.7b). Although pyruvate absorbs strongly in the ultraviolet region, which limits its use in absorbance-detected AUC experiments, saturating (i.e., mM) concentrations of pyruvate were able to be employed in this study, since the substrate is nonfluorescent and thus did not contribute to the signal derived from the Alexa Fluor 488-labeled enzyme. Subsequent sedimentation equilibrium studies were then conducted at high picomolar to low nanomolar enzyme concentrations in the absence and presence of pyruvate to show that the dimer-monomer dissociation constant ($K_D^{2\to1}$) was 33 nM for the apoenzyme but 20-fold tighter in the presence of pyruvate (i.e., $K_D^{2\to1} = 1.6$ nM). The dimerization affinity determined by AU-FDS was subsequently validated using enzyme kinetics assays that determined the concentration dependence on the specific activity of *S. aureus* DHDPS (Burgess et al. 2008). This was made possible since the *S. aureus* DHDPS monomer is significantly less active than the dimer (Burgess et al. 2008). The Burgess et al. (2008) study was one of the first to employ AU-FDS to quantify a high-affinity interacting system and provide an example of both NUTS and BOLTS applications of the technique (Sect. 4.3). Subsequent studies have also employed the platform to quantify high-affinity interactions, including self-associating proteins (Wowor et al. 2011; Zhao et al. 2012), antibody-antigen interactions (Kroe and Laue 2009; Zhao et al. 2014b), and protein-RNA interactions (Husain et al. 2012).

4.6.2 Studies in Complex Backgrounds

The use of absorbance-detected AUC provides a convenient method for measuring sedimentation. While this is a powerful technique for purified samples, the nonspecific nature of protein absorption makes absorbance-detected AUC inappropriate in complex or heterogeneous sample backgrounds. As discussed in Sect. 4.3, the BOLTS application of AU-FDS makes it possible to characterize the hydrodynamic properties of a fluorescently labeled macromolecule in biological fluids, such as blood serum and cell lysates (Table 4.1). As will be seen below, these more complex solutions can change the behavior of the macromolecule of interest, and so observations made using simpler buffers may not predict the true behavior of biomolecules in vivo. However, when assessing the behavior of protein drugs (such as monoclonal antibodies) or studying protein aggregation linked to common age-related diseases (such as systemic amyloidosis), it is critical to understand how proteins behave in biological fluids.

4.6.2.1 Behavior of Protein Therapeutics in Blood Serum

The importance of studying biomolecules in biological fluids was highlighted by Demeule et al. (2009) who showed that proteins can form different quaternary structures in plasma than they do in phosphate-buffered saline (PBS). Demeule et al. (2009) reasoned that while much in vitro work is conducted to assess the properties of protein drugs, little is known of their properties in the bloodstream, due in part to the expense and difficulty of conducting animal and clinical studies. One such drug is omalizumab, an anti-IgE monoclonal antibody used in the treatment of asthma (Liu et al. 1995). Earlier work had shown that omalizumab formed different complexes with IgE when the omalizumab/IgE molar ratio was varied

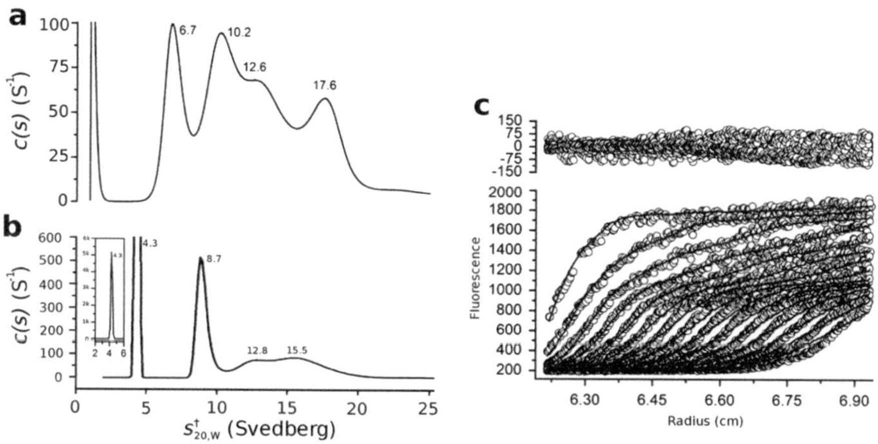

Fig. 4.8 AU-FDS analysis of antibody complexes in serum. *c(s)* distributions for omalizumab-IgE complexes in (**a**) PBS and (**b**) serum where $s^{\dagger}_{20,w}$ is the apparent corrected sedimentation coefficient. The distribution in serum shows substantially more monomers (8.7 S) than oligomers, compared to the distribution in PBS. The 4.3 S species (panel (**b**) inset) is attributed to albumin. (**c**) Data (*symbols*) and fits (*lines*) for the *c(s)* distribution best fit in serum (refer to panel (**b**)). Residuals are shown above plotted as a function of radial position (Figure adapted from Demeule et al. (2009))

(Liu et al. 1995). Demeule and colleagues therefore set out to employ AU-FDS to further explore these antibody-antigen interactions. Omalizumab was labeled with Alexa Fluor 488, mixed in PBS buffer with IgE in equimolar concentrations, and then added to human serum. The hydrodynamic properties of the mixture in serum (Fig. 4.8b) compared to the PBS control (Fig. 4.8a) were determined by sedimentation velocity studies. The *c(s)* distribution analyses showed that substantial differences exist between the omalizumab-IgE complexes in buffer and serum, with higher levels of monomer (8.7 S species) observed in the more complex background of serum. The 4.3 S species (Fig. 4.8b) is attributed to albumin, which has been shown to bind endogenous fluorescent molecules, highlighting that AU-FDS data generated in serum must be carefully interpreted. The nonlinear regression least squares output for the *c(s)* distribution in serum showed that the residuals are systematic and larger than the random noise (Fig. 4.8c), indicating that the *c(s)* distribution model does not perfectly model the data. The bulk density and viscosity of serum were employed in these analyses to determine the apparent corrected sedimentation coefficient ($s^{\dagger}_{20,w}$) of the various species. Although the authors acknowledge that this may produce inaccuracies, the resulting $s^{\dagger}_{20,w}$ values actually matched those determined in PBS. This study was one of the first to provide proof of concept that the hydrodynamic properties of fluorescently labeled macromolecules can be determined in complex biological backgrounds using the AU-FDS.

4.6.2.2 Aggregation of Serum Proteins Linked to Systemic Amyloidosis

The nature of complex formation in biological fluids is of particular interest in the study of amyloid diseases, such as Alzheimer's disease, Parkinson's disease, Huntington's disease, and systemic amyloidosis, since it is yet to be established whether protein aggregates observed in these diseases are pathogenic or merely symptomatic (Swart et al. 2014). In the case of senile systemic amyloidosis (SSA), the primary cause of death for 70 % of supercentenarians (Coles and Young 2012), the disease is manifested by the aggregation of transthyretin (TTR), which normally functions as a tetramer to transport thyroid hormone in the bloodstream. It is unknown how TTR transitions between a physiologically active soluble tetramer to a disease-associated amyloid-like aggregate, but it is generally assumed that the first step involves dissociation of the tetramer (Coles and Young 2012). To examine the aggregation of TTR, Kingsbury et al. (2008) conducted sedimentation velocity experiments in the AU-FDS using recombinant TTR labeled with fluorescein that was subsequently added to human serum. Similar to the aforementioned studies of omalizumab (Sect. 4.6.2.1), the *c(s)* distribution best fit of TTR in serum resulted in large systematic residuals (Fig. 4.9a). This once again indicates that the *c(s)* distribution model (Schuck 2000; Schuck et al. 2002) does not appropriately account for the nonideality in the complex background of serum. Nevertheless, the resulting *c(s)* distribution (at a confidence interval of 68 %) yielded a single peak with a sedimentation coefficient of 2.41 S, which is consistent with the size found by boundary midpoint analysis (2.45 S) (Fig. 4.9b). Kingsbury et al. (2008) therefore argue that although investigating protein aggregation in complex solutions is somewhat confounded by current analysis approaches, AU-FDS is a valuable platform to study protein aggregation ex vivo. However, the outputs using current

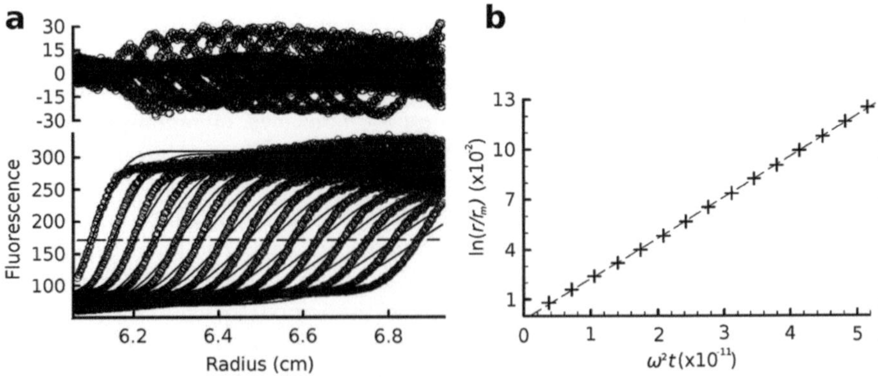

Fig. 4.9 AU-FDS studies of transthyretin aggregation in serum. (**a**) Data (*symbols*) and fits (*lines*) resulting from the *c(s)* distribution best fit of TTR in serum resulting in a single peak with sedimentation coefficient of 2.41 S (distribution not shown). Residuals are plotted above as a function of radial position. (**b**) Boundary midpoint analysis for the data shown in panel (a) yields a single species of 2.45 S (Figure adapted from Kingsbury et al. (2008))

data analysis methods, such as *c(s)* distribution analysis, should be observed in a more qualitative rather than quantitative manner.

Kingsbury et al. (2012) subsequently set out to explore a second aggregating system, namely, immunoglobulin light chain, which is associated with another form of systemic amyloidosis known as light chain amyloidosis (AL). AL is the most common form of systemic amyloidosis and the most life threatening, particularly when cardiomyopathy presents (Sanchorawala 2006). It is caused by unregulated light chain production by plasma cells and, if left untreated, can induce pathogenic changes in many systems including the hepatic, digestive, peripheral nervous, and cardiovascular system (Sanchorawala 2006). In the study by Kingsbury et al. (2012), serum was collected from patients suffering from systemic amyloidosis and classed into one of four clinical groups, namely, (i) no amyloidosis, (ii) SSA, (iii) AL without cardiomyopathy (AL), and (iv) AL with cardiomyopathy (AL-CMP). A high concentration of albumin was used as a negative control. As adopted in the TTR study described earlier (Kingsbury et al. 2008), fluorescein was used as the tracer for BOLTS experiments in the AU-FDS. The fluorophore was simply added to the serum samples collected from individual patients. The resulting sedimentation profiles of the four groups showed that the samples could be clearly differentiated into two types, designated type 1 or type 2. Type 1 samples displayed only one boundary with a sedimentation coefficient of 2.1–3.3 S, comparable to the fluorescein-bound albumin control (2.3 S) (Fig. 4.10a, c). However, type 2 samples were heterogeneous with two major high-molecular-weight complexes (HMWCs) with sedimentation coefficients of 5.8 S and 10.8 S (Fig. 4.10b, d). The additional inverted slow-moving boundary observed in the type 2 sample (Fig. 4.10b) is consistent with the Johnston-Ogston effect, which is a sedimentation artifact arising from nonideal boundaries (Johnston and Ogston 1946). All four clinical groups displayed the type 1 profile, but type 2 patterns were only observed in samples displaying amyloidosis, suggesting that HMWCs are a hallmark of disease. Furthermore, HMWCs were detected in over half of the AL-CMP samples and less often in AL and SSA, correlating with AL-CMP being the most life-threatening form. This supports the notion that HMWCs may play a role in pathogenesis. Interestingly, the addition of recombinant TTR to SSA patient serum increased the levels of HMWCs, suggesting that the HMWCs observed in systemic amyloidosis are associated with TTR. As with the study by Kingsbury et al. (2008) reported earlier, the *c(s)* distribution model provides a poor description of the data. This is likely to be due to time-dependent spreading of the boundary resulting from the high-density gradient presenting in the centrifuge due to the complex nature of serum. Although this prevented the determination of accurate molecular weights for the HMWCs, the sedimentation coefficients derived via Lamm equation modeling approximated to that determined by boundary midpoint migration. Despite these limitations, AU-FDS has allowed the first observations of high-molecular-weight complexes in the serum of patients suffering systemic amyloidosis.

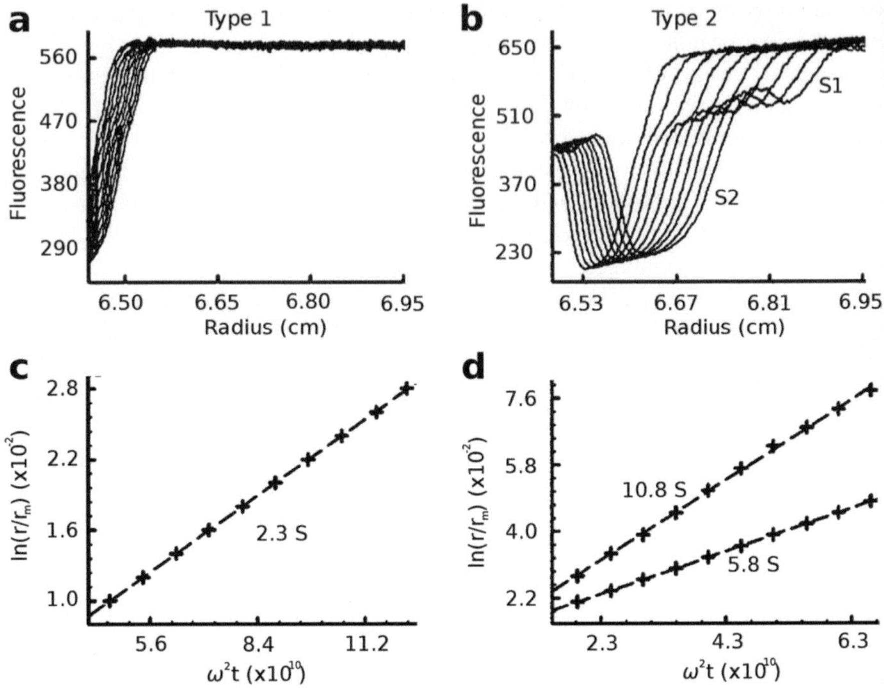

Fig. 4.10 AU-FDS analysis of high-molecular-weight complexes in serum from patients with systemic amyloidosis. Fluorescence intensity is plotted as a function of radial position for (**a**) type 1 and (**b**) type 2 samples, with the resulting midpoint boundary analyses shown in panels (**c**) and (**d**), respectively. Based on the sedimenting boundaries shown, patient samples can be grouped into type 1 or type 2 (Figure adapted from Kingsbury et al. (2012))

4.6.2.3 Aggregation of Huntingtin in Cell Lysates

AU-FDS has also been employed to study protein aggregation of huntingtin (Htt) protein associated with Huntington's disease (Olshina et al. 2010). This is an autosomal dominant condition caused by the addition of repeated CAG codons in exon 1 of the huntingtin gene, which results in poly-glutamine (poly-Q)-rich mutant huntingtin proteins (Landles and Bates 2004). Olshina et al. (2010) investigated Htt aggregation in the AU-FDS by studying Htt containing 46 glutamine repeats (Htt46Q) as a GFP fusion protein in aqueous buffer and also in mouse neuroblastoma (Neuro2a) cell lysates. In buffer, the recombinant Htt46Q-GFP construct was shown to exist as a 2.4 S monomer that aggregated over time to form 100 to 6000 S complexes. By contrast, AU-FDS studies of Htt46Q-GFP in Neuro2a cell lysates harvested at different transfection time points showed that three distinct species are present, namely, 2.3 S monomer, 140 S soluble oligomer, and 320,000 S inclusion bodies (Fig. 4.11). Interestingly, the relative amount of 2.3 S monomer decreases over time, while the proportion of 320,000 S inclusion bodies increase

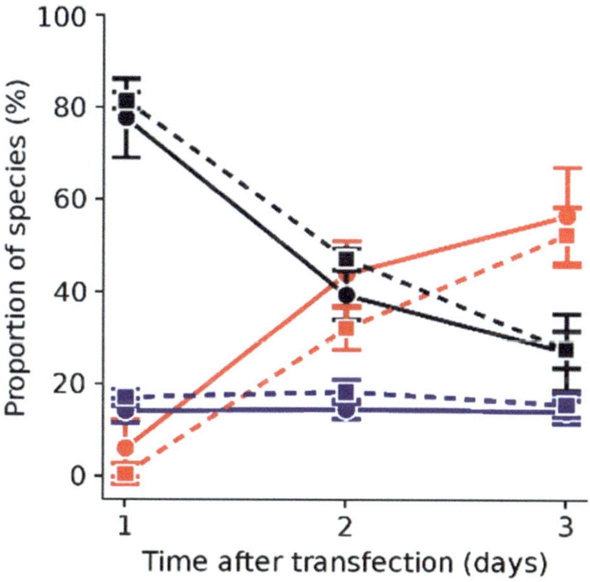

Fig. 4.11 AU-FDS experiments of huntingtin aggregates in cell lysates. Proportions of different species of Htt46Q-GFP in Neuro2a cell lysates are plotted as a function of transfection time. *Black*, monomer (2.3 S); *blue*, oligomer (140 S); and *red*, inclusion bodies (320,000 S). *Dotted* and *solid lines* indicate the absence and presence, respectively, of Hsc70 (Figure adapted from Olshina et al. (2010))

and the amount of 140 S oligomer remains constant (Fig. 4.11). This suggests that the 140 S species represents a rate-limiting factor in aggregation. Studies were also conducted in the presence of Hsc70, which is a member of the heat shock protein 70 (Hsp70) chaperone family that is known to co-localize with inclusion bodies and lower Htt toxicity (Warrick et al. 1999). AU-FDS experiments conducted in Neuro2a cell lysates containing overexpressed Hsc70 show that the relative proportion of soluble oligomers decrease while the proportion of inclusion bodies increase (Fig. 4.11, solid lines). This suggests that Hsp70 reduces Htt toxicity by facilitating the conversion of soluble oligomers into inclusion bodies. Accordingly, this study yielded critical insights into the molecular mechanisms underpinning pathogenesis in Huntington's disease.

4.6.3 Identifying Enzyme Complexes

In addition to the aforementioned NUTS and BOLTS applications of AU-FDS, Wang et al. (2012) have recently employed the technology to explore the size and composition of enzyme complexes functioning in translation from the model organism, *Saccharomyces cerevisiae*. By tagging specific proteins or RNA with GFP and performing a pulldown with FLAG-labeled large ribosomal subunit protein RPL25A or poly(A)-binding protein PAB1, Wang et al. were able to identify a new 77 S monosomal translation complex. Akin to Western blotting, this AU-FDS application showed that the complex is comprised of the 80S ribosome, mRNA, PAB1, and eukaryotic initiation factors, eIF4E and eIF4G (Wang et al.

2012). Additional experiments conducted in a separate study demonstrate that the complex also contains eRF1, SLF1, SSD1, PUB1, and SBP1 (Zhang et al. 2014). Further, Wang et al. (2012) probed the mechanisms by which environmental stress inhibits translation. This study showed that glucose starvation leads to a substantial reduction of 77 S complex abundance, an effect that is partially rescued by mutations in eIF4E and eIF3b or by deletion of PAT1 but not by mutations in PAB1. This suggests that glucose starvation acts through eIF4E. By contrast, amino acid deprivation, osmotic stress, and heat shock are not rescued, suggesting they do not act primarily through this complex (Wang et al. 2012).

4.7 Conclusions

The recent development of the fluorescence detection system for the analytical ultracentrifuge (AU-FDS) has provided significant advances to the hydrodynamic and thermodynamic analyses of macromolecular systems. This platform technology provides *sensitivity* for the measurement of low-abundance proteins and the quantitation of tight biomolecular interactions (NUTS applications), as well as *selectivity* for the specific detection of a biomolecule of choice in complex biological backgrounds (BOLTS applications). In this chapter, we have described studies of tight self-associating enzymes (DHDPS), protein therapeutics (omalizumab), and amyloid-like aggregating proteins (TTR, immunoglobulin light chain, and huntingtin) to demonstrate NUTS and BOLTS applications of the AU-FDS. We also describe the use of the technology for proteomic investigations, such as defining the composition of the 77 S monosomal translation complex. This highlights the great diversity of the platform in this, the post-genome era, where the emphasis has switched from the delineation of genome structure and the characterization of isolated gene products to the qualitative and quantitative measurement of gene product interaction networks in complex in vivo-like backgrounds. However, several challenges remain for the advancement of the AU-FDS, including the development of multiwavelength excitation laser sources to permit the use of a broader range of fluorescent probes and advancement of analytical tools to circumvent nonideality observed in complex biological fluids. Nevertheless, the AU-FDS provides researchers with a cutting edge tool for contemporary applications of macromolecular characterization in solution.

References

Bailey MF, Angley LM, Perugini MA (2009) Methods for sample labeling and meniscus determination in the fluorescence-detected analytical ultracentrifuge. Anal Biochem 390(2):218–220

Burgess BR, Dobson RCJ, Bailey MF, Atkinson SC, Griffin MDW, Jameson GB, Parker MW, Gerrard JA, Perugini MA (2008) Structure and evolution of a novel dimeric enzyme from a clinically-important bacterial pathogen. J Biol Chem 283(41):27598–27603

Campbell RE, Tour O, Palmer AE, Steinbach PA, Baird GS, Zacharias DA, Tsien RY (2002) A monomeric red fluorescent protein. Proc Natl Acad Sci U S A 99(12):7877–7882
Cole JW, Lary JW, Moody TP, Laue TM (2008) Analytical ultracentrifugation: sedimentation velocity and sedimentation equilibrium. In: Correia JJ, Detrich H (eds) Methods in cell biology, vol 84. Elsevier, San Diego, pp 143–179
Coles LS, Young RD (2012) Supercentenarians and transthyretin amyloidosis: the next frontier of human life extension. Prev Med 54:S9–S11
Demeler B (2005) UltraScan a comprehensive data analysis software package for analytical ultracentrifugation experiments. In: Scott DJ, Harding SE, Rowe AJ (eds) Modern analytical ultracentrifugation: techniques and methods. Royal Society of Chemistry, Cambridge, pp 210–229
Demeler B, van Holde KE (2004) Sedimentation velocity analysis of highly heterogeneous systems. Anal Biochem 335(2):279–288
Demeule B, Shire SJ, Liu J (2009) A therapeutic antibody and its antigen form different complexes in serum than in phosphate-buffered saline: a study by analytical ultracentrifugation. Anal Biochem 388(2):279–287
Dogovski C, Atkinson S, Dommaraju S, Dobson RCJ, Hor L, Hutton CA, Gerrard JA, Perugini MA (2009) Lysine biosynthesis in bacteria: an unchartered pathway for novel antibiotic design. In: Doelle HW, Rokem S (eds) Encyclopedia of life support systems, vol 11. Encyclopedia of Life Support Systems (EOLSS). Developed under the Auspices of the UNESCO, EOLSS Publishers, Paris, pp 116–136
Dogovski C, Atkinson SC, Dommaraju SR, Downton M, Hor L, Moore S, Paxman JJ, Peverelli MG, Qiu TW, Reumann M, Siddiqui T, Taylor NL, Wagner J, Wubben JM, Perugini MA (2012) Enzymology of bacterial lysine biosynthesis. In: Ekinci D (ed) Biochemistry. InTech Open Access Publisher, Rijeka, pp 225–262
Dogovski C, Gorman MA, Ketaren NE, Praszkier J, Zammit LM, Mertens HD, Bryant G, Yang J, Griffin MD, Pearce FG, Gerrard JA, Jameson GB, Parker MW, Robins-Browne RM, Perugini MA (2013) From knock-out phenotype to three-dimensional structure of a promising antibiotic target from *Streptococcus pneumoniae*. PLoS ONE 8(12):e83419
Gerrard JA, Hutton CA, Perugini MA (2007) Inhibiting protein-protein interactions as an emerging paradigm for drug discovery. Mini Rev Med Chem 7(2):151–157
Giebeler R (1992) The optima XL-A: a new analytical ultracentrifuge with a novel precision absorption optical system. In: Harding SE, Rowe AJ, Horton JC (eds) Analytical ultracentrifugation biochemistry and polymer science. Royal Society of Chemistry, Cambridge, pp 16–25
Hellman LM, Zhao C, Melikishvili M, Tao X, Hopper JE, Whiteheart SW, Fried MG (2011) Histidine-tag-directed chromophores for tracer analyses in the analytical ultracentrifuge. Methods 54(1):31–38
Hochuli E, Bannwarth W, Döbeli D, Gentz R, Stüber D (1988) Genetic approach to facilitate purification of recombinant proteins with a novel metal chelate adsorbent. Nat Biotech 6(11):1321–1325
Howlett GJ, Minton AP, Rivas G (2006) Analytical ultracentrifugation for the study of protein association and assembly. Curr Opin Chem Biol 10(5):430–436
Husain B, Mukerji I, Cole JL (2012) Analysis of high-affinity binding of protein kinase R to double-stranded RNA. Biochemistry 51(44):8764–8770
Hutton CA, Perugini MA, Gerrard JA (2007) Inhibition of lysine biosynthesis: an evolving antibiotic strategy. Mol Biosyst 3(7):458–465
Johnston JP, Ogston AG (1946) A boundary anomaly found in the ultracentrifugal sedimentation of mixtures. Trans Faraday Soc 42:789–799
Kingsbury JS, Laue TM, Klimtchuk ES, Theberge R, Costello CE, Connors LH (2008) The modulation of transthyretin tetramer stability by cysteine 10 adducts and the drug diflunisal: direct analysis by fluorescence-detected analytical ultracentrifugation. J Biol Chem 283(18):11887–11896

Kingsbury JS, Laue TM, Chase SF, Connors LH (2012) Detection of high-molecular-weight amyloid serum protein complexes using biological on-line tracer sedimentation. Anal Biochem 425(2):151–156

Kobayashi K, Ehrlich SD, Albertini A, Amati G, Andersen KK, Arnaud M, Asai K, Ashikaga S, Aymerich S, Bessieres P, Boland F, Brignell SC, Bron S, Bunai K, Chapuis J, Christiansen LC, Danchin A, Debarbouille M, Dervyn E, Deuerling E, Devine K, Devine SK, Dreesen O, Errington J, Fillinger S, Foster SJ, Fujita Y, Galizzi A, Gardan R, Eschevins C, Fukushima T, Haga K, Harwood CR, Hecker M, Hosoya D, Hullo MF, Kakeshita H, Karamata D, Kasahara Y, Kawamura F, Koga K, Koski P, Kuwana R, Imamura D, Ishimaru M, Ishikawa S, Ishio I, Le Coq D, Masson A, Mauel C, Meima R, Mellado RP, Moir A, Moriya S, Nagakawa E, Nanamiya H, Nakai S, Nygaard P, Ogura M, Ohanan T, O'Reilly M, O'Rourke M, Pragai Z, Pooley HM, Rapoport G, Rawlins JP, Rivas LA, Rivolta C, Sadaie A, Sadaie Y, Sarvas M, Sato T, Saxild HH, Scanlan E, Schumann W, Seegers JF, Sekiguchi J, Sekowska A, Seror SJ, Simon M, Stragier P, Studer R, Takamatsu H, Tanaka T, Takeuchi M, Thomaides HB, Vagner V, van Dijl JM, Watabe K, Wipat A, Yamamoto H, Yamamoto M, Yamamoto Y, Yamane K, Yata K, Yoshida K, Yoshikawa H, Zuber U, Ogasawara N (2003) Essential *Bacillus subtilis* genes. Proc Natl Acad Sci U S A 100(8):4678–4683

Kroe RR, Laue TM (2009) NUTS and BOLTS: applications of fluorescence-detected sedimentation. Anal Biochem 390(1):1–13

Landles C, Bates GP (2004) Huntingtin and the molecular pathogenesis of Huntington's disease. EMBO Rep 5(10):958–963

Laue TM (1996) Optical systems of the XLA ultracentrifuge. Applications data note for Spinco division of Beckman Instruments. Beckman Instruments Inc, Palo Alto

Laue TM (2004) Analytical ultracentrifugation: a powerful 'new' technology in drug discovery. Drug Discov Today Technol 1(3):309–315

Laue TM, Stafford WF (1999) Modern applications of analytical ultracentrifugation. Annu Rev Biophys Biomol Struct 28:75–100

Laue TM, Anderson AL, Weber BW (1997) Prototype fluorescence detector for the XLA analytical ultracentrifuge. SPIE Proc 2985:196–204

Liu J, Lester P, Builder S, Shire SJ (1995) Characterization of complex formation by humanized anti-IgE monoclonal antibody and monoclonal human IgE. Biochemistry 34(33):10474–10482

Lyons DF, Lary JW, Husain B, Correia JJ, Cole JL (2013) Are fluorescence-detected sedimentation velocity data reliable? Anal Biochem 437(2):133–137

MacGregor IK, Anderson AL, Laue TM (2004) Fluorescence detection for the XLI analytical ultracentrifuge. Biophys Chem 108(1-3):165–185

Olshina MA, Angley LM, Ramdzan YM, Tang J, Bailey MF, Hill AF, Hatters DM (2010) Tracking mutant huntingtin aggregation kinetics in cells reveals three major populations that include an invariant oligomer pool. J Biol Chem 285(28):21807–21816

Pettikiriarachchi A, Gong L, Perugini MA, Devenish RJ, Prescott M (2012) Ultramarine, a chromoprotein acceptor for Förster resonance energy transfer. PLoS One 7(7):e41028

Polling S, Hatters DM, Mok Y (2013) Size analysis of polyglutamine protein aggregates using fluorescence detection in an analytical ultracentrifuge. Methods Mol Biol 1017:59–71

Romanini DW, Cornish VW (2012) Protein labelling: playing tag with proteins. Nat Chem 4(4):248–250

Sanchorawala V (2006) Light-chain (AL) amyloidosis: diagnosis and treatment. Clin J Am Soc Nephrol 1:1331–1341

Schachman HK (1959) Ultracentrifugation in biochemistry. Academic, New York

Schmidt B, Rappold W, Rosenbaum V, Fischer R, Riesner D (1990) A fluorescence detection system for the analytical ultracentrifuge and its application to proteins, nucleic acids, and viruses. Colloid Polym Sci 268(1):45–54

Schuck P (2000) Size distribution analysis of macromolecules by sedimentation velocity ultracentrifugation and Lamm equation modeling. Biophys J 78:1606–1619

Schuck P, Perugini MA, Gonzales NR, Howlett GJ, Schubert D (2002) Size-distribution analysis of proteins by analytical ultracentrifugation: strategies and application to model systems. Biophys J 82(2):1096–1111

Stafford WF, Sherwood PJ (2004) Analysis of heterologous interacting systems by sedimentation velocity: curve fitting algorithms for estimation of sedimentation coefficients, equilibrium and rate constants. Biophys Chem 108(1–3):231–243

Svedberg T, Pedersen KO (1940) The ultracentrifuge. Clarendon, Oxford

Swart C, Haylett W, Kinnear C, Johnson G, Bardien S, Loos B (2014) Neurodegenerative disorders: dysregulation of a carefully maintained balance? Exp Gerontol 58:279–291

Tsien RY (1998) The green fluorescent protein. Annu Rev Biochem 67:509–544

van Holde KE, Hansen JC (1998) Analytical ultracentrifugation from 1924 to present: a remarkable history. Chemtracts Biochem Mol Biol 11:933–941

Vistica J, Dam J, Balbo A, Yikilmaz E, Mariuzza RA, Rouault TA, Schuck P (2004) Sedimentation equilibrium analysis of protein interactions with global implicit mass conservation constraints and systematic noise decomposition. Anal Biochem 326(2):234–256

Wang X, Zhang C, Chiang Y, Toomey S, Power MP, Granoff ME, Richardson R, Xi W, Lee DJ, Chase S, Laue TM, Denis CL (2012) Use of the novel technique of analytical ultracentrifugation with fluorescence detection system identifies a 77S monosomal translation complex. Protein Sci 21(9):1253–1268

Warrick JM, Chan HYE, Gray-Board GL, Chai Y, Paulson HL, Bonini NM (1999) Suppression of polyglutamine-mediated neurodegeneration in Drosophila by the molecular chaperone HSP70. Nat Genet 23(4):425–428

Wilmann PG, Petersen J, Pettikiriarachchi AZ, Rossjohn J, Buckle AM, Smith SC, Olsen S, Perugini MA, Devenish RJ, Prescott M (2005) The 2.1 Å crystal structure of the far-red fluorescent protein HcRed: inherent conformational flexibility of the chromophore. J Mol Biol 349(1):223–237

Wowor AJ, Yu D, Kendall DA, Cole JL (2011) Energetics of SecA dimerization. J Mol Biol 408(1):87–98

Yphantis DA, Lary JW, Stafford WF, Liu S, Olson PH, Hays DB, Moody TP, Ridgeway T, Lyons D, Laue T (1994) On line data acquisition for the Rayleigh interference optical system of the analytical ultracentrifuge. In: Schuster TM, Laue TM (eds) Modern analytical ultracentrifugation: acquisition and interpretation of data for biological and synthetic polymer systems. Birkhauser, Boston, pp 209–226

Zhang C, Wang X, Park S, ChiangY XW, Laue TM, Denis CL (2014) Only a subset of the PAB1-mRNP proteome is present in mRNA translation complexes. Protein Sci 23(8):1036–1049

Zhao C, Hellman LM, Zhan X, Bowman WS, Whiteheart SW, Fried MG (2010) Hexahistidine-tag-specific optical probes for analyses of proteins and their interactions. Anal Biochem 399(2):237–245

Zhao H, Berger AJ, Brown PH, Kumar J, Balbo A, May CA, Casillas E Jr, Laue TM, Patterson GH, Mayer ML, Schuck P (2012) Analysis of high-affinity assembly for AMPA receptor amino-terminal domains. J Gen Physiol 139(5):371–388

Zhao H, Lomash S, Glasser C, Mayer ML, Schuck P (2013) Analysis of high affinity self-association by fluorescence optical sedimentation velocity analytical ultracentrifugation of labeled proteins: opportunities and limitations. PLoS One 8(12):e83439

Zhao H, Ma J, Ingaramo M, Andrade E, MacDonald J, Ramsay G, Piszczek G, Patterson GH, Schuck P (2014a) Accounting for photophysical processes and specific signal intensity changes in fluorescence-detected sedimentation velocity. Anal Chem 86(18):9286–9292

Zhao H, Mayer ML, Schuck P (2014b) Analysis of protein interactions with picomolar binding affinity by fluorescence-detected sedimentation velocity. Anal Chem 86(6):3181–3187

Chapter 5
The Multiwavelength UV/Vis Detector: New Possibilities with an Added Spectral Dimension

Engin Karabudak and Helmut Cölfen

Abstract The multiwavelength (MWL) detector is a new type of absorption detector for AUC. The commercial absorption detector of the Beckman Coulter XL-A AUC can only handle a single wavelength per scan with the possibility to scan at maximum 3 wavelengths, whereas MWL-AUC can handle all the wavelengths in the UV/Vis region at one time. The result is impressive since now a full spectral dimension is added to each single scan. In this chapter, we are explaining development history, instrumentation, and future perspective of MWL-AUC.

Keywords Analytical ultracentrifuge • UV/Vis spectroscopy • Separation • Instrumentation • Detector • Proteins • Multiwavelength detector • Online detection

5.1 Introduction

Analytical ultracentrifugation (AUC) allows the determination of important physicochemical quantities like sedimentation and diffusion coefficients, molar mass, size and shape, as well as the stoichiometries and equilibrium constants of interacting systems. Its power is in its ability to fractionate the sample, enabling the investigation even of most complex mixtures. However, every AUC experiment requires the detection of the relevant analytes, which in turn calls for a number of parallel available optical detectors to maximize the range of molecules.

Traditionally, AUC is equipped with refractive index (Schlieren and Rayleigh interference) as well as UV/Vis detectors. The latter enables the detection of the radial concentration profile at one wavelength. The commercially available Beckman Coulter XL-I is equipped with UV/Vis and Rayleigh interference optics.

E. Karabudak
Chemistry Department, Izmir Institute of Technology, TR-35430 Izmir, Turkey

H. Cölfen (✉)
Department of Chemistry, University of Konstanz, Universitätsstr. 10, D-78457 Constance, Germany
e-mail: Helmut.Coelfen@uni-konstanz.de

S. Uchiyama et al. (eds.), *Analytical Ultracentrifugation*,
DOI 10.1007/978-4-431-55985-6_5

A fluorescence detector also is available as a retrofit for this instrument (MacGregor et al. 2004). For nanoparticles, special turbidity detectors were developed, which allow the detection of very broad particle size distributions using a rotor velocity profile (gravitational sweep technique) (Müller 1989; Mächtle 1999). Fluorescence and turbidity detectors extend the range of applications for AUC considerably. However, especially in the field of nanoparticles, size-dependent optical properties are observed, such as bandgap absorbance for semiconductors or the plasmon resonance for metal nanoparticles. The investigation of such samples would benefit very much from the detection of UV/Vis absorption spectra while the sample is fractionated by the ultracentrifugal field. The same is true for mixtures of samples with different chromophores, regardless if nanoparticle or (bio)polymer. This was the scientific motivation for the development of a multiwavelength (MWL)[1] UV/Vis detector, which allows the detection of a full spectrum instead of a single wavelength.

5.2 The Multiwavelength Detector

The multiwavelength (MWL) detector is a new type of absorption detector for AUC. The commercial absorption detector of the Beckman Coulter XL-A AUC can only handle a single wavelength per scan with the possibility to scan at maximum 3 wavelengths, whereas MWL-AUC can handle all the wavelengths in the UV/Vis region at one time. The result is impressive since now a full-spectral dimension is added to each single scan as shown in Fig. 5.1. Each component in multicomponent samples can thus easily be tracked by MWL-AUC. Data of MWL-AUC are conveniently observed as a movie of sedimentation that includes all

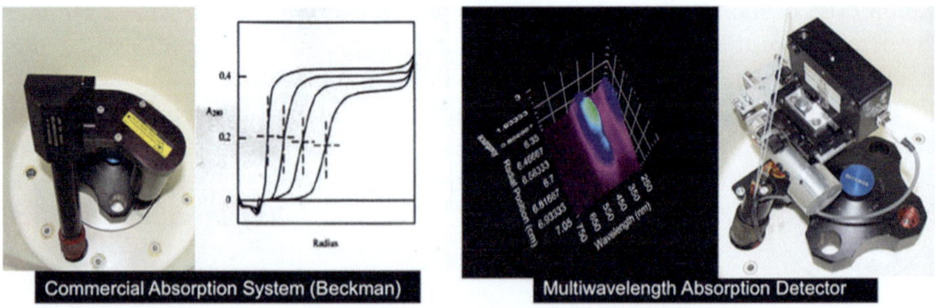

Fig. 5.1 Commercial UV/Vis absorption detector of the Beckman Coulter XL-I and the multiwavelength absorption detector (Reproduced from Karabudak (2009) with permission)

[1]Please note that besides MWL, also the abbreviation MWA for multiwavelength absorption detector is used in the literature. Both abbreviations are equivalent.

Table 5.1 Comparison of commercial XL-A and open-source MWL-AUC second-generation detector

	Commercial absorption system (Beckman)	Multiwavelength detector
Data	Single wavelength per radial point	Full UV/Vis spectrum per radial point
Technology	Photomultiplier tube (1990)	CCD
Cell scan time	1.5 min with 50 μm steps	45 s with 50 μm steps
Minimum motor step size	10 μm	100 nm
Software	Commercial	Open source

the wavelengths of the UV/Vis region, which is an advantage for multicomponent samples.

It is important to compare MWL-AUC and the commercial XL-A AUC in order to explain the novelty of MWL-AUC. First of all, MWL-AUC costs about $50.000 including the basis preparative ultracentrifuge, whereas the commercial XL-A ultracentrifuge costs about a factor of 10 more. In addition, MWL-AUC is an open-source project in the framework of the open AUC project (Coelfen et al. 2010). Nothing is confidential and anyone can get the designs and can build his own detector by downloading the construction plans for free from the open AUC website (http://wiki.bcf2.uthscsa.edu/openAUC/wiki/WikiStart). Its detector technology is also different; MWL-AUC is using CCD technology, whereas the XL-A uses a photomultiplier tube. The CCD technology allows for a very fast readout of the CCD pixels in the ms time range yielding a full UV/Vis spectrum in ms. One single wavelength radial scan takes about 1.5 min in the XL-A and the full wavelength scan about 45 s in the MWL-AUC. The minimum detector step size is less than 1 μm in MWL-AUC and 10 μm in the XL-A. The comparison can be seen in Table 5.1.

In this chapter, we will present the history of the development, optical parts, mechanical parts, software, and electronics of MWL-AUC, and we will give examples of chemical, biological, and industrial applications of MWL-AUC.

5.3 History of MWL-AUC Development

The commercial Beckman Coulter Optima XL-A has an optical system using a Xenon flash lamp with a maximum flash frequency of 100 Hz. White light from this lamp is passed to a toroidal diffraction grating and the system sends the now monochromatic light with pre-selected wavelength via the measurement cell to a photomultiplier tube (PMT) (Giebeler 1992). This optical setup is based on the 1990 technology. Technological developments in the last two decades decreased the price of spectrometers, counter cards, analog to digital converters, and other computer-based technologies significantly. In addition, CCD technology became broadly commercially available. All these new, cheaper, faster, and more precise technologies do not exist in the commercial XL-I AUC.

MWL-AUC was developed as a cooperation project AUC 2004 with BASF SE in the Cölfen lab at the Max Planck Institute of Colloids and Interfaces in Potsdam, Germany, between 2002 and 2009. With the agreement of BASF SE, the detector could be made open source (http://wiki.bcf2.uthscsa.edu/openAUC/wiki/WikiStart). Throughout the years, two different generations of MWL detectors were developed which are described below. The important characteristic of the two detector generations is that they are based on optical UV/Vis fibers which are prone to intensity decrease in the important UV range due to fiber bending and chemical reactions caused by high-energy UV light (fiber solarization). A third-generation MWL optics has been developed by Spin Analytical for the new centrifugal fluid analyzer (CFA) generation of AUC, which will be soon available commercially. This MWL detector is based on a mirror optics, thus eliminating chromatic aberration and, more importantly, avoiding the use of optical fibers (Laue and Austin in press).

5.3.1 First-Generation MWL Detectors

The first-generation MWL was developed on the basis of feeding the light into the AUC vacuum chamber and also feeding it out again to the spectrometer after it has passed the measurement cell. The design is shown in Fig. 5.2. The light from a Xenon flash lamp was coupled into the optical fiber, which had a diameter of 600 μm in order to provide as much light as possible at the detector. However, the large fiber diameter resulted in poor spatial resolution. The light exiting the fiber was collimated to a parallel light beam[2] and then reflected by 90° via a prism or mirror and then passed the cell (Fig. 5.2 left). Then the light was collimated by a slit-lens assembly taken from an XL-A AUC containing only one 10 mm lens and passed the 25 μm slit which defines the radial resolution of the system. Then it was finally coupled into a 1000 μm optical fiber, which transported the light to a CCD array spectrometer. The whole optics in the vacuum chamber was fixed to an arm, which allowed for radial scanning of the AUC cell. The data acquired using this system clearly showed what was possible by adding a spectral dimension to AUC; however it also suffered from a number of problems (Bhattacharyya 2006). First of all, this MWL system produced noisy spectra, which prevented further hydrodynamic analysis. In addition, the intensity was very low in the UV region.

To overcome some of these limitations, the design was modified in how light was focused into the detection fiber (Fig. 5.2 right, elements 4–6; a photo of this setup

[2]Note that the light can only be made parallel for one wavelength due to the wavelength dependence of the refractive index (chromatic aberration).

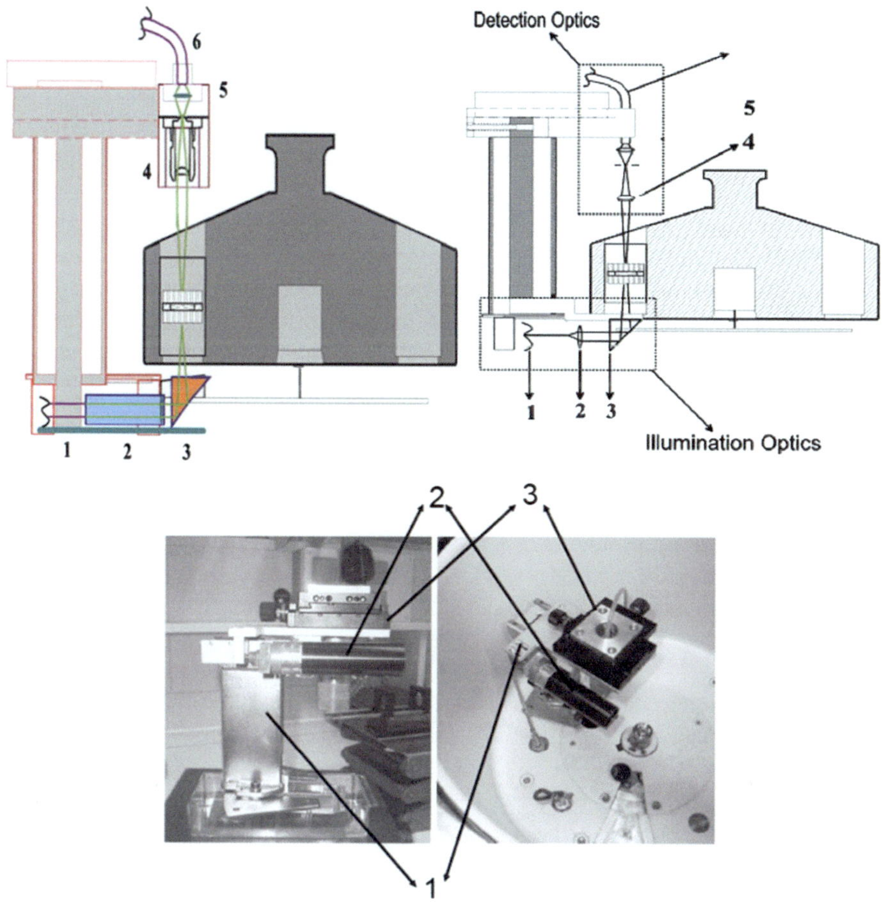

Fig. 5.2 First-generation MWL detectors
Left: Very first experimental MWL optics setup. (*1*) 600 μm patch fiber, UV/Vis (Ocean optics). (*2*) The collimating lens system (self-built) $f = 20.6$ mm biconvex. (*3*) 90 °quartz prism. (*4*) Slit-lens assembly (from the XL-A) only one lens $f = 10$ mm biconvex. (*5*) Focusing (OZ Optics) $f = 15$ mm. 6–1000 μm patch fiber, UV-Vis (Ocean Optics). *Right*: Modified detector arm. The detector arm. 1, 2, and 3 are the same as in the earlier setup. (*4*) Biconvex lens $f = 20$ mm. (*5*) Slit of dimension 25 μm. (*6*) Biconvex lens $f = 10$ mm (*7*) 600 μm patch fiber (Ocean optics). The light path is also shown schematically. *Lower*: Photograph of the arm. (*1*) Detector arm, (*2*) stepping motor (0.1 μm resolution at up to 4 mm/s), (*3*) detection unit with X-Y positioning screws. Right: Mounted arm in the AUC chamber (Images taken from Bhattacharyya (2006) with permission)

is shown in Fig. 5.2 lower panel). (Bhattacharyya 2006; Bhattacharyya et al. 2006). This setup improved the intensity in the UV region and decreased the noise some. Furthermore, the system became more stable (Bhattacharyya 2006).

5.3.2 Second-Generation MWL Detectors

The second-generation detector design was significantly improved concerning light intensity and optical quality, since the light after the cell was not anymore coupled into a fiber after passing a 25 μm slit for the optical resolution (Fig. 5.2) and then passed another 25 um slit at the spectrometer entrance for the wavelength resolution, but now the spectrometer was directly mounted on the detector arm (Figs. 5.3 and 5.4) (Karabudak 2009). This eliminated one slit in the optical path, which significantly improved the light intensity as now the 25 μm slit of the spectrometer is simultaneously used for the definition of the radial as well as of the spectrometer wavelength resolution. This is the open AUC second-generation detector design (downloadable for free using the open AUC website (http://wiki.bcf2.uthscsa.edu/openAUC/wiki/WikiStart)), which is in use since 2008 up to now and which since then has been further improved by gradual changes. The most notable of them are (1) the programming of a new control and data acquisition software by Johannes Walter based on Labview and 64 bit technology (Walter et al. 2014), which eliminated the occasional timing problems of the first MWL generation 32 bit MWL control and data acquisition software (Karabudak 2009; Bhattacharyya 2006; Bhattacharyya et al. 2006), and (2) the alternative development of an external hardware solution, which takes over all time critical multiplexing steps by Nanolytics (Pearson et al. 2015).

The second-generation optical system uses a high-power Xe flash lamp L-9456-12 from Hamamatsu Photonics GmbH, which has a flash rate of 530 Hz. This is

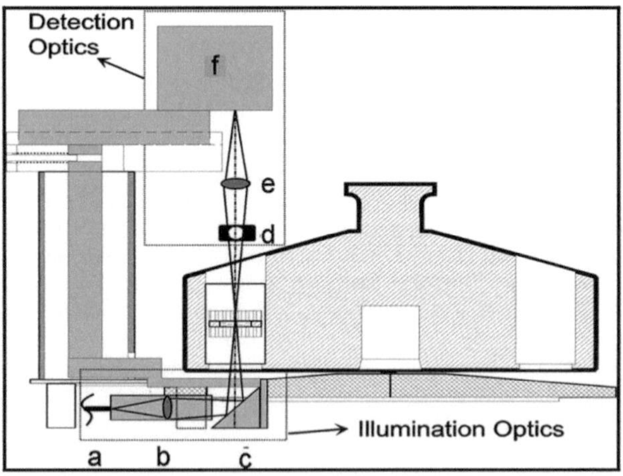

Fig. 5.3 Schematics of the second-generation MWL detector arm (Taken from Karabudak (2009) with permission) (**a**) 600 μm patch fiber UV/Vis (Ocean Optics). (**b**) The collimating lens system (self-built), $f = 20.6$ mm biconvex. (**c**) 90° quartz prism. (**d**) Iris diaphragm for reducing light intensity. (**e**) Focusing biconvex lens (40 mm). (**f**) Spectrometer. The light path is also shown schematically

about five times faster than that of the Beckman Coulter XL-A allowing for faster flashing which increases the detection speed. The light from the lamp is passed from the fiber (Fig. 5.3a) through a collimating lens (Fig. 5.3b). The (almost) parallel light is then reflected by a 90° mirror (Fig. 5.3c). Afterward, the light passes the sample compartment and reaches an adjustable iris (Fig. 5.3d). The iris is used for adjusting the total intensity since the light in the visible is so intense that it maxes out the spectrometer, while that in the UV is significantly reduced as compared to the initial white light spectrum from the flash lamp. Finally light reaches a collimating lens, which images the cell and focuses the light into the 25 μm entrance slit of the Ocean Optics USB2000 spectrometer. The USB2000 spectrometer uses CCD technology, which has one fixed wavelength for each of the pixels. These pixels can be read out in a few milliseconds, yielding 2000 data points, which allows for averaging the data from several pixels at each wavelength. Newer spectrometers have 3000 pixel CCD chips.

5.3.3 *Mechanical Parts of Second-Generation MWL-AUC*

There are two main mechanical parts of the open AUC MWL-AUC (Strauss et al. 2008). Firstly, the detector arm is the main mechanical part of MWL-AUC as shown in Fig. 5.4. This arm carries the spectrometer, the stepping motor, lenses, iris, and quartz prism. The second mechanical part of MWL-AUC is the vacuum feedthrough. The vacuum feedthrough includes the electronic feedthrough and optical feedthrough. The electronic feedthrough is used for the USB signal from

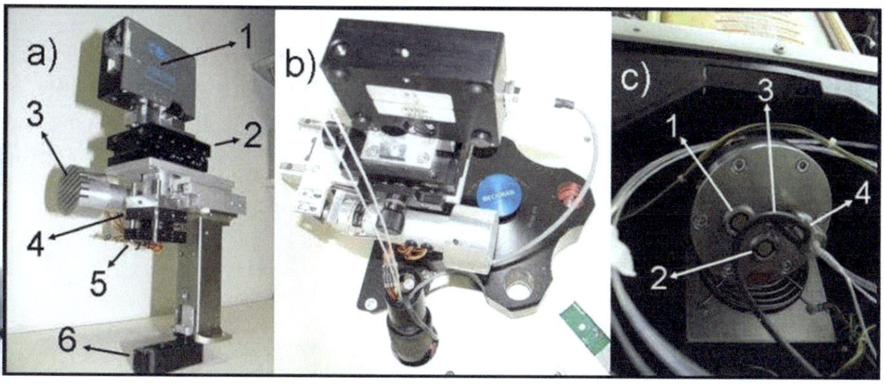

Fig. 5.4 Photographs of second-generation MWL-AUC (Taken from (Karabudak 2009) with permission).(**a**) Photograph of the detector arm: *1* spectrometer. *2* table with the possibility of X-Y movement. *3* step motor. *4* lens (40 mm biconvex). *5* iris *6* 90° quartz prism. (**b**) The arm fitted in the centrifuge. (**c**) Photograph of the vacuum feedthrough: *1* Electronic feedthrough for step motor. *2* Electronic feedthrough for spectrometer. *3* Electronic connection for TTL pulse for rpm measurements. *4* Optical feedthrough for fibers

and to the spectrometer, step motor, and Transistor-Transistor Logics (TTL) pulses for speed measurements. The optical feedthrough is used for fiber connections. All these mechanical parts are mounted into a preparative ultracentrifuge. In order to mount the feedthrough into the preparative machine, the heat sink of the preparative machine needs to be modified. This requires drilling holes into the heat sink of the preparative ultracentrifuge as shown and discussed in the open AUC publication (Coelfen et al. 2010) and the user loses the Beckman Coulter warranty. A more elegant solution is to take an XL-A heat sink and then use the hole for the flash lamp to plug in the vacuum feedthrough as shown in Fig. 5.4c. This allows maximum flexibility and more importantly allows the modification of any XL-A AUC into a MWL-AUC and back again in about an hour. To accommodate swapping, a slit was cut into the thermal shroud to allow mounting of the detector arm.

5.3.4 Electronics of MWL-AUC

There are four main electronic parts of MWL-AUC. First, an optical sensor is used to time the rotor revolution over speeds ranging from 1000 rpm to 60,000 rpm. A TTL timing pulse generated by light reflected from a spot on the rotor allows measurement of this speed and even allows the system to follow steep speed ramps for very polydisperse samples (Mächtle 1999; Bhattacharyya et al. 2006). The speed is measured by a national instrument digital counter/timer card. This card measures the TTL pulse for speed determination and triggers the flash lamp and spectrometer when the cell of interest passes the optical path. Second, there is a USB connection for the USB spectrometer. This connection collects the spectral data and sends it to the computer. A computer is used to control the digital counter/timer card, USB connector, and data collection. The third electronic part is the USB spectrometer. This spectrometer is triggered by the counter card and sends the digital signal to the computer via the USB connection. The fourth part is the stepping motor, which is controlled by the MWL-AUC control software. As soon as data are taken at a given radial position, this motor moves the detector arm to the next radial position at the radial specified step interval.

5.3.5 Software of MWL-AUC

Initially, the MWL detector was programmed using Labview programming language in a 32 bit version. The Labview control program was controlling the timer/counter card as well as the movement of the step motor. In addition, the software triggers the flash lamp and spectrometer and collects the data from the spectrometer. The software also calculates the absorption and plots it. Furthermore, the software stores the data on the computer hard disk. The initial program allowed for radial scanning at constant speed and for speed ramps but suffered from occasional timing problems of the trigger pulses. This was overcome with the 64 bit program (Walter et al. 2014) as well as the omega device (Pearson et al. 2015).

5.4 MWL-AUC in the Scientific Literature

Although the first publication about the MWL detector (first generation) was published in 2006 (Bhattacharyya et al. 2006), nine years before the time of writing of the present manuscript, not many publications have appeared using the MWL-AUC. The reason for this is that the MWL detector was only available in a few dedicated laboratories and was still under continuous development. Additionally, data evaluation by powerful evaluation packages like the UltraScan (Gorbet et al. 2015) or Sedanal (Walter et al. 2015) became available only very recently. The situation has, therefore, now improved. The MWL detector has a quality comparable or superior to the Beckman Coulter XL-A in the visible and even in the UV (Walter et al. 2014; Strauss et al. 2008; Gorbet et al. 2015) and is freely available from the open AUC website (http://wiki.bcf2.uthscsa.edu/openAUC/wiki/WikiStart). In addition, it can be reversibly mounted into any XL-A AUC (after cutting of a slit into the can for the detector arm). These improvements will very likely lead to more operational MWL detectors. In addition, the CFA will soon be available, which will have a third-generation MWL detector installed (Laue and Austin in press).

While the first publication described the basic MWL design and first measurements, it already contains first measurements, which qualify the MWL as multi-sensitivity turbidity detector due to the proportionality of the scattering intensity with the particle radius to the power of six and the wavelength of light to the inverse power of four (Bhattacharyya et al. 2006). Large particles scatter light a lot and can be detected at higher wavelengths than the weaker scattering small particles, which can be detected at short wavelengths with higher sensitivity. Due to the spectral dimension, an optimum wavelength can be found for each particle size, which significantly enhances the dynamic range of turbidity detection

Strauss et al. analyzed the first prototype of the second-generation MWL-AUC (Strauss et al. 2008). The performance of the optical setup MWL-AUC was tested, and the result was compared with the commercial XL-A showing that MWL wavelength accuracy and radial resolution are comparable with the XL-A. In addition wavelength dependent noise levels were compared and showed that the MWL was better than the XL-A in the visible range and worse in the UV as a result of UV light attenuation by the optical fiber. This publication is important because it showed for the first time that MWL-AUC is comparable with the commercial detector but has the advantage of adding the spectral dimension to the data. The advantage of this spectral dimension becomes obvious already in the raw data. Figure 5.5 shows a sedimentation experiment of an industrial and commercial product with MWL-AUC (Karabudak et al. 2010a). The product is gelatin-coated β-carotene composite particles, which are industrially used as food colorants.

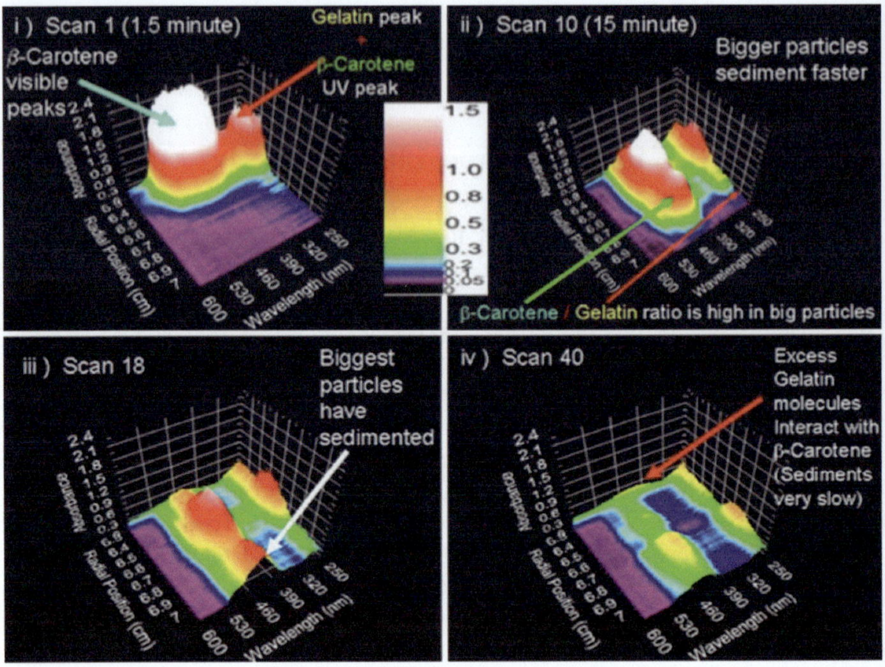

Fig. 5.5 Three-dimensional plots of the raw data from a band sedimentation experiment with β-carotene detected with the MWL detector (Taken from Karabudak et al. (2010a) with permission)

The raw data from this experiment in Fig. 5.5 show that the product contains multiple species as labeled. Carotene absorbs in the visible around 460 nm and gelatin in the UV at 280 nm. The fastest-sedimenting species mainly consists of carotene, while the second fastest species contains carotene and significantly more gelatin than the fastest-sedimenting species. Finally free gelatin and gelatin interacting with β-carotene sediment slowly. Not only could the raw data show the number and spectral characteristics of the components. Spectral analysis of the fractionated species showed that the previously assumed structure of the product was not correct. MWL-AUC showed that H-aggregates and J-aggregates are formed in different species and that a previously published literature structure (Auweter et al. 1999) is not correct.

In another study, scientists extensively studied the interaction of single-walled carbon nanotubes (SWCNTs) with surfactants (Backes et al. 2010a) and determined the surfactant density of SWCNTs. Also surfactants were tested for SWCNT absorption (Backes et al. 2010a). This kind of study is not possible with the commercial XL-A, since each of the SWCNT species has its spectral fingerprint. Therefore, simultaneous fractionation and size determination as well as spectral detection for each of the fractionated species are necessary.

When a carbon nanotube, surfactant, and intercalant system were analyzed by MWL-AUC (Karabudak et al. 2010b), the results showed that previously published results on this system were incorrect (Backes et al. 2010b). The MWL-AUC experiments revealed that the intercalant is oxidized and sediments independent of the carbon nanotubes. This finding changed the previously assumed function of the intercalant.

In a recent study, Walter et al. (2014) studied polydisperse nanoparticles successfully. An impressive example was the MWL experiment on Au nanoparticles, which shows size- and shape-dependent UV/Vis spectra due to the surface plasmon resonance. Polydisperse Au nanorods (Fig. 5.6a) were subjected to MWL-AUC. The

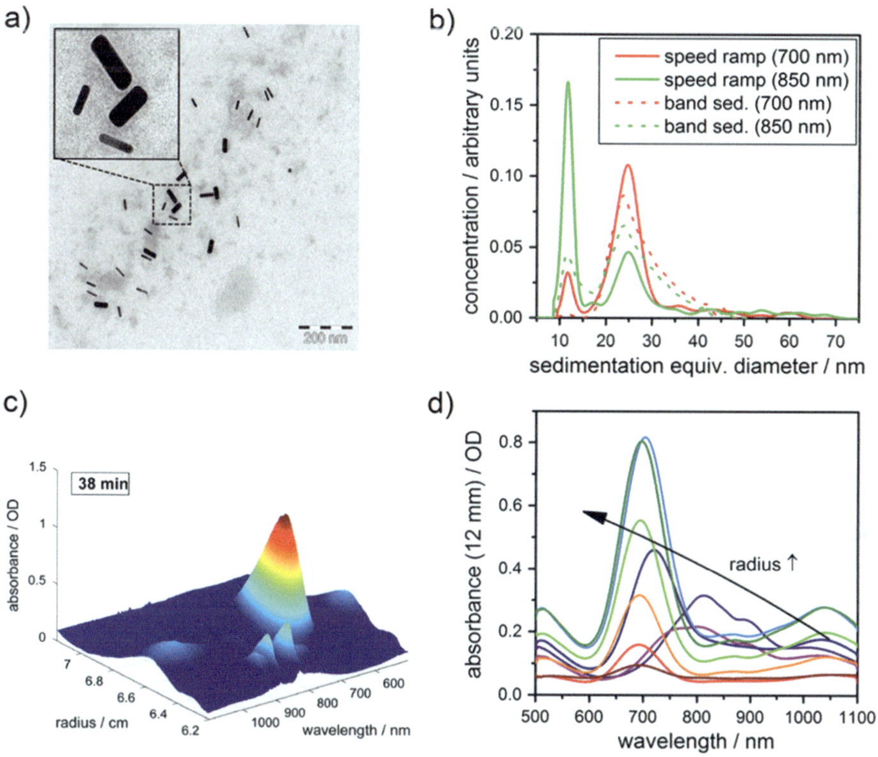

Fig. 5.6 (**a**) TEM image of a gold nanorod mixture clearly showing the two main species. (**b**) Extinction weighted particle size distribution of two mixtures of the same nanorod species gained at a speed ramp and band sedimentation experiment evaluated at 700 and 850 nm. (**c**) Multiwavelength spectra of gold nanorods sedimenting at 2 krpm in a direct band centrifugation experiment as a function of the radial distance from the axis of rotation for a scan taken after 38 min. The three-dimensional data surface is clearly visible. (**d**) Absorbance data of gold nanorods recorded in the radial dimension from 6.3 to 7.1 cm with 1 mm increment (*purple* to *blue* to *green* to *red* and *arrow* as a guide to the eyes) for the same snapshot (**c**) after 38 min in the band centrifugation experiment. A higher radial position results in an increased sedimentation equivalent diameter, which causes a redshift in the absorption spectrum due to the particle size- and shape-dependent surface plasmon resonance (Taken from Walter et al. (2014) with permission)

particle size distributions (Fig. 5.6b) showed wavelength-dependent differences in the intensity, which is expected from the size-dependent UV/Vis spectra.

A scan containing raw data (Fig. 5.6c) shows the complexity of the three-dimensional data surface and directly allows to reveal the spectral properties of the different species in the mixture. From this complex three-dimensional surface, the spectra at different radii can be extracted and show the change of the UV/Vis spectra with location and size of the nanoparticles. Due to these advantageous properties, MWL-AUC was applied for other complex mixtures like polyfluorene-coated semiconductor particles (de Roo et al. 2014).

A very recent report showed that application of the Sedanal software to multiwavelength data allows for an easy and successful deconvolution of individual spectra of components in a mixture (Fig. 5.7) (Walter et al. 2015). In this example, blue silica nanoparticles were mixed with Au nanoparticles. The spectra are shown in Fig. 5.7a. MWL-AUC (Fig. 5.7b) shows that two different components can be seen with distinctly different spectra. Deconvolution of the mixture spectra yields the spectra of the individual components in good quality (Fig. 5.7d), and the sedimentation coefficient distributions of the individual components could also be deconvoluted (Fig. 5.7c). The wiggle in the sedimentation coefficient distribution is an as yet unexplained artifact.

Another exciting and recently reported application is the use of MWL-AUC to detect spectra of nucleating species (Voelkle et al. 2015). Overlaying the two reactants of a crystallizing species in an AUC cell by the technique of synthetic boundary crystallization ultracentrifugation (Borger et al. 2000) allows to initiate the reaction in a sharp boundary formed upon speeding up the ultracentrifuge. Since one reactant is consumed within seconds, further nucleation and growth are quenched and the formed early nucleation and growth species get then fractionated by the ultracentrifugal field and can then be detected. Since the particle size resolution of the AUC is in the Angström range for such small species (Colfen and Pauck 1997), valuable information about the early species in a crystallization reaction can be obtained. Addition of the spectral dimension by MWL-AUC now significantly increases the information content by the UV/Vis spectra of the different species. An example for Ag nucleation was recently reported (Voelkle et al. 2015).

The last and most recent example discussed in this MWL overview is the application of MWL-AUC for biopolymer samples, which is certainly one of the most important applications for AUC. MWL-AUCs using detectors of the second generation have the problem of UV attenuation as discussed above, which significantly decreases the data quality in this important wavelength range. Nevertheless, a recent study on BSA-DNA mixtures shows that proteins can be investigated by MWL-AUC as well, yielding unsurpassed information, which cannot be obtained with the XL-A ultracentrifuge as a direct comparison of MWL and XL-A data showed (Pearson et al. 2015; Gorbet et al. 2015). Figure 5.8 shows the raw data for the BSA mixture with two different DNA fragments.

From these data, already three species can be identified. The fastest-sedimenting species coded with yellow in Scan 10 and 25 with absorption around 260 nm is likely a DNA species. A second following species absorbs broadly from 240 to

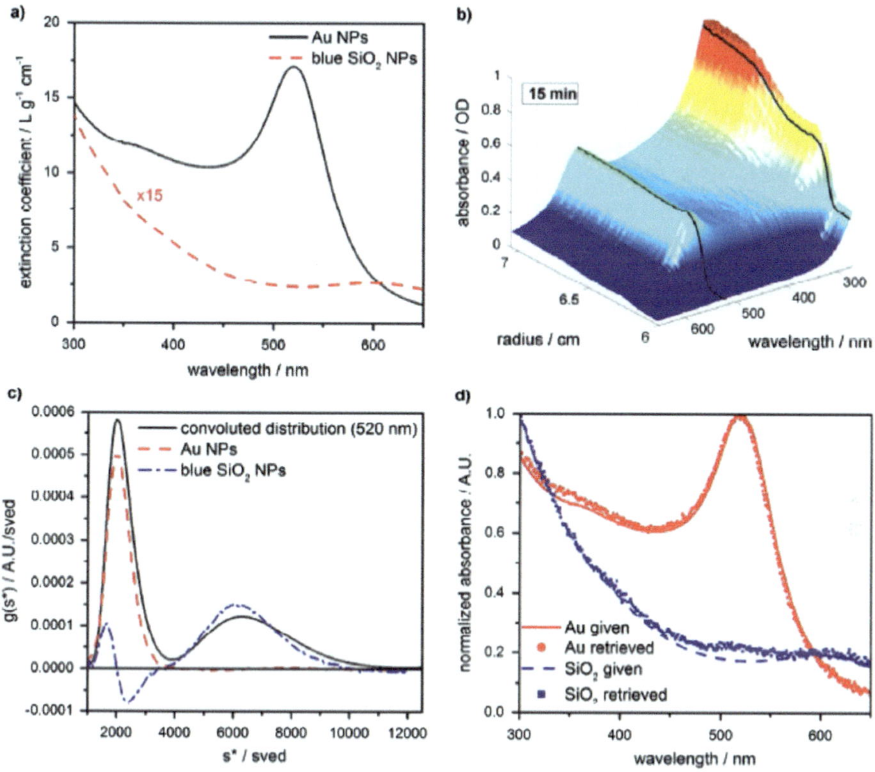

Fig. 5.7 (**a**) Extinction coefficients of the silica and gold NPs. The extinction coefficient of the silica NPs was multiplied by 15 to be better recognizable. (**b**) Multiwavelength spectra of the silica-gold mixture sedimenting at 4 krpm in a sedimentation velocity experiment as a function of the radial distance from the axis of rotation. The scan was taken after 15 min. Two different wavelengths corresponding to the spectral features of the two species were highlighted with black lines as a guide to the eye. (**c**) Convoluted sedimentation coefficient distributions at 520 nm and deconvoluted sedimentation coefficient distributions of gold and silica NPs. 329 wavelengths values were used for the MWL analysis. The distributions were not normalized for the sake of clarity. (**d**) Extinction spectra of the two species measured using a benchtop UV/Vis spectrometer as well as the extinction data reproduced from the MWL-AUC experiment. Spectra were normalized to a maximum value of one to be comparable (Taken from Walter et al. (2015) with permission)

280 and thus likely contains protein and DNA and the same is true for the slowest-sedimenting species in Fig. 5.8.

Evaluation of this experiment with UltraScan 3.0, which is now able to fully analyze MWL data, reveals much more detail about the sample mixture.

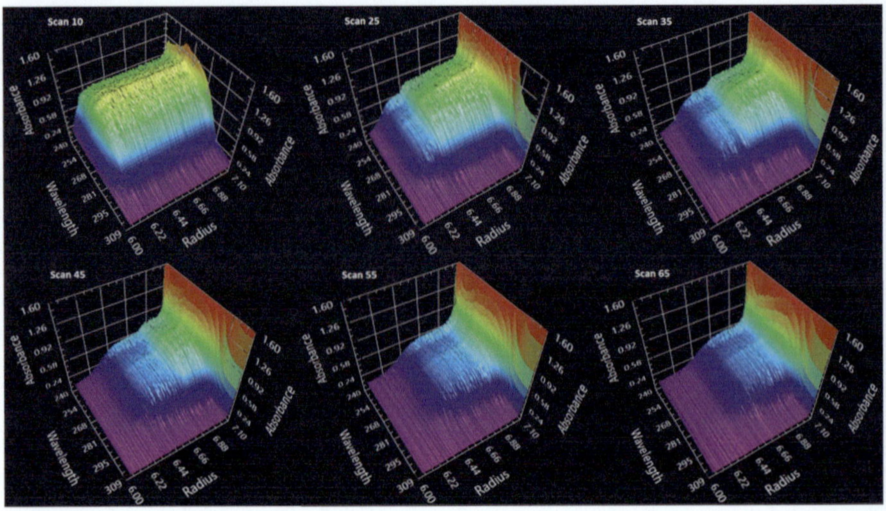

Fig. 5.8 Time series images of three-dimensional sedimentation absorbance data for a DNA-BSA mixture from open-source MWL recorded with 50 μm radial step size. The intensity in OD units is dependent on the extinction coefficient at each wavelength. The absorbance spectra of the species in solution are evident across the wavelength range. The *yellow* coded species at ca. 270 nm across the spectra is indicative of a more rapidly sedimenting species, in this case a larger DNA fragment (Image reproduced from Pearson et al. (2015) with permission)

Two-dimensional spectrum analysis (2DSA) shows BSA monomers and dimers as well as a 208 base pair DNA sedimenting at almost the same speed like the BSA. Only the additional spectral dimension makes it possible to distinguish these almost equally fast-sedimenting species. The faster sedimenting 12 S species turned out to be 2811 base pair DNA and the fastest-sedimenting species at low concentration were uncut DNA plasmids. This example shows in an impressive way what can be learned from MWL-AUC experiments with a sophisticated analysis as is possible with UltraScan (Fig. 5.9).

Other analysis methods implemented into UltraScan are also possible with MWL data, which was demonstrated for the van Holde-Weischet method, which is especially attractive since it needs no assumptions about the sample (Gorbet et al. 2015).

If the UV/Vis spectra of the individual components in the mixture are known, like in this case those for BSA and DNA, it is possible to deconvolute the MWL data of mixtures of these components into the contributions of each component to the detected mixture UV/Vis spectrum. It is then possible to average the individual sedimentation velocity profiles of each component in the mixture for each

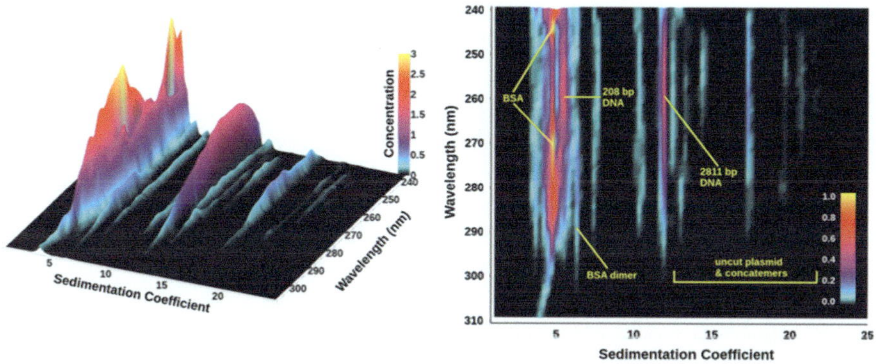

Fig. 5.9 *Left:* Three-dimensional view of the sedimentation profile as a function of wavelength for a 50:50 DNA-BSA mixture. The protein absorbance spectrum at 4.3 s (*two yellow peaks*) can be clearly distinguished from the DNA peak with absorbance maximum around 258 nm. Minor species can be identified based on their spectrum. *Right:* Projection view of the 2DSA-MC sedimentation profile as a function of wavelength for the 50:50 DNA-BSA mixture. Remarkably, the protein absorbance spectrum at 4.3 s (*two yellow peaks*) can be clearly distinguished from the adjacent DNA peak with absorbance maximum around 258 nm, despite the proximity of the peaks (4.5 s vs. 5.2 s). Minor species can be identified based on their spectrum. The *straight lines* attest to the high resolution and robustness of this approach to fit multiwavelength data (each wavelength is separately analyzed) (Taken from Gorbet et al. (2015) with permission)

wavelength according to the wavelength-dependent extinction coefficient profile and arrive at two spectrally decomposed sedimentation velocity profiles for BSA and DNA individually (Gorbet et al. 2015). This spectral decomposition works extremely well as was demonstrated for different BSA-DNA mixing ratios. If now the sedimentation velocity profiles are evaluated for the two components, the data in Fig. 5.10 are obtained.

Figure 5.10 shows that the individual sedimentation coefficient distributions are obtained for DNA as well as for BSA in a very high quality showing all species present in the mixture. For the XL-A, such high-quality analysis is not possible (Fig. 5.10). In addition the MWL decomposition showing the sedimentation coefficient plotted versus the shape in terms of the frictional ratio reveals the sedimentation coefficients of the individual species in the complex mixture together with their frictional ratio/shape. The XL-A analysis, even when performed for the two relevant wavelengths, lacks this resolution clearly showing the advantage of the spectral dimension in the MWL-data.

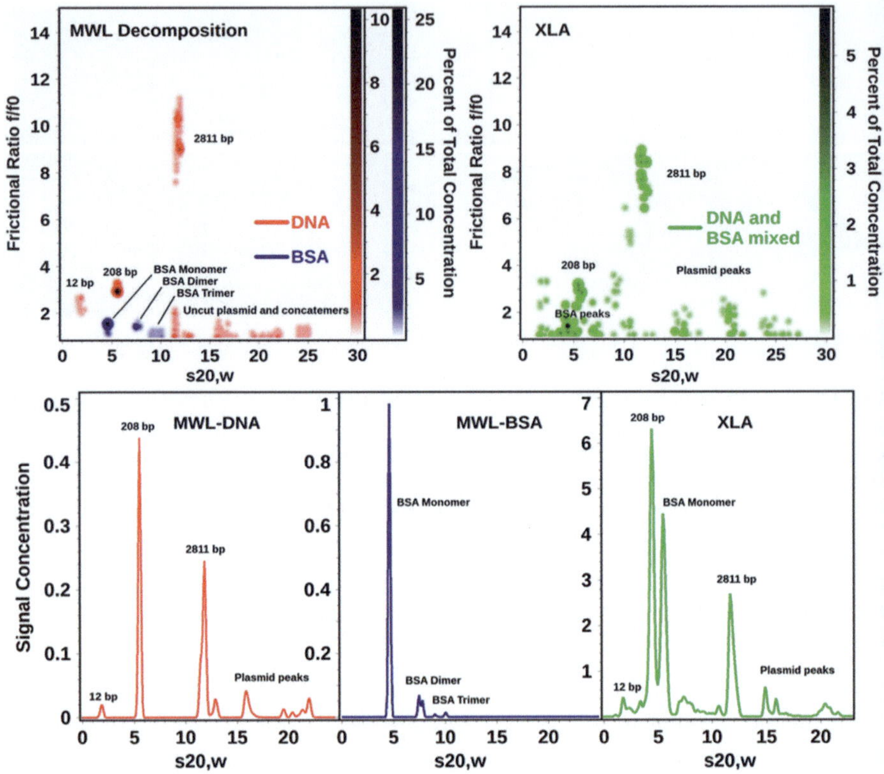

Fig. 5.10 Global genetic algorithm Monte Carlo analysis of decomposition results obtained from six different DNA and BSA mixtures analyzed on the open AUC MWL instrument (*top left*) and the dual wavelength results obtained from the Beckman Coulter XL-A (*right panel*). The separate decomposition results for DNA (*red*) and BSA (*blue*) are combined in the left panel pseudo-three-dimensional plot to illustrate the exceptional separation achieved by spectral decomposition which even separates species with nearly identical sedimentation coefficients (the two major species sedimenting near 5 S). This approach demonstrates the superior resolution obtained from MWL analysis compared to the global 2-wavelength analysis performed on the Beckman Coulter XL-A (*right panel*, *green*). Lower panel: differential distributions from the same data shown above (*red*, decomposition for DNA; *blue*, decomposition for BSA; *green*, non-separated XLA data for DNA-BSA mixtures globally fitted to genetic algorithm – Monte Carlo analysis) (Taken from Gorbet et al. (2015) with permission)

5.5 Conclusions

In this chapter, we discussed the development as well as design and applications of the MWL detector. This detector has a huge advantage over existing AUC detectors by the addition of a spectral dimension to the hydrodynamic data from AUC. MWL-AUC has now reached a data quality, which allows its application for the wide range of samples from (bio)polymers to nanoparticles with a data quality comparable or superior to that of the Beckman Coulter XL-I instrument. Fiber solarization is still an issue in the second-generation MWL detectors, although it could be demonstrated that even with attenuated UV intensity, a high-quality analysis of biological samples is possible (Gorbet et al. 2015). Already the MWL raw data contain a significant amount of information. This information content can be dramatically increased if sophisticated data analysis packages like UltraScan or Sedanal are used, which were recently adapted for the analysis of MWL data. Therefore, everything necessary for the successful application of MWL-AUC is now available. The second-generation open AUC detector construction plans can be freely downloaded from the open AUC website (http://wiki.bcf2.uthscsa.edu/openAUC/wiki/WikiStart). This detector can be mounted in any XL-A after introduction of a cut for the detector arm in the can and can be reversibly exchanged with the XL-A optics. Importantly, the new CFA AUC will have a third-generation (Laue and Austin in press) MWL detector, which works with mirror optics and thus avoids the light attenuation in the UV in the optical fibers as well as chromatic aberration problems. The data quality of this detector will be superior to the fiber-based MWL designs. For all these reasons, a wide use of MWL-AUC can be expected in the near future. Although only a few publications using MWL-AUC exist so far, they all demonstrate the superior information content of these measurements by the added spectral dimension. Therefore, we expect that this detector will enable measurements, which have so far not been possible by AUC, and that MWL-AUC will help to solve important scientific problems in the future.

Acknowledgments We thank a number of people who helped to make MWL-AUC what it is today. First of all, we acknowledge funding of the AUC 2004 project by BASF SE, which made the start of this project possible. Dr. Walter Mächtle, Dr. Lars Börger, and Dr. Wendel Wohlleben, all from BASF SE, are thanked for the fruitful cooperation over the years of development. Dipl. Ing. Johannes Walter (University of Erlangen) is thanked for the new data acquisition software. We also thank Dr. Kristian Schilling (Nanolytics) for discussions and exchange of ideas during detector development. Prof. Dr. Borries Demeler (University of Texas, Health Center, at San Antonio) is thanked for adapting the UltraScan evaluation software to MWL data, which now allows unsurpassed evaluations using MWL-AUC as well as for a long-standing cooperation in this area. We also thank all students and coworkers who worked and work in the Cölfen lab on MWL development, namely, Patricyja Maciejewska, Saroj Bhattacharyya, Akif Gülsün, Basri Cicek, Dirk Haffke, Joe Pearson, Dr. Holger Strauss, and Dr. Karel Planken.

References

Auweter H et al (1999) Supramolecular structure of precipitated nanosize beta-carotene particles. Angew Chem Int Ed 38:2188–2191. doi:10.1002/(sici)1521-3773(19990802)38:15<2188::aid-anie2188>3.0.co;2-#

Backes C et al (2010a) Determination of the surfactant density on SWCNTs by analytical ultracentrifugation. Chem Eur J 16:13176–13184. doi:10.1002/chem.200903461

Backes C et al (2010b) Nanotube surfactant design: the versatility of water-soluble perylene bisimides. Adv Mater 22:788–802. doi:10.1002/adma.200902525

Bhattacharyya SK (2006) Development of detectors for the analytical ultracentrifuge, PhD thesis, Universität Potsdam
Bhattacharyya SK et al (2006) Progress in colloid and polymer science: analytical ultracentrifugation VIII, Springer, pp 9–22
Borger L, Colfen H, Antonietti M (2000) Synthetic boundary crystallization ultracentrifugation: a new method for the observation of nucleation and growth of inorganic colloids and the determination of stabilizer efficiencies. Colloids Surf A Physicochem Eng Asp 163:29–38. doi:10.1016/s0927-7757(99)00427-6
Coelfen H et al (2010) The open AUC project. Eur Biophys J Biophys Lett 39:347–359. doi:10.1007/s00249-009-0438-9
Colfen H, Pauck T (1997) Determination of particle size distributions with angstrom resolution. Colloid Polym Sci 275:175–180. doi:10.1007/s003960050068
de Roo T et al (2014) A direct approach to organic/inorganic semiconductor hybrid particles via functionalized polyfluorene ligands. Adv Funct Mater 24:2714–2719. doi:10.1002/adfm.201304036
Giebeler R (1992) In: Harding SE, Rowe AJ, Horton JC (eds) Analytical ultracentrifugation in biochemistry and polymer science. Royal Society of Chemistry, Cambridge, pp 16–31
Gorbet G, Pearson J, Demeler A, Cölfen H, Demeler B (2015) Next generation AUC (2): multi-wavelength data analysis adds a spectral dimension to hydrodynamic information. Methods Enzymol 562:27–47
http://wiki.bcf2.uthscsa.edu/openAUC/wiki/WikiStart
Karabudak E (2009) PhD thesis; Development of MWL-AUC / CCD-C-AUC / SLS-AUC detectors for the analytical ultracentrifuge PhD thesis, Max Planck Institute of Colloids and Interfaces
Karabudak E, Wohlleben W, Colfen H (2010a) Investigation of beta-carotene-gelatin composite particles with a multiwavelength UV/vis detector for the analytical ultracentrifuge. Eur Biophys J Biophys Lett 39:397–403. doi:10.1007/s00249-009-0412-6
Karabudak E et al (2010b) A universal ultracentrifuge spectrometer visualizes CNT-intercalant-surfactant complexes. ChemPhysChem 11:3224–3227. doi:10.1002/cphc.201000504
Laue T, Austin B (in press) The CFA analytical ultracentrifuge architecture. Analytical ultracentrifugation: instrumentation software and application
MacGregor IK, Anderson AL, Laue TM (2004) Fluorescence detection for the XLI analytical ultracentrifuge. Biophys Chem 108:165–185. doi:10.1016/J.Bpc.2003.10.018
Mächtle W (1999) High-resolution, submicron particle size distribution analysis using gravitational-sweep sedimentation. Biophys J 76:1080–1091
Müller H (1989) Automated determination of particle-size distributions of dispersions by analytical ultracentrifugation. Colloid Polym Sci 267:1113–1116
Pearson J et al (2015) Next generation AUC (1): multi-wavelength detectors for the analytical ultracentrifuge add a spectral dimension to AUC instrumentation. Methods Enzymol 562:q–26
Strauss HM et al (2008) Performance of a fast fiber based UV/Vis multiwavelength detector for the analytical ultracentrifuge. Colloid Polym Sci 286:121–128. doi:10.1007/s00396-007-1815-5
Voelkle C, Gebauer D, Cölfen H (2015) High-resolution insights into the early stages of silver nucleation and growth. Faraday Discuss 179:59–77. doi:10.1039/C4FD00269E
Walter J et al (2014) Multidimensional analysis of nanoparticles with highly disperse properties using multiwavelength analytical ultracentrifugation. ACS Nano 8:8871–8886. doi:10.1021/nn503205k
Walter J et al (2015) Simultaneous analysis of hydrodynamic and optical properties using analytical ultracentrifugation equipped with multiwavelength detection. Anal Chem 87:3396–3403

Chapter 6
SEDANAL: Model-Dependent and Model-Independent Analysis of Sedimentation Data

Peter J. Sherwood and Walter F. Stafford

Abstract SEDANAL (Stafford and Sherwood, Analysis of heterologous interacting systems by sedimentation velocity: curve fitting algorithms for estimation of sedimentation coefficients, equilibrium and kinetic constants. Biophys Chem 108:231–243, 2004) is a suite of routines that are used to analyze data from the analytical ultracentrifuge and other types of centrifugal fluid analyzers. It can handle data from both sedimentation velocity and sedimentation equilibrium experiments. Two general approaches are used: (1) model independent and (2) model dependent. The model-independent modules are based on the time-derivative method for sedimentation velocity and BioSpin for sedimentation equilibrium data. The model-dependent modules use several curve fitting techniques to fit user-specified models both to the sedimentation velocity and sedimentation equilibrium data. SEDANAL allows the global analysis of data from multiple runs and multiple optical systems and of absorbance data from multiwavelength instruments. Models are specified in the Model Editor module of SEDANAL. This chapter describes the various modules and routines of SEDANAL. The Model Editor especially is described in detail.

Keywords SEDANAL • DCDT • BioSpin • Sedimentation analysis • Sedimentation velocity • Analytical ultracentrifugation • Model Editor • Molecular hydrodynamics • Thermodynamics • Nonideality • Interacting systems

6.1 Introduction

SEDANAL has evolved considerably since it was first introduced in 2004 (Stafford and Sherwood 2004) as a program for fitting reversibly interacting hetero-associating systems. In what follows we describe the main features of the latest

---

P.J. Sherwood
Interactive Technology, Inc., 20 Woodside Glen Ct., Oakland, CA 94602, USA

W.F. Stafford (✉)
Boston Biomedical Research Institute, c/o 12 Francis Ave, Cambridge, MA 02138, USA
e-mail: stafford@sedanal.org

S. Uchiyama et al. (eds.), *Analytical Ultracentrifugation*,
DOI 10.1007/978-4-431-55985-6_6

version of SEDANAL. It comprises several modules for time-derivative analysis as well as direct fitting to solutions of the Lamm equation and fitting to the sedimentation equilibrium equations. The Model Editor, which allows specification of nearly any arbitrary molecular interaction model, has been intercalated into the body of the main program. The Model Editor is described in detail. SEDANAL now has the ability to read and process data from multiwavelength instruments, deconvolute concentration profiles of each component, and extract spectra from each of the separated components.

6.2 Modules of SEDANAL

6.2.1 Preprocessor

The preprocessor can load XL-A/I absorbance, intensity, interference, and fluorescence scan files or multiwavelength binary files from a centrifuge equipped with multiwavelength optics. The preprocessor is used to identify the meniscus and base and the range to fit. For interference data, it also allows the removal of time-dependent vertical jitter and fringe jumps in the raw data. For intensity data the vertical jitter can also be removed and the scans converted to pseudo-absorbance data. The multiwavelength data is also an intensity data and is also converted to pseudo-absorbance data for subsequent processing. The preprocessed data are stored in a "run" file (*.abr) that contains the original raw data along with all the adjustments as metadata. When the program reads a "run" file for a particular type of analysis, it reads the raw data from the "run" file and applies the adjustments on the fly as it reads in the run file. The preprocessor can also write out XL-I/A files from a run file either as the original data or as the adjusted data. The preprocessor, given the extinction spectra of the components, will deconvolute the concentration profiles of each of the constituent components and write out a corresponding set of XL-A/I files in units of mass concentration versus radius for each time point.

6.2.2 Fitter

The Fitter determines the best fit of velocity or equilibrium data to the Lamm equation for diffusion and sedimentation, combined with equations for chemical equilibrium or kinetics for both ideal and nonideal systems (Stafford and Sherwood 2004). The Fitter is based on the method of Todd and Haschemeyer (1981). With the Fitter, SEDANAL allows either least squares (L^2 norm) or robust (least sum of absolute values) (L^1 norm) curve fitting of sedimentation velocity data to sedimentation models using the finite element solutions to the Lamm equation developed by Claverie et al. (1975). For least squares fitting, either the Levenberg-

Marquardt or the simplex algorithms can be used, and for robust fitting, the simplex algorithm is used. The Fitter has been expanded since 2004 to include fitting of sedimentation equilibrium data (unpublished "A fitting program for global analysis of general hetero- and self-associating systems by sedimentation equilibrium," Lausanne AUC Meeting, 2005 and Mukhopadhyay et al. 2005) for noninteracting and self- and hetero-associating systems. Nonideality can be accommodated through second and third virial coefficients including the first-order cross terms. See also Chap. 7 of this book for a description of the equilibrium Fitter.

6.2.3 *Simulator*

The simulator generates solutions to the Lamm equation using the finite element method, combined with chemical equilibrium or kinetics equations generated by the Model Editor. The Fitter uses the simulator to generate the sedimentation patterns for each set of guesses and then compares them with the data to compute the residuals as described previously (Stafford and Sherwood 2004). The simulator can output synthetic data in XL-A/I format for further use to test models with SEDANAL or other software. It is useful for designing experiments and to test hypotheses.

6.2.4 *DCDT and WDA*

Time-derivative and wide distribution analysis of velocity data. Time-derivative (DCDT) analysis is described in previous publications (Stafford 1992, 1997, 2000, 2004). Wide distribution analysis (WDA) also has been described previously (Stafford and Braswell 2004).

In DCDT analysis, initially, the program arbitrarily chooses a subset of scans centered near the middle of the run and presents a tentative plot of both the time derivative, dc/dt, vs. s* in the left-hand window and a plot of g(s*) vs. s* in the right-hand window. The user is then allowed to choose a different subset of scans depending on the desired range of s* to be analyzed. The g(s*) curve can also be plotted on a log scale as s*g(s*) vs. ln(s*) to allow better zero level adjustment. The g(s*) curve can be considered as a snapshot of the concentration gradient at a particular time during the run.

Under the DCDT menu, one may select either standard g(s*) analysis (Stafford 1992) or wide distribution analysis (Stafford and Braswell 2004) (WDA). WDA can be used to analyze either single-speed data or to analyze multispeed data using the equilibrium method to establish the speed change protocol. The WDA curves are plotted on a log scale as s*g(s*) vs. ln(s*) to accommodate a wide range of sedimentation coefficient values in a single, multispeed run. It can easily accommodate a range of 1.0 S to 250,000 S depending on the speeds used during the run. Data from several radii can be overlapped and averaged to span a very wide range of sedimentation coefficients in a single plot. SEDANAL version 6.37 has a vastly improved interface and analysis method compared to previous versions.

N.B. There are at least two ways to plot time-derivative $g(s*)$ data, either as $g(s^*)$ vs. s^* or as $s^*g(s^*)$ vs. $ln(s^*)$. The integral of either of these plots gives the concentration as a function of s^* or $ln(s^*)$, respectively, so that the integral between the meniscus and the plateau of either of these plots gives the plateau concentration.

$$c_p = \int_{\text{meniscus}}^{\text{plateau}} g(s^*)ds^* = \int_{\text{meniscus}}^{\text{plateau}} s^*g(s^*)dln(s^*) \tag{6.1}$$

This means that if one plots the data on a log s^* scale, one must multiply $g(s^*)$ by s^* so that one plots $s^*g(s^*)$ vs. $ln(s^*)$) to allow computation of relative amounts of material from the areas under the peaks. A $ln(s^*)$ plot allows the analysis of a wide range of s^* values from a single run – either multispeed or single speed – and is used in wide distribution analysis (WDA) to display the entire range of s^* values observable in a given run. The noise distribution on the $s^*g(s^*)$ scale is uniform with respect to s^* and does not blow up near the meniscus as it does on the $g(s^*)$ scale.

6.2.5 BioSpin

BioSpin calculates point-by-point molecular-weight averages for sedimentation equilibrium data using the original code by Roark and Yphantis (1969).

6.2.6 Chemical Equilibrium Calculator

In addition, SEDANAL has a built-in calculator for homogeneous chemical equilibrium and one for homogeneous chemical kinetics allowing one to simulate the kinetics of an arbitrary reaction scheme that can be represented in the Model Editor.

6.2.7 Model Editor and Equation Editor

Two important parts of the FITTER are the MODEL EDITOR and the EQUATION EDITOR. These allow users to specify a reaction scheme of either chemical equilibrium or kinetics and a set of default constraints on the parameters to be fitted. The specified scheme and constraints are applied during fitting. The constraints can be modified on the Fitter control screen.

For example, tubulin (Tb) binds stepwise to stathmin (St) to form StTb and $StTb_2$. This model would be stored by the MODEL EDITOR as two chemical reactions, involving four species. The binding is rapid, so these will be in rapidly reversible equilibrium. Figure 6.1 shows the MODEL EDITOR tab where the reaction scheme is input. Figure 6.2 shows how species and stoichiometries and

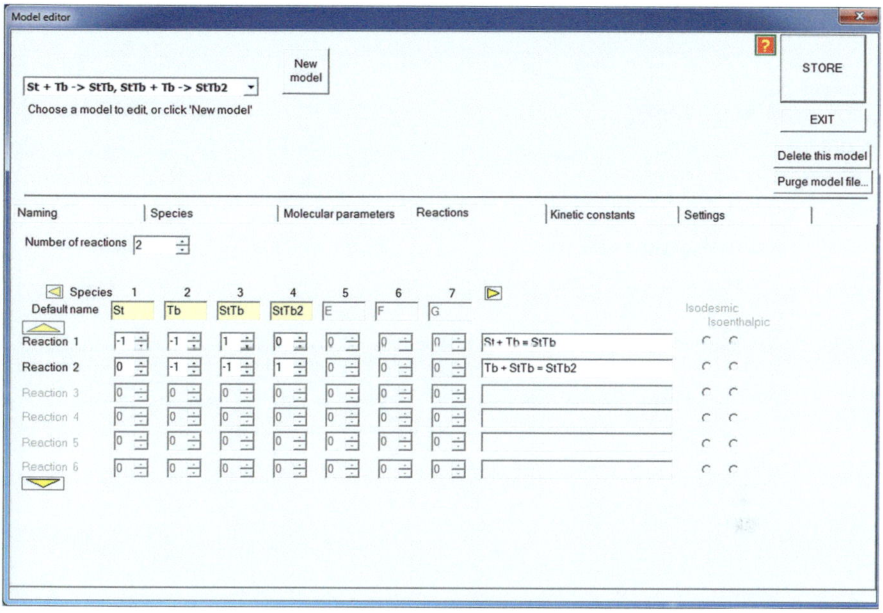

Fig. 6.1 Model Editor reactions tab

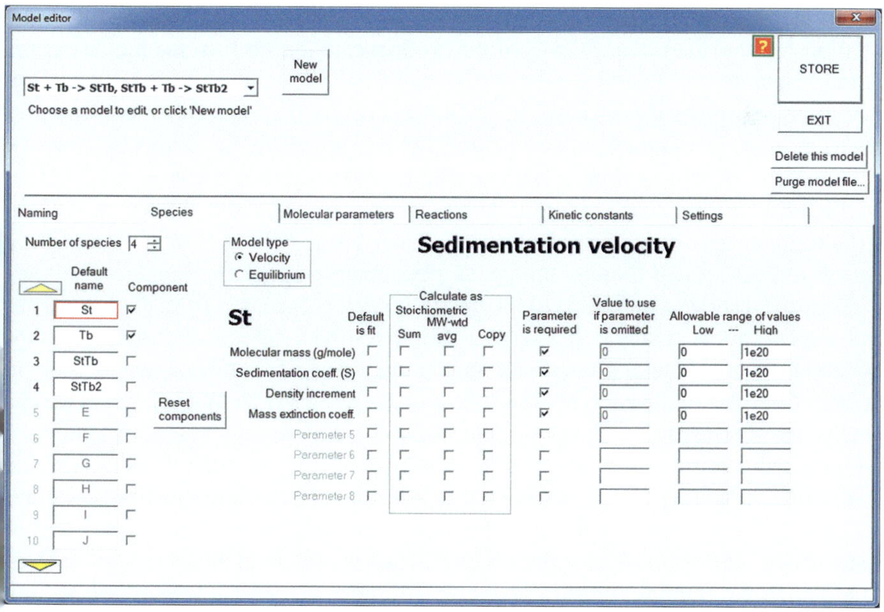

Fig. 6.2 Model Editor species tab

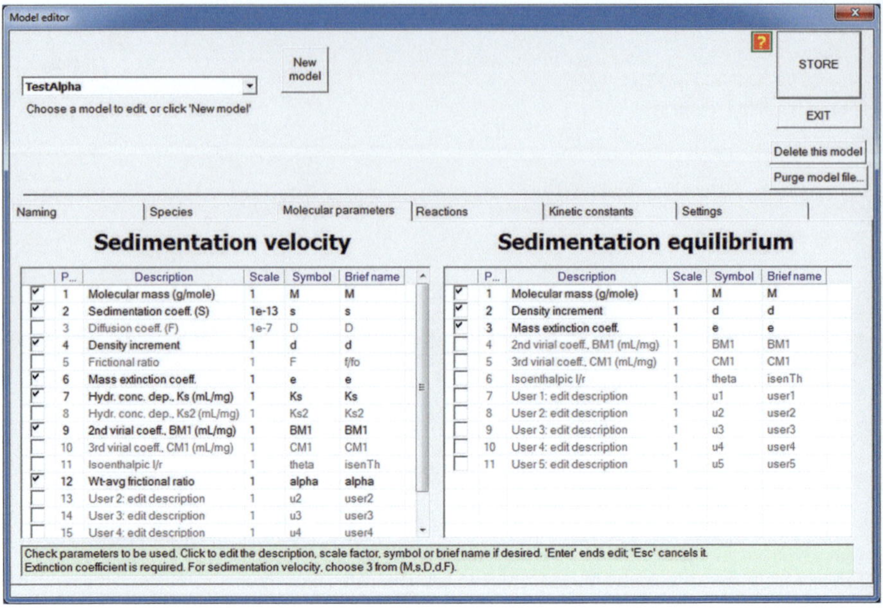

Fig. 6.3 Model Editor molecular parameters tab

fitting parameters are entered. For example, by selecting "SUM" we establish the relationship that the molar masses of the products are the sums of the molar masses of the reactants.

Figure 6.3 shows the MODEL EDITOR's molecular parameters screen. The user chooses a set of parameters to be fitted for the model (e.g., molecular mass or frictional coefficient, sedimentation coefficient, density increment, and extinction coefficient). In this case, two nonideality parameters (Ks and BM1), as well as an additional user-defined parameter (α), have been chosen. User-defined fitting parameters are related to other molecular parameters through the Equation Editor.

Figure 6.4 shows the EQUATION EDITOR, which is accessed from the Fitter control screen. In this example, the system comprises three independent species (components), with no interactions (reactions) among them. Therefore, there will be no information supplied by the Model Editor concerning stoichiometric relationships among the three species. If we suspect that this solution may contain monomers, dimers, and tetramers that are not in equilibrium, we can enter FORTRAN-like statements indicating those stoichiometric relations. For example, we would enter $M(2) = 2.0 * M(1)$ to require that the molar mass of species 2 be 2 times that of species one, and similarly for the species 3, we would enter $M(3) = 4.0 * M(1)$ to establish the required constraints during fitting. The control screen would look as shown in Fig. 6.5. The yellow background indicates that those values are not independent and depend on other parameters according to the relationships specified in the Equation Editor. This would also reduce the number of independent fitting

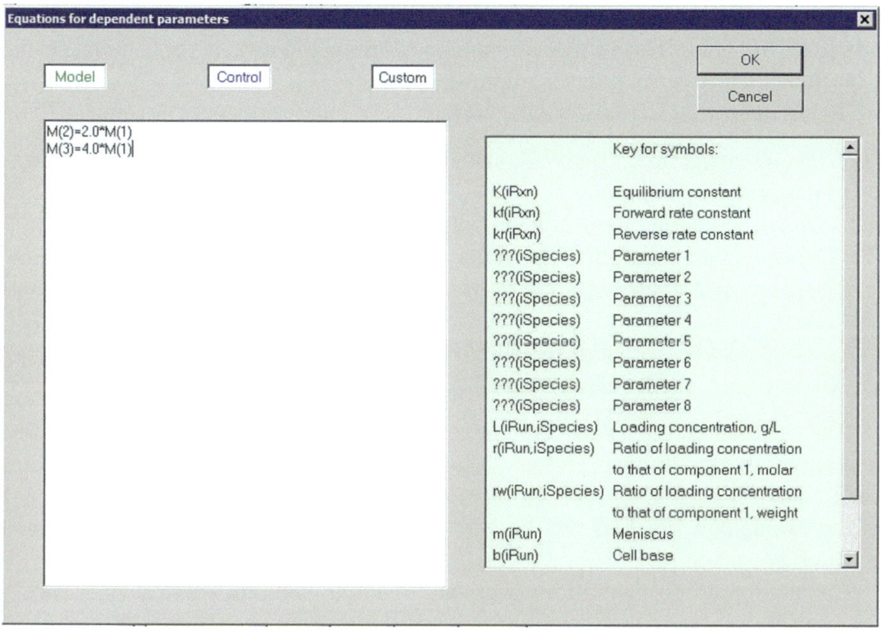

Fig. 6.4 Equation Editor window accessed from the control screen

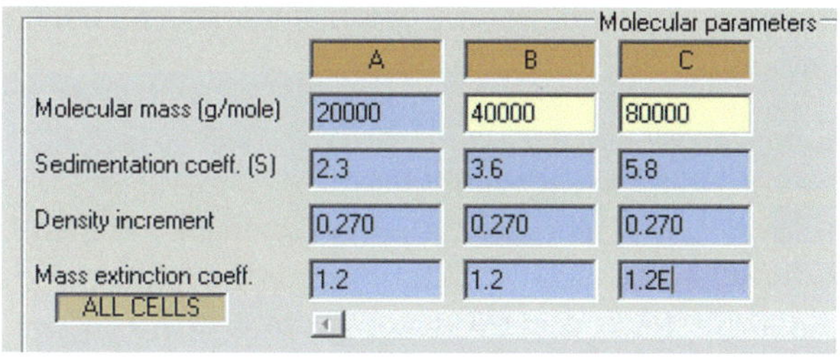

Fig. 6.5 Control screen: molecular parameters in yellow background are specified by the Equation Editor

parameters by 2. The right-hand panel in Fig. 6.4 lists the possible variables that can be used. The variable names are case sensitive.

For each fit, the user indicates which parameters should be allowed to vary during the fit and which will be held constant. Data from the MODEL EDITOR and the EQUATION EDITOR constrain the parameters (a simple example of a model constraint is that a dimer's molar mass is twice that of the monomer). During the fitting procedure, the independent parameters are varied in accord with the fit method (e.g., Levenberg-Marquardt), and the dependent parameters calculated from the independent parameters using the constraints.

The Fitter for sedimentation velocity data works by simulating a set of scans at the same elapsed times as the experimental scans. Both the simulated and experimental scans are subtracted in pairs to remove time-independent systematic errors in the data. The root mean square difference between the simulated and experimental difference scans is minimized.

To simulate a set of scans for the current set of independent parameters:

(1) Dependent parameters are calculated from independent ones, using the EQUATION EDITOR's output.
(2) A small sedimentation-diffusion time step is made.
(3) Dependent concentrations from the independent ones, using the MODEL EDITOR's stoichiometry and the current values for equilibrium and/or rate constants.
(4) Steps (2) and (3) are repeated until the time of a scan is reached.
(5) Steps (2)–(4) are repeated until all scans have been simulated.

The calculation of the dependent parameters uses an interpreter, which maintains a symbol table containing current values for all defined variables, and evaluates each equation in order. Variables that are required by the simulation (such as M in the example in Fig. 6.3) will be used for sedimentation, diffusion, or chemical equilibria or kinetics.

Calculation of concentrations uses one of three methods, chosen by the user: (a) an analytic solution, (b) equilibrium (only) by successive approximation of the components' concentrations, and (c) integration of the rate equations.

In the case of (b), the dependent concentrations are calculated from the components' concentrations using equilibrium equations from the model file (see below).

Analytic solutions are built into SEDANAL for:

Isomerization $A = A^*$ (kinetically limited)
Isomerization $A = A^*$ (rapid equilibrium)
Monomer-dimer, self-association $2A = A_2$ (kinetically limited)
Monomer-dimer, self-association $2A = A_2$ (rapid equilibrium)
Hetero-dimerization $A + B = C$ (kinetically limited)
Hetero-dimerization $A + B = C$ (rapid equilibrium)
Isodesmic, indefinite self-association $A- > A2- > A3- > A3\ldots$ (rapid equilibrium)
Isoenthalpic, indefinite self-association $A- > A2- > A3- > A3\ldots$ (rapid equilibrium)

Calculation of the equilibrium concentrations is carried out by successive approximation using the Newton-Raphson method.

For slow reactions, numerical integration of the rate equations is accomplished by either of two methods, chosen by the user: (a) Bulirsch-Stoer (BulSt) (Teukolsky et al. 2007) and (b) semi-implicit Euler extrapolation (SEulEx) (Teukolsky et al. 2007). The rate equations are a set of simultaneous first-order ordinary differential equations (ODE), and the right-hand sides are evaluated by a special-purpose interpreter, using the stoichiometric matrix provided by the MODEL EDITOR.

As an example for fitting sedimentation velocity data, suppose the model is A + B = C and C + B = D, and it is kinetically limited (the reactions are slow compared to the sedimentation times). Let's call it "ABCD." Then the user will have given guesses for the forward and reverse rate constants k_f and k_r for each reaction. There is no analytic solution; so the rate equations will be integrated over the discrete sedimentation time step used in the numerical solution to the Lamm equation. There are two rate equations. It is convenient to use moles of reaction (x_i) as the dependent variable.

$$\dot{x}_1 = k_{f,1}\mathrm{A} \cdot \mathrm{B} - k_{r,1}\mathrm{C}$$
$$\dot{x}_2 = k_{f,2}\mathrm{B} \cdot \mathrm{C} - k_{r,2}\mathrm{D}$$

At each integration time step, Δt, we start with the concentrations of all four species from the sedimentation time step. Numerically integrating the system $x_1(t)$ and $x_2(t)$ over the sedimentation time step gives x_1 and x_2, the moles of reaction during that time step. We then update all the species concentrations according to the stoichiometric matrix.

The information needed to compute the rate for each reaction ($\dot{x}_i$) is contained in a data file `ModelInfo.txt` maintained by the user using the MODEL EDITOR. Here is part of the data stored for the "ABCD" model. Each model has a part for fitting sedimentation velocity experiments and a part for sedimentation equilibrium experiments:

```
"A+B=C;C+B=D"  "ABCD"  ""        // name and alternate names of the
                                    model
...
  2   4   4   3   2             // reactions, species, params/species
                                    (vel),
                                // params/species (eq), components
  1  2                          // species numbers of components
"A"                             // default name for species 1
"B"                             // default name for species 2
"C"                             // default name for species 3
"D"                             // default name for species 4
"A + B = C"    1    0           // name, group, analytic solution for
                                // its group, for reaction 1
"B + C = D"    1    0           // name, group, analytic solution for
                                // its group, for reaction 2
  -1   -1    1    0             // stoichiometry for reaction 1
   0   -1   -1    1             // stoichiometry for reaction 2
...
// Independent parameters for sedimentation velocity run
K(1) kf(1) kr(1) K(2) kf(2) kr(2) M(1) M(2) s(1) s(2) s(3) s(4)
d(0,1) d(0,2) e(0,1) e(0,2)
// Dependent parameters for sedimentation velocity run (6,210)
M(3)=M(1)+M(2)
M(4)=M(2)+M(3)
d(0,3)=(M(1)*d(0,1)+M(2)*d(0,2))/(M(1)+M(2))
d(0,4)=(M(2)*d(0,2)+M(3)*d(0,3))/(M(2)+M(3))
```

```
e(0,3)=(M(1)*e(0,1)+M(2)*e(0,2))/(M(1)+M(2))
e(0,4)=(M(2)*e(0,2)+M(3)*e(0,3))/(M(2)+M(3))
...
// Equilibrium information
Cons 1      2     -1      1      0
Cons 2     -1      1      0      1
Rxn 1 Species 3      1      0   |      1      1     -1      0
Rxn 2 Species 4      1      1   |      1      2      0     -1
dc 1/ 1      1      0      0   |      0      0      0      0
dc 1/ 2      0      0      0   |      0      0      0      0
dc 2/ 1      0      0      0   |      0      0      0      0
dc 2/ 2      1      0      0   |      0      0      0      0
dc 3/ 1      1      1      0   |      0      1      0      0
dc 3/ 2      1      1      0   |      1      0      0      0
dc 4/ 1      1      1      1   |      0      2      0      0
dc 4/ 2      2      1      1   |      1      1      0      0
```

The dependent parameters are calculated from the equations shown, plus any that the user has added for a particular fit using the EQUATION EDITOR.

Similarly, certain species are chosen as independent ("components"); at equilibrium, the remaining species may be calculated from these using the stoichiometry and equilibrium constants. The equations for doing this are stored in the model file, encoded in shown above in the "`Rxn` *r* `Species` *s*" lines.

6.2.7.1 Determination of the Components

From the representation of the chemical reactions as a stoichiometric matrix, SEDANAL determines which species' concentrations must be specified in order to be able to create any feasible set of concentrations of all species.

This problem is one of determining a basis for a vector space. The vector space is the set of all possible concentration vectors, constrained by the chemical equilibria. A basis is a set of vectors that spans the vector space from which any vector in the space can be expressed as a linear combination.

In the case of no reactions, the number of components, n_c, is the same as the number of species, n_s: i.e., $n_c = n_s$, because all the species are independent.

If there are chemical reactions, the interdependence of the concentrations reduces the number of components, n_c, and is less than the number of species, n_r. Assuming there are n_r reactions, all linearly independent (e.g., not multiples of one another), $n_c = n_s - n_r$.

Each reaction generates a row in the stoichiometric matrix $\boldsymbol{\nu}$; the values in the row are the number of moles of each species produced by the reaction. For example, if there are 5 species, A, B, C, D and E, and one of the reactions is $A + 3B \rightleftharpoons 2C$, the stoichiometric matrix will have a row $\begin{pmatrix}-1 & -3 & 2 & 0 & 0\end{pmatrix}$. The stoichiometric matrix will have n_r rows and n_s columns.

The n_r rows of $\boldsymbol{\nu}$ will be supplemented by $n_s - n_r$ unit vectors for the n_c components. We must determine which of the species to select as components in order for the n_s rows of this supplemented matrix to be a basis for the concentration vector space.

The test for whether a set of n vectors is a basis for an n-dimensional vector space is just whether they are linearly independent. This can in turn be determined by finding the rank of the matrix whose columns are the n vectors. The rank is calculated by doing Gaussian elimination (conversion to a row-echelon matrix) and counting the number of pivots.

The overall process consists of trying each possible unit vector and adding those that increase the rank of the matrix to the basis set (i.e., choosing them as components).

For example, suppose the species are A, B, C, and D, with the two-step equilibria:

$$\mathrm{A + B} \overset{\mathrm{K_1}}{\rightleftharpoons} \mathrm{C}$$

$$\mathrm{B + C} \overset{\mathrm{K_2}}{\rightleftharpoons} \mathrm{D}$$

The stoichiometric matrix is $\boldsymbol{\nu} = \begin{pmatrix} -1 & -1 & 1 & 0 \\ 0 & -1 & -1 & 1 \end{pmatrix}$. In practice, we use the transpose of $\boldsymbol{\nu}$, $\boldsymbol{\nu}^T$, so we can think of the columns as vectors. The transpose of the 4×2 stoichiometric matrix itself, $\boldsymbol{\nu}^T = \begin{pmatrix} -1 & 0 \\ -1 & -1 \\ 1 & -1 \\ 0 & 1 \end{pmatrix}$, has a rank of 2.

To add A (species 1) as a component, we append the vector $\begin{pmatrix} 1 \\ 0 \\ 0 \\ 0 \end{pmatrix}$; the 4×3 matrix $\begin{pmatrix} -1 & 0 & 1 \\ -1 & -1 & 0 \\ 1 & -1 & 0 \\ 0 & 1 & 0 \end{pmatrix}$ has a rank of 3.

Since we need a matrix with rank 4 (the number of species), we do not yet have a basis. So we try adding B (species 2) as a component by appending the vector $\begin{pmatrix} 0 \\ 1 \\ 0 \\ 0 \end{pmatrix}$.

The 4×4 matrix $\begin{pmatrix} -1 & 0 & 1 & 0 \\ -1 & -1 & 0 & 1 \\ 1 & -1 & 0 & 0 \\ 0 & 1 & 0 & 0 \end{pmatrix}$ has a rank of 4, so we have found a basis.

The species we used, A and B, are the components.

In this case, any two of A, B, C, and D could have been chosen as components.

6.2.7.2 Determining Equations for Equilibrium

We want to solve for the concentrations of all species in a homogeneous solution with multiple chemical reactions at equilibrium, in a closed system. The solution is assumed to be ideal; that is, activity coefficients are unity, and activities can be replaced with concentrations.

For n_s species and n_r independent chemical reactions, there are $n_c = n_s - n_r$ components and n_c independent conservation relations. Each of the n_r mass-action equilibria is of the form (ν_{ij} = stoichiometric number for reaction i, species j)

$$K_i = \prod_{j=1}^{n_s} c_j^{\nu_{ij}} = \left(\prod_{\substack{j=1\\ j\neq k}}^{n_s} c_j^{\nu_{ij}}\right) \cdot c_k^{\nu_{ik}}$$

or

$$1 = K_i^{-1} \prod_{j=1}^{n_s} c_j^{\nu_{ij}} = K_i^{-1}\left(\prod_{\substack{j=1\\ j\neq k}}^{n_s} c_j^{\nu_{ij}}\right) \cdot c_k^{\nu_{ik}} \tag{6.2}$$

To express each of the non-component concentrations in terms of component concentrations, the n_r equilibrium expressions are used to eliminate each non-component concentration. This is done by solving the equilibrium expression for reaction i (i.e., Eq. 6.2) for component c_k

$$c_k^{\nu_{ik}} = K_i \prod_{\substack{j=1\\ j\neq k}}^{n_s} c_j^{-\nu_{ij}}$$

$$c_k = K_i^{1/\nu_{ik}} \prod_{\substack{j=1\\ j\neq k}}^{n_s} c_j^{-\nu_{ij}/\nu_{ik}} \tag{6.3}$$

and substituting for c_k in the equilibrium expressions for all the other reactions.

To represent expressions such as the above, we use a scalar p and vectors of rational values (fractions) $\mathbf{q}$ and $\mathbf{r}$. We write such an expression as

$$p \prod_{j=1}^{n_r} K_j^{r_j} \prod_{j=1}^{n_s} c_j^{q_j} \tag{6.4}$$

Using this notation, (1) becomes

$$1 = p \prod_{j=1}^{n_r} K_j^{r_j} \prod_{j=1}^{n_s} c_j^{q_j}$$

with $p = 1$; $r_i = -1$ and $r_1 = r_2 = \ldots = r_{i-1} = r_{i+1} = \ldots = r_{n_r} = 0$; and $q_1 = \nu_{i1}, q_2 = \nu_{i2}, \ldots, q_{n_s} = \nu_{in_s}$ and (2) may be rewritten

$$c_k = p \prod_{j=1}^{n_r} K_j^{r_j} \prod_{j=1}^{n_s} c_j^{q_j}) \tag{6.5}$$

with $p = 1$; $r_k = 1/\nu_{ik}$ and $r_1 = r_2 = \ldots = r_{k-1} = r_{k+1} = \ldots = r_{n_r} = 0$; and $q_1 = -\nu_{i1}/\nu_{ik}$, $q_2 = -\nu_{i2}/\nu_{ik}$, ..., $q_{k-1} = -\nu_{ik-1}/\nu_{ik}$, $q_k = 0$, $q_{k+1} = -\nu_{i(k+1)}/\nu_{ik}$, ..., $q_{n_s} = -\nu_{in_s}/\nu_{ik}$.

If reaction i has been solved for non-component k, with an expression for c_k represented by $(p', \mathbf{q}', \mathbf{r}')$, the substitution of c_k into reaction j, with an equilibrium expression represented by $(p, \mathbf{q}, \mathbf{r})$, the resulting expression is represented by $(p'', \mathbf{q}'', \mathbf{r}'')$, which are related by

$$p'' = p \cdot p'^{q_k} \qquad r_i'' = r_i - (r_i'/q_k')^{q_k} \qquad q_i'' = q_i - (q_i'/q_k')^{q_k} \tag{6.6}$$

Example 1: $A + B = C$ $\quad n_r = 2 \quad n_s = 4 \quad n_c = n_s - n_r = 2$
$B + C = D$ $\quad$ Components chosen as A and B

$$K_1 = \frac{C}{A \cdot B} \qquad C = K_1 A \cdot B$$

$$1 = K_1 A \cdot B \cdot C^{-1} \quad p' = 1 \quad r_1' = 1 \quad r_2' = 0 \quad q_1' = q_2' = 1 \quad q_3' = -1 \quad q_4' = 0$$

$$K_2 = \frac{D}{B \cdot C}$$

$$1 = K_2 B \cdot C \cdot D^{-1} \quad p = 1 \quad r_1 = 0 \quad r_2 = 1 \quad q_1 = 0 \quad q_2 = q_3 = 1 \quad q_4 = -1$$

The first line gives C in terms of components A and B; substituting for the non-component C in the second equation gives

$$1 = K_1 K_2 A \cdot B^2 \cdot D^{-1}$$

This final expression is represented by $p'' = 1$, $r_1'' = r_2'' = 1$, $q_1'' = 1$, $q_2'' = 2$, $q_3'' = 0$, and $q_4'' = -1$. Using the relations (4), with $k = 3$, $q_k = 1, and\ q_k' = -1$ give

$$p'' = p \cdot p'^{q_k} = 1 \cdot 1^1 = 1$$
$$r_1'' = r_1 - (r_1'/q_3')^{q_3} = 0 - (1/-1)^1 = 1$$
$$r_2'' = r_2 - (r_2'/q_3')^{q_3} = 1 - (0/-1)^1 = 1$$

$$q''_1 = q_1 - (q'_1/q'_3)^{q_3} = 0 - (1/-1)^1 = 1$$
$$q''_2 = q_2 - (q'_2/q'_3)^{q_3} = 1 - (1/-1)^1 = 2$$
$$q''_3 = q_3 - (q'_3/q'_3)^{q_3} = 1 - (-1/-1)^1 = 0$$
$$q''_4 = q_4 - (q'_4/q'_3)^{q_3} = -1 - (0/-1)^1 = -1$$

which agree with the above values. The concentrations of C and D computed from the component concentrations A and B are stored as

Model file: `Rxn 1 Species 3   1   0   |   1   1  -1   0`
Means: $1 = K_1^1\ K_2^0\ c_1^1\ c_2^1\ c_3^{-1}\ c_4^0$ or $\mathrm{C} = K_1 \mathrm{A} \cdot \mathrm{B}$
Model file: `Rxn 2 Species 4   1   1   |   1   2   0  -1`
Means: $1 = K_1^1\ K_2^1\ c_1^1\ c_2^2\ c_3^0\ c_4^{-1}$ or $\mathrm{D} = K_1 K_2 \mathrm{A} \cdot \mathrm{B}^2$

Using the degrees of reaction x_i to express concentrations in terms of initial concentrations

$$\mathrm{A} = \mathrm{A}_0 - x_1 \qquad \mathrm{C} = \mathrm{C}_0 + x_1 - x_2$$
$$\mathrm{B} = \mathrm{B}_0 - x_1 - x_2 \qquad \mathrm{D} = \mathrm{D}_0 + x_2$$

$$2\mathrm{A} - \mathrm{B} + \mathrm{C} = 2(\mathrm{A}_0 - x_1) - (\mathrm{B}_0 - x_1 - x_2) + \mathrm{C}_0 + x_1 - x_2 = 2\mathrm{A}_0 - \mathrm{B}_0 + \mathrm{C}_0$$
$$-\mathrm{A} + \mathrm{B} + \mathrm{D} = -(\mathrm{A}_0 - x_1) + \mathrm{B}_0 - x_1 - x_2 + \mathrm{D}_0 + x_2 = -\mathrm{A}_0 + \mathrm{B}_0 + \mathrm{D}_0$$

These $n_c = 2$ conservation relations are stored in `ModelInfo.txt` as

`Cons 1    2   -1    1    0` meaning $2\mathrm{A} - \mathrm{B} + \mathrm{C} = \textit{constant}$
`Cons 2   -1    1    0    1` meaning $-\mathrm{A} + \mathrm{B} + \mathrm{D} = \textit{constant}$

The Jacobian is stored in the model file as

dc 1/ 1	1	0	0	0	0	0	0	$\partial \mathrm{A}/\partial \mathrm{A} = 1$
dc 1/ 2	0	0	0	0	0	0	0	$\partial \mathrm{A}/\partial \mathrm{B} = 0$
dc 2/ 1	0	0	0	0	0	0	0	$\partial \mathrm{B}/\partial \mathrm{A} = 0$
dc 2/ 2	1	0	0	0	0	0	0	$\partial \mathrm{B}/\partial \mathrm{B} = 1$
dc 3/ 1	1	1	0	0	1	0	0	$\partial \mathrm{C}/\partial \mathrm{A} = K_1 \mathrm{B}$
dc 3/ 2	1	1	0	1	0	0	0	$\partial \mathrm{C}/\partial \mathrm{B} = K_1 \mathrm{A}$
dc 4/ 1	1	1	1	0	2	0	0	$\partial \mathrm{D}/\partial \mathrm{A} = K_1 K_2 \mathrm{B}^2$
dc 4/ 2	2	1	1	1	1	0	0	$\partial \mathrm{D}/\partial \mathrm{B} = K_1 K_2 \mathrm{A} \cdot \mathrm{B}$

Example 2: $\mathrm{A} + \mathrm{B} = \mathrm{AB}$ $n_r = 2$ $n_s = 4$ $n_c = n_s - n_r = 2$
$2\mathrm{AB} = (\mathrm{AB})_2$ Components chosen as A and B
Species numbering is 1: A; 2: AB; 3: B; 4: $(\mathrm{AB})_2$.

$$K_1 = \frac{\mathrm{AB}}{\mathrm{A} \cdot \mathrm{B}} \qquad \mathrm{AB} = K_1 \mathrm{A} \cdot \mathrm{B}$$
$$1 = K_1 \mathrm{A} \cdot \mathrm{B} \cdot \mathrm{AB}^{-1} \qquad p' = 1 \quad r'_1 = 1 \quad r'_2 = 0$$
$$q'_1 = q'_3 = 1 \quad q'_2 = -1 \quad q'_4 = 0$$

$$K_2 = \frac{(\mathrm{AB})_2}{\mathrm{AB}^2} \qquad (\mathrm{AB})_2 = K_2\mathrm{AB}^2$$

$$1 = K_2\mathrm{AB}^2 \cdot (\mathrm{AB})_2^{-1} \qquad p = 1 \qquad r_1 = 0 \quad r_2 = 1$$

$$q_1 = 0 \quad q_2 = 2 \quad q_3 = 0 \quad q_4 = -1$$

The first line gives AB in terms of components A and B; substituting for the non-component AB in the second equation gives

$$1 = K_2(K_1\mathrm{A} \cdot \mathrm{B})^2 \cdot (\mathrm{AB})_2^{-1} = K_1^2 K_2 \mathrm{A}^2 \cdot \mathrm{B}^2 \cdot (\mathrm{AB})_2^{-1}$$

This final expression is represented by $p'' = 1$, $r_1'' = r_2'' = 1$, $q_1'' = 2$, $q_3'' = 2$, $q_2'' = 0$, $q_4'' = -1$.

The concentrations of AB and AB_2 computed from the component concentrations A and B are stored in the model file (`ModelInfo.txt`) as

Model file: `Rxn 1 Species 2   1   0   |   1  -1   1   0`
Means: $1 = K_1^1\, K_2^0\, c_1^1\, c_2^{-1}\, c_3^1\, c_4^0$ or $\mathrm{AB} = K_1\mathrm{A} \cdot \mathrm{B}$
Model file: `Rxn 2 Species 4   2   1   |   2   0   2  -1`
Means: $1 = K_1^2\, K_2^1\, c_1^2\, c_2^0\, c_3^2\, c_4^{-1}$ or $(\mathrm{AB})_2 = K_1^2 K_2 \mathrm{A}^2 \cdot \mathrm{B}^2$

Using the degrees of reaction x_i to express concentrations in terms of initial concentrations

$$\mathrm{A} = \mathrm{A}_0 - x_1 \qquad \mathrm{AB} = \mathrm{AB}_0 + x_1 - 2x_2$$

$$\mathrm{B} = \mathrm{B}_0 - x_1 \qquad (\mathrm{AB})_2 = (\mathrm{AB})_{2,0} + x_2$$

$$-\mathrm{A} + \mathrm{B} = -(\mathrm{A}_0 - x_1) + \mathrm{B}_0 - x_1 = -\mathrm{A}_0 + \mathrm{B}_0$$

$$\tfrac{1}{2}\mathrm{A} + \tfrac{1}{2}\mathrm{AB} + (\mathrm{AB})_2 = \tfrac{1}{2}(\mathrm{A}_0 - \mathrm{x}_1) + \tfrac{1}{2}(\mathrm{AB}_0 + \mathrm{x}_1 - 2\mathrm{x}_2) + (\mathrm{AB})_{2,0} + \mathrm{x}_2$$

$$= \tfrac{1}{2}\mathrm{A}_0 + \tfrac{1}{2}\mathrm{AB}_0 + (\mathrm{AB})_{2,0}$$

These conservation relations are stored in `ModelInfo.txt` as

`Cons 1  -1   0    1  0` meaning $-\mathrm{A} + \mathrm{B} = \mathit{constant}$
`Cons 2  1/2 1/2 0  1` meaning $\tfrac{1}{2}\mathrm{A} + \tfrac{1}{2}\mathrm{AB} + (\mathrm{AB})_2 = \mathit{constant}$

The Jacobian is stored in the model file as

`dc 1/ 1`	1	0	0	|	0	0	0	0	$\partial\mathrm{A}/\partial\mathrm{A} = 1$
`dc 1/ 3`	0	0	0	|	0	0	0	0	$\partial\mathrm{A}/\partial\mathrm{B} = 0$
`dc 2/ 1`	1	1	0	|	0	0	1	0	$\partial\mathrm{AB}/\partial\mathrm{A} = K_1\mathrm{B}$
`dc 2/ 3`	1	1	0	|	1	0	0	0	$\partial\mathrm{AB}/\partial\mathrm{B} = K_1\mathrm{A}$
`dc 3/ 1`	0	0	0	|	0	0	0	0	$\partial\mathrm{B}/\partial\mathrm{A} = 0$
`dc 3/ 3`	1	0	0	|	0	0	0	0	$\partial\mathrm{B}/\partial\mathrm{B} = 1$
`dc 4/ 1`	2	2	1	|	1	0	2	0	$\partial(\mathrm{AB})_2/\partial\mathrm{A} = 2K_1^2 K_2 \mathrm{A} \cdot \mathrm{B}^2$
`dc 4/ 3`	2	2	1	|	2	0	1	0	$\partial(\mathrm{AB})_2/\partial\mathrm{B} = 2K_1^2 K_2 \mathrm{A}^2 \cdot \mathrm{B}$

6.2.7.3 Representation of the Jacobian

In the `Equilibrium information` section of the model file are the conservation relations (e.g., 2A − B + C = *constant*), equations for computing non-component species' concentrations from component species' at chemical equilibrium, and the partial derivatives $\frac{\partial c_i}{\partial c_j}$, which are the elements of the $n_s \times n_c$ Jacobian matrix, **J** (number of components $n_c = n_s - n_r$). The Jacobian is used for sedimentation equilibrium. These are equations of the form (cf. Eq. 6.5)

$$\frac{\partial c_i}{\partial c_j} = p \prod_{i=1}^{n_r} K_i^q \prod_{j=1}^{n_s} c_j^r$$

In the model file, each $\frac{\partial c_i}{\partial c_j}$ is represented by a `dc` $i/\ j$ line, which has p and n_r qs to the left of the bar and n_s rs to the right. For example

```
dc 4/ 2 2 1 1 | 1 1 0 0
```

means

$$\frac{\partial c_4}{\partial c_2} = 2\,K_1^1\,K_2^1\,c_1^1\,c_2^1\,c_3^0\,c_4^0 = 2\,K_1\,K_2\,c_1\,c_2$$

6.2.7.4 Determination of the Conservation Relations

The conservation relations will be the last n_c rows of

$$\mathbf{L}^{-1}\mathbf{Pa} = \mathbf{L}^{-1}\mathbf{Pa_0}$$

$\mathbf{L}^{-1}\mathbf{P}$ are found from the relation $\mathbf{P}\boldsymbol{\nu}^{\mathrm{T}} = \mathbf{LU}$, in which $\boldsymbol{\nu}$ is the $n_r \times n_s$ stoichiometric matrix; **P** is an $n_s \times n_s$ permutation matrix; **L** is an $n_s \times n_s$ lower-triangular matrix, with 1s on the diagonal; and **U** is an $n_s \times n_r$ upper-echelon matrix. $\mathbf{a}(t)$ is the vector of time-dependent species amounts (not concentrations), and $\mathbf{a_0}$ the initial species amounts (i.e., $\mathbf{a}(0)$).

For example, suppose a monomer, A, and dimer, A_2, are in the chemical equilibrium:

$$2A \rightleftharpoons A_2 \qquad K_{eq} = \frac{(A_2)}{(A)^2}$$

Then $n_r = 1, n_s = 2,$

$$\boldsymbol{\nu} = \begin{pmatrix} -2 & 1 \end{pmatrix} \quad \text{and} \quad \boldsymbol{\nu}^{\mathrm{T}} = \begin{pmatrix} -2 \\ 1 \end{pmatrix}$$

To find $\mathbf{L}$ that puts $\boldsymbol{\nu}^{\mathrm{T}}$ into upper-echelon form, we reduce $\boldsymbol{\nu}^{\mathrm{T}}$ to upper-echelon form using row operations, which gives

$$\mathbf{L} = \begin{pmatrix} 1 & 0 \\ -\frac{1}{2} & 1 \end{pmatrix} \quad \text{and} \quad \mathbf{U} = \begin{pmatrix} -2 \\ 0 \end{pmatrix}$$

Since no row transpositions are necessary during the reduction,
$\mathbf{P} = \mathbf{I} = \begin{pmatrix} 1 & 0 \\ 0 & 1 \end{pmatrix}$. Next, $\mathbf{L}^{-1} = \begin{pmatrix} 1 & 0 \\ \frac{1}{2} & 1 \end{pmatrix}$. Finally,

$$\begin{aligned} \mathbf{L}^{-1}\mathbf{Pa} &= \begin{pmatrix} 1 & 0 \\ \frac{1}{2} & 1 \end{pmatrix} \begin{pmatrix} A \\ A_2 \end{pmatrix} = \begin{pmatrix} A \\ \frac{1}{2}A + A_2 \end{pmatrix} \\ &= \begin{pmatrix} 1 & 0 \\ \frac{1}{2} & 1 \end{pmatrix} \begin{pmatrix} A_0 \\ A_{2,0} \end{pmatrix} = \begin{pmatrix} A_0 \\ \frac{1}{2}A_0 + A_{2,0} \end{pmatrix} \end{aligned}$$

There is one component, so the $n_c \times n_s$ conservation matrix is the last row of $\mathbf{L}^{-1}\mathbf{Pa}$, $(\frac{1}{2}\ 1)$, and the conservation relation is

$$\tfrac{1}{2}A + A_2 = \tfrac{1}{2}A_0 + A_{2,0} \quad \text{or} \quad A + 2A_2 = A_0 + 2A_{2,0}$$

which we think of as conservation of monomer.

In case the permutation matrix $\mathbf{P}$ is not the identity matrix $\mathbf{I}$, the species are renumbered to make it so.

6.2.7.5 Determination of the Jacobian

The $\dfrac{\partial c_i}{\partial c_j}$ is needed for all species i, for each component species j. When i is a component, $\dfrac{\partial c_i}{\partial c_j} = \delta_{ij}$ (Kronecker delta). When i is not a component, we start with the expression for c_i (Eq. 6.3), substitute for all c_k, where k is not a component, to get c_i in terms of only component concentrations. The Model Editor has derived these using Eq. 6.6 and stored them in the model file. Then the partial derivative is just replacing c_j^q with qc_j^{q-1}, and this is also in the model file.

Using the reactions in Example 2 (page 94), we have the representation of the non-component species AB and $(AB)_2$ in terms of components $c_1 \equiv$ A and $c_3 \equiv$ B as

Model file: `Rxn 1 Species 2    1    0    |    1   -1    1    0`
Means: $1 = K_1^1\ K_2^0\ c_1^1\ c_2^{-1}\ c_3^1\ c_4^0$ or $AB = K_1 A \cdot B$
Model file: `Rxn 2 Species 4    2    1    |    2    0    2    -1`
Means: $1 = K_1^2\ K_2^1\ c_1^2\ c_2^0\ c_3^2\ c_4^{-1}$ or $(AB)_2 = K_1^2 K_2 A^2 \cdot B^2$

For instance, $\frac{\partial c_4}{\partial c_3}$ has $p = 2$ (from derivative of c_3^2), $q_1 = 2$, $q_2 = 1$, $r_1 = 2$, $r_2 = 0$, $r_3 = 1$, $r_4 = 0$, or $2K_1^2K_2c_1^2c_2^0c_3^1 = 2K_1^2K_2c_1^2c_3$.

6.2.7.6 Methods of Calculation

The reaction rates during numerical integration, the non-component species' concentrations at equilibrium, and the Jacobian are each computed using the model file data and a different special-purpose interpreter in Fortran, shown below.

```
forwardRate=kf(iRxn); reverseRate=kr(iRxn)
DO iSpecies=1,nSpecies
  s=stoich(iSpecies,iRxn)
  DO i=1,-s
   forwardRate=forwardRate*initialConc(iSpecies) ! species
  END DO                                         ! appears on LHS
  DO i=1,s
   reverseRate=reverseRate*initialConc(iSpecies) ! species
  END DO                                         ! appears on RHS
END DO
ReactionRate=forwardRate-reverseRate
```

```
DO iSpecies=1,nSpecies
  ! conc is the initial concentration
  DO iRxn=1,nRxns
    IF(stoich(iSpecies,iRxn)/=0) &
      conc(iSpecies) = conc(iSpecies) &
      + rxnDegree(iRxn)*stoich(iSpecies,iRxn)
  END DO
END DO
```

```
DO iComp=1,nComponents
  DO jComp=1,nComponents
    DO iSpecies=1,nSpecies
      prod=pF(iSpecies,jComp)
      DO iRxn=1,nRxns
        prod=prod*(Keq(iRxn)**rF(iSpecies,jComp,iRxn))
      END DO ! iRxn
      DO jSpecies=1,nSpecies
        prod=prod*(concEq(jSpecies)**qF(iSpecies,jComp,jSpecies))
      END DO ! jSpecies
      fCol(iSpecies)=prod
    END DO ! iSpecies

    IF(.NOT.homogeneous) THEN
      fCol(1:nSpecies)=fCol(1:nSpecies)*dxdc(1:nSpecies)
    END IF
    J(iComp,jComp) = &
      DOT_PRODUCT(LConserv(iComp,1:nSpecies),fCol(1:nSpecies))

  END DO ! jComp
END DO ! iComp
```

6.2.8 *Wide Distribution Analysis*

The wide distribution analysis (WDA) in SEDANAL has been improved with better methods of differentiation, averaging, and interpolation. It can be used to analyze either single-speed runs or multispeed runs to cover an extremely wide range of sedimentation coefficients. The main advantage of this method over other approaches is that a very wide range, from less than 1S to over 250,000 S, can be achieved in a single run when the speed is ramped from 3,000 RPM to 60,000 RPM. Methods that depend on the finite element solutions to the Lamm equation have difficulty handling the very large molecules that have very small diffusion coefficients and typically cannot handle data collected over several speeds in a single run. This is not a limitation with the WDA method.

6.2.8.1 Advantages of WDA vs. DCDT

An advantage of this method over the DCDT method is that all the scans from the entire run are used in the calculation of the distribution function; so the user doesn't have to choose which scans to display; moreover, no scans are ignored as in other methods. Results are plotted on a log scale as $s^*g(s^*)$ vs. $ln(s^*)$: this preserves that area under the curve so that the integral $\int_{s_1}^{s_2} s^*g(s^*)dln(s^*)$ gives the concentration between s_1 and s_1 (Stafford and Braswell, 2004); it also allows the calculation of the weight-average sedimentation coefficient for baseline separated peaks or for the whole sample, assuming nothing very large has sedimented before taking the first scans.

The time-derivative (DCDT) method uses a subset of scans to give essentially a snapshot of the sedimenting boundary at a particular time during the run and is plotted on a linear s* scale (Stafford 1992). On the other hand, since the wide distribution analysis (WDA) method includes all the scans from a run, it gives a sedimentation pattern spanning the entire range of s values observable for that run, the log scale accommodating the wide range of s values observable. In addition, WDA has a multispeed capability allowing extremely wide ranges of s value to be observed in a single run (Stafford and Braswell 2004). In both methods use of the time derivative eliminates the need to fit for the systemic background offset at each radial position in each scan.

6.2.8.2 Multiwavelength Analysis

After the MWL data have been read into the preprocessor, we have a large array of scans at each wavelength for every time point. If we have M wavelengths and N components in our mixture, at each radial point, we can represent this system as a linear system in matrix notation:

$$\mathbf{a} = \mathbf{Ec} \tag{6.7}$$

where **a** is an M element column vector of absorbances measured at each of M wavelengths, **E** is an M × N matrix of extinction coefficients for each of N components at M wavelengths, and **c** is an N element column vector of concentrations. This is a highly overdetermined system that could be solved exactly if there were no noise on the data. In fact we could obtain an exact solution using only N wavelengths if we had N components giving N equations in N unknowns. However, in the presence of noise, we can obtain the best unbiased estimate of **c** in the least squares sense using data from all the wavelengths by the following manipulations of the matrices:

$$\hat{\mathbf{c}} = (\mathbf{E}^T\mathbf{E})^{-1}\mathbf{E}^T\mathbf{a} \tag{6.8}$$

where $\hat{\mathbf{c}}$ is a column vector containing the N best linear unbiased estimates of the concentrations of the N components.

E is an M × N matrix where M is the number of wavelengths and N is the number of components in the mixture. Each column element in the matrix **E** is the value of the extinction coefficient at each wavelength, and each row corresponds to the total contribution to the absorbance at each wavelength.

6.2.8.3 Extracting Component Extinction Spectra

If individual peaks in either the DCDT or WDA distributions are sufficiently well separated, the extinction spectra of each species can be extracted from the distribution functions by integrating over small ranges of sedimentation coefficient over which only one component is contributing to the signal.

From DCDT we have:

$$a(\lambda) = \int_{s_1}^{s_2} \left(\frac{\partial a(\lambda)}{\partial s^*}\right) ds^* \tag{6.9}$$

and from WDA, we have:

$$a(\lambda) = \int_{s_1}^{s_2} \left(\frac{\partial a(\lambda)}{\partial ln(s^*)}\right) dln(s^*) \tag{6.10}$$

where $a(\lambda)$ is the absorbance at a particular wavelength, λ, and then the extinction spectrum, $E(\lambda)$, is just

$$E(\lambda) = \frac{a(\lambda)}{lc} \tag{6.11}$$

where c is the known concentration of the component under consideration, expressed in the appropriate concentration units, and l is the path length of the centerpiece used in the experiment (Walter et al. 2015).

6.2.9 *Summary*

We have presented a description of the software package SEDANAL including several new features since its initial presentation in 2004 (Stafford and Sherwood 2004). The additions since then include the ability to fit sedimentation equilibrium data (See chapter 7 of this volume); the incorporation of the Model Editor, for which we have presented a full description, into the main program; and the incorporation of an Equation Editor into the fitting control screen so that relationships between parameters can be established outside the Model Editor. This is especially useful, for example, for establishing stoichiometric relationships between components in noninteracting systems. Multiwavelength capability has been added to allow analysis of data from multiwavelength instruments. A procedure has been implemented to deconvolute concentrations of components given known extinction spectra covering up to more than 2000 wavelengths, as well as the ability to determine extinction spectra of unknown components that exhibit discrete, well-separated peaks in DCDT and/or WDA distribution curves. SEDANAL can also fit directly to multiwavelength data given the extinction spectra of the components without the need to deconvolute first. SEDANAL can be downloaded from the SEDANAL web site http://sedanal.org/.

"In everything, no matter what it may be, uniformity is undesirable. Leaving something incomplete makes it interesting, and gives one the feeling that there is room for growth."

- Yoshida Kenko Essays in Idleness (14th Century). The Tsurezuregusa of Kenko

References

Claverie JM, Dreux H, Cohen R (1975) Sedimentation of generalized systems of interacting particles. I. Solution of systems of complete Lamm equations. Biopolymers 14(8):1685–1700

Mukhopadhyay S, Langsetmo K, Stafford WF, Henry GD, Baleja JD, Sarkar S (2005) Identification of a region of fast skeletal troponin T required for stabilization of the coiled-coil formation with troponin I. J Biol Chem 280:538–547

Stafford WF (1992) Boundary analysis in sedimentation transport experiments: a procedure for obtaining sedimentation coefficient distributions using the time derivative of the concentration profile. Anal Biochem 203:295–300

Stafford WF (1997) Sedimentation velocity spins a new weave for an old fabric. Curr Opin Biotechnol 8:14–24

Stafford WF (2000) Analysis of reversibly interacting macromolecular systems by time derivative sedimentation velocity. Methods Enzymol 323:302–325

Stafford WF (2004) Boundary analysis in sedimentation velocity experiments. Methods Enzymol 240:478–501

Stafford WF, Braswell E (2004) Sedimentation velocity, multi-speed method for analyzing polydisperse solutions. Biophys Chem 108:273–279

Stafford WF, Sherwood PJ (2004) Analysis of heterologous interacting systems by sedimentation velocity: curve fitting algorithms for estimation of sedimentation coefficients, equilibrium and kinetic constants. Biophys Chem 108:231–243

Teukolsky S, Flannery BP, Vetterling WT, Press WH (2007) Numerical recipes: the art of scientific computing, 3rd edn. Cambridge University Press, Cambridge/New York. ISBN:978-0-521-88068-8

Todd GP, Haschemeyer RH (1981) General solution to the inverse problem of the differential equation of the ultracentrifuge. Proc Natl Acad Sci U S A 78(11):6739–6743

Walter J, Sherwood PJ, Lin W, Segets D, Stafford WF, Peukert W (2015) Simultaneous analysis of hydrodynamic and optical properties using analytical ultracentrifugation equipped with multiwavelength detection. Anal Chem 87(6):3396–3403

Roark, DE, Yphantis DA (1969) Studies of self-associating systems by equilibrium ultracentrifugation. Ann N Y Acad Sci 164, 245–278.

Chapter 7
SEDANAL: Global Analysis of General Hetero- and Self-Associating Systems by Sedimentation Equilibrium

Walter F. Stafford and Peter J. Sherwood

Abstract Algorithms have been developed for the analysis of sedimentation equilibrium data by fitting to arbitrary reaction schemes. These have been implemented within the framework of a larger program called SEDANAL, which until inclusion of equations for fitting equilibrium data was capable of treating only sedimentation velocity data (Stafford and Sherwood, Biophys Chem 108:231–243, 2004) A predecessor to this program, called NONSIM (first used by Margossian and Stafford, Biochemistry, 1982), forms the basis for the algorithms used in SEDANAL. Fitting to the equilibrium equations is carried out by minimization with respect to either the L1 norm (average absolute value of the residuals) using the simplex method of Nelder and Mead (1965) or the L2 norm using the Levenberg-Marquardt method. In cases involving more than one macromolecular component (i.e., hetero-associations or self-associating systems involving nonparticipating species), conservation of mass is invoked, and weight fractions of the components become fitting parameters. Thus, it is possible to fit directly for the weight fraction of incompetent species in a self-associating system without resorting to further mathematical treatment of apparent equilibrium constants. Global fitting of data spanning multiple speeds and loading concentrations (and multiple optical systems) allows the determination of both equilibrium constants for the interacting species and the weight fractions of the several components. Because the weight fraction of components must remain constant upon dilution while the distribution of individual species will vary characteristically with local concentration according to the law of mass action, these types of mixed systems can be resolved as long as data from a sufficiently wide range of loading concentrations and speeds can be combined in a global fit. A large array of arbitrary reaction schemes can be represented using the MODEL EDITOR, which is part of SEDANAL.

W.F. Stafford (✉)
Boston Biomedical Research Institute, c/o 12 Francis Ave Cambridge, MA 02138, USA
e-mail: wstafford3@walterstafford.com

P.J. Sherwood
Interactive Technology, Inc. 20 Woodside Glen Ct., Oakland, CA 94602, USA

S. Uchiyama et al. (eds.), *Analytical Ultracentrifugation*,
DOI 10.1007/978-4-431-55985-6_7

Keywords Equilibrium analytical ultracentrifugation • Heterologous interacting systems • Self-associating systems • Curve fitting • Arbitrary reaction schemes • SEDANAL

7.1 Introduction

Programs for global analysis of heterologous interacting systems had not been widely available until the introduction of the program HeteroAnalysis by Jim Cole in 2004 at the University of Connecticut (Cole 2004).

The program NONLIN (Johnson et al. 1981) has become the standard for the analysis of self-associating systems, but cannot be rigorously applied to interacting systems comprised of more than one macromolecular component. Until the introduction of HeteroAnalysis, there had been no widely available software for the global analysis of multicomponent interacting macromolecular systems. At about the same time, Peter Schuck incorporated fitting of heterologous systems into his program SEDPHAT (Vistica et al. 2004).

The software package SEDANAL, described here, in use for sedimentation equilibrium analysis since 2004 (Mukhopadhyay et al. 2005), can be used for the analysis of either sedimentation equilibrium or sedimentation velocity data obtained from general interacting systems. An account of the sedimentation equilibrium fitter was presented at the Lausanne AUC meeting in 2005 (Stafford 2005, unpublished)

The MODEL EDITOR, part of SEDANAL, maintains a user-defined and editable database of reaction schemes that is used by the fitter in SEDANAL.

Fitting is carried out by at least two methods: least squares (L2) or least average absolute value (L1) fitting using either the simplex algorithm of Nelder and Mead or the least squares routine of Levenberg and Marquardt (Levenberg 1944; Marquardt 1963) to minimize the fitting function with respect to the parameters of interest. These may include loading concentrations, molar masses, association equilibrium constants, and non-ideality coefficients.

7.2 Theoretical Background

The model for ideal sedimentation equilibrium data is

$$S_i(\xi) = \delta_i + \sum_{j=1}^{n_s} \epsilon_{i,j} A_{i,j} \exp(\sigma_j \xi) \tag{7.1}$$

where $S_i(\xi)$ is the noise-free signal measured with a particular optical system from channel i (a channel of data being data obtained from a single solution column), with $i = 1, 2, 3, \ldots, m$; m is the number of data sets; n_s is the number of macromolecular species; and ϵ_{ij} is the mass extinction coefficient for species j in the ith optical system. σ_j is the buoyant reduced molar mass of species j, defined by Yphantis (1964) as

$$\sigma_j \equiv M_j(1 - \bar{v}_j\rho)\omega^2/RT \tag{7.2}$$

M_j and $\bar{v}_j$ are the molar mass and partial specific volume of species j, ξ is the radial coordinate defined by

$$\xi \equiv r^2/2$$

and $A_{i,j}$ is an integration factor defined below (Eq. 7.6ff.); δ_i is a constant, radially independent, vertical offset in the signal for data set i.

7.2.1 *Radially Dependent Baseline Subtraction*

For most systems that we observe, the measured signal, $S_{\text{obs}}(\xi)$, is composed of the true signal, $S_i(\xi)$, plus a time-independent background signal, $S_b(\xi)$, that is a function of radius, arising from non-compensating inhomogeneities in the optical system and cell components:

$$S_{\text{obs}}(\xi) = S_i(\xi) + S_b(\xi) \tag{7.3}$$

The optical inhomogeneities, represented by $S_b(\xi)$, are most serious for interference optics, pseudo-absorbance (intensity), and fluorescence measurements, but are often present and have to be dealt with when using the absorbance optical system as well. The most straightforward way to deal with the baseline inhomogeneities is to perform a blank run with water vs. water at the same speeds as those of the experimental run and to subtract those from the experimental run to obtain $S_i(\xi)$. It is assumed in the following discussion that such an appropriate baseline pattern, $S_b(\xi)$, has been determined and subtracted from the observed signal before fitting. This radially dependent baseline subtraction is accomplished in the preprocessor of SEDANAL by loading a baseline "blank" scan after loading the equilibrium scan. The preprocessor performs an interpolated baseline subtraction to produce a corrected equilibrium scan that is used for the curve fitting. Optical baseline subtraction is usually when fitting all types of data, but is especially important for interference, fluorescence, and intensity data, which have considerable background components to their respective signals. Blank subtraction with absorbance data may not be required if the optics (windows and lenses) are free of dirt and imperfections.

7.2.2 *Conservation of Mass Relations*

SEDANAL fits for the loading concentration of each component. Accurate determination of loading concentration requires accurate knowledge of the radius of the meniscus and base in order to perform the conservation of mass integration. With this caveat in mind, we proceed to derive the conservation of mass relationships used by SEDANAL. The conservation of mass calculation is especially sensitive to the radius of the base of the cell, since the concentration is highest in that region and increasing exponentially. One complication is that gelation or pelleting at the base may alter the effective base radial position. This is not a problem for accurate determination of equilibrium constants or stoichiometries. Accurate knowledge of the loading concentration is required only if data sets obtained on the same sample have different loading concentrations or speeds. For example, if we have data from one cell at three different speeds, we should be able to require that the loading concentration be the same for all three data sets. However if the base radius is not known accurately at each speed, we cannot reduce the fitting parameters by making the loading concentrations equal. Fitting for the base radius does not solve this problem, since it requires adding another fitting parameter, which defeats the purpose.

A general strategy for fitting equilibrium data from interacting systems is outlined below.

7.2.2.1 One Species

The simplest system is a single-component, single-species system. Conservation of mass requires that the integral over the equilibrium mass distribution be the same as that over the initial mass distribution, allowing us to write:

$$\int_{\xi_m}^{\xi_b} c_0 \, d\xi = \int_{\xi_m}^{\xi_b} A e^{\sigma \xi} \, d\xi \tag{7.4}$$

$$c_0(\xi_b - \xi_m) = \frac{A}{\sigma}\left(e^{\sigma \xi_b} - e^{\sigma \xi_m}\right) \tag{7.5}$$

$$A = \frac{c_0 \sigma (\xi_b - \xi_m)}{\left(e^{\sigma \xi_b} - e^{\sigma \xi_m}\right)} \tag{7.6}$$

where c_0 is the loading concentration and the pre-exponential factor:

$$A = \frac{c_0 \sigma (\xi_b - \xi_m)}{\exp(\sigma \xi_b) - \exp(\sigma \xi_m)}$$

7.2.2.2 Monomer-Oligomer

The simplest single-component, self-associating system is the two-species, monomer-oligomer equilibrium system.

$$n\mathrm{B} = \mathrm{B}_n$$

For a two-species, ideal monomer-n-mer (n is the degree of association) self-associating system, let $c(\xi)$ be the total concentration of macromolecule in mass concentration units and $K_{1,n} = \mathrm{c}_n/\mathrm{c}_1{}^n$ the association constant (in mass concentration units) for the monomer-n-mer reaction. Then $c(\xi) = c_1(\xi) + c_n(\xi)$ becomes

$$c(\xi) = A_1 \exp(\sigma_1 \xi) + K_{1,n} A_1^n \exp(\sigma_n \xi) \tag{7.7}$$

and $\sigma_n = n\sigma_1$, which follows from Eq. 7.2 and the assumption that there is no volume change upon association, so that $\bar{v}_n = \bar{v}_1$. Since $c_n = K_{1,n} c_1^n$, the relationship between the loading concentration , c_0, and A_1 is given by

$$\int_{\xi_m}^{\xi_b} c_0(\xi)\, d\xi = A_1 \int_{\xi_m}^{\xi_b} \exp(\sigma_1 \xi)\, d\xi + K_{1,n} A_1^n \int_{\xi_m}^{\xi_b} \exp(n\sigma_1 \xi)\, d\xi \tag{7.8}$$

and carrying out the integration we arrive at

$$c_0(\xi_b - \xi_m) = \frac{A_1}{\sigma_1}\left[\exp(\sigma_1 \xi_b) - \exp(\sigma_1 \xi_m)\right] + \frac{K_{1,n} A_1^n}{n\sigma_1}\left[\exp(n\sigma_1 \xi_b) - \exp(n\sigma_1 \xi_m))\right] \tag{7.9}$$

and rearranging a little

$$c_0 = A_1 \left[\frac{\exp(\sigma_1 \xi_b) - \exp(\sigma_1 \xi_m)}{\sigma_1(\xi_b - \xi_m)}\right] + K_{1,n} A_1^n \left[\frac{\exp(n\sigma_1 \xi_b) - \exp(n\sigma_1 \xi_m))}{n\sigma_1(\xi_b - \xi_m)}\right] \tag{7.10}$$

This polynomial in A_1 can be solved for A_1 given c_0 and $K_{1,n}$, either by quadratic formula for $n = 2$ or by iteration for $n > 2$.

7.2.2.3 Hetero-Association

The simplest hetero-association is a bimolecular reaction between two dissimilar species, say, A and B (species 1 and 2).

$$\mathrm{A} + \mathrm{B} = \mathrm{C} \qquad K_{1,2} = c_C/(c_A c_B)$$

For the total concentration, c_t, at each point we have

$$c_t(\xi) = A_1 \exp(\sigma_A \xi) + A_2 \exp(\sigma_B \xi) + K_{1,2} A_1 A_2 \exp(\sigma_C \xi)$$

The radial distributions of components 1 (total A) and 2 (total B) are

$$c_A(\xi) = A_1 \exp(\sigma_A \xi) + \frac{M_A}{M_C} K_{1,2} A_1 A_2 \exp(\sigma_C \xi) \tag{7.11}$$

$$c_B(\xi) = A_2 \exp(\sigma_B \xi) + \frac{M_B}{M_C} K_{1,2} A_1 A_2 \exp(\sigma_C \xi) \tag{7.12}$$

leading to the following relations for their conservation of mass

$$\int_{\xi_m}^{\xi_b} c_A(\xi) d\xi = A_1 \int_{\xi_m}^{\xi_b} \exp(\sigma_A \xi) d\xi + \frac{M_A}{M_C} K_{1,2} A_1 A_2 \int_{\xi_m}^{\xi_b} \exp(\sigma_C \xi) d\xi$$

$$\int_{\xi_m}^{\xi_b} c_B(\xi) d\xi = A_2 \int_{\xi_m}^{\xi_b} \exp(\sigma_B \xi) d\xi + \frac{M_B}{M_C} K_{1,2} A_1 A_2 \int_{\xi_m}^{\xi_b} \exp(\sigma_C \xi) d\xi$$

These two simultaneous equations can be solved for A_1 and A_2, given loading concentrations, $c_{A,0}$ and $c_{B,0}$, of the two components.

$$c_{A,0} = A_1 \left[\frac{\exp(\sigma_A \xi_b) - \exp(\sigma_A \xi_m)}{\sigma_A(\xi_b - \xi_m)} \right] + \left(\frac{M_A}{M_C} \right) (K_{1,2} A_1 A_2) \left[\frac{\exp(\sigma_C \xi_b) - \exp(\sigma_C \xi_m)}{\sigma_C(\xi_b - \xi_m)} \right] \tag{7.13}$$

$$c_{B,0} = A_1 \left[\frac{\exp(\sigma_B \xi_b) - \exp(\sigma_B \xi_m)}{\sigma_B(\xi_b - \xi_m)} \right] + \left(\frac{M_B}{M_C} \right) (K_{1,2} A_1 A_2) \left[\frac{\exp(\sigma_C \xi_b) - \exp(\sigma_C \xi_m)}{\sigma_C(\xi_b - \xi_m)} \right] \tag{7.14}$$

7.2.3 *Self-Associating System with Incompetent or Adventitious Species*

A monomer-oligomer system with either incompetent (i.e., nonparticipating) monomer or incompetent oligomer is a two-component system with three macromolecular species. The adventitious species could also be an impurity having no relation with the monomer-oligomer system. Since the incompetent species does not participate in the reaction, it is an additional component, and, therefore, upon dilution the weight fraction of incompetent species in the mixture will be invariant and we can write:

$$c(\xi) = \left[A_1 \exp(\sigma_1 \xi) + K_{1,n} A_1^n \exp(n\sigma_1 \xi)\right] + A_{\text{inc}} \exp(\sigma_{\text{inc}} \xi) \tag{7.15}$$

Invoking conservation of mass, we have:

$$\int_{\xi_m}^{\xi_b} c(\xi)\, d\xi = \int_{\xi_m}^{\xi_b} \left[A_1 \exp(\sigma_1 \xi) + K_{1,n} A_1^n \exp(n\sigma_1 \xi)\right] d\xi \quad + \quad \int_{\xi_m}^{\xi_b} A_{\text{inc}} \exp(\sigma_{\text{inc}} \xi)\, d\xi \tag{7.16}$$

And we can write:

$$c_{0,total} = c_{0,1,N} + c_{0,\text{inc}} = f_{1,n}\, c_0 + (1 - f_{1,n}) c_0 \tag{7.17}$$

where $f_{1,n}$ is the weight fraction of the monomer-oligomer component and $(1 - f_{1,n})$ is the weight fraction of the incompetent component. This leads to the following two equations:

$$c_{0,1,N} = A_1 \left[\frac{\exp(\sigma_1 \xi_b) - \exp(\sigma_1 \xi_m)}{\sigma_1 (\xi_b - \xi_m)}\right] + K_{1,n} A_1^n \left[\frac{\exp(\sigma_n \xi_b) - \exp(\sigma_n \xi_m)}{\sigma_n (\xi_b - \xi_m)}\right] \tag{7.18}$$

and

$$c_{0,\text{inc}} = A_{\text{inc}} \left[\frac{\exp(\sigma_{inc} \xi_b) - \exp(\sigma_{inc} \xi_m)}{\sigma_{inc} (\xi_b - \xi_m)}\right] \tag{7.19}$$

The pre-exponential factors A_i are obtained by solving Eq. 7.19 for the incompetent species given $c_{0,\text{inc}}$ and by solving Eq. 7.18 for the monomer-oligomer system given $c_{0,1,n}$. Note that the weight fraction of incompetent species in the initial mixture remains constant upon dilution, and likewise the weight fraction of the monomer-oligomer system remains constant, while the weight fractions of monomer and oligomer species participating in the monomer-oligomer reaction change in accordance with the law of mass action. It is this difference in the behavior of the components of the mixture that allows us to resolve them and to fit directly for $f_{1,n}$ and f_{inc} and $K_{1,n}$ (where $f_{\text{inc}} = 1 - f_{1,n}$).

The pre-exponential factors are often recast in terms of the concentration at a reference radius, r_{ref}, often chosen as the first data point, but can be any convenient position, such as the meniscus. In that case, the concentration, $c_{i,\text{ref}}$, at that point is given by $c_i(\xi_{\text{ref}}) = A_i \exp(\sigma_i \xi_{\text{ref}})$ and we have that

$$c_i(\xi) = c_i(\xi_{\text{ref}}) \exp(\sigma_i(\xi - \xi_{\text{ref}})).$$

7.2.4 *Non-ideality*

This section describes the algorithm implemented in SEDANAL for the treatment of non-ideality in polydisperse systems.

The concentration of each species, i, varies according to:

$$c_i(\xi) = c_i(\xi_m)\, e^{\sigma_i(\xi - \xi_m) - \tau} \qquad i = 1, \ldots, n_s \tag{7.20}$$

The non-ideality is contained in τ ($\tau = 0$ for the ideal case):

$$\tau = \sum_{k=1}^{n_s} \left(2BM1_k \cdot c_k + \frac{3}{2} CM1_k \cdot c_k^2 \right) \tag{7.21}$$

In SEDANAL's implementation, the c_k, and the virial coefficients $BM1_k$ and $CM1_k$ are in molar units.

The concentrations $c_k(\xi_m)$ are related by mass action, since the system is in chemical equilibrium at every radius. A conservation relation may be written for each of the n_c components.

Since the loading concentration of each component is spread out over the entire cell at equilibrium, the conserved quantities are:

$$\mathbf{q} = \mathbf{C} \int_{\text{cell}} \mathbf{c}\, dV \tag{7.22}$$

where $\mathbf{q}$ is a vector of n_c quantities (i.e., concentrations × cell volume), $\mathbf{c}(\xi)$ is a vector of concentrations at ξ, and $\mathbf{C}$ is a *conservation matrix* so that:

$$\int_{\text{cell}} \mathbf{C}\,\mathbf{c}\, dV = \int_{\text{cell}} \mathbf{C}\,\mathbf{c}_0\, dV$$

For a sector-shaped cell of constant thickness h, the volume element is:

$$dV = h \cdot d\theta \, d\xi$$

where $d\xi = r\,dr$. The volume of this cell between meniscus and base is:

$$V \equiv h \int_{\theta_1}^{\theta_2} \int_{\xi_m}^{\xi_b} d\theta \, d\xi = h(\theta_2 - \theta_1)\,(\xi_b - \xi_m)$$

If the initial (loading) concentration vector is $\mathbf{c_0}$, and of species i is $c_{i,0}$, initially

$$\mathbf{q} = \mathbf{C}\,\mathbf{c_0} V$$

and

$$q_k = V \sum_{i=1}^{n_s} C_{ki}\, c_{i,0} = h(\theta_2 - \theta_1)(\xi_b - \xi_m) \sum_{i=1}^{n_s} C_{ki}\, c_{i,0} \tag{7.23}$$

and at any time, because $\mathbf{c}$ is independent of θ, integration of Eq. 7.22 over θ gives

$$\mathbf{q} = h(\theta_2 - \theta_1)\mathbf{C} \int_{\xi_m}^{\xi_b} \mathbf{c}\, d\xi \tag{7.24}$$

or, for component species k,

$$q_k = h(\theta_2 - \theta_1) \sum_{i=1}^{n_s} C_{ki} \int_{\xi_m}^{\xi_b} c_i \, d\xi \tag{7.25}$$

The concentrations vary according to Eq. 7.20.

Combining Eqs. 7.24 and 7.20 gives

$$q_k = h(\theta_2 - \theta_1) \sum_{i=1}^{n_s} C_{ki}\, c_i(\xi_m) \int_{\xi_m}^{\xi_b} e^{\sigma_i(\xi - \xi_m) - \tau} d\xi \tag{7.26}$$

$$(\xi_b - \xi_m) \sum_{i=1}^{n_s} C_{ki}\, c_{i,0} = \sum_{i=1}^{n_s} C_{ki}\, c_i(\xi_m) \int_{\xi_m}^{\xi_b} e^{\sigma_i(\xi - \xi_m) - \tau}\, d\xi \tag{7.27}$$

Defining $\mathbf{x}$ as the vector whose components are

$$x_i = c_i(\xi_m) \int_{\xi_m}^{\xi_b} e^{\sigma_i(\xi - \xi_m) - \tau}\, d\xi \tag{7.28}$$

and

$$\mathbf{x_0} = (\xi_b - \xi_m)\,\mathbf{c_0}$$

converts Eq. 7.27 to

$$\mathbf{Cx} = \mathbf{Cx_0} \tag{7.29}$$

The nonideal term τ is a function of $c_i(\xi_m)$.

The main iteration is over the solutions of 7.20 and 7.29 using fixed values of σ_i and virial coefficients. One main iteration consists of using a set of

$$z_i \equiv x_i / c_i(\xi_m) = \int_{\xi_m}^{\xi_b} e^{\sigma_i(\xi-\xi_m)-\tau}\, d\xi \tag{7.30}$$

to solve Eq. 7.29 for $\mathbf{x}$, as described in the following paragraph. The z_i are recomputed for the next iteration. The process is complete when no concentrations are significantly changed by an iteration.

The $n_s + 1$ equations represented by Eq. 7.29 and the definition of τ are solved simultaneously for c_i and τ, at each ξ. That means finding the c_i and τ that satisfy both Eqs. 7.29 and 7.21.

The method of solution is to define a set of

$$f_i \equiv c_i(\xi_m)\, e^{\sigma_i(\xi-\xi_m)} \cdot \exp(-\tau) - c_i \qquad i = 1, \ldots, n_s$$

Since there is no analytical solution to the function $\mathbf{f} = 0$, the solution is achieved by iteration. The procedure is to find the zeros of the f_i using the multidimensional Newton's method for a given τ at each step. We need the Jacobian of $\mathbf{f}$:

$$\frac{\partial \mathbf{f}}{\partial \mathbf{c}} = \begin{pmatrix} \frac{\partial f_1}{\partial c_1} & \frac{\partial f_1}{\partial c_2} & \cdots & \frac{\partial f_1}{\partial c_{n_s}} \\ \frac{\partial f_2}{\partial c_1} & \frac{\partial f_2}{\partial c_2} & \cdots & \frac{\partial f_2}{\partial c_{n_s}} \\ & \vdots & \ddots & \\ \frac{\partial f_{n_s}}{\partial c_1} & \frac{\partial f_{n_s}}{\partial c_2} & \cdots & \frac{\partial f_{n_s}}{\partial c_{n_s}} \end{pmatrix} \tag{7.31}$$

$$\frac{\partial f_i}{\partial c_j} = c_i(\xi_m)\, e^{\sigma_i \xi} \cdot \exp(-\tau) \cdot \left(-2BM1_j - 3CM1_j \cdot c_j\right) - \delta_{ij} \tag{7.32}$$

using $\frac{\partial \tau}{\partial c_j} = 2BM1_j + 3CM1_j \cdot c_j$, and δ_{ij} the Kronecker delta defined by

$$\delta_{i,j} = \begin{cases} 1 \text{ when } i = j \\ 0 \text{ when } i \neq j. \end{cases}$$

Given the value of the function (**f**) and its Jacobian $\left(\frac{\partial \mathbf{f}}{\partial \mathbf{c}}\right)$, the solution is computed using library function `nwnleq` (Dennis and Schnabel 1996) (using the "double dogleg" global strategy).

This solution gives a set of c_i for all the species at the given radius and a value for τ at that radius, which are used to recompute the z_i for the next main iteration.

After the main iterations are complete, the concentrations determine the signal at each radial position according to

$$y(\xi) = y_0 + \sum_{k=1}^{n_s} c(\xi)_k M_k \epsilon_k \tag{7.33}$$

where c_k is the molar concentration of species k, M_k is its molar mass, and ϵ_k is its mass extinction coefficient.

7.3 Fitting to a Heterologous Interacting System

As an example of a heterologous system, consider troponin I HR domain interacting with a troponin T fragment called TnT(172–241) to form a 1:1 complex (Mukhopadhyay et al. 2005), which interacts according to the following chemical equation:

$$\text{TnI(HR)} + \text{TnT(172–241)} = \text{TnI(HR)} \cdot \text{Tn(172–241)}$$

Figure 7.1 shows SEDANAL's control screen for the global fit to nine data sets (three speeds and three loading concentrations). The molar mass of each component species is known and held constant (as indicated by the blue background). Windows with yellow background indicate values that are derived from other parameters. For example, the molar mass of the complex is equal to the sum of the molar masses of the reactants and is automatically entered according to the stoichiometric relationships established by either the MODEL EDITOR or the Equation Editor. The column of numbers labeled "molar ratio" with yellow background indicate that a single ratio of components is used for all cells, appropriate in this case because the samples were part of dilution series in which the ratio of TnI(HR) to Tn(172–241) was the same for all dilutions. We float (fit for) the ratio but require that it be the same for all dilutions, and we float the loading concentrations.

In general it's a good idea to float the loading concentrations even if you think you know them because there are errors in pipetting and in the values of the extinction coefficients themselves. We are most interested in the association equilibrium constant and so allow its value to float. Error analysis can be done by several methods (see below). In this case, the F-statistics method was used to compute the 95 % confidence limits for the association constant K_{eq}. Figure 7.2 shows the fitting screen at convergence after several iterations of the Levenberg-

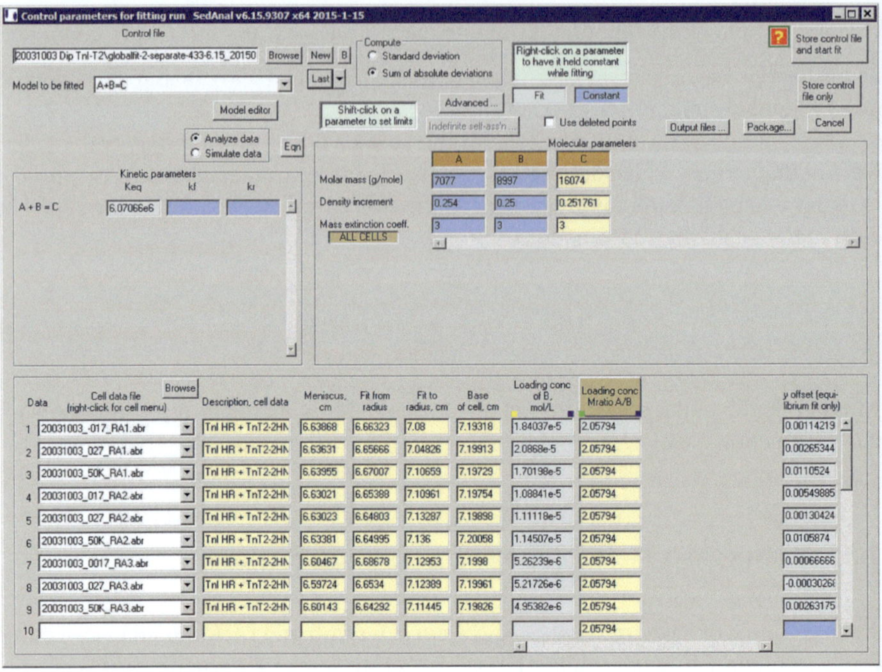

Fig. 7.1 TnI(HR)-TnT(172–241) hetero-association control screen showing the floating parameters in *gray*, the parameters held constant in *blue*. Values shown in *yellow* are derived from the other parameters on the screen. For example, the molar mass of C, shown in *yellow*, is computed as the sum of the molar masses of A and B according to the model. Under the column "Loading conc. Molar ratio" the *yellow* entries have been set equal to the top value, which is being allowed to float during the fitting

Marquardt procedure. In this example there were three loading concentrations run at three speeds. The molar masses of TnI and TnT(172–241) are known and, therefore, are held constant during the fit.

7.3.1 *Fitting Strategy*

In general, one fits to the form given by Eq. 7.1. For example consider three species:

$$c(\xi) = A_1 \exp(\sigma_1 \xi) + A_2 \exp(\sigma_2 \xi) + A_3 \exp(\sigma_3 \xi) \tag{7.34}$$

The buoyant molar masses σ_i can be determined whether the species are interacting or not by specifying a three-component, noninteracting model and ignoring the pre-exponential factors. After the number of species and their buoyant molar masses

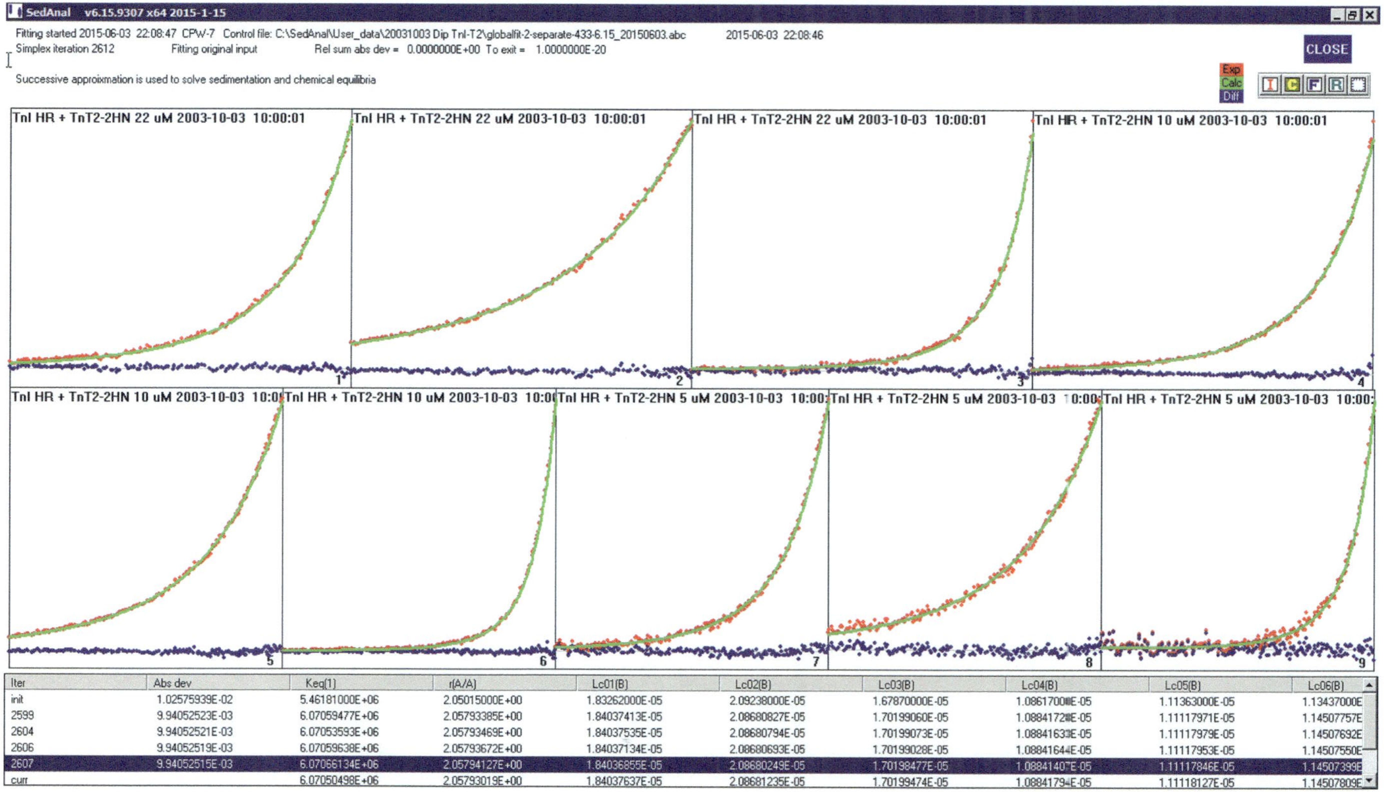

Fig. 7.2 TnI(HR)-TnT(172–241) hetero-association fitting screen. showing the data in *red*, the best fit in *green* and the residuals in *blue*

have been determined, those values can be held constant, while various association models consistent with those buoyant molar masses can be tested.

The pre-exponential factors, A_i, will have different interpretations depending on the association scheme being modeled. To be more explicit, for a noninteracting system, the A_i are related directly to the loading concentrations and weight fractions of the components present. For an interacting system, the pre-exponential factors are related to the loading concentrations according to conservation of mass and contain information about the values of the equilibrium constants. However, the actual values of the A_i obtained by curve fitting do not depend on the model, but the model specifies the relationships that must hold between the pre-exponential factors for that model. SEDANAL uses the information contained in the stoichiometric matrix supplied by the MODEL EDITOR to interpret the information contained in the values of the A_i. It should be pointed out that one cannot distinguish an associating system from a noninteracting system from a run at single speed and single loading concentration. Multiple speeds and loading concentrations are required to uniquely establish a model and reveal any incompetent species.

7.3.2 BIOSPIN

BIOSPIN is a model-independent analysis method first published in 1969 by Roark and Yphantis. Their original code has been incorporated into SEDANAL, and a GUI front end added to display number, weight, and *z*-average sigmas as a function of local cell concentration, as well as two-species plots and local radius plots. Plots of two charge-independent molar mass moments can also be displayed. BIOSPIN also outputs the so-called charge-independent sigmas, σ_{y1} and σ_{y2}, that are independent of the second and higher virial coefficients. The reader is referred to the original papers for more information (Roark and Yphantis 1971; Yphantis and Roark 1972).

7.3.3 Error Analysis

Confidence limits for the fitted parameters can be determined by any of three methods in SEDANAL: (1) bootstrap with replacement, (2) Monte Carlo, or (3) F-statistics (Sherwood and Stafford 2016).

7.3.4 Summary

We have described a routine and strategy for fitting sedimentation equilibrium data that has been implemented in SEDANAL. SEDANAL can treat noninteracting systems and hetero- and self-associating systems with or without incompetent (i.e., nonparticipating) components. It can fit multiple data sets at multiple wavelengths and multiple optical systems. It can fit to any arbitrary association model that can be described in the MODEL EDITOR. Parameter error analysis can be carried out by any of three ways: bootstrap (with replacement), Monte Carlo, or F-statistics methods. BIOSPIN, the model-independent equilibrium sedimentation analysis program originally developed by Dennis Roark and David Yphantis, has been incorporated into SEDANAL with a GUI interface. SEDANAL is available for download from http://sedanal.org/.

References

Cole JL (2004) Analysis of heterogeneous interactions. Meth Enzymol 348:212–232

Dennis JE, Schnabel, RB (1996) Numerical methods for unconstrained optimization and nonlinear equations. SIAM, Philadelphia

Johnson ML, Correia JJ, Yphantis DA, Halvorson HR (1981) Analysis of data from the analytical ultracentrifuge by nonlinear least-squares techniques. Biophys J 36(3):575–588

Levenberg K (1944) A method for the solution of certain non-linear problems in least squares. Q Appl Math 2:164–168

Margossian SM, Stafford WF (1982) Calcium-induced dimerization of troponin-c. J Biol Chem 257(3):1160–1165

Marquardt D (1963) An algorithm for least-squares estimation of nonlinear parameters. SIAM J Appl Math 11:431–441

Mukhopadhyay S, Langsetmo K, Stafford WF, Henry GD, Baleja JD, Sarkar S (2005) Identification of a region of fast skeletal troponin T required for stabilization of the coiled-coil formation with troponin I. J Biol Chem 280(1):538–547

Nelder J, Mead R (1965) A simplex method for function minimization. Comp J 7:308–313

Roark DE, Yphantis DA (1969) Studies of self-associating systems by equilibrium ultracentrifugation. Ann NY Acad Sci 164:245–728

Roark DE, Yphantis DA (1971) Equilibrium centrifugation of nonideal systems. The Donnan effect in self-associating systems. Biochemistry 10(17):3241–3249

Sherwood PJ, Stafford WF (2016) SEDANAL: model-dependent and model-independent analysis of sedimentation data. In: Uchiyama S et al (eds) Analytical ultracentrifugation. Springer, Tokyo

Stafford WF (2005) A fitting program for global analysis of general hetero- and self- associating systems by sedimentation equilibrium. In: Lausanne AUC meeting (unpublished)

Stafford WF, Sherwood PJ (2004) Analysis of heterologous interacting systems by sedimentation velocity: curve fitting algorithms for estimation of sedimentation coefficients, equilibrium and kinetic constants. Biophys Chem 108:231–243

Vistica J, Dam J, Balbo A, Yikilmaz E, Mariuzza RA, Rouault TA, Schuck P (2004) Sedimentation equilibrium analysis of protein interactions with global implicit mass conservation constraints and systematic noise decomposition. Anal Biochem 326(2):234–256

Yphantis DA (1964) Equilibrium ultracentrifugation of dilute solutions. Biochemistry 3:297–317

Yphantis DA, Roark DE (1972) Equilibrium centrifugation of nonideal systems. Molecular weight moments for removing the effects of nonideality. Biochemistry 11(16):2925–2934

Chapter 8
Analytical Ultracentrifugation Data Analysis with UltraScan-III

Borries Demeler and Gary E. Gorbet

Abstract The current status of the UltraScan-III (US3) data analysis software suite is described. An overview of the US3 concepts, software layout, the data workflows and US3 components is presented, followed by a discussion of the analysis methods and their applications. Also described are visualization modules for analysis results, US3's utilities and simulation tools, as well as the collaboration environments for online data and result exchange.

Keywords Analytical ultracentrifugation • UltraScan • Lamm equation modeling • Two-dimensional spectrum analysis • Genetic algorithms • Multi-wavelength • AUC • Supercomputing • Simulations • Sedimentation velocity experiments

8.1 Introduction

UltraScan-III (US3, http://www.ultrascan.uthscsa.edu) is a free and open source, multi-platform software suite designed to provide high-performance and high-throughput data analysis and modeling of hydrodynamic data. The UltraScan-III project grew out of the requirement to support the needs of the analytical ultracentrifugation core facility at the University of Texas Health Science Center in San Antonio. This facility owns multiple analytical ultracentrifugation (AUC) instruments and has several hundred users, many of whom collaborate on joint projects which necessitates sharing data online. As is discussed below, US3 addresses many challenges posed by this large, multi-user environment through the use of a relational MySQL database with a web-based interface, termed the UltraScan Laboratory Information Management System (USLIMS) (Demeler 2009). Foremost, US3 allows researchers to achieve an unsurpassed level of accuracy and resolution in their data analysis. US3 is designed to provide maximum flexibility in formulating a great variety of custom analysis models and optimization

B. Demeler (✉) • G.E. Gorbet
Department of Biochemistry, The University of Texas Health Science Center at San Antonio, San Antonio, TX 78229, USA
e-mail: demeler@biochem.uthscsa.edu

S. Uchiyama et al. (eds.), *Analytical Ultracentrifugation*,
DOI 10.1007/978-4-431-55985-6_8

approaches, while at the same time offering an intuitive interface that is easy to learn and use. The software currently supports data from the Beckman XLA/I UV/visible absorbance and intensity detector, the Rayleigh interference detector, the Aviv fluorescence detector, and from two recently developed multi-wavelength detectors (Bhattacharyya et al. 2006; Schilling 2014). Both single- and multi-speed experiments are supported. The software is highly configurable so that repetitive actions can be largely automated to speed up routine analysis and accelerate discovery, without compromising rigor, accuracy, and flexibility in the functionality. In recent years, increasing availability of high performance computing (HPC) and network infrastructure has opened up new avenues for biophysical modeling and analysis of sedimentation data. It is now possible to use computationally demanding, parallelized fitting approaches (Demeler and Brookes 2008; Brookes et al. 2006, 2010a; Brookes and Demeler 2007, 2008) based on whole boundary models using an adaptive space-time finite element solution for the underlying flow equation (Cao and Demeler 2005, 2008). The solution built into US3 is able to simulate self- and hetero-associating reactions, including kinetic rate constants (Demeler et al. 2010), supports solvent compressibility, co-sedimenting solutes and gradient formation, as well as concentration dependency of s and D. The parallel methods programmed into US3 provide significantly higher accuracy and resolution than conventional approaches, which are limited by traditional desktop or laptop computers where high-resolution analysis is impractical and time consuming. US3 also allows the user to process many datasets in parallel, greatly improving throughput and time savings (Memon et al. 2013). This is particularly critical for the new multi-wavelength data format, where datasets for several hundred wavelengths must be evaluated from each channel. Additional performance gains are realized from streamlined and automated workflows available through the networked science gateway and offered by the Extreme Science and Discovery Environment (XSEDE, funded by the National Science Foundation in the USA) and from the Partnership for Advanced Computing in Europe (PRACE). These workflows are accessed through efficient grid middleware implementations that allow investigators to distribute jobs to multiple supercomputer clusters simultaneously (Memon et al. 2014). US3 aims to provide a comprehensive and robust analysis environment for all hydrodynamic analysis. In addition to sedimentation data analysis, UltraScan offers the SOMO Solution Modeler with comprehensive facilities for hydrodynamic modeling (Brookes et al. 2010b, c; 2013), further discussed in Chap. 10. The integration of remote HPC resources in UltraScan and the exchange of research data and analysis results are accomplished through the USLIMS, and the Apache Airavata grid middleware that manages the communication with the HPC clusters (Marru et al. 2011). US3 adheres to the OpenAUC data standard (Cölfen et al. 2010), which provides significantly higher storage and I/O efficiency than traditional Beckman formatted ASCII files, increasing data loading and network transfer speed. The OpenAUC standard offers database associations between related data elements which improves accuracy and automation. Below, the most important features of US3 are discussed. US3 is an ongoing project with many collaborators and contributors. A wiki is available on the UltraScan website that provides many resources for users and developers (UltraScan III).

8.2 UltraScan-III Components

The US3 software consists of several local, online, and remote components. A multi-platform desktop binary (Linux/X11, Windows XP/7/8, Macintosh OS-X) is used to import and edit experimental data, visualize results, create analysis reports, and to provide access to many utilities and simulation programs. The user has the option to use US3 for the analysis of data stored locally, or stored in the online database. The latter is required if data are submitted to a remote supercomputer for analysis, the former is intended for situations when Internet is temporarily unavailable, or the user chooses to work without a database back-end. Not all analysis methods are suitable for local analysis, some require supercomputer capability. Computationally demanding routines are all multi-threaded to take advantage of modern multi-core architectures. When data are stored in the database, they are accessible to any authorized user from any Internet location. In addition to functioning as the preferred data storage for the desktop component, this database is also at the core of the online UltraScan Laboratory Information Management System (USLIMS, http://www.uslims3.uthscsa.edu). The user interacts with the USLIMS through a web browser. The USLIMS offers the user remote access to analysis reports and metadata, an administrative interface, and the online submission system for remote supercomputer analysis. The remote analysis is performed by a parallelized MPI routine running the ASTFEM codes (Cao and Demeler 2005, 2008), which is installed on multiple XSEDE resources in the USA, and on Juropa at the Forschungszentrum Jülich, available to European users. Compute cycles in the USA are offered for free to all users through a community account, which is supported through an NSF/XSEDE allocation grant to one of the authors (BD). The final component constitutes the Airavata Science Gateway infrastructure developed at Indiana University. It is responsible for managing all analysis requests from the USLIMS and for sending them to the selected remote supercomputer, and for moving input data and results between supercomputers and the database. All data transfers and communications between database, supercomputer, and Airavata are ssl-encrypted to protect the data from unauthorized access. To facilitate collaboration among users, each desktop installation can be configured to access multiple databases. Database usernames and passwords are stored encrypted on disk and are decrypted with a single master password, specific for each user's installation. The master password is carried in memory as long as the US3 application is open. For the duration of the session, each transaction with a remote database is authenticated by MySQL stored procedures, requiring a one-time sign-on with the master password. To assure data ownership integrity, the US3 database maintains a hierarchy of user levels, depending on the user's role in the database, which in order of decreasing permission level can be *superuser*, *administrator*, *analyst*, *investigator*, or *unprivileged*. The desktop version of US3 honors these levels and prevents unauthorized access to data from other users of the database. To facilitate collaboration and exchange of result data through the USLIMS website, the user can choose to share their data with other users registered on the same USLIMS instance, regardless of their user level. In order to access experimental data other than results, the user must have *analyst* or higher permission set. Below this level, data stored in the database cannot be processed with US3, and they can only be viewed in the USLIMS instance.

8.3 UltraScan-III Concepts

An important change in US3 compared to earlier versions is the emphasis on sedimentation velocity (SV) data and the preference of intensity over absorbance data. The main reason sedimentation equilibrium (SE) experiments historically were of interest is the relative simplicity of their analysis. SE models are based on analytical exponential functions instead of finite element solutions of differential equations, and only a single scan at the equilibrium stage of the experiment is needed for each speed or concentration. In addition, equilibrium columns are generally less than 3 mm high, so very little data needs to be modeled, greatly reducing memory requirements. This also means that, compared to SV experiments, only limited information content is available, significantly decreasing the confidence an investigator can have in the results (Demeler et al. 2010). Furthermore, systematic time invariant noise (TI) subtraction is impossible for equilibrium data, since scans are by definition time-invariant. Due to the availability of high-performance computing and high-resolution SV analysis methods in US3, the perceived advantage of simpler models for SE experiments is no longer relevant. As a consequence, the US3 user is encouraged to measure SV experiments instead. Even if an equilibrium experiment is planned, it can still be treated as a velocity experiment by collecting also the data during the approach-to-equilibrium period of the experiment, taking advantage of additional information in the data and using SV analysis methods to interpret the data. Users wishing to analyze legacy equilibrium data are able to do this with the previous version (UltraScan-II), which includes an extensive analysis suite for SE experimental data, and is still available for download from our website. Another important emphasis in US3 is on the replacement of absorbance data (ABS) by intensity data (INT). In ABS, a reference scan is subtracted from the sample scan, thereby convoluting stochastic noise from the reference scan with the stochastic noise from the sample scan. This leads to a $\sqrt{2}$ increase in the stochastic noise signal. Historically, this degradation of experimental data was tolerated because this subtraction also eliminated the majority of TI noise contributions that are present in intensity data, although not completely, since cell windows may have different TI contributions for each channel, and they are not eliminated by reference subtraction, but instead compounded. In US3, efficient algorithms exist to remove both TI and radially invariant (RI) noise contributions from velocity data (Schuck and Demeler 1999), hence providing a superior dataset and higher confidence in the analysis. Additional advantages of intensity measurements include the fact that the reference sector can be used for a low-concentration sample (<0.5 OD in the Beckman XL-A). Importantly, the design of the experiment is critical for its success. This does not only relate to the optical quality of the data acquired (which is entirely dependent on the maintenance status of the machine), but extends also to the speed of the experiment, which affects the amount of data available for analysis, and the relative sedimentation and diffusion signal contained in the data. Important factors also include sample preparation, buffer selection, and sample concentration. A failure to optimize these parameters is never remedied by applying a sophisticated analysis available in UltraScan. These considerations are further discussed in reference (Demeler 2010). All basic models and optimization algorithms available in US3 provide a general description of SV experiments suitable for most experimental conditions and will always converge to the global minimum. A requirement is that the underlying data behave ideally and are not impacted by systematic instrument errors, and do not exhibit concentration dependent non-ideality or change composition mid-run due to a chemical instability, such as pressure dependence, degradation or time-dependent aggregation processes, or gradient formation. Advanced models are available to handle those special cases but they require additional user input.

8.4 The UltraScan Analysis Workflow

8.4.1 Overview

An overview of the general workflow in US3 is illustrated in Fig. 8.1. In step 1, the user enters a project request together with related information such as the solution details, images for absorbance scans, gel pictures, and experimental designs. The solution details describe the analytes, their partial specific volumes and extinction coefficients, and buffer components, which US3 uses to predict the viscosity and density of the solution. Buffers also specify the pH and compressibility of the solution. Next, experimental data are acquired and imported into the database and associated with the solution information and other ancillary metadata (step 2). In step 3, these data are retrieved to a PC where the data are edited. The edit profiles are stored in the database (step 4) and analyzed locally or on a supercomputer to obtain Lamm equation models (step 5). Next, additional data analysis can be performed locally, and all models and results are visualized. All analysis results and visualizations are deposited in the database (step 6) where a report can later be retrieved from the USLIMS website by the user (step 7). A detailed and updated flowchart for the analysis of SV experiments is available on our website: http://www.ultrascan3.uthscsa.edu/sed-veloc-flowchart.php

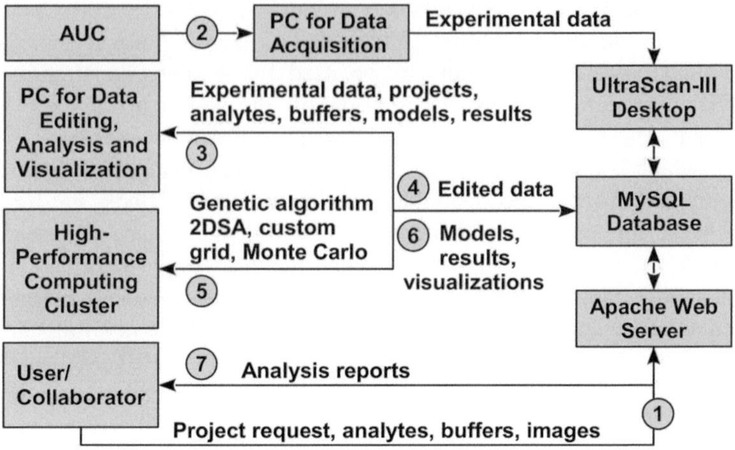

Fig. 8.1 UltraScan-III data flow. The order of steps performed is indicated by numbers (see text)

8.4.2 Importing Experimental Data

Before data can be analyzed with US3 modules, the data must be converted into the binary OpenAUC standard data format (Cölfen et al. 2010). Experimental data from all supported detectors and instruments can be converted by US3. The OpenAUC format is very efficient, scaling the precision of the data type to the accuracy of the detector that was used for data acquisition. In this step, data belonging to an experiment are separated into a unique cell, channel, and wavelength combination. Each combination is referred to as a *triple*. For example, triple 3/B/280 is a dataset containing all scans from the sample channel of cell 3, acquired at 280 nm. Channels are assigned letters A-H, supporting up to eight channels per cell, with "A" referring to the innermost reference channel, and "H" the outermost sample channel. For a multi-wavelength experiment, the number of triples for each channel equals the number of wavelengths acquired (see Fig. 8.2). Additional data relationships are then established: An experiment is first associated with an investigator (the data owner), an instrument operator, the laboratory, instrument and optical system used for acquisition, an experimental project description, and a rotor and rotor calibration (discussed below), the date of the experiment, average temperature, comments, and a run protocol. For each triple, a centerpiece also needs to be selected. The centerpiece geometry, together with the rotor stretching factor (provided from a stored rotor calibration), is later used to calculate a precise position for the bottom of any cell, which is needed for one of the boundary conditions in the solution of the Lamm equation. Geometries of all common centerpieces and previously measured stretching factors for rotors available commercially have been calibrated and entered into US3 for reference, but users can upload their own calibrations. For each triple, a solution must also be defined. A solution is composed of one or more analytes and a buffer, and both are entered by the user. Where possible, partial specific

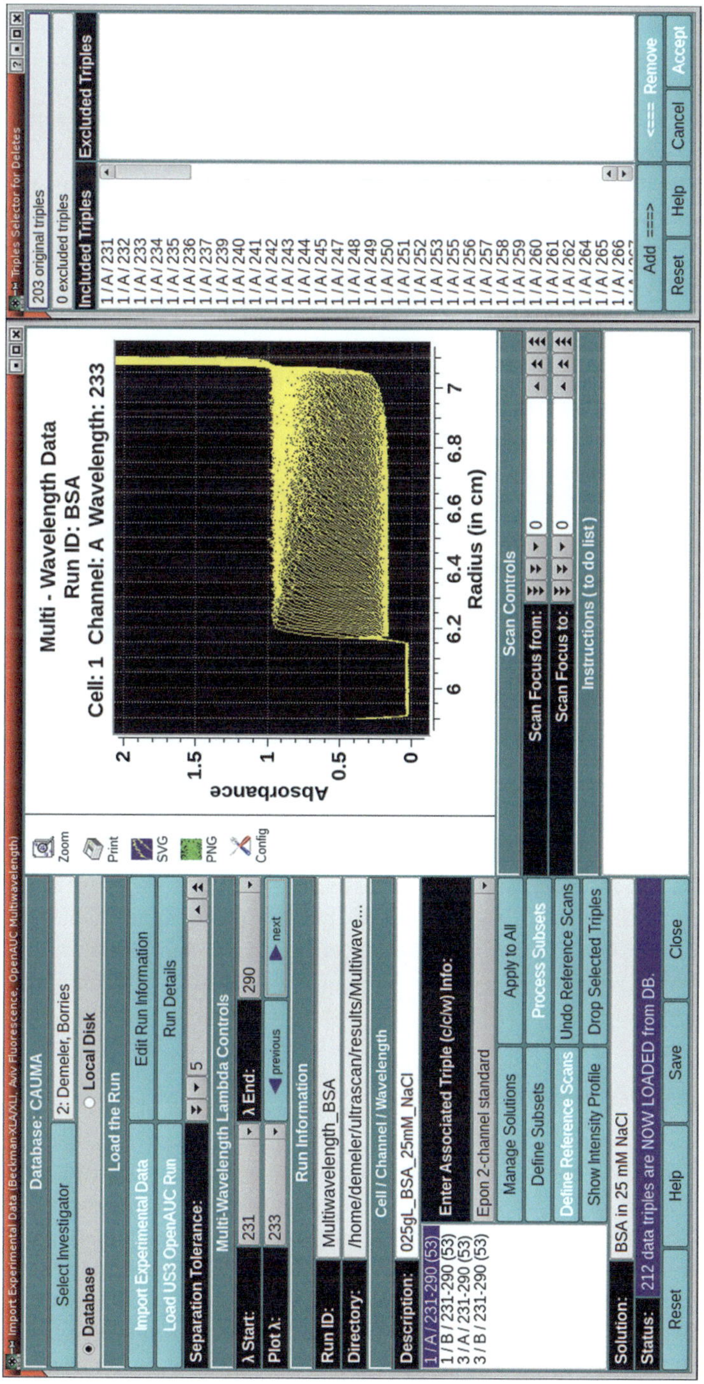

Fig. 8.2 *Left*: Import data dialog with a loaded multi-wavelength experiment for BSA with four channels and 53 wavelengths each. *Right*: The user can exclude selected triples from the data

volumes, molecular weight, and extinction coefficients are estimated automatically by US3 from sequence, or they can be specified by the user. Buffers are composed of buffer components whose density and viscosity increments are directly provided by US3, analogous to the Sednterp software (Laue et al. 1992). Once defined, a solution provides density, viscosity, absorption spectra, extinction coefficient(s), and estimates for the partial specific volumes of the analytes. This information is used in all analyses implemented in US3 to correct all results automatically to standard conditions (water at 20 °C). For temperature correction, US3 assumes an aqueous solution unless a manual correction is specified. This way, experiments performed under different solution conditions can be compared directly, and global fitting of sedimentation data is greatly facilitated (e.g., for organic solvents). Newer multi-wavelength analytical ultracentrifugation (AUC) instruments can produce a time state file, which records the temperature, rotor speed, vacuum, and other system diagnostics throughout the run at short time increments ($\sim$1 s). This information is also stored in binary format and uploaded to the database and provides details about the speed profile, which can be used to more accurately simulate multiple speed steps and acceleration during the run, as well as additional run diagnostics. Finally, an XML file is produced which stores all metadata for the experiment. For multi-speed experiments, special care needs to be taken to correct for the shift in meniscus and bottom of cell position due to different rotor stretching. This is handled automatically by the rotor calibration routine discussed below. In addition, due to the radial shift, TI noise must be determined for each speed separately. Once uploaded, any subsequent analysis of these data is now inextricably linked to the metadata associated with the experiment. While changes to the primary data are possible, such changes would invalidate any derived results and violate their integrity. Therefore, US3 will enforce a deletion of all analysis results if any changes are made to the primary data. Importantly, by storing all associated run details with the experimental data in a relational database at the time of data acquisition, important details about the experiment are associated with the experiment and the user can easily retrieve them later during manuscript preparation. Especially for multi-user facilities, this practice greatly reduces errors and automates the analysis process.

8.4.3 Editing Experimental Data

The next step requires data to be edited. During this process, experimental data are prepared for analysis. An edit profile (stored in XML) is generated, which identifies the meniscus position, the radial data range, any excluded scans, an estimated plateau position needed for analysis methods such as time derivative and second moment, and allows automatic removal of spikes in the data resulting from failed lamp flashes. Each triple can be associated with one or more edit profiles. Multiple profiles are possible to allow investigators to easily evaluate the effect of different editing strategies (for example, exclusion of different scans, selecting different radial ranges), which can be of interest to diagnose whether samples show time dependent changes during the run, are sensitive to pressure effects, or contain aggregates visible only in early scans of the experiment. After defining an edit profile, a dataset can be analyzed with any analysis available in US3, including remote supercomputer analysis.

8.4.4 *Data Refinement*

After editing, the user is tasked with the removal of TI and RI noise that may contribute to the experimental data and to find an optimal meniscus position. Intensity and interference data generally contain noticeable TI and RI noise components, while fluorescence and absorbance data contain less. To remove TI and RI noise, data need to be simultaneously modeled with the intrinsic sedimentation and diffusion transport. By using a degenerate, high-resolution two-dimensional model over all possible s and D coefficients present in the experimental data, the intrinsic sedimentation and diffusion transport will be optimally represented, resulting in an uncorrelated noise determination which can then be subtracted from the experimental data (Schuck and Demeler 1999). This process is accomplished in three steps: In the first step, the sedimentation coefficient range is estimated either from an enhanced van Holde–Weischet analysis (Demeler and van Holde 2004), or, for cases where the data contain a significant amount of time-invariant noise, with the time derivative method (Stafford 1992). A single two-dimensional spectrum analysis (2DSA) (Brookes et al. 2010a) with TI noise removal is then fitted over the determined range. The initial fit not only removes a first-order estimation of the TI noise but also baseline offsets common for all scans. In the next step, the 2DSA is iterated with typically 10–30 meniscus positions in the vicinity of the graphically determined meniscus position during editing. At the same time, RI and TI noise contributions are re-fitted to obtain less correlated noise components. This results in an optimally noise-corrected fit for each fitted meniscus position. Next, the root mean square deviations (RMSD) from each fit is plotted against the meniscus position and fitted to a second-order polynomial (see Fig. 8.3). The lowest RMSD position is used to update the meniscus position in the associated edit profile. Depending on the resolution of the meniscus fit, this position does not necessarily correspond to a position previously fitted. To obtain an uncorrelated TI, RI noise profile for the new position, a final 2DSA with TI and RI noise analysis is performed. This last analysis is performed with a maximum of ten iterative refinement steps to obtain the most optimal solution. The TI, RI noise vectors obtained in this last fit are then subtracted from the data in any subsequent data analysis, allowing the user to omit further noise analysis. By maintaining the instrument well, and using intensity mode for UV/visible data collection, the remaining stochastic noise should then be minimal and random. While the noise vectors from the last step are to be preferred, all noise vectors from previous steps are stored in the database and can be evaluated and applied instead, if desired.

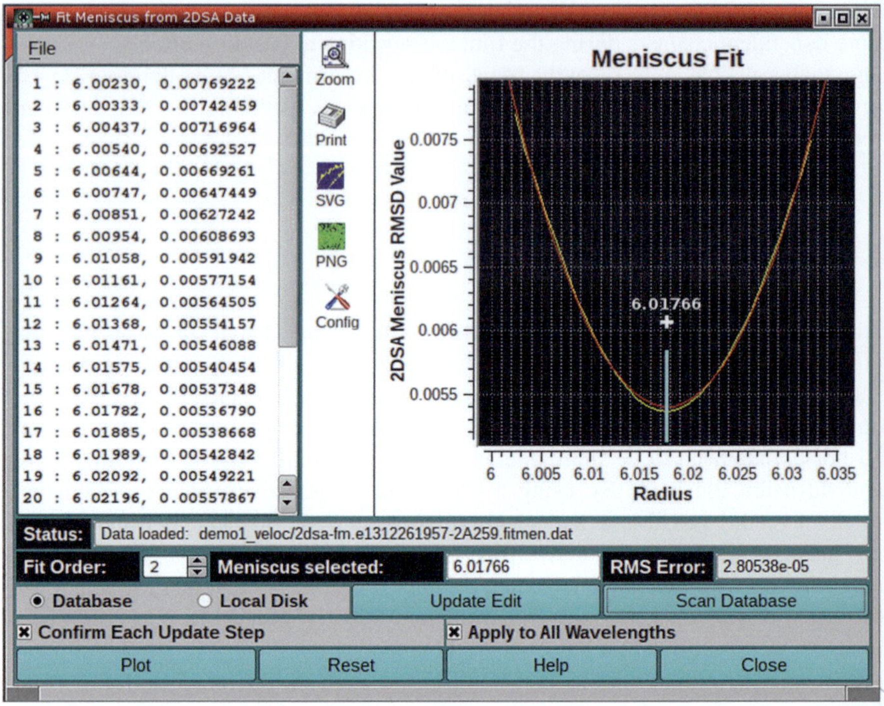

Fig. 8.3 US3 dialog for the meniscus fitter. RMSD values from 2DSA Model files are plotted against meniscus position to identify the optimal meniscus position at the lowest RMSD

8.4.5 *Remote Supercomputer Analysis*

The steps described in Sect. 8.4.4 can either be performed with the desktop version of the 2DSA analysis using a local computer (Fig. 8.4) or by submission through the USLIMS system to a remote compute cluster. In the latter case, the user will log into their Apache web account on their institutional USLIMS instance and select their edited data from the LIMS3 database. The data is then submitted to a remote cluster for analysis (see Fig. 8.5). This process is handled by the Apache Airavata middleware (Marru et al. 2011), currently operating out of Indiana University. A record for each submission is created in the GFAC database, which allows multiple users to submit multiple jobs synchronously from different instances. The Airavata middleware then stages the job(s) on the requested resource's local queuing system and registers the job as started in the generic factory (GFAC) database. A daemon (gridctrl.php) continually monitors the contents of the GFAC database and updates the USLIMS queue viewer where the user can track progress. Once the job starts, status information is sent via UDP to a second daemon (listen.php), which updates each running job in the queue viewer with status details. Once completed, the resulting models are deposited in the LIMS3 database, and the job is marked as

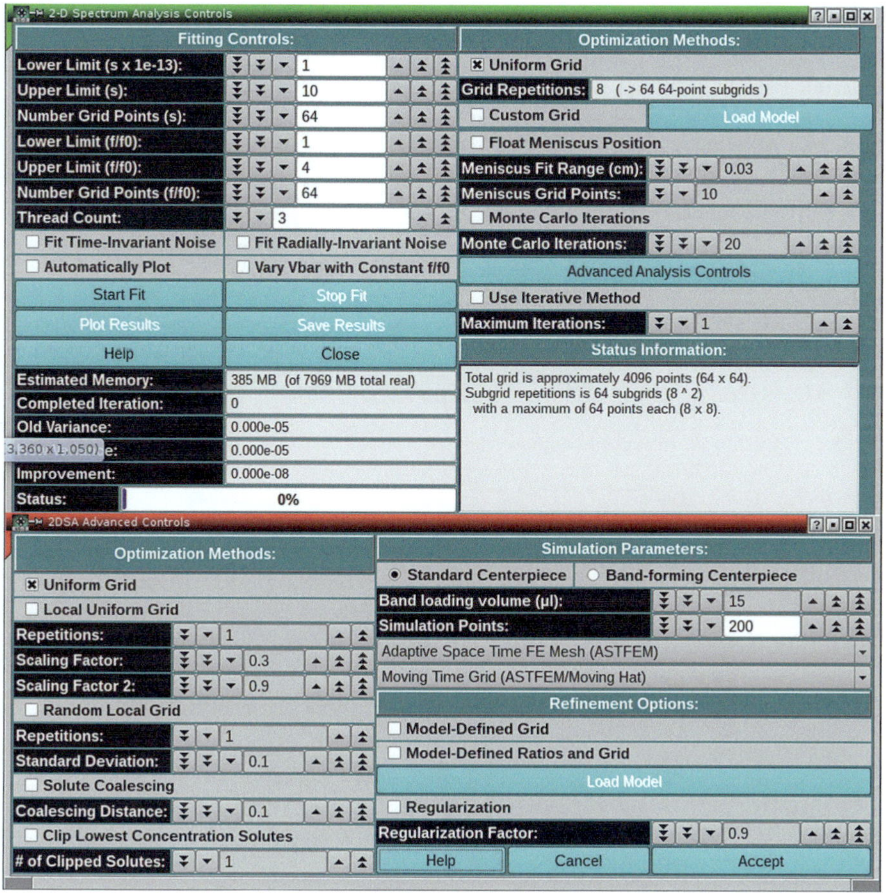

Fig. 8.4 *Top*: 2DSA analysis control window for the UltraScan-III desktop version. *Bottom*: Advanced controls. The desktop version is multi-threaded to support multi-core architectures

completed in the GFAC database and deleted by the grid control daemon. This daemon also notifies the user per e-mail of job completion and updates the queue viewer. Completed jobs are permanently stored in the LIMS database, where the results can be accessed both by the US3 desktop version and through the LIMS system.

8.4.6 Advanced Data Analysis

Once refinement is completed, all subsequent analysis can be performed without additional noise processing or optimization of the boundary conditions. This

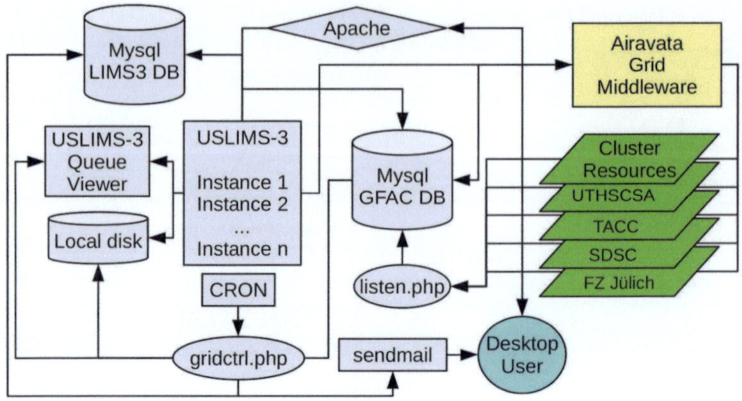

Fig. 8.5 Data flow during supercomputer analysis

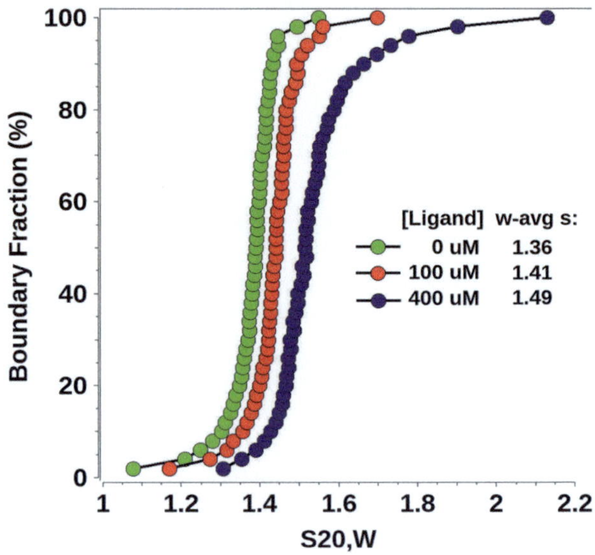

Fig. 8.6 Diffusion-corrected van Holde–Weischet integral s-value distributions provide very sensitive comparisons. In this case, multiple ligand concentrations are measured for a weakly interacting system

reduces calculation time for all subsequent analysis. A range of additional analysis methods are available in US3 which can now be applied. For example, to compare diffusion-corrected sedimentation profiles for multiple samples, an enhanced van Holde–Weischet analysis (Demeler and van Holde 2004) (vHW) can be performed. Combined diffusion-corrected integral G(s) distributions available from the vHW analysis provide a very sensitive way to compare the sedimentation profiles from multiple datasets, even when the changes are very slight (see Fig. 8.6). For

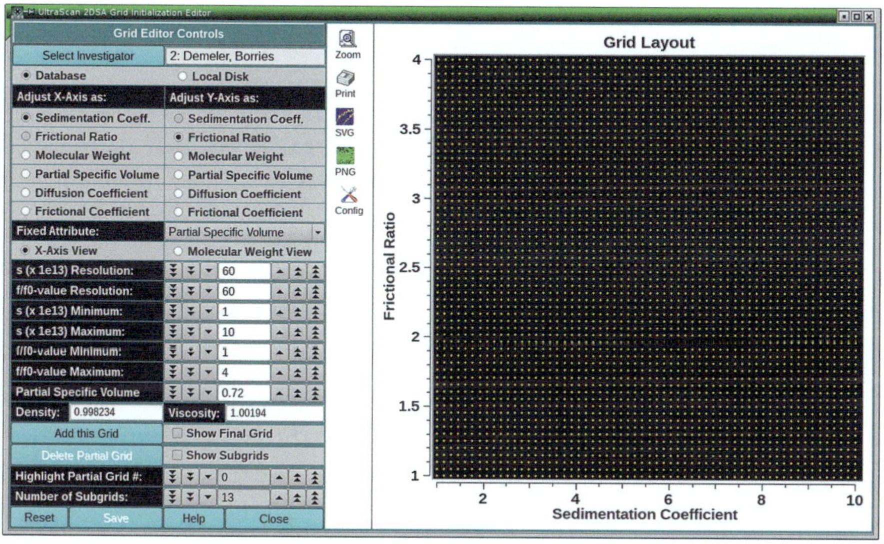

Fig. 8.7 The custom 2DSA grid initialization editor is used to define custom grids that can be fitted with the 2DSA analysis

heterogeneous samples, a 2DSA-Monte Carlo analysis is used to obtain molecular weight and anisotropy distributions, emphasizing intrinsic sedimentation signals while simultaneously attenuating the stochastic noise contributions. For pauci-disperse systems, a parsimonious regularization using the genetic algorithm (GA) optimization method can be used to eliminate non-essential species from the solution without degrading the quality of the fit by applying Occam's razor (Brookes and Demeler 2007). The GA result can be further refined with a Monte Carlo analysis to provide statistical evaluations for all parameters fitted for each identified species and to test the reliability of the fit. A recent addition to US3 is the Custom Grid (CG) method (Fig. 8.7), which allows the user to define the two-dimensional grid analyzed by the 2DSA analysis in terms of any two hydrodynamic parameters that define the sedimentation and diffusion process: s, D, anisotropy, molar mass, partial specific volume, and frictional coefficient when a third parameter is available from an independent measurement. This approach provides great flexibility and allows mixed grids with different parametrizations to be combined. For example, when fitting a DNA–protein associating system, free DNA, free protein, and DNA/protein complex each have a different partial specific volume that can be accommodated by individual custom subgrids to more accurately describe the molar mass distributions present in a mixture. Should molar mass be available from sequence or a mass spectrometry experiment, it can be fixed in the CG analysis and anisotropy and partial specific volume can be fitted (Demeler et al. 2014). Likewise, when anisotropy is available from electron microscopy or crystal structure, it can be held fixed, and heterogeneity in partial specific volume and molar mass can be fitted with the CG analysis. A special case where oligomerization leads to a predictable anisotropy change can be addressed well by the parametrically constrained spectrum analysis (PCSA (Gorbet et al. 2014)). It allows the user to find the best functional parametrization for the two-dimensional parameter space to constrain the solution to a uni-valued function where only a single frictional ratio matches a single sedimentation coefficient. Arbitrary functional forms can be defined in this method to accommodate any distribution function. GA optimization can be initialized with a manual model, or with the results from any 2DSA, Monte Carlo, CG or PCSA analysis. A second-moment analysis is also available in US3 to provide a diagnostic for samples that are not at chemical equilibrium and change sedimentation behavior throughout the experiment. The second moment analysis reports a weight average sedimentation coefficient for each scan, independent of time.

8.4.7 Global Analysis

US3 offers true global fitting where multiple SV datasets can be fitted simultaneously to a single model with either the 2DSA or GA analysis. Due to the large memory requirements of fitting combined datasets, global fitting is only available on remote supercomputers. In either method, the assumption is made that multiple experiments represent the same sample. The samples can be measured either at the same or different speeds and can be at multiple concentrations. The underlying assumption is that they contain the same set of solutes but not necessarily at the same concentration. By fitting multiple datasets globally, additional signal can be obtained. For example, if a sample is measured at a slow speed, diffusion signal is favored, while a fast rotor speed improves resolution of the sedimentation information. By globally fitting both to a single model, the optimized signal from both experiments for either transport process is combined to provide a more reliable model. The number of models generated from a global fit is $2n + 1$, where n is the number of datasets. The first model represents the best fit global model for all datasets. For each dataset, two more models are generated: The first model contains the same set of solutes found in the global model with identical ratios for each solute maintained from the best-fit global model, but scaled to the total concentration of each dataset. The second model contains the same solutes but re-adjusted in partial concentration to optimally match each dataset. The latter model may have one or more solutes from the global model set to zero concentration. This situation could arise when a reversibly self-associating system was measured at multiple concentrations, and the ratio of the oligomers changes as a function of solute concentration, or aggregates appear in a high concentration sample. The comparison of RMSD for each model therefore serves as a reliable diagnostic for the absence or presence of reversible association. If the composition does not change as a function of concentration, both models will produce similar RMSD values, and mass action is absent. For samples with appreciable mass action occurring, only the model with the adjusted ratios of concentrations will fit well. For samples that are true replicates of the same sample, at the same concentration, all three types of models will have similar RMSD values.

8.4.8 Models

The core analysis in US3 is based on modeling SV data using Lamm equations solved by the finite element method. One set of optimization methods in US3 attempts to find the best fitting model consisting of linear combinations of Lamm equation solutions representing any solutes present in the experimental data. Each solute has a set of properties, which include partial specific volume, partial concentration, $s_{20,W}$ and $D_{20,W}$ coefficients, concentration dependency factors for s and D, and extinction coefficients. A second set of optimization methods uses GA to fit models containing two or more solutes as well as reaction terms, such as equilibrium constants, rate constants, and stoichiometry (Demeler et al. 2010). A model also contains details about co-sedimenting solutes, such as gradient forming materials which affect flow of any other solutes as a function of time and radius. From the three hydrodynamic parameters stored for each solute, other parameters can be derived, such as molar mass, anisotropy or frictional ratio, and frictional coefficients. This model structure is used universally in US3 to communicate information between all analysis and visualization modules, including: (1) output from any optimization method based on finite element modeling. Such a model will always contain a reference to the method that was used to determine the model, as well as a root mean square deviation (RMSD) of the fit; (2) user-defined models, either for simulation or to fit the model's parameters directly; (3) input models for the refinement or initialization of a subsequent method, and as a custom grid; (4) a model structure can be visualized with several modules in UltraScan; (5) multiple models can be combined to create global models used for global analysis. Thus, a model serves as a well-defined structure for different methods and modules in US3.

8.4.9 Visualization

US3 contains several powerful model visualization capabilities. First, the user can simulate any model using the US3 simulation routine, either by defining a custom instrument setting or by comparison of an experimental data set with a fitted model. In the latter case, all settings from the experiment are inherited, and an overlay is generated. If time- or radially invariant noise is part of the model, it can be subtracted from the dataset and the experimental and simulated data can be shown overlaid, and all hydrodynamic parameters from the model can be plotted in two dimensions (Fig. 8.8, upper left). The distribution of solutes and their partial concentrations can be visualized in three dimensions by using pseudo-three-dimensional plots or true three-dimensional plots as shown in Fig. 8.9. Here, the user has the option

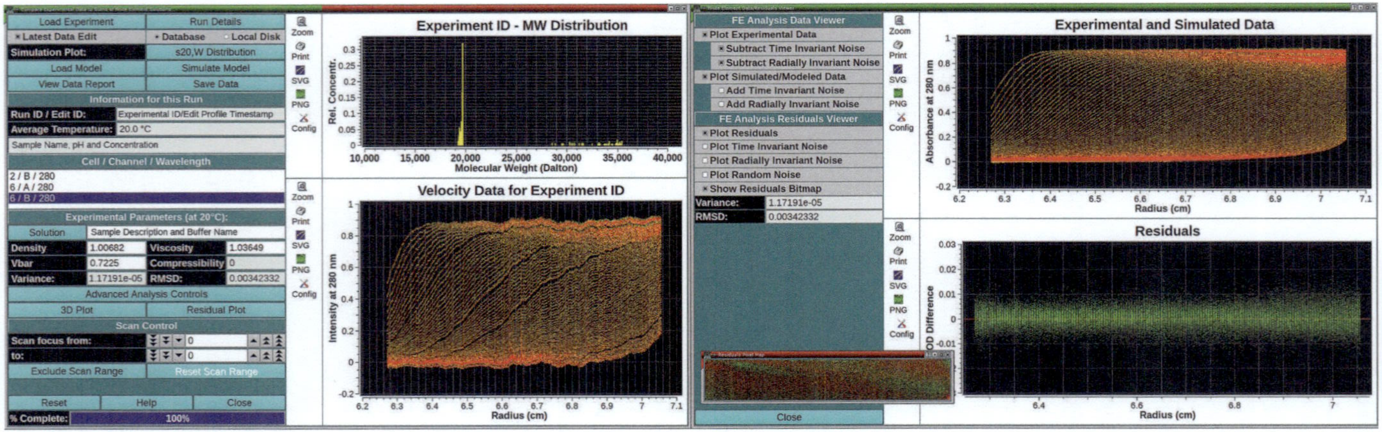

Fig. 8.8 Visualization of experimental intensity SV data and modeled simulations in US3. Experimental data (*yellow*) overlaid by simulations (in *red*) including (*lower left*) and after subtraction of systematic noise components (*upper right*), stochastic noise (lower right), residual bitmap (*low center*) and molar mass distribution information, which can be displayed for all hydrodynamic parameters and transformations (*upper left*). Systematic noise components can be displayed in the lower right panel (*not shown*), and various controls provide flexibility in the display and selection of data

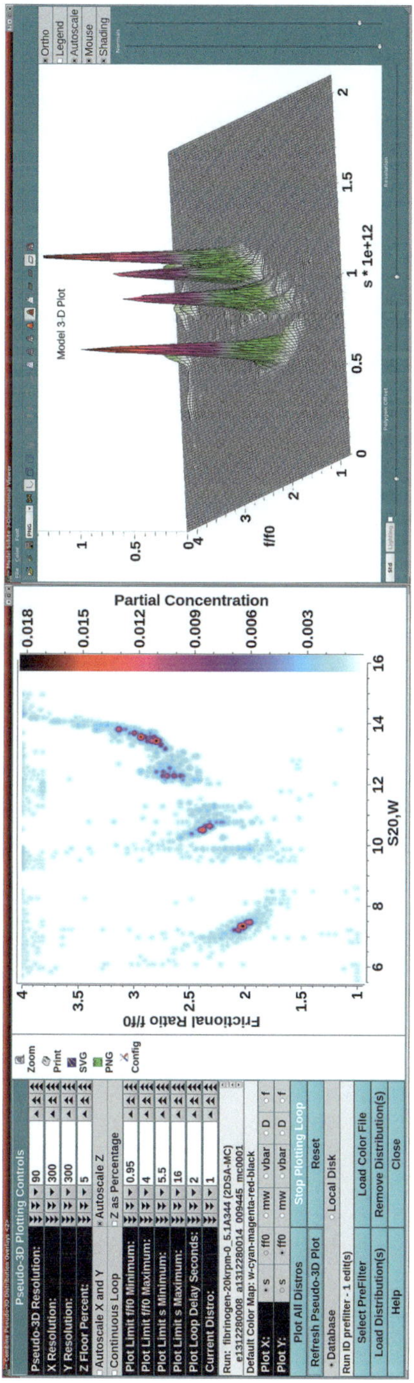

Fig. 8.9 Model visualization options (the same multicomponent 2DSA Monte Carlo model is shown in both viewers) *Left*: The pseudo-three-dimensional viewer. Low concentration or noise points are shown in light cyan. *Right*: A flexible three-dimensional viewer module based on the Qt framework is integrated in US3. Both viewers provide a flexible choice of visualizing all hydrodynamic parameter combinations for x- and y-axes to accommodate the Custom Grid method

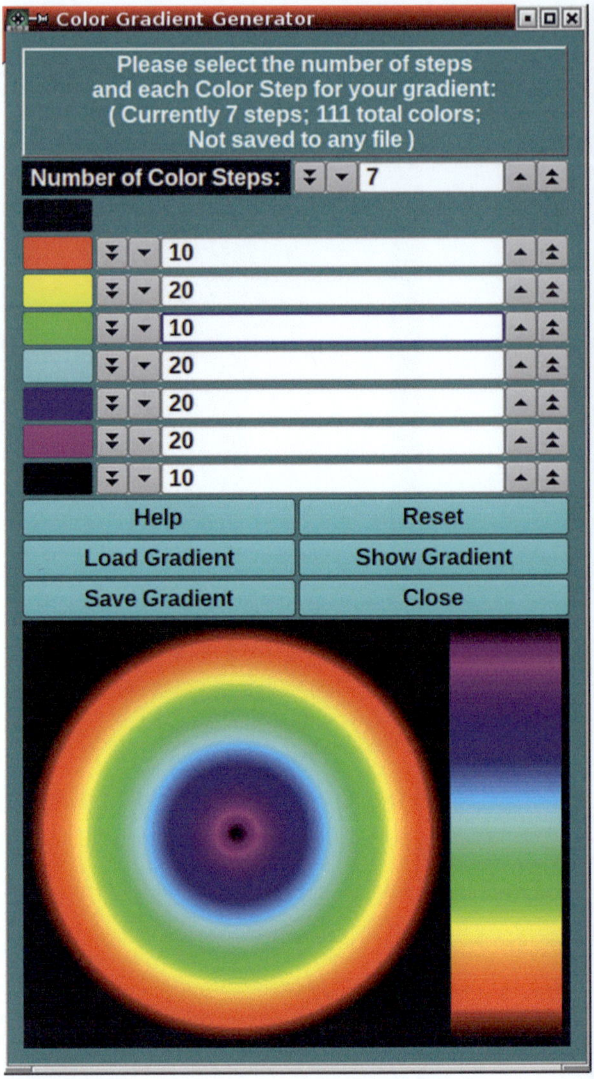

Fig. 8.10 Gradient editor

to select any two of the variable hydrodynamic parameters for the X- and Y-axis. A gradient editor (see Fig. 8.10) provides a convenient interface to define custom color gradients, which can be used to differentially color the three-dimensional plots. In the PCSA analysis, an arbitrary equation is used to specify a constraint within the two-dimensional hydrodynamic parameter space to be probed. The analysis will provide a heatmap of the root mean square deviation (RMSD) values for each variation of the parametrization, providing a visual feedback of the solution's error surface (Fig. 8.11). Two-dimensional combination graphs can be created both from enhanced van Holde–Weischet analyses (see Fig. 8.6) or any of the whole boundary

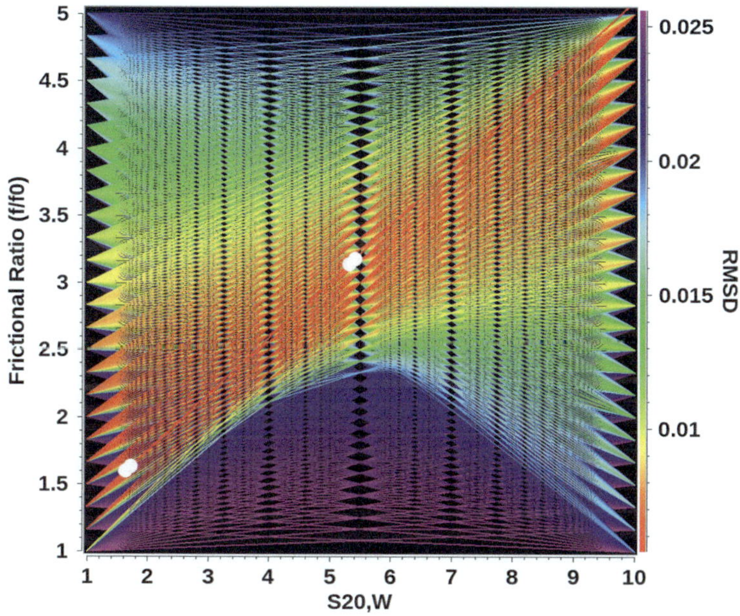

Fig. 8.11 PCSA error surface heat map for SV data from a mixture of lysozyme and a 208 bp DNA fragment using a straight line parametrization. The white dots show the solutes found for the line that best fits the SV data. The color gradient indicates the RMSD value for each line

modeling programs (Fig. 8.12). The former program is well suited to compare the results from multiple experiments where a parameter like the solute concentration, the ionic strength or the pH of the buffer is modulated, or mutants and wildtype are compared, or a titration of a binding partner is followed. The latter program provides detailed views of all hydrodynamic parameters and is well suited to compare results from multiple analysis methods for the same sample. For multi-wavelength data, either three-dimensional plots or movies can be used to visualize the data. Three-dimensional plots are appropriate for simultaneously viewing spectral and hydrodynamic sample properties in a single plot (Fig. 8.13). Movies can be created by showing individual pseudo-three-dimensional plots (Fig. 8.9, left panel), where each frame represents the next wavelength in a sequence of wavelengths. To enable customization of visualization output in US3, each graph has a configuration button that allows customization of every plot element, such as axes, labels, legends, lines, symbols, colors, grid lines, plot canvas, and graphing items.

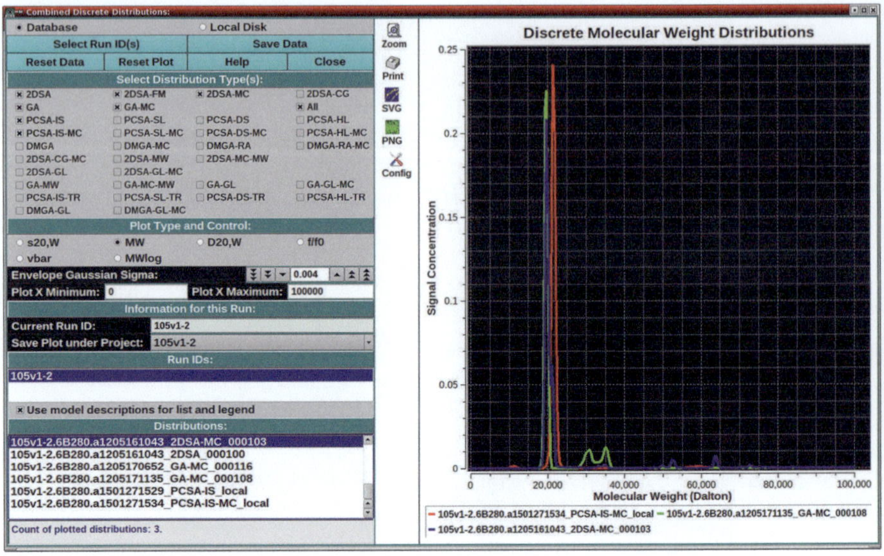

Fig. 8.12 Discrete molecular weight distributions for a 19.7 kDa protein analyzed with Monte Carlo for 2DSA, PCSA and GA

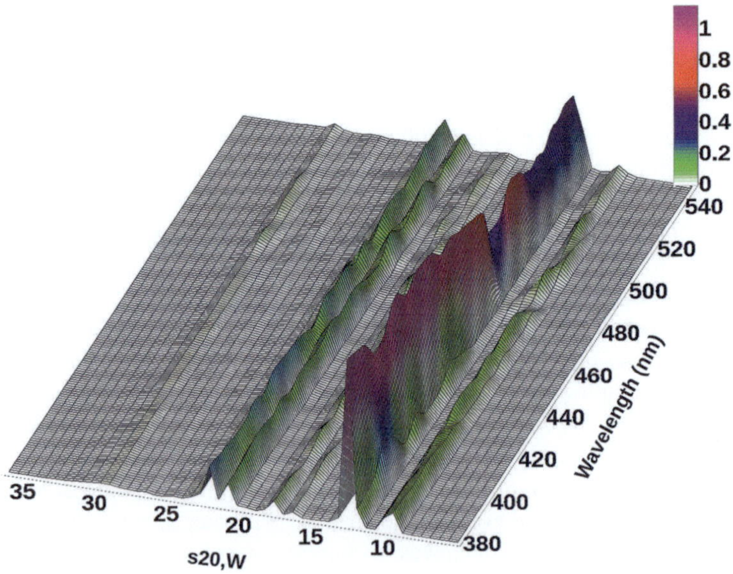

Fig. 8.13 Three-dimensional view of a global multi-wavelength analysis showing the relative absorbance for different hydrodynamic species for each fitted wavelength (Credits: Robert Whetten, German Plascencia, UTSA)

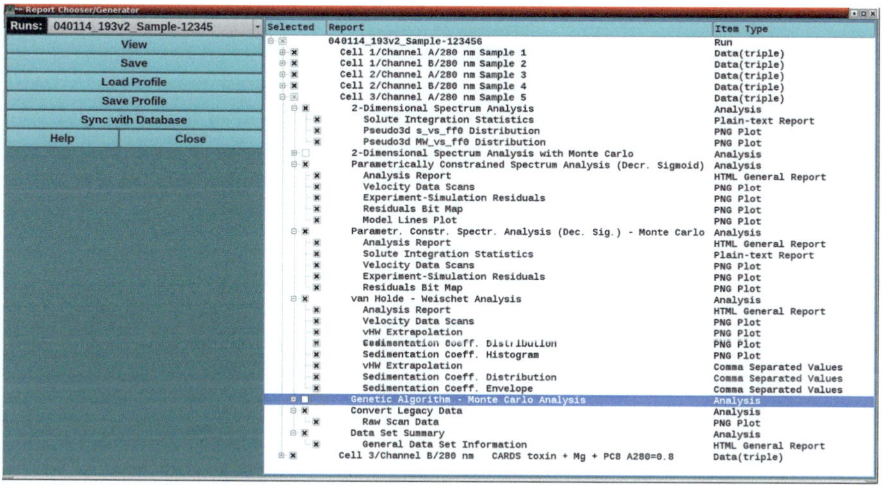

Fig. 8.14 Report Generator. Individual report items from each triple can be included into a PDF-formatted report, and the selection profile can be stored

8.4.10 Reports

Completed analysis results are captured and saved to disk and database. Each analysis generates a selection of graphs, analysis report records, and portable spreadsheets of the resulting data. The user can retrieve all result documents either from the USLIMS or through the desktop version. In the desktop version, a hierarchy of completed analysis methods and record items is presented for each experiment. The user can create a custom report by selecting desired items from the tree and then either generate a pre-formatted PDF document of all selected items or print it directly. If a particular selection profile is used repeatedly, the user can save this profile and re-apply it to other experiments to regenerate the desired report selection. The user can also migrate report records to a new computer by synchronizing the local storage with the database (Fig. 8.14). In the USLIMS, a report can be created dynamically from any record stored in the database. After selecting the experiment, any triple in the experiment provides a link to all report records belonging to this triple, sorted by analysis method, which the user can view or download. Graphics are provided both in PNG compressed bitmaps and also in scalable vector graphics format, suitable for post-editing at arbitrary resolution.

8.5 Simulation Programs

US3 offers a range of simulation modules to assist the user in designing experiments and interpreting results. These include a graphical finite element modeling program for the simulation of SV data from any model stored locally or in the database.

All parameters and boundary conditions for the run can be modified using an instrument control panel (rotor speed, length of run, number of scans, scan delay, rotor acceleration rate, temperature), as well as data parameters such as noise levels for TI, RI and stochastic noise components, selection of boundary or band-forming centerpieces, meniscus and cell bottom positions. Pre-defined buffers can be chosen to apply viscosity and density corrections. Multi-speed experiments can be simulated by defining speed profiles. Finite element solutions can be generated using multiple discretization schemes (radius: adaptive space, fixed Claverie mesh, moving hat, specified mesh file, or finite volume meshes needed for advanced simulations, time: fixed or adaptive). All profiles can be saved and applied to future simulations. The ASTFEM/ASTFVM solutions programmed in US3 can produce simulations for arbitrary advanced models, including reacting systems with kinetics, concentration dependent non-ideality, co-sedimenting solutes and solvent compressibility. Models for such experiments can be created with the US3 model editor. Simulations can be shown in accelerated time as movies. US3 has an equilibrium simulation program that predicts the time it takes to reach equilibrium based on molecular weight, rotor speed, centerpiece position, and column height. Multiple speed steps can be simulated. A self-association simulator can be used to predict relative concentrations of individual oligomers in a reversible self-associating monomer *n*-mer *m*-mer system as a function of concentration, where the equilibrium constants for each association reaction can be supplied by the user. Finally, US3 offers two hydrodynamic calculators. The first takes as input the molar mass, partial specific volume and the solution conditions, as well as an axial ratio to predict s, D, f, f/f_0, and the dimensions of the two axes of a prolate and oblate ellipsoid, and for a long rod model and a sphere. The second module takes two of the three parameters molar mass, s and D to predict axial ratios and dimensions for the same two ellipsoids and the long rod model, and predicts the Stokes radius, f/f_0, and the remaining hydrodynamic parameters.

8.6 Utilities

The US3 software offers a number of utilities that support sedimentation analysis. First, an export module can be used to convert OpenAUC formatted data to traditional Beckman ASCII file format. For intensity data, the user has the choice to export either intensity or pseudo-absorbance data. This makes US3 fully backward compatible with other analysis packages and offers a conversion path between traditional Optima XLA/I acquired data and newer multi-wavelength instruments collecting data in binary OpenAUC format. A second utility is used to calibrate rotor stretching. This calibration allows US3 to predict the precise displacement of the cell bottom due to stretching of the rotor as a function of speed and will calculate an exact position of the cell bottom at any speed, since an exact centerpiece geometry is also measured and stored for each centerpiece in US3. The exact bottom position is critical for a correct solution of the Lamm equation in all whole boundary fitting

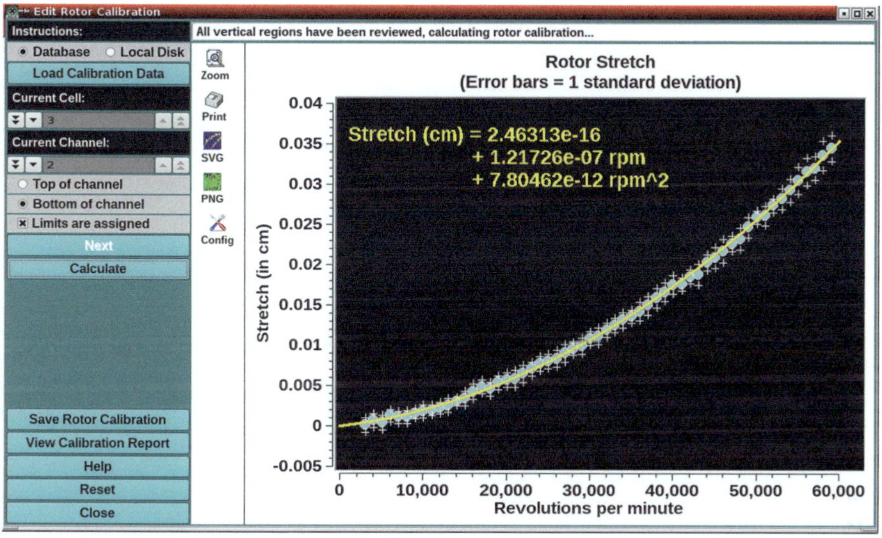

Fig. 8.15 Rotor Calibration program showing a stretch calibration for a Beckman An60Ti 4-hole rotor

programs, especially for multi-speed experiments. First, a calibration experiment is performed where an empty two- or six-channel centerpiece is placed into each rotor hole. All channels are radially scanned in intensity mode. This records positions for the edges of each channel in the centerpiece. The first scan of all cells, including the counterbalance, is collected at 3000 rpm. Subsequently, the rotor is accelerated in 1000 rpm increments to maximum speed, pausing every 1000 rpm and scanning all cells and channels, generating a total of 57 scans for each cell and channel (47 for the 8-hole rotor). The data are cropped to each edge of a centerpiece channel. For each speed, the center of the edge is determined by the calibration algorithm, and the average displacement for each speed increment is calculated and plotted against rotor speed. These data are then fitted to a second-order polynomial, and the baseline is adjusted such that the displacement at zero rpm is zero. A fit for a typical rotor calibration is shown in Fig. 8.15. The stored calibrations for each rotor are read automatically based on the associations made for each cell of each experiment every time finite element calculations are made. This mechanism provides a more accurate alternative to the introduction of another fitting parameter for the bottom of the cell position. A configuration utility offers flexible configurations for personal preferences such as fonts, color schemes, advanced interface options, debug levels, default file locations, and for database connectivity options, including passwords. Individual database dialogs provide access to tables which may be stored locally or in a remote database. These dialogs allow the user to retrieve and edit existing records, and create new records for investigator information, buffer files, analytes, solutions, experimental data, edit profiles, models, noise files, projects, rotors, and rotor calibrations. Database users can use the data management tool to synchronize any computer with the contents of a remote database, which makes US3 data inherently portable and independent of location. The data managed with these dialogs will always be synchronized with the USLIMS information. US3 also features a complete online help menu that covers each routine with a context-specific help module.

References

Bhattacharyya SK et al (2006) Development of fast fiber based UV-vis multiwavelength detector for an ultracentrifuge. Progr Colloid Polym Sci 131:9–22

Brookes EH, Demeler B (2007) Parsimonious regularization using genetic algorithms applied to the analysis of analytical ultracentrifugation experiments. In: GECCO '07 proceedings of the 9th annual conference on genetic and evolutionary computation. ACM, New York, pp 361–368. ISBN: 978-1-59593-697-4. doi:10.1145/1276958.1277035

Brookes EH, Demeler B (2008) Parallel computational techniques for the analysis of sedimentation velocity experiments in UltraScan. Colloid Polym Sci 286:138–148

Brookes EH, Boppana RV, Demeler B (2006) Computing large sparse multivariate optimization problems with an application in biophysics. In: Proceedings of the SC 2006 conference. ACM/IEEE. doi:10.1109/SC.2006.18 E-ISBN: 0-7695-2700-0 Print ISBN: 0-7695-2700-0 INSPEC Accession Number: 9343095

Brookes EH, Cao W, Demeler B (2010a) A two-dimensional spectrum analysis for sedimentation velocity experiments of mixtures with heterogeneity in molecular weight and shape. Eur Biophys J 39:405–414

Brookes E, Demeler B, Rosano C, Rocco M (2010b) The implementation of SOMO (SOlution MOdeller) in the UltraScan analytical ultracentrifugation data analysis suite: enhanced capabilities allow the reliable hydrodynamic modeling of virtually any kind of biomacromolecule. Eur Biophys J 39(3):423–435

Brookes E, Demeler B, Rocco M (2010c) Developments in the US-SOMO bead modeling suite: new features in the direct residue-to-bead method, improved grid routines, and influence of accessible surface area screening. Macromol Biosci 10(7):746–753. doi:10.1002/mabi.200900474

Brookes E, et al (2013) US-SOMO cluster methods: year one perspective. XSEDE '13. In: Proceedings of the conference on extreme science and engineering discovery environment: gateway to discovery, at San Diego, California 01/2013. doi:10.1145/2484762. 2484815. ISBN: 9781450321709

Cao W, Demeler B (2005) Modeling analytical ultracentrifugation experiments with an adaptive space-time finite element solution of the Lamm equation. Biophys J 89(3):1589–1602

Cao W, Demeler B (2008) Modeling analytical ultracentrifugation experiments with an adaptive space-time finite element solution for multi-component reacting systems. Biophys J 95(1):54–65

Cölfen H et al (2010) The open AUC project. Eur Biophys J 39(3):347–359

Demeler B (2009) High-resolution modeling of hydrodynamic experiments with UltraScan. http://www.XSEDE.org/gateways/projects.php?id=77

Demeler B (2010) Methods for the design and analysis of sedimentation velocity and sedimentation equilibrium experiments with proteins. Curr Protoc Protein Sci 07: Unit 7.13. doi:10.1002/0471140864.ps0713s60 PMCID: PMC4547541

Demeler B, Brookes EH (2008) Monte Carlo analysis of sedimentation experiments. Colloid Polym Sci 286:129–137

Demeler B, van Holde KE (2004) Sedimentation velocity analysis of highly heterogeneous systems. Anal Biochem 335:279–288

Demeler B et al (2010) Characterization of reversible associations by sedimentation velocity with UltraScan. Macromol Biosci 10(7):775–782. PMID: 20486142

Demeler B et al (2014) Characterization of size, anisotropy, and density heterogeneity of nanoparticles by sedimentation velocity. Anal Chem 86(15):7688–7695

Gorbet G et al (2014) A parametrically constrained optimization method for fitting sedimentation velocity experiments. Biophys J 106:1741–1750

Laue TM, Shah BD, Ridgeway TM, Pelletier SL (1992) Computer-aided interpretation of analytical sedimentation data for proteins. In: Harding SE, Rowe AJ, Horton JC (eds) Analytical ultracentrifugation in biochemistry and polymer science. Royal Society of Chemistry, Cambridge, pp 90–125

Marru S, et al (2011) Apache Airavata: a framework for distributed applications and computational workflows. In: GCE '11 Proceedings of the 2011 ACM workshop on Gateway computing environments, pp 21–28, ACM New York, NY, USA. Apache Airavata: http://airavata.apache.org/

Memon S, et al (2013) Improvements of the UltraScan scientific gateway to enable computational jobs on large-scale and open-standards based cyberinfrastructures. In: XSEDE '13 Proceedings of the conference on extreme science and engineering discovery environment: gateway to discovery No. 39, ACM New York, ISBN: 978-1-4503-2170-9, doi:10.1145/2484762.2484800

Memon S, et al (2014) Advancements of the UltraScan scientific gateway for open standards-based cyberinfrastructures. Concurr Computat Pract Exper, Wiley. doi: 10.1002/cpe.3251

Schilling K (2014) A multiwavelength capable detector for the analytical ultracentrifuge (Nanolytics, Potsdam, Germany). Personal Communication

Schuck P, Demeler B (1999) Direct sedimentation analysis of interference optical data in analytical ultracentrifugation. Biophys J 76(4):2288–2296

Stafford WF (1992) Boundary analysis in sedimentation transport experiments: a procedure for obtaining sedimentation coefficient distributions using the time derivative of the concentration profile. Anal Biochem 203(2):295–301

UltraScan-III wiki: http://wiki.bcf2.uthscsa.edu/ultrascan3/

Part III
Hydrodynamic Modeling

Chapter 9
Introduction: Calculation of Hydrodynamic Parameters

Olwyn Byron

Abstract This introduction considers the approaches to the calculation of hydrodynamic (and related) parameters described in detail in the following Chaps. 10, 11 and 12 (Chap. 10, US-SOMO; Chap. 11, the HYDRO suite; and Chap. 12, BEST). Starting with a description of what hydrodynamic modelling is and why it is useful, the first part of this chapter then presents 12 equations as a very basic tutorial in the hydrodynamic computations underlying the majority of the methodology that is then summarised in the subsequent section on current approaches in both rigid body and flexible modelling. The pros and cons of these approaches are then given before a few concluding remarks and an outlook.

Keywords Hydrodynamic modelling • Bead modelling • Boundary element modelling • Electrostatic-hydrodynamic analogy modelling • Sedimentation coefficient • Diffusion coefficient • Macromolecular hydration • Rigid body modelling • Flexible body modelling

9.1 Introduction

One might expect that because this chapter and the subsequent three chapters it seeks to introduce are in a book about analytical ultracentrifugation, the reader would be a paid-up member of the AUC and hydrodynamics club, with corresponding data for their macromolecular system and a clear idea of the utility of hydrodynamic modelling as an interpretative tool. However, in the era of accessing book chapters in isolation, this chapter starts by assuming no such background whatsoever, in the hope and expectation of encouraging new members to sign up!

Most macromolecular systems function in (aqueous or otherwise) solution, but much structural understanding of these systems is based on high-resolution coordinates determined crystallographically or as homology models. And yet it is well recognised that many macromolecules have so far failed to crystallise at all or

---

O. Byron (✉)
School of Life Sciences, College of Medical, Veterinary and Life Sciences, University of Glasgow, Glasgow G12 8QQ, Scotland, UK
e-mail: olwyn.byron@glasgow.ac.uk

S. Uchiyama et al. (eds.), *Analytical Ultracentrifugation*,
DOI 10.1007/978-4-431-55985-6_9

in such a way as to yield diffraction data at a satisfactory resolution, although this is expected to be much less of a limitation to high-resolution structure determination with the advent of free electron lasers and the possibility of determining atomic structures for macromolecules that form micro- (Boutet et al. 2012) or nanocrystals at best. Additionally, many molecules and complexes are too large for routine structural determination via NMR spectroscopy (often a viable alternative to X-ray crystallography), and there are entire classes of macromolecules that are poorly suited to high-resolution structural study with one or other method (e.g. polymers or intrinsically disordered proteins). Low- or meso-resolution solution methods offer valuable insights to macromolecular structure in these instances *and* for systems that *have* been successfully characterised with X-ray crystallography, since it is now accepted that the dilute solution conformation can be significantly different from that adopted in the crystal lattice (Nakasako et al. 2001; Smolle et al. 2006; Trewhella et al. 1988; Vestergaard et al. 2005).

Hydrodynamic modelling entails computation of experimentally determinable hydrodynamic and related parameters (hereafter HARPs) for some form of macromolecular model. If the computed and experimental parameters agree to within some acceptable limit, the model is consistent with the solution macromolecular state. The information content of the single-value parameters concerned (e.g. translational diffusion coefficient, intrinsic viscosity, radius of gyration, etc.) is insufficient to permit *determination* of solution structures *de novo* but, especially when used in combination, can lend strong support to (or rule out) the proposed model.

What kinds of measureable parameters can be computed by these methods? This depends on the particular software that is used. In addition to the sedimentation coefficient (s) and translational diffusion coefficient (D_t), two of the principal experimentally determined parameters at the heart of this book, are the many other complementary parameters, a comprehensive (if slightly outdated) list of which is given in Table 1 of Byron (2008).

In the next three sections of this chapter, the most widely used approaches for macromolecular hydrodynamic computation will be described in some detail. These are summarised in Table 9.1. Emre Brookes and Mattia Rocco's chapter on US-SOMO (Brookes et al. 2010a, b) (Chap. 10) describes in detail its use for computing HARPs for models constructed directly from atomic resolution structures (or homology models) via either (1) the *SoMo* (Rai et al. 2005) or *AtoB* (Byron 1997) methods combined with the algorithms for HARP computation developed by José García de la Torre and collaborators over many years (Chap. 11) or the newer *Zeno* (Kang et al. 2004) algorithm (Brookes and Rocco, Sect. 10.3) or (2) the BEST algorithm of Sergio Aragon (2004, 2011) (Chap. 12 where the theoretical basis for this approach is fully described). US-SOMO has, since its inception, developed to embrace the hugely complementary modelling possibilities afforded by small-angle X-ray and neutron scattering (SAXS and SANS, respectively); this is not covered in any detail by Brookes and Rocco in Chap. 10, but clear reference is made to complementary published work in which it is described. The chapter by José García de la Torre (Chap. 11) on the HYDROxxx suite of programs (for rigid

Table 9.1 Summary of freely available software for the computation of HARPs for macromolecular models

Software	References	URL	Overview of capability
HYDRO++	García de la Torre et al. (2007)	http://leonardo.inf.um.es/macromol/programs… …/hydro++/hydro++.htm	Computation of HARPs from multi-sphere models that have been generated externally
US-SOMO	Brookes et al. (2010a, b)	http://www.somo.uthscsa.edu/	Constructs bead models from PDB files using either the *SoMo* or *AtoB* algorithm. HARPs are computed for these models using either the García de la Torre and Bloomfield "supermatrix inversion" method or the *Zeno* method. An interface to BEST computation on a cluster (below) is also offered. Molecular flexibility can be evaluated with the DMD tool
Zeno	Kang et al. (2004)	http://web.stevens.edu/zeno/	Computes selected HARPs for any molecule by enclosing it in a sphere from whose internal surface a series of random walks is launched. The fraction of these walks that reaches the molecule determines the HARP value. *Zeno* is also available in US-SOMO
HYDROPRO	Ortega et al. (2011b)	…/hydropro/hydropro.htm	Constructs shell models from PDB files and computes HARPs for shell models of decreasing sphere size, extrapolating values to the case of infinitely small spheres to arrive at the final outputs. Also can construct and compute HARPs for bead-per-atom/residue models
BEST	Aragon (2004, 2011)	http://esmeralda.sfsu.edu/	Constructs triangular patchwork surface models from PDB files and computes HARPs for models of decreasing triangle size, extrapolating values to the case of infinitely small triangles to arrive at the final outputs

(continued)

Table 9.1 (continued)

Software	References	URL	Overview of capability
HYDRONMR	García de la Torre et al. (2000)	. . . /hydronmr/hydronmr.htm	Shell modelling as for HYDROPRO but additionally computes the rotational correlation time (τ_c) and NMR relaxation times (T_1 and T_2) and for each residue
HYDROMIC	García de la Torre et al. (2001)	. . . /hydromic/hydromic.htm	Shell modelling as for HYDROPRO but constructs models from three-dimensional reconstruction (e.g. SPIDER or MRC) files
HYDROSUB	García de la Torre and Carrasco (2002)	. . . /hydrosub/hydrosub.htm	Shell modelling as for HYDROPRO but shell models are constructed from ellipsoids and/or cylinders that are positioned to represent the modelled system
HYDROPIX	García de la Torre (2001a)	. . . /hydropix/hydropix.htm	Shell modelling as for HYDROPRO but the shell model is based on a solid geometrical shape that is generated by an ancillary program MAKEPIXB
MULTIHYDRO	García de la Torre et al. (2005)	. . . /multihydro/multihydro1c.htm	Generates a number of conformations of a given bead model for which solution parameters are then computed with HYDRO++
HYDFIT	Ortega et al. (2011a)	. . . /hydfit/hydfit.htm	Searches for the best fit structure generated by MULTIHYDRO by comparing the computed solution parameter landscape with experimentally determined values
MONTEHYDRO	García de la Torre et al. (2005)	. . . /montehydro-4/montehydro-4.htm	A Monte Carlo method is used to generate different conformations of models comprising beads joined by conceptually flexible connectors for which solution parameters are computed and reported as the MC conformational average
SIMUFLEX	García de la Torre et al. (2009)	. . . /simuflex-4/simuflex-4.htm	Directly simulates the internal dynamics of a bead-and-connector model using a Brownian dynamics simulation engine

body modelling), MONTEHYDRO (García de la Torre et al. 2005) and SIMUFLEX (García de la Torre et al. 2009) (respectively, for Monte Carlo and Brownian dynamics modelling of flexible systems) and HYDFIT (Ortega et al. 2011a) for rigid body modelling of multiple conformations of a given model follows next. While the discrete molecular dynamics (DMD (Dokholyan et al. 1998; Ding and Dokholyan 2006)) module of US-SOMO permits the application of a type of MD to gain some understanding about the conformers that are consistent with HARPs, flexibility is best modelled with the Brownian dynamics approach in SIMUFLEX. The final chapter, by Sergio Aragon (Chap. 12), describes the boundary element (BE) algorithm for computation of HARPs for models whose surface comprises a triangular patchwork constructed from atomic coordinates (e.g. PDB files). HARPs are computed for models of decreasing triangle size, extrapolating values to the case of infinitely small triangles to arrive at the final outputs, in a manner analogous to the bead/shell-model concept implemented in the HYDROPRO (Ortega et al. 2011b) method and other programs in the HYDRO suite of García de la Torre and colleagues.

9.2 Hydrodynamics 101: A Simple Tutorial

Imagine having determined a sedimentation coefficient (s) and/or translational diffusion coefficient (D_t) for a macromolecule for which you have very limited other structural data apart from the molecular weight (M, determined via, e.g., sedimentation equilibrium or known from the primary structure). What can these parameters tell you? Immediately, you can assess them in terms of what they would tell you about your macromolecule if it was a sphere and was "anhydrous". As described by Uchiyama and Arisaka in Chap. 1, the Svedberg equation

$$s = \frac{M(1-\overline{v}\rho)}{N_A f} \tag{9.1}$$

relates s to the frictional coefficient f via M, the buoyancy factor ($(1-\overline{v}\rho)$ where $\overline{v}$ is the partial specific volume of the molecule and ρ the solvent density), and Avogadro's number (N_A). Stokes' law

$$f_0 = 6\pi\eta_0\sigma \tag{9.2}$$

describes the dependence of the frictional coefficient of a sphere (f_0) on its radius σ and the solvent viscosity η_0. Knowing that the volume of a sphere with the same M and $\overline{v}$ as the molecule is

$$V_a = \frac{M\overline{v}}{N_A} \tag{9.3}$$

an expression for σ can be introduced to Eq. (9.2), and the resultant expression for f_0 can be substituted into Eq. (9.1), i.e.

$$s = \frac{M\left(1 - \overline{v}\rho\right)}{N_A 6\pi\eta_0}\left(\frac{4\pi N_A}{3M\overline{v}}\right)^{\frac{1}{3}} \tag{9.4}$$

This is the maximum possible sedimentation coefficient a molecule could have – any contribution to the frictional behaviour of the molecule arising from the deviation of its shape from a sphere (which is the most hydrodynamically efficient shape) is neglected, as are any hydration effects. What is meant by hydration in the context of hydrodynamic modelling is discussed briefly in Sect. 9.3.1 and more fully by Aragon in Sect. 12.3. Simplistically, in order to obtain agreement between computed (anhydrous) values for s and their (solvated) experimental counterparts, 0.3–0.4 g water/g protein is included in hydrodynamic models for proteins. A very full account of the various ways in which hydrodynamic hydration is modelled is given in section 5 of Byron (2008).

Nonsphericity and hydration will both decrease the sedimentation coefficient. Interpreting s in this way depends on a knowledge of $\overline{v}$, which in turn is difficult to measure but can be computed (e.g. from the $\overline{v}$ of constituent residues with, e.g., SEDNTERP (http://sednterp.unh.edu/) (Laue et al. 1992)). But for some macromolecules, the number and type of constituent residues are not well known (e.g. glycoproteins purified from animal serum), making computation of $\overline{v}$ unreliable. The sedimentation coefficient also depends on M, but this can normally be determined with far less uncertainty than $\overline{v}$. The translational diffusion coefficient, on the other hand, does not suffer from this problem since, from the Stokes-Einstein equation for a sphere,

$$D_t = \frac{RT}{N_A f} \tag{9.5}$$

Hence, D_t should be the better parameter of the two for hydrodynamic modelling studies, assuming it can be determined from sedimentation velocity experiments with the same precision that is achievable for the measurement of s.

But we must be able to do better than this. Some molecular systems can be modelled by general triaxial ellipsoids (Harding 1982), for which the exact frictional behaviour is also well known, but many cannot be satisfactorily represented with such regular solid shapes. Instead, it is usual to represent the macromolecule as an assembly of elements for which the solution behaviour is well defined and to derive equations that satisfactorily describe their collective interaction with the solution environment. At this point, it is useful to mention the two extremes of solvent behaviour at the macromolecular surface, termed "stick" and "slip". In the "stick" boundary condition, the solvent at the macromolecular surface has zero velocity and sticks to or moves with the macromolecule. This is in contrast to the alternative "slip" condition at the molecule/solvent boundary where the component of solvent velocity perpendicular to the molecular surface is zero, but the tangential component is unconstrained and so the solvent slips past the molecular surface. The stick boundary condition is appropriate for macromolecules that are far larger than the solvent molecules, whereas the slip condition better represents the HARPs of much smaller molecules whose size is comparable to that of the solvent (see Aragon, Sect. 12.2).

Considering the case of the frictional interaction of a macromolecule moving in a solution (directly relevant to the translational diffusion coefficient (D_t) and s measured by AUC), the general expression relating the frictional force **F** on the molecule to its velocity **u** is, for a sphere,

$$\mathbf{F} = -f\mathbf{u} \tag{9.6}$$

Kirkwood (1949, 1954) established the first general theory of "irreversible" (e.g. transport) processes in solutions of macromolecules and devised equations that facilitated the computation of D_t, intrinsic viscosity $[\eta]$ and parameters determined by electric birefringence. Macromolecules were represented by three-dimensional arrays of N spherical beads of radius σ_i ($i = 1$ to N) in a solvent of viscosity η_0. The force exerted on the ith bead by the solvent is the product of the frictional coefficient (f_i) for that bead and the velocity of that bead relative to the solvent:

$$\mathbf{F}_i = -f_i\,(\mathbf{u}_i - \mathbf{v}_i) \tag{9.7}$$

where $f_i = 6\pi\eta_0\sigma_i$ and $\mathbf{u}_i$ is the velocity of the ith bead while $\mathbf{v}_i$ is the velocity the solvent would have at the centre of that bead were that bead absent from the system. This is an equation of frictional drag but neglects the motion of the other elements in the model (beads $j = 1$ to N; $j \neq i$) that perturbs the solvent flow pattern. Oseen (1927) and Burgers (1938) had already derived a correction for this perturbation for stick boundary conditions, as follows:

$$\mathbf{F}_i = -f_i\left(\mathbf{u}_i - \mathbf{v}_i^0\right) - f_i\sum_{j=1}^{N}\mathbf{T}_{ij}\mathbf{F}_j \tag{9.8}$$

where $\mathbf{v}_i^0$ is the velocity the solvent would have at the centre of the ith bead were *all other* beads to be absent – i.e. the unperturbed velocity of the solvent – and $\mathbf{T}_{ij}$ is the hydrodynamic interaction tensor, which is at the heart of hydrodynamic bead modelling computations. A tensor is a geometric object that describes a relationship between vectors (geometric quantities with magnitude and direction, e.g. **F, u, v**), scalars (real numbers with magnitude only, e.g. f) and other tensors (e.g. **I** below). When a vector force is applied to a material system and the response of that system to the force is a vector that does not point in the same direction as the original force, then that response is described by a matrix of numbers instead of a scalar. This matrix is a tensor. Oseen (1927) described the hydrodynamic interaction tensor as follows:

$$\mathbf{T}_{ij} = \frac{1}{8\pi\eta_0 R_{ij}}\left(\mathbf{I} + \frac{\mathbf{R}_{ij}\mathbf{R}_{ij}}{R_{ij}^2}\right) \tag{9.9}$$

where $\mathbf{R}_{ij}$ is the distance vector between the centres of beads i and j, R_{ij} is the distance and $\mathbf{I}$ is the unit tensor:

$$\mathbf{I} = \begin{bmatrix} 1 & 0 & 0 \\ 0 & 1 & 0 \\ 0 & 0 & 1 \end{bmatrix} \tag{9.10}$$

Equation (9.9) is equivalent to Eq. (9.2) in Aragon, Chap. 12, and Eq. (9.8) is an approximate form of the exact integral formulation of hydrodynamics for infinitesimal surface elements (Eq. 9.1 in Aragon, Chap. 12), one of the unique features of which is that either stick, slip or mixed boundary conditions can be addressed, while bead modelling is currently limited to stick boundary conditions.

In order to compute a measureable parameter such as D_t or s for a hydrodynamic model, the hydrodynamic interaction equation (Eq. 9.8), which is actually a system of N linear equations with $3N$ unknowns, has to be solved. This is usually done by inversion of a coefficient supermatrix of N^2 blocks with dimension 3×3. The computational time taken for this supermatrix inversion follows an approximate cubic dependence on the number of elements comprising the model (i.e. N^3). Since the advent of hydrodynamic modelling, computers have become unimaginably fast so that this dependence on N^3 (and its associated memory requirement) is a consideration only for very large molecules or complexes comprising many elements.

There are limitations to the expression for the hydrodynamic interaction tensor (Eq. 9.9) in the case of bead modelling: it does not take into account the finite volume (and thus the radius) of the beads; it is restricted to beads of equal radius and the beads cannot overlap. Since this tensor does not take into account the bead radius, it is not possible for it to "sense" any bead overlap. This instead becomes an issue when the tensor does take into account the finite bead size, as is the case for the hydrodynamic interaction tensor devised by Rotne and Prager (1969) and Yamakawa (1970) for *overlapping* beads of *equal* radius σ:

$$\mathbf{T}_{ij} = \frac{1}{6\pi\eta_0\sigma}\left(\left(1 - \frac{9R_{ij}}{32\sigma}\right)\mathbf{I} + \frac{3\mathbf{R}_{ij}\mathbf{R}_{ij}}{32\sigma R_{ij}}\right) \tag{9.11}$$

whereas the Oseen tensor (Eq. 9.9) was extended by García de la Torre and Bloomfield (1977) to account for *non-overlapping* beads of *differing* radii:

$$\mathbf{T}_{ij} = \frac{1}{8\pi\eta_0 R_{ij}}\left(\mathbf{I} + \frac{\mathbf{R}_{ij}\mathbf{R}_{ij}}{R_{ij}^2} + \frac{\sigma_i^2 + \sigma_j^2}{R_{ij}^2}\left(\frac{\mathbf{I}}{3} - \frac{\mathbf{R}_{ij}\mathbf{R}_{ij}}{R_{ij}^2}\right)\right) \tag{9.12}$$

There remains no tensor to describe the hydrodynamic interaction between *overlapping* beads of *differing* radii. There have been successive incremental developments of the theory underlying and formulation of the hydrodynamic interaction tensor. One notable correction to it is known as the "volume correction" (García de la Torre and Rodes 1983) which overcomes a deficiency that becomes obvious when hydrodynamic bead models are dominated by a small number of beads whose volume is comparable to that of the entire macromolecule. The correction term (important in the computation of $[\eta]$ and rotational diffusion coefficient (D_r), only) is proportional to the volume of the constituent beads. Regardless of any of the corrections mentioned here, bead modelling hydrodynamic interaction tensors usually assume stick boundary conditions and are approximations to what is an infinite series of powers in the inverse of R_{ij}.

The Oseen and Burgers tensor (Eq. 9.9), however, gives the exact hydrodynamic interaction between two points on a molecular surface and is the starting point for the computation of HARPs by boundary element (BE) modelling (Aragon, Chap. 12) under stick boundary conditions.

9.3 Current Approaches

9.3.1 Rigid Body Modelling

The HYDRO suite of programs (see García de la Torre, Chap. 11) for the computation of HARPs follows on from the pioneering work by García de la Torre and colleagues from the late 1970s onwards (see, e.g., García de la Torre and Bloomfield (1981)). The first in the suite was HYDRO (García de la Torre et al. 1994), the forerunner of the currently used version HYDRO++ (García de la Torre et al. 2007), a general-purpose program that enables the calculation of HARPs for rigid macromolecules and colloidal particles that can be represented as bead models. Hydrodynamic parameters are computed by HYDRO++ by solving the equation for frictional drag with hydrodynamic interaction (Eq. 9.8) as described above for any particle that can be represented as a three-dimensional bead array. The user generates and supplies to HYDRO++ the Cartesian coordinates and radii of the composite beads which must include some volume to account for hydrodynamic hydration (see below for more on hydration and section 5 of Byron (2008) for a very full consideration of the topic).

Bead model coordinates are easily derived when (1) the model is based on atomic coordinates from a crystal or NMR structure (in which case the use of HYDROPRO ((Ortega et al. 2011b), below), US-SOMO ((Brookes et al. 2010a, b), below) or BEST ((Aragon 2004; 2011), below) is more appropriate) or (2) electron microscopy density maps (when HYDROMIC ((García de la Torre et al. 2001), below) can be used) or (3) the particle can be reliably represented by a geometric shape which can, in turn, be defined by an equation and populated with spheres by

HYDROPIX ((García de la Torre 2001a), below) or (4) *AtoB* (Byron 1997) is used to construct a bead model *de novo* (see, e.g., Byron (2008)).

US-SOMO (Brookes and Rocco, Chap. 10) offers a choice of utilising the same hydrodynamic interaction tensors as HYDRO++ (as originally implemented in the BEAMS suite (Spotorno et al. 1997)) or the alternative *Zeno* (Kang et al. 2004) method with two alternative methods for bead model construction. The so far more frequently utilised of these is the *SoMo* algorithm (Rai et al. 2005) that generates medium-resolution bead models from the atomic coordinates of biomacromolecules by placing a bead of volume equal to the sum of the constituent atom volumes at the centre of mass of the main-chain segment of each residue (or equivalent) and a second bead at a defined position for the side-chain segment depending on its chemical characteristic. The resultant models comprise about one-quarter the number of atoms in the original Protein Data Bank (PDB, (Berman et al. 2000)) file. The volumes of the beads are increased in order to include water of hydration for particular residues according to the data of Kuntz and Kauzmann (1974) for proteins (and other data for different kinds of biomacromolecules; see Brookes and Rocco, Chap. 10). Overlaps between beads are removed in either a hierarchical or a synchronous process that tries to maintain the topography of the original outer surface and so maximises the reliability of the subsequently computed HARPs (Brookes et al. 2010a, b; Rai et al. 2005). Because *SoMo* also determines which beads of the resultant model are solvent exposed (and thus contribute to the frictional interaction with solvent), buried beads can be excluded from core hydrodynamic calculations, increasing the size limit and speed with which hydrodynamic computations can be completed. US-SOMO also accepts as input models of the type generated from SAXS or SANS data by the *ab initio* dummy atom or residue modelling programs DAMMIF (Franke and Svergun 2009) and GASBOR (Svergun et al. 2001). This is particularly useful in applying, e.g., s as a restraint on *ab initio* modelling: if s and/or D_t computed for a (hydrated) dummy atom/residue model disagrees with the experimentally determined value, the model is likely to be incorrect.

Less frequently utilised is the *AtoB* (Byron 1997) algorithm that conceptually superimposes a three-dimensional grid of user-defined resolution onto the molecular structure and places one bead at the centre (or centre of mass) of each cubic element of the grid with a volume corresponding to the atoms contained within that cube. *AtoB* is useful for the construction of (appropriately hydrated) bead models for very large molecular complexes in order to economise on CPU time in subsequent hydrodynamics computations – especially if many conformations of the molecule are to be assessed after, e.g., discrete molecular dynamics (DMD) calculations (see below and Brookes and Rocco, Chap. 10).

Probably the most utilised of the HYDRO suite of programs is HYDROPRO (Ortega et al. 2011b) which (in one mode) constructs shell models, composed of very small beads, from atomic coordinates (from PDB files) and computes HARPs for shell models of decreasing sphere size (and increasing sphere number), extrapolating the resultant values to the case of infinitely small spheres to generate the finally reported values for subsequent comparison with, e.g., experimentally determined parameters. The primary hydrodynamic model on which the shell model is built is characterised by an adjustable parameter, the radius of its constituent beads (the atomic element radius (AER)). The choice of this radius was important in providing a basis for empirically adjusting the anhydrous HARPs computed by HYDROPRO to values that agreed with those experimentally observed. The currently recommended value for globular macromolecules is 2.9 Å. This adjustment is historically and usually referred to as hydration, but, as pointed out by Halle and Davidovic (2003) and expounded at length by Aragon in Sect. 12.3, the residence time of water molecules at the surface of a protein is of the order of 50 ps, regardless of the chemical nature of the residue side chain. There is, thus, unlikely to be a layer of water that associates or moves with a sedimenting (or tumbling) protein. And the increase in experimental f (or decrease in s) compared with that computed for a hydrodynamic model is likely to arise instead from a difference in the viscosity of the first solvation layer, compared with the bulk solvent. This in turn is the result of the rough and dynamic nature of the protein surface where the side-chain atoms "jostle" water molecules in this solvation layer. It is a fortunate coincidence that assuming a number of "bound" waters nearly exactly compensates for this local viscosity change effect (Halle and Davidovic 2003). Interestingly, the value of 2.9 Å, currently recommended as the AER in HYDROPRO, is equal to the sum of a typical protein constituent atom van der Waals radius (see, e.g., Tsai et al. (1999)) and 1.1 Å, which happens to be the thickness of the "hydration layer" utilised by BEST (see next paragraph).

While surface modelling by HYDROPRO with default settings that limit the maximum number of shell beads used offers a computational time advantage over programs that convert the entire atomic resolution structure to beads (e.g. US-SOMO, albeit recognising the exclusion from *SoMo* models of non-surface beads that do not contribute to the frictional interaction with the solvent), this advantage is lost in the need to repeat the HARP computations for a series of surfaces comprising increasingly small beads prior to the extrapolation to zero bead size. For macromolecules that can be represented in *SoMo* or *AtoB* with up to 2000 exposed beads (the default maximum number of shell beads in HYDROPRO), HYDROPRO is slower than *SoMo* or *AtoB* followed by supermatrix inversion computation or *Zeno* computation, although HYDROPRO can be operated in a "one-bead-per-residue" mode which is extremely fast for moderately sized proteins for which this number of beads is not too large. For extremely large models, the computing time in US-SOMO can be reduced by using *AtoB* with a suitably large grid size to decrease the number of beads comprising a given model.

BEST (BE modelling under stick boundary conditions) (Aragon 2004, 2011) is conceptually similar to HYDROPRO in that the surface of the macromolecule is discretised, in order to facilitate the solution of the integral form of Eq. (9.8), not by dividing its volume into beads but instead by covering it with a patchwork of N very small triangles. From this, it computes HARPs, for surfaces comprising triangles of increasing number and decreasing size, and extrapolates the HARPs to the case of infinitely small triangles in order to obtain values for comparison with experimentally determined counterparts. Any differences in approaches to discretisation used by BE and the bead/shell-model concept implemented in HYDROPRO and related programs should largely vanish upon extrapolation to zero element size. In order to match computed HARPs with experimental values, BEST applies a default 1.1 Å uniform layer of conceptual hydration (see Aragon, Sect. 12.3) for globular monomeric proteins. This 1.1 Å does not reflect the actual thickness of the solvation layer that is perturbed by the residue side chains but is simply the magnitude required to adjust the HARPs computed with stick boundary conditions to a level that agrees with a set of experimentally determined values. Some multimeric proteins may require a higher level of "hydration" (for reasons proposed by Aragon, Chap. 12), and HARPs for small rigid molecules, whose size is comparable to that of the solvent molecules, are best computed with slip boundary conditions since their atoms perturb the solvent very little and it is free to simply "slip" past. An important difference between BE and bead modelling is that BE modelling requires no approximations to account for bead overlaps or the "volume correction". Therefore, computed HARPs are very precise. But BEST is computationally more intensive, and, while US-SOMO offers an interface to BEST, it currently does this only via cluster access. However, a Windows OS 64-bit command line version of BEST that runs on a single fast processor with 8 GB RAM is available separately (see Aragon, Chap. 12).

In HYDRONMR (García de la Torre et al. 2000), the rotational diffusion tensor ($\mathbf{D}_{rr}$) and the coordinates of the centre of diffusion are combined with the atomic coordinates of the experimental particle to compute the rotational correlation time (τ_c) and the NMR relaxation times (T_1 and T_2) for each residue. As for HYDROPRO (Ortega et al. 2011b), calculations in HYDRONMR are based on a shell model and the assumption is made that relaxation stems only from the modulation of dipolar couplings and chemical shift anisotropy by global tumbling. Bernadó and colleagues (2002) noticed that if the AER was optimised to maximise the agreement between calculated and experimental T_1/T_2 ratios, it became diagnostic of "problems" with the molecular system: larger values indicative of oligomerisation or aggregation and smaller values stemming from models that do not adequately describe solution molecular conformation.

In the absence of atomic resolution coordinates, HYDROMIC (García de la Torre et al. 2001) can construct bead models from 3D reconstruction (e.g. SPIDER or MRC) files generated from cryo-electron microscopy data. It assigns constituent voxels to the particle according to a selected threshold and calculates their Cartesian coordinates. The voxels are then converted to beads, yielding a primary hydrodynamic model for subsequent shell modelling for the computation of hydrodynamic and related parameters. If even cryo-EM data are lacking for a system of interest, HYDROSUB (García de la Torre and Carrasco 2002) can be used to generate shell models of ellipsoids and/or cylinders from which models based on user-defined parameters such as subunit dimensions and coordinates of subunit centres of mass and polar angles that define the orientation of the major symmetry axis can be composed and for which solution parameters can be computed. Last in this sequence of decreasing model "resolution" is HYDROPIX (García de la Torre 2001a) wherein the shell model, for which solution parameters are computed, is generated from a solid geometrical shape which is constructed according to a user-supplied formula by the ancillary program MAKEPIXB. In its original inception, *AtoB* could also be used to construct *de novo* a bead model to represent *any* three-dimensional shape via a combination of geometric operations (including, e.g., add a new bead; delete a bead; move a bead or a subset of beads by a defined distance in x-, y- and/or z-space; rotate a bead or a subset of beads about a defined origin or axis; create a circular array of defined radius of touching beads; expand a bead or a subset of beads). This functionality will shortly be reintroduced to *AtoB* together with an alternative "drag-and-drop" tool in the GUI (E. Brookes, personal communication). The coordinates of the finalised model can then be introduced to US-SOMO or HYDRO++ for hydrodynamic computation.

A number of different conformations of a given bead model can be evaluated in a single run of the program MULTIHYDRO (García de la Torre et al. 2005) which produces the conformers and ports them to whichever of the HYDRO suite of programs is to be used to compute the solution parameters. The results are then evaluated by the program HYDFIT (Ortega et al. 2011a) which, like an earlier program *Rayuela* by Nöllmann and colleagues (2004, 2005), searches for the best fit structure by comparing the computed solution parameter landscape with the experimentally determined values. A similar tool (model classifier; see Brookes and Rocco, Sect. 10.4) is incorporated into US-SOMO for selection of the best-fitting models when HARPs for, e.g., a range of conformations have been computed.

The *Zeno* (Kang et al. 2004) method can be used to compute f, electrostatic capacity, $[\eta]$, intrinsic conductivity and electrical polarisability of arbitrarily shaped objects. The electrostatic analogy used by *Zeno* to compute HARPs does not generate tensor values of translational diffusion and is an approximate method but a good one. From f, D_t can be directly computed utilising the Stokes-Einstein relation ($D_t = kT/f$). *Zeno* encloses the test object (i.e. a macromolecule, presented to the program in its properly hydrated form, e.g., as a *SoMo* or *AtoB* model) in a sphere from whose internal surface it launches a series of random walks which eventually (after a number of steps) either reach the molecular surface or return to the sphere surface at which point the walk is either terminated or restarted. Computed parameters are determined from the fraction of random walk trajectories that reach the molecule surface. The procedure computes the electrostatic capacity and electrostatic polarisability of a perfect conductor having the same size and shape as the model. From the electrostatic capacity, f can be computed and has been shown to be accurate within 1 %. From the electrostatic polarisability, $[\eta]$ can be computed to within 2–3 % (Mansfield and Douglas 2008). The method can be used on bead models or atomic structures defined as bead models (e.g. utilising van der Waals radii). Importantly, the bead models may contain overlaps, and the individual beads can be arbitrarily sized, allowing high-resolution structures to be processed. Additionally, the required computation time scales *linearly* with the number of random walk trajectories, the number of beads (N) or the molecular volume (Kang et al. 2004), as opposed to cubically (N^3) as in methods solving the system of equations for frictional drag with hydrodynamic interaction (e.g. the HYDRO programs, US-SOMO or BEST) making the *Zeno* computation of f

for high-resolution bead models relatively fast and feasible. The *Zeno* method is available in US-SOMO (see Brookes and Rocco, Sect. 10.3).

9.3.2 Flexible Body Modelling

The methods so far summarised compute solution parameters for rigid models. What about flexible molecules? There are (at least) four approaches to calculating HARPs for flexible systems. In one, Monte Carlo rigid body (MCRB) modelling (with the program MONTEHYDRO (García de la Torre et al. 2005)), the flexibility is approximated by using an MC method to generate many different (random) conformations of models comprising beads joined by conceptually flexible connectors (characterised by potentials), and solution parameters for the MC sample are computed. This sample comprises models whose averaged solution properties agree with the experimentally determined values. Because the result is an average over conformations, the MCRB approach is appropriate for the evaluation of overall properties like s, D_t, $[\eta]$, R_g and scattering form factors, but it does not model the *internal* dynamics of the system.

An alternative exploration of conformational space is afforded by the discrete molecular dynamics (DMD) approach offered within US-SOMO (see Brookes and Rocco, Sect. 10.5) that, like the MCRB approach, also does not model the internal dynamics of the system but instead allows the generation of numerous conformers of a starting model, delivered by changing the conformation of model segments presented to the program as having the potential to be flexible. The starting model is a PDB file, for which residues that are to remain static (i.e. are non-flexible) are identified. Control parameters for the DMD simulation that generates different conformations of the flexible regions of the model include the Andersen thermostat temperatures, durations, time intervals and number of models to be generated for the relax and run phases of the simulation. The recommended Andersen thermostat temperature is 0.5 kcal/mol/kB where most proteins will not unfold or deviate much from native state. This temperature corresponds to 251 K, although the temperature in these simulations generally does not correspond to the physical temperature at which, e.g., hydrodynamics measurements are made. Since water is not explicitly defined in DMD simulations, the system will not freeze, and there will be sufficient sampling of conformational dynamics near the native states. At higher thermostat temperatures, fluctuations will have larger amplitudes, and the protein might unfold. This is useful in the search for atomic resolution models that are consistent with HARPs for partially disordered proteins. A starting (perhaps fully folded model or actual structure) can be presented to the DMD interface and successively more unfolded versions of the putative partially disordered region generated. These can then be evaluated by computation of their HARPs.

BEST has also been used to model HARPs for flexible macromolecules, again by using carefully selected MD force fields, volumes, temperatures, pressures, salt concentrations and time frames. The precise simulations performed were fully atomistic with discrete water molecules using the AMBER (Perlman et al. 1995) MD package. HARPs averaged over typically 3000 structures captured from an MD simulation have been shown to provide excellent agreement with experimental data for the case of a monoclonal antibody (see Aragon, Sect. 12.5.3 and references cited therein). MONTEHYDRO, US-SOMO-DMD and BEST-MD share a common approach to model and HARP generation: they generate a set of conformations from a starting model and compute HARPs for these, reporting averaged HARPS according to user preference.

The fourth method considered here is a Brownian dynamics (BD) approach delivered by the program SIMUFLEX (García de la Torre et al. 2009) which directly simulates the internal dynamics of a bead and connector model by generating the macromolecular trajectory using Brownian dynamics simulation (with full consideration of hydrodynamic interactions) and then analysing this trajectory to extract the macromolecular HARPs. Unlike the MD simulations used in combination with BEST, the more approximate BD method of SIMUFLEX is not fully atomistic, nor does it use discrete water molecules. It has the advantage, however, of much shorter computation times.

All four approaches are relatively new in the field of hydrodynamic modelling and so examples of their use are few. There is an excellent tutorial mini-review on MONTEHYDRO and also SIMUFLEX that serves as a good starting point (García de la Torre et al. 2010), and the DMD and BEST tools within US-SOMO are explained in the on-line manual.

9.4 Pros and Cons of Current Approaches

Each of the programs previewed in this introduction and described much more fully in the following three chapters has their strengths and weaknesses. There is no such thing as the perfect hydrodynamic modelling program since different molecular systems are better suited by different programs.

For instance, HYDRO++ (García de la Torre et al. 2007), the descendant of the first freely available hydrodynamic bead modelling program HYDRO, appears not to be used as much now as it was prior to the release of programs such as HYDROPRO and US-SOMO for modelling directly from PDB files. However, there remain systems that can best (or only) be described by very simple bead models, for example, dumbbell-shaped polyelectrolyte brush particles (Hoffmann et al. 2008). That said, the same authors (Hoffmann et al. 2009) later chose to replace very simple (four-)bead models for colloidal clusters with shell models generated by and evaluated by HYDROPRO, taking advantage of the concomitant increase in precision of the HARP computations and the more reliable procedure for mimicking hydration.

Of all the currently available approaches to hydrodynamic modelling, HYDROPRO is the least demanding and will accept any properly formatted PDB file as an input, while US-SOMO is more demanding, needing an internal "coding" of each residue comprising a biomacromolecule to properly translate it into a bead model. While codes are provided for a large number of commonly encountered residues for proteins, nucleic acids, carbohydrates and some lipids, detergents and prosthetic groups, the list is far from being exhaustive, and coding for new residues can be demanding. Furthermore, based on this coding, US-SOMO will check any given input model for missing atoms in coded residues or breaks in the sequence, warning the user and requiring remedial action. However, approximate methods are available in US-SOMO to cope with either non-coded residues or missing atoms within coded residues. This level of scrutiny is not exerted by HYDROPRO, which can therefore generate HARPs that lack precision if the user has not realised that the input PDB is incomplete in some way compared with the experimental macromolecule.

Both HYDROPRO and BEST require extrapolations of parameters computed for shell models with spheres of decreasing radius (HYDROPRO) or for surfaces with triangles of decreasing area (BEST). On occasion, the computed parameters can lie surprisingly far from the line of best fit, and the extrapolation can require the application of statistical treatments in order to gain validity (see, e.g., Brookes and Rocco, Chap. 10). For this reason, it is important to visually inspect the extrapolation prior to finalisation of the result; this is implemented in the US-SOMO BEST interface (see Brookes and Rocco, Sect. 10.6).

Both HYDROPRO and BEST treat hydration water as a uniform layer, while US-SOMO uses a differential hydration scheme that reflects in some way the chemical and hydrodynamic interaction between the macromolecular surface and the solvent. While the uniform layer of HYDROPRO and BEST is a *mimic* of hydration, the layer thickness is a parameter that can be adjusted to render the computed HARPs comparable with the experimentally determined values for a particular category of molecule (e.g. in the case of BEST: monomeric compared with multimeric). This adjustment is not required for the hydration apportioned by US-SOMO where it is as effective for monomeric or multimeric models. That said, if a correct "hydration" level could be determined for BEST modelling, it should provide unparalleled precision thereafter. Hydration in the context of hydrodynamic modelling is currently a topic of intense discussion amongst the main protagonists, and those of us who benefit from their efforts can expect some form of agreement to be reached in the near future.

BEST definitely offers the most precise computation of D_r and $[\eta]$, since it does not require the "volume correction" that plagues bead modelling (see Sect. 9.2 above and García de la Torre and Carrasco (1998)). But BEST is computationally more intensive and therefore currently less practical for the evaluation of many conformations or very large molecules. In this instance, *Zeno* becomes attractive since the time taken for its computations depends linearly in the number of beads used in an input model.

Interestingly, a recent comparison of the hydrodynamic modelling methods available in US-SOMO and HYDROPRO shows that *SoMo* slightly overestimates s and D_t, whereas BEST slightly underestimates them and HYDROPRO even more so (Rocco and Byron 2015). But *AtoB* with a 5 Å grid size performed better, giving computed values that more closely matched the experimentally determined parameters. This notwithstanding, the overall best performance was delivered by bead models generated with *SoMo without* the removal of bead overlaps, for which HARPs were computed using *Zeno*.

Whereas MONTEHYDRO, US-SOMO DMD and BEST-MD provide an exploration of conformational space and essentially deal in conformational averages, flexibility can also be modelled for an individual bead-and-connector model with SIMUFLEX (García de la Torre et al. 2009) using its explicit Brownian dynamics engine.

9.5 Concluding Comments and Outlook

Some questions remain: what is the sensitivity of solution parameters to the conformations on which they report? How much can an individual parameter tell us about molecular conformation? A rapid answer to this is that within limits a single parameter can be used to support or discard a potential molecular model. But a far more satisfactory outcome is arrived at when a model has to reproduce a combination of solution parameters (e.g. MULTIHYDRO or the model classifier in US-SOMO).

How much detail in a model is reasonable or necessary? All of the programs described in this chapter generate s in the form of the ratio (or inverse thereof) of f/η_0. In Sect. 9.2, the dependence of s on other experimental parameters (such as η_0, M, T, $\bar{v}$ and ρ) is explored briefly. Each of these parameters has an error associated with it that will contribute to uncertainty in s. The same will apply to other HARPs. Therefore, any comparison between the various computational procedures in terms of the numerical difference in their predictions should be done with this accumulated uncertainty in mind (see García de la Torre (2001b) for a thorough analysis of error propagation in HARPs). In Chap. 11 (García de la Torre), an important conclusion is reported: the difference in precision in the computation of HARPs for bead models generated from atomic coordinates and those generated on the basis of one bead per residue is comparable with the experimental errors. And in Chap. 12 (Aragon), it is reported that the precision in transport coefficients computed with BE methods (compared with exact results for shapes for which exact or very precise values are available (e.g. ellipsoids of revolution, toroids, etc.)) is 1 %, whereas it is 2 % for rotational parameters and $[\eta]$ (comparable with the precision reported for *Zeno*). This precision is better than the accuracy with which most HARPs can be experimentally determined.

In the field of small-angle X-ray and neutron scattering (SAXS, SANS), a publicly accessible repository (called SASBDB, http://www.sasbdb.org/) of experimental data and low-resolution models derived therefrom has been established. This is in response to requests by the SAXS/SANS community and a recommendation made by the wwPDB Small-Angle Scattering Task Force (Trewhella et al. 2013). Given the complementary nature of SAXS/SANS and hydrodynamics, the AUC community should consider whether it could usefully contribute to a project such as SASBDB so that models consistent with hydrodynamic calculation can be made available to the wider community for further evaluation.

Acknowledgements Thanks to Sergio Aragon, Emre Brookes, José García de la Torre and Mattia Rocco for helpful discussions, preview of their chapter sections during the writing of this introduction and comments on this section prior to its finalisation.

References

Aragon S (2004) A precise boundary element method for macromolecular transport properties. J Comp Chem 25(9):1191–1205

Aragon SR (2011) Recent advances in macromolecular hydrodynamic modeling. Methods 54(1):101–114

Berman HM, Westbrook J, Feng Z, Gilliland G, Bhat TN, Weissig H, Shindyalov IN, Bourne PE (2000) The protein data bank. Nucleic Acids Res 28(1):235–242

Bernadó P, Åkerud T, García de la Torre J, Akke M, Pons M (2002) Combined use of NMR relaxation measurements and hydrodynamic calculations to study protein association. Evidence for tetramers of low molecular weight protein tyrosine phosphatase in solution. J Am Chem Soc 125:916–923

Boutet S, Lomb L, Williams GJ, Barends TRM, Aquila A, Doak RB, Weierstall U, DePonte DP, Steinbrener J, Shoeman RL, Messerschmidt M, Barty A, White TA, Kassemeyer S, Kirian RA, Seibert MM, Montanez PA, Kenney C, Herbst R, Hart P, Pines J, Haller G, Gruner SM, Philipp HT, Tate MW, Hromalik M, Koerner LJ, van Bakel N, Morse J, Ghonsalves W, Arnlund D, Bogan MJ, Caleman C, Fromme R, Hampton CY, Hunter MS, Johansson LC, Katona G, Kupitz C, Liang M, Martin AV, Nass K, Redecke L, Stellato F, Timneanu N, Wang D, Zatsepin NA, Schafer D, Defever J, Neutze R, Fromme P, Spence JCH, Chapman HN, Schlichting I (2012) High-resolution protein structure determination by serial femtosecond crystallography. Science 337(6092):362–364

Brookes E, Demeler B, Rocco M (2010a) Developments in the US-SOMO bead modeling suite: new features in the direct residue-to-bead method, improved grid routines, and influence of accessible surface area screening. Macromol Biosci 10(7):746–753

Brookes E, Demeler B, Rosano C, Rocco M (2010b) The implementation of SOMO (SOlution MOdeller) in the UltraScan analytical ultracentrifugation data analysis suite: enhanced capabilities allow the reliable hydrodynamic modeling of virtually any kind of biomacromolecule. Eur Biophys J 39(3):423–435

Burgers JM (1938) On the motion of small particles of elongated form, suspended in a viscous liquid. In: Burgers JM, Jaeger FM, Houwink R, Van Nieuwenberg CJ, Saal RNJ (eds) Second report on viscosity and plasticity. Nordemann, Amsterdam, p 209

Byron O (1997) Construction of hydrodynamic bead models from high-resolution X-ray crystallographic or nuclear magnetic resonance data. Biophys J 72(1):408–415

Byron O (2008) Hydrodynamic modeling: the solution conformation of macromolecules and their complexes. In: Correia JJ, Detrich HW (eds) Methods in cell biology, vol 84. Elsevier, San Diego, pp 327–373

Ding F, Dokholyan NV (2006) Emergence of protein fold families through rational design. PLoS Comput Biol 2(7):e85

Dokholyan NV, Buldyrev SV, Stanley HE, Shakhnovich EI (1998) Discrete molecular dynamics studies of the folding of a protein-like model. Fold Des 3(6):577–587

Franke D, Svergun DI (2009) DAMMIF, a program for rapid ab-initio shape determination in small-angle scattering. J App Cryst 42(2):342–346

García de la Torre J (2001a) Building hydrodynamic bead–shell models for rigid bioparticles of arbitrary shape. Biophys Chem 94(3):265–274
García de la Torre J (2001b) Hydration from hydrodynamics. General considerations and applications of bead modelling to globular proteins. Biophys Chem 93(2–3):159–170
García de la Torre J, Bloomfield VA (1977) Hydrodynamic properties of macromolecular complexes. I Translation. Biopolymers 16(8):1747–1763
García de la Torre J, Bloomfield VA (1981) Hydrodynamic properties of complex rigid biological macromolecules: theory and applications. Q Rev Biophys 14(1):81–139
García de la Torre J, Carrasco B (1998) Intrinsic viscosity and rotational diffusion of bead models for rigid macromolecules and bioparticles. Eur Biophys J 27:549–557
García de la Torre J, Carrasco B (2002) Hydrodynamic properties of rigid macromolecules composed of ellipsoidal and cylindrical subunits. Biopolymers 63:163–167
García de la Torre J, Rodes V (1983) Effects from bead size and hydrodynamic interactions on the translational and rotational coefficients of macromolecular bead models. J Chem Phys 79(5):2454–2460
García de la Torre J, Navarro S, Lopez Martinez MC, Díaz FG, Lopez Cascales JJ (1994) HYDRO: a computer program for the prediction of hydrodynamic properties of macromolecules. Biophys J 67(August):530–531
García de la Torre J, Huertas ML, Carrasco B (2000) HYDRONMR: prediction of NMR relaxation of globular proteins from atomic-level structures and hydrodynamic calculations. J Magn Reson 147:138–146
García de la Torre J, Llorca O, Carrascosa JL, Valpuesta JM (2001) HYDROMIC: prediction of hydrodynamic properties of rigid macromolecular structures obtained from electron microscopy images. Eur Biophys J 30(6):457–462
García de la Torre J, Ortega A, Pérez Sánchez HE, Hernández Cifre JG (2005) MULTIHYDRO and MONTEHYDRO: conformational search and Monte Carlo calculation of solution properties of rigid or flexible bead models. Biophys Chem 116(2):121–128
García de la Torre J, del Rio G, Ortega A (2007) Improved calculation of rotational diffusion and intrinsic viscosity of bead models for macromolecules and nanoparticles. J Phys Chem B 111(5):955–961
García de la Torre J, Hernandez Cifre JG, Ortega A, Rodríguez Schmidt R, Fernandes MX, Pérez Sánchez HE, Pamies R (2009) SIMUFLEX: Algorithms and tools for simulation of the conformation and dynamics of flexible molecules and nanoparticles in dilute solution. J Chem Theory Comput 5(10):2606–2618
García de la Torre J, Ortega Á, Amorós D, Rodriguez Schmidt R, Hernández Cifre JG (2010) Methods and tools for the prediction of hydrodynamic coefficients and other solution properties of flexible macromolecules in solution. A tutorial minireview. Macromol Biosci 10(7):721–730
Halle B, Davidovic M (2003) Biomolecular hydration: from water dynamics to hydrodynamics. Proc Natl Acad Sci U S A 100(21):12135–12140
Harding SE (1982) A computer program for evaluating the hydrodynamic parameters of a macromolecule in solution for any given value of its axial dimensions. Comput Biol Med 12(2):75–80
Hoffmann M, Lu Y, Schrinner M, Ballauff M, Harnau L (2008) Dumbbell-shaped polyelectrolyte brushes studied by depolarized dynamic light scattering. J Phys Chem B 112(47):14843–14850
Hoffmann M, Wagner CS, Harnau L, Wittemann A (2009) 3D Brownian diffusion of submicron-sized particle clusters. ACS Nano 3(10):3326–3334
Kang EH, Mansfield ML, Douglas JF (2004) Numerical path integration technique for the calculation of transport properties of proteins. Phys Rev E 69(3):031918
Kirkwood JG (1949) The statistical mechanical theory of irreversible processes in solutions of macromolecules (visco-elastic behaviour). J Chem Phys 68(7):649–660
Kirkwood JG (1954) The general theory of irreversible processes in solutions of macromolecules. J Polym Sci 12(1):1–14

Kuntz ID, Kauzmann W (1974) Hydration of proteins and polypeptides. In: Anfinsen CB, Edsall JT, Richards FM (eds) Advances in protein chemistry, vol 28. Academic, New York, pp 239–345

Laue TM, Shah DD, Ridgeway TM, Pelletier SL (1992) Computer-aided interpretation of analytical sedimentation data for proteins. In: Harding SE, Rowe AJ, Horton JC (eds) Analytical ultracentrifugation in biochemistry and polymer science. Royal Society of Chemistry, Cambridge, pp 90–125

Mansfield ML, Douglas JF (2008) Improved path integration method for estimating the intrinsic viscosity of arbitrarily shaped particles. Phys Rev E 78(4):046712

Nakasako M, Fujisawa T, Adachi S-i, Kudo T, Higuchi S (2001) Large-scale domain movements and hydration structure changes in the active-site cleft of unligated glutamate dehydrogenase from *Thermococcus profundus* studied by cryogenic X-ray crystal structure analysis and small-angle X-ray scattering. Biochemistry 40(10):3069–3079

Nöllmann M, Stark WM, Byron O (2004) Low-resolution reconstruction of a synthetic DNA Holliday junction. Biophys J 86(5):3060–3069

Nöllmann M, Stark W, Byron O (2005) A global multi-technique approach to study low-resolution solution structures. J App Cryst 38:874–887

Ortega A, Amorós D, García de la Torre J (2011a) Global fit and structure optimization of flexible and rigid macromolecules and nanoparticles from analytical ultracentrifugation and other dilute solution properties. Methods 54(1):115–123

Ortega A, Amorós D, García de la Torre J (2011b) Prediction of hydrodynamic and other solution properties of rigid proteins from atomic- and residue-level models. Biophys J 101(4):892–898

Oseen CW (1927) Neuere Methoden und Ergebnisse in der Hydrodynamik. In: Hilb E (ed) Mathematik und ihre Anwendungen in Monographien und Lehrbuchern. Academisches Verlagsgellschaft, Leipzig, p 337

Perlman D, Case D, Caldwell J, Ross W, Cheathamm T III, Debold S, Ferguson D, Seibel G, Kollman G (1995) AMBER, a package of computer programs for applying molecular mechanics, normal mode analysis, molecular dynamics and free energy calculations to simulate the structural and energetic properties of molecules. Comput Phys Commun 91:1–41

Rai N, Nöllmann M, Spotorno B, Tassara G, Byron O, Rocco M (2005) *SOMO* (*SO*lution *MO*deler): differences between x-ray and NMR-derived bead models suggest a role for side chain flexibility in protein hydrodynamics. Structure 13(5):723–734

Rocco M, Byron O (2015) Computing translational diffusion and sedimentation coefficients: an evaluation of experimental data and programs. Eur Biophys J 44:417–431

Rotne J, Prager S (1969) Variational treatment of hydrodynamic interaction in polymers. J Chem Phys 50(11):4831–4837

Smolle M, Prior AE, Brown AE, Cooper A, Byron O, Lindsay JG (2006) A new level of architectural complexity in the human pyruvate dehydrogenase complex. J Biol Chem 281(28):19772–19780

Spotorno B, Piccinini L, Tassara G, Ruggiero C, Nardini M, Molina F, Rocco M (1997) BEAMS (BEAds Modelling System): a set of computer programs for the generation, the visualisation and the computation of the hydrodynamic and conformational properties of bead models of proteins. Eur Biophys J 25(5/6):373–384

Svergun DI, Petoukhov MV, Koch MHJ (2001) Determination of domain structure of proteins from X-ray solution scattering. Biophys J 80(6):2946–2953

Trewhella J, Carlson VAP, Curtis EH, Heidorn DB (1988) Differences in the solution structures of oxidised and reduced cytochrome C measured by small-angle X-ray scattering. Biochemistry 27(4):1121–1125

Trewhella J, Hendrickson WA, Kleywegt GJ, Sali A, Sato M, Schwede T, Svergun DI, Tainer JA, Westbrook J, Berman HM (2013) Report of the wwPDB Small-Angle Scattering Task Force: data requirements for biomolecular modeling and the PDB. Structure 21(6):875–881

Tsai J, Taylor R, Chothia C, Gerstein M (1999) The packing density in proteins: standard radii and volumes. J Mol Biol 290(1):253–266

Vestergaard B, Sanyal S, Roessle M, Mora L, Buckingham RH, Kastrup JS, Gajhede M, Svergun DI, Ehrenberg M (2005) The SAXS solution structure of RF1 differs from its crystal structure and is similar to its ribosome bound cryo-EM structure. Mol Cell 20(6):929–938

Yamakawa H (1970) Transport properties of polymer chains in dilute solution: hydrodynamic interaction. J Chem Phys 53(1):435–443

Chapter 10
Calculation of Hydrodynamic Parameters: US-SOMO

Emre Brookes and Mattia Rocco

Abstract Although it now offers a number of complementary applications for data processing and multiresolution modeling of (bio)macromolecules in solution, the UltraScan SOlution MOdeler (US-SOMO) suite of programs was initially developed for the computation of their solution properties starting from atomic-resolution structures and their comparison with experimental data. In this chapter, we will give a brief overview of the basic principles behind the hydrodynamic parameter computational methods available in US-SOMO and outline their operation.

Keywords Diffusion coefficient • Sedimentation coefficient • Intrinsic viscosity • Bead modeling • Multiresolution modeling • Solution structure

10.1 Introduction

US-SOMO (http://somo.uthscsa.edu/) was initially started as a graphical user interface (GUI) within the analytical ultracentrifugation data analysis program UltraScan (Demeler 2005) for the SOlution MOdeler (*SoMo*) method developed by the Rocco and Byron labs (Rai et al. 2005). The previously published *AtoB* grid method (Byron 1997) was also available in US-SOMO from its initial release (Brookes et al. 2010a). Both methods were developed for the computation of the hydrodynamic parameters starting from high-resolution structures of (bio)macromolecules using different bead modeling procedures, trying to avoid some of the drawbacks present in other approaches. Since then, it has grown to include other methods such as *Zeno* (http://www.stevens.edu/zeno/; Kang et al. 2004) and *BEST* (http://esmeralda.

E. Brookes
Department of Biochemistry, University of Texas Health Science Center at San Antonio, San Antonio, TX, USA
e-mail: emre@biochem.uthscsa.edu

M. Rocco (✉)
Biopolimeri e Proteomica, IRCCS AOU San Martino-IST, Istituto Nazionale per la Ricerca sul Cancro, Genova, Italy
e-mail: mattia.rocco@hsanmartino.it

S. Uchiyama et al. (eds.), *Analytical Ultracentrifugation*,
DOI 10.1007/978-4-431-55985-6_10

sfsu.edu; Aragon 2004), the former directly implemented, the latter operating on a supercompute cluster through a dedicated interface (see also Chap. 12). Small-angle scattering (SAS) data analysis and simulation modules have been subsequently added (Brookes et al. 2010b, 2013a), and discrete molecular dynamics (DMD) procedures (Ding and Dokholyan 2006; Dokholyan et al. 1998) for the expansion of conformational space when dealing with flexibility issues have been implemented, again operating on a supercompute cluster. While the overarching goal of the US-SOMO suite is to provide a full toolbox for the multiresolution modeling of (bio)macromolecules, in this chapter we will deal only with the features relating to hydrodynamic computation. Recent literature summarizing the other US-SOMO capabilities is available (Brookes et al. 2012; Rocco and Brookes 2014).

10.2 Operational Principles of the Bead Modeling Methods

Two bead modeling approaches are available in US-SOMO, *SoMo* (Rai et al. 2005) and *AtoB* (Byron 1997), with the computations carried out, in their original implementation, by solving a system of n linear equations with $3n$ unknowns using the coefficients "supermatrix" inversion (*SMI*) procedure (Brookes et al. 2010a; García de la Torre and Bloomfield 1981; Spotorno et al. 1997). The *SoMo* method is based on a direct correspondence between the structural elements of a (bio)macromolecule and the beads used to represent it. Appropriately positioned beads of different radii are used, but since in the *SMI* procedure the hydrodynamic parameters are computed using the Rotne-Prager-Yamakawa hydrodynamic interaction tensor as modified by García de la Torre and Bloomfield (1981), valid for assemblies of variable-sized beads only if they do not overlap (see below), overlaps must be removed after the initial set of beads is defined. For proteins, a distinction is made between side- and main-chain segments, each represented with a bead as the default option. Two alternatives are available for representing the latter: the main chain of each nth residue (N-CA-C-CO)$_n$ or the peptide bond between the nth and (nth + 1) residues (CA-C-CO)$_n$N$_{(n+1)}$. The second is the default option, because it reduces the chances of overlaps between the main- and side-chain beads. The initial spatial location of each bead is chosen according to the nature of the segment it represents. For the main chain (peptide bond) and for the hydrophobic and nonpolar side chains, the bead is placed at the center of mass of the atoms involved, while for polar and charged side chains, the bead is located toward the end of the side chain. Similar rules are employed for the sugar units in carbohydrates and for the sugars and bases forming the nucleotides in RNA/DNA. Prosthetic groups are likewise treated. The anhydrous volume of each bead is defined by the sum of the anhydrous atomic volumes of the atoms it represents, taken from literature analyses of crystallographic data (Perkins 1986; Tsai et al. 1999; Nadassy et al. 2001; Voss and Gerstein 2005). Alternatively, volumes can be calculated from structural models using dedicated software (*e.g.*, the 3 V Contact Volume Calculator: http://3vee.molmovdb.org/volumeCalc.php; Voss and Gerstein 2010).

The *AtoB* method relies instead on a cubic grid approach to "assign" atoms to a particular bead (Byron 1997). The initial volume of each bead is then simply calculated by summing up all the assigned atom volumes, and the position of each bead is defined either at the center of mass of its constituent atoms or the center of the cubelet. The overlap removal problem also applies to the *AtoB* method. The resolution of the model is controlled by the chosen size of the grid spacing.

A key, common aspect of the two bead modeling methods available in US-SOMO is a more realistic treatment of the water of hydration, in contrast to the uniform expansion of the model or to the addition of a uniform shell on the model surface as utilized by other approaches (*e.g.*, see Chaps. 11 and 12).

In the *SoMo* direct correspondence method, a number of water molecules are assigned to each bead, based on the theoretical, statistical hydration values determined by Kuntz and Kauzmann (1974) for each residue using NMR freezing. The volume of these water molecules is taken to be different from that of bulk water molecules, on the basis of crystallographic studies (Gerstein and Chothia 1996). Although this representation of the hydration effect as "bound" water molecules is not correct in principle, it turns out that it compensates quite well for the real physical effects involving changes in local viscosity and density at the protein/water interface (Halle and Davidovic 2003) (see also Rocco et al. (2012) and Chap. 12). This procedure can in general be applied to other types of biomacromolecules (*e.g.*, nucleic acids, carbohydrates, etc.). Operationally, the volume of the theoretically bound water molecules is then added to each corresponding bead, thereby accounting for the local variation in hydration. This approach is also implemented in the revised *AtoB* grid method available in US-SOMO, where water molecules are assigned to atoms within residues. Currently (May 2016), the waters/atoms assignment is provided for amino acid and carbohydrate residues only, but experienced users can define their own values for other residues modifying the *somo.residue* lookup table (see below).

Another innovation in both the *SoMo* and *AtoB* methods is a prescreening of the (bio)macromolecule to identify buried and exposed patches. This information is then associated with atoms/residues-representing beads, which are then labeled as being either buried or exposed. A further distinction is also made between exposed main- and side-chain segments. This information is utilized to greatly reduce the number of beads that are subsequently included in hydrodynamic computations using the *SMI* procedure, because only the beads that contribute to the surface frictional interaction with solvent are then considered.

The steps required to generate bead models in the *SoMo* and *AtoB* methods are illustrated in Fig. 10.1. In *SoMo*, the accessible surface area (ASA) is first determined, assigning each main and side chain as being either buried or exposed (the colors used refer to the nature of the placed residues; see Fig. 10.1 legend). The beads corresponding to exposed side chains are subsequently placed (A $\rightarrow$ B). The overlaps between these beads are then removed (B $\rightarrow$ C), first fusing together beads that overlap by more than a preset threshold and then proportionally reducing the bead radii either hierarchically (the couple with the largest overlap first and then the others) or synchronously (the radii of all overlapping beads are reduced

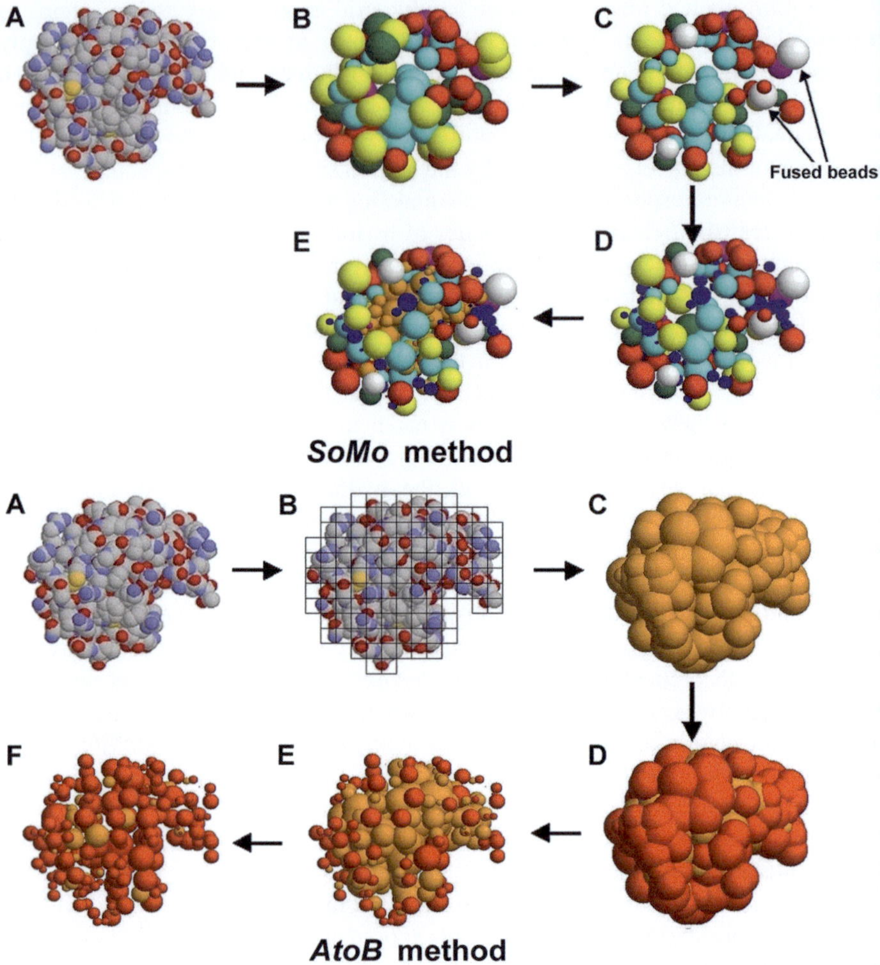

Fig. 10.1 Schematic representation of the *SoMo* (*top*) and *AtoB* (*bottom*) bead model generation methods. A test protein atomic structure is shown in space-filling mode in step A in both procedures. The monodimensional grid visualized in step B of the *AtoB* method panel is in practice a three-dimensional "cage." The steps in both procedures (A → B, etc.) are described in detail in the text. The color coding in the *SoMo* B → E steps is *blue*, main chain; *cyan*, hydrophobic; *magenta*, nonpolar; *red,* polar; *yellow*, basic; *green*, acidic; *white*, fused beads; and *orange*, buried beads. The color coding in the *AtoB* steps D → F is *orange*, buried beads, and *red*, exposed beads

by a percentage of their original radius, and the procedure is repeated until no overlaps remain). An important procedure is implemented in this step to preserve the original surface as much as possible: while their radii are reduced, the bead centers are moved outwardly along a line connecting them to the center of mass of the (bio)macromolecule by an equal amount ("outward translation," OT). In the subsequent step (C → D), the main-chain exposed beads (blue) are placed and their overlaps removed using one of the procedures described above but without the OT (this choice is made because the peptide bond segments do not usually protrude from the protein surface as most of the exposed side chains do). In the last step (D → E), buried residue beads (orange) are placed and their overlaps removed, again without OT. An ASA screen is then performed again on the final bead model, because some beads might have changed their exposed/buried status during the overlap removal procedure.

In *AtoB*, a cubic grid of a selected spacing is first placed on the original structure and atoms are "assigned" to cubes (A → B). In a single step, all beads are generated summing up the hydrated volumes of the atoms in each cube, and beads are placed according to the centering method chosen (B → C). An ASA screen is then performed (C → D; orange, buried; red, exposed). Overlaps are subsequently removed first in the exposed subset (D → E) using preferentially the synchronous procedure (default; the hierarchical procedure is also available), with OT. The same procedure is then applied to the buried subset, without OT, and the entire set is rescreened for ASA, resulting in many more beads becoming exposed (E → F).

Very recently, on the basis of an extensive investigation of the performance of the main available methods/programs used to compute the hydrodynamic parameters starting from atomic-resolution structures, it was found that the best results could be obtained by utilizing *SoMo*-type bead models *without* removing the overlaps between them and using the *Zeno* method for the computations (Rocco and Byron 2015). This approach is now directly available within US-SOMO (see Sect. 10.3).

10.3 The US-SOMO Main GUI Interface and Option Settings

10.3.1 PDB Function Area

The first button ("Select Lookup Table") allows the user to change the main reference file containing all the information necessary to properly recognize each residue and the atoms within it (the automatically uploaded default file is shown in the corresponding field) (see Fig. 10.2). The main lookup table and the other tables necessary for its construction can be edited from the top bar pull-down menu ("Lookup Tables"). The proper coding of each residue is a fundamental step in hydrodynamic bead model generation in US-SOMO (as well as for SAS computations, not dealt with here), and the tables contain the atomic radii, hydration numbers, SAXS/SANS coefficients, and the atoms to bead conversion/bead positioning rules.

Although advanced editors are available within US-SOMO (see Figs. 10.3 and 10.4), coding for atoms/residues and assignment to beads are not simple operations, as they entail knowledge of several physicochemical properties. The hybridization state of each non-H atom (see Tsai et al. 1999) and its related properties (*i.e.*,

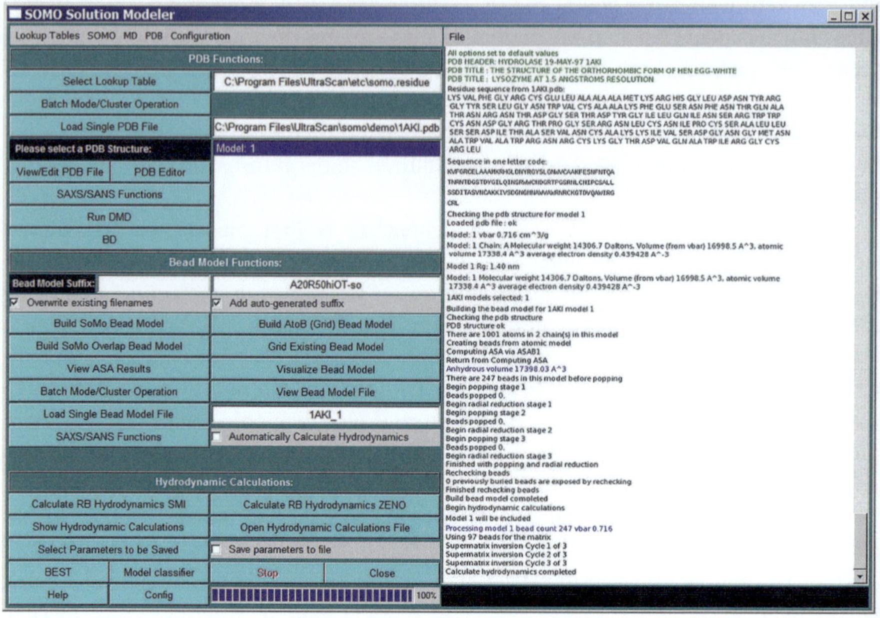

Fig. 10.2 The US-SOMO main panel GUI. The *left side* of the window is divided into three subpanels: PDB Functions, Bead Model Functions, and Hydrodynamic Calculations. The *right-side* panel reports on structure loading/verification, modeling, and calculation progress (Shown with a reduced font size are the steps in the processing of the 1AKI.pdb RNase A structure)

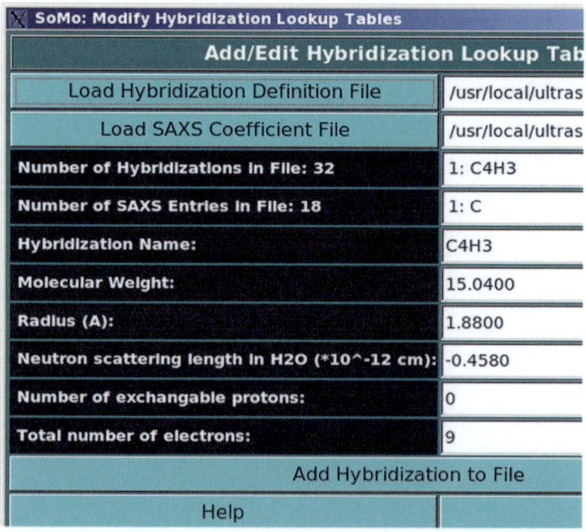

Fig. 10.3 The "Add/Edit Hybridization Lookup Table" (*left panel*) and "Add/Edit Atom Lookup Table" (*right panel*) modules of US-SOMO

Fig. 10.4 The "Add/Edit Residue Lookup Table" module of US-SOMO

molecular weight including the H atoms attached to it, radius, etc.) are defined in a first table (default, *somo.hybrid*, currently containing 42 entries; see Fig. 10.3, left panel). Since in PDB files each type of atom (*e.g.*, C, O, N) can have many different "names" (*e.g.*, C1, OG, N3), a second table is built where the atom names present in the PDB entries are linked to the proper hybridization and associated parameters (default, *somo.atom*, currently containing 629 entries; see Fig. 10.3, right panel). Both tables are connected to a third basic table containing the SAXS coefficients (default, *somo.saxs_atoms*; editor not shown). Finally, the residues making up a (bio)macromolecule are stored in the main lookup table (default, *somo.residue*, currently containing 122 residues, including all standard and some nonstandard amino acids, ribo- and deoxyribonucleosides/nucleotides, and carbohydrates, plus some lipids, detergents, and various prosthetic groups). In Fig. 10.4, the editor module for the residue lookup table is shown. A detailed description of these procedures is provided in the US-SOMO help files, accessible by pressing the "Help" button located at the bottom of each GUI module (see also the Supplementary Information of Brookes et al. (2010a, b)). It is important to emphasize that for reliable results, all atoms/residues present in the sample for which experimental data are collected must also be present in the structural model used for the hydrodynamic parameter computations. However, to avoid the complicated task of encoding new residues, skipping noncoded atoms/residues or approximate methods to represent them are provided, the latter being now the default option. A warning message will appear if noncoded atoms/residues are found in a structure, and the user can proceed with the approximate method or chose a different option. Both are controlled from the "PDB" pull-down menu in the top bar. If skipping is chosen (not recommended), the user is warned about the risks of underestimating the molecular weight (*mw*) of the model and of miscalculating its partial specific volume (*psv*; both parameters are needed to compute the sedimentation coefficient from the computed translational frictional coefficient of the bead model, and the *mw* is needed for the computation of the intrinsic viscosity [η]). If the correct *mw* and *psv* are available, the user can enter them in the appropriate US-SOMO modules (see below). This of course would not take into account the lack of friction due to the skipped residue(s). The approximate "automatic bead builder" instead at least partially compensates for it and will roughly define a single "side-chain" bead for each noncoded residue. This procedure is based on an "average" volume for each atom (with an "average" *mw* and hydration number), from which a global volume (and *mw*) is calculated. An "average" radius for each atom is also provided for the ASA routines. The bead is then placed at the center of mass of all the atoms within the noncoded residue, and an "average" *psv* is assigned to it. All these "average" values can be modified in the Miscellaneous Options panel (see below), allowing the user to tune them to the type of noncoded residue (*e.g.*, amino acid, sugar, nucleotide, etc.). As with the "skip" option, if available "true" *mw* and *psv* values should be entered anyway in the appropriate US-SOMO modules. Likewise, in the more common case of when incomplete (but coded) residues are present in the PDB file, the default option is to use an approximate method to generate and place a bead. In this case, since the residue is encoded, *mw* and *psv* are computed as for complete structures. If the missing atom(s) are not marked in the *somo.residue* table as needed to position the bead, the approximation will lead to a "normal" bead, indistinguishable from what would be obtained for a complete residue. Otherwise, the level of the approximation will depend on the number and position of the missing atom(s). As long as there is even a single atom belonging to a coded residue, a bead representing it can be generated. Again, a warning message pops-up if incomplete residues are found in a structure, and the user can proceed or chose another option, like stopping or skipping the whole residue (not recommended), by selecting it in the "PDB" pull-down menu. Of course, there is no cure in US-SOMO for totally missing residues: the users are urged to complete their structures using external methods (*e.g.*, ROBETTA, http://robetta.bakerlab.org/ (Kim et al. 2004); I-TASSER, http://zhanglab.ccmb.med.umich.edu/I-TASSER/ (Roy et al. 2010); MODELLER, https://salilab.org/modeller/ (Eswar et al. 2006)). Missing atoms within coded protein residues can be added by WHATIF (http://swift.cmbi.ru.nl/servers/html/index.html; Vriend 1990).

PDB files can be individually loaded using the "Load Single PDB file" button or in batch mode (the latter will open a new window with advanced functions; see Sect. 10.5). When NMR-style files are opened, all models present are listed in the field provided, and either individual or multiple/all models can then be selected for further operations. Each structure is automatically visualized upon loading using RasMol (http://www.bernstein-plus-sons.com/software/rasmol/; Sayle and Milner-White 1995).

PDB files can be viewed in text mode and manually edited by pressing the "View/Edit PDB Files" button. Alternatively, an advanced PDB editor is also available, including cut/splice capabilities and the possibility to extract individual models from NMR-style files or to create NMR-style files from single models. These two functions are particularly useful as a complement to the DMD utility (Ding and Dokholyan 2006; Dokholyan et al. 1998) (accessed by pressing the "Run DMD" button, see Sect. 10.5), *e.g.*, to splice generated multiple conformations of a connecting segment between two static domains. The DMD utility will not be dealt with in detail in this chapter.

SAXS/SANS functions allowing computations directly on the atomic structure can also be accessed from this area (not dealt with here; see Brookes et al. (2012, 2013a); Rocco and Brookes (2014)). A Brownian Dynamics (BD) module (in preparation) will be also available in the future for the hydrodynamic parameter computation for flexible/partially disordered structures.

The Miscellaneous Options menu (Fig. 10.5, left-side panel), in addition to the "Average Parameters for Automatic Bead Builder" settings, contains the *psv* ("vbar") controls. The *psv* can be automatically computed from the composi-

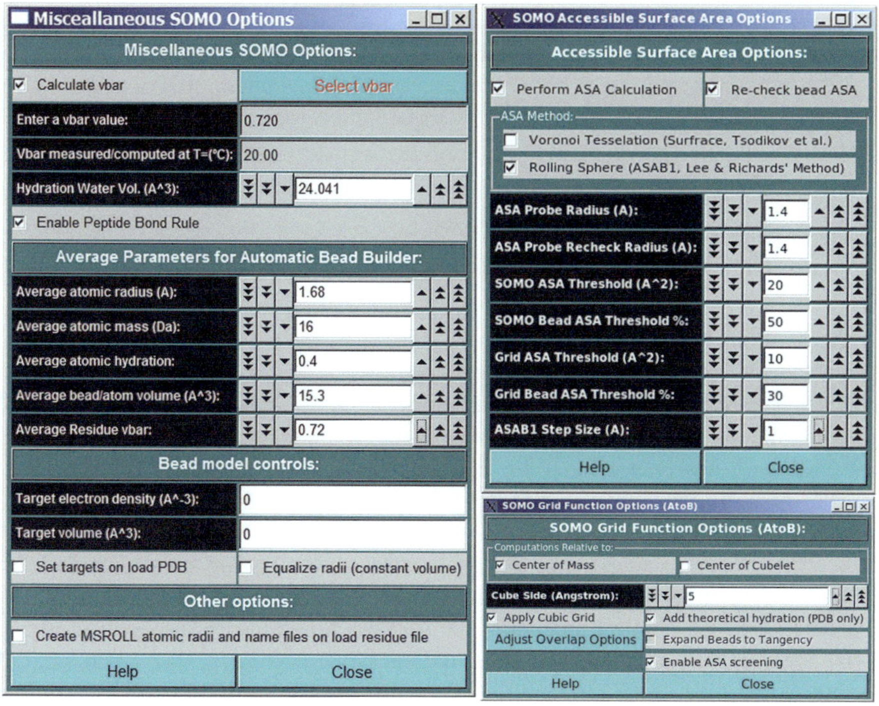

Fig. 10.5 The US-SOMO "Miscellaneous Options" (left-side panel), "Accessible Surface Area Options" (*top right-side panel*), and "Grid Functions Options" (AtoB) (*bottom right-side panel*) modules

tion using the matching between the residues in the PDB file and those in the *somo.residue* lookup table ("Calculate vbar" checkbox selected), it can be uploaded from a database by pressing the "Select vbar" button, or it can be manually entered in the "Enter a vbar value" field. In the latter case, the "vbar measured/computed at T = (°C)" field should also be updated (default, 20 °C). Thus, if the temperature entered is different from 20 °C, the program can then recalculate a proper *psv* at the standard 20 °C T to which all hydrodynamic parameters are standardized by default (see also the Hydrodynamic Computations Options module).

Another important entry in this module is the volume assigned to the hydration waters, controlled in the "Hydration Water Vol (A^3)" field (default, 24.041 Å^3; Gerstein and Chothia 1996). This is the volume that will be added to the sum of the anhydrous atom volumes for each water molecule assigned to a bead.

The "Enable Peptide Bond Rule" checkbox controls if this rule is used by the *SoMo* method. With it, the peptide bond segment is used for the main-chain beads of a protein structure. These beads are thus positioned at the center of gravity of the $(CA\text{-}C\text{-}O)_n\text{-}(N)_{(n+1)}$ atoms, except when PRO is the $(n+1)$ residue. In this case, the peptide bond bead is positioned at the center of gravity of the $(CA\text{-}C\text{-}O)_n$ atoms. Additional rules control the generation of the OXT bead and of the first N atom at the beginning of each protein chain. All these rules are controlled by "special" residues in the *somo.residue* table. To gain total control over the positioning, volumes and masses of every bead, the "Enable Peptide Bond Rule" checkbox should be deselected (default, selected, but if breaks are found in a chain, it is disabled). The "Bead Model Controls" (for SAS work) and "Other options" (relating to BEST operations, see Sect. 10.6) sections will not be dealt with here.

10.3.2 BEAD Model Function Area

In this section, new bead models can be generated from selected PDB structures according to one of the three methods available, *SoMo* (without overlaps), *AtoB* (also without overlaps), and *SoMo* with overlaps (see Fig. 10.2). The various menus with the options and settings in the bead generation routines are accessible from the "SOMO" pull-down menu in the top bar.

The ASA options are controlled by the "Accessible Surface Area Options" module (Fig. 10.5, top right-side panel). By default, the "Perform ASA Calculation" and "Re-check bead ASA" checkboxes are selected, allowing the assignment of each bead in the final model to either an exposed or buried status. The hydrodynamic computations with the *SMI* procedure can then be carried out on the exposed beads subset only, greatly reducing the computational load (see Sect. 10.2). The default method is the Lee and Richards (1971) rolling sphere algorithm ("ASAB1"), but a Voronoi tessellation method ("Surfracer"; Tsodikov et al. 2002) is also available. The ASA probe radii can be independently set for the original structure and for the resulting bead model (default, both 1.4 Å). The "SOMO ASA threshold (A^2)" and the "Grid ASA threshold (A^2)" fields control the levels above which a main or side chain will be considered exposed to the solvent in the *SoMo* method and above which primary beads will be considered exposed to the solvent in the *AtoB* (Grid) method, respectively (defaults, 20 and 10 Å^2, respectively). The "SOMO bead ASA threshold (%)" and "Grid Bead ASA Threshold (%)" fields set the minimum % of the surface of a bead that must be accessible to reclassify as exposed a bead previously considered to be buried in the *SoMo* and *AtoB* methods, respectively (defaults, 50 % and 30 %, respectively). Finally, the "ASAB1 step size (A)" field defines the increment between the 2D slices, to be integrated, in which the structure (or the model) is subdivided in the rolling sphere method (default, 1 Å).

The options for the *AtoB* grid method can be seen in Fig. 10.5, bottom right-side panel. The positioning method can be either the center of mass of the atoms assigned to each bead or to the center of the cubelet. The grid size can be set here (default, 5 Å). "Apply Cubic Grid" allows the grid procedure to be executed (default, active). It could be deselected to allow the use of the Grid module for overlap removal of a previously loaded bead model. The "Add theoretical hydration (PDB only)" checkbox will enable the addition of the theoretically bound water molecules volume to those of the atoms assigned to a bead. The "Adjust Overlap Options" button will open the *AtoB* overlap reduction options module (see below and Fig. 10.6). Finally, the "Enable ASA screening" checkbox will allow the user to select/deselect that routine (default, selected). The other checkbox controls a function still under development.

The overlap reduction routines have several options that can be accessed from two dedicated modules, one for the *SoMo* and the other for the *AtoB* methods (see Fig. 10.6). A common "overlap cutoff" field, which determines the level of precision in computing the overlaps between beads (default, 0.001 in the model units) is present at the top. Each module then has three different sections, dealing with the overlaps between exposed side-chain beads only, between main- and side-chain beads, and between buried beads for the *SoMo* method, while the distinctions are made between exposed grid beads and buried grid beads for the *AtoB* method. For the latter, in case no ASA screen is selected, there is a specific panel for the overlap reduction settings. All the options visible in Fig. 10.6 are common in the three

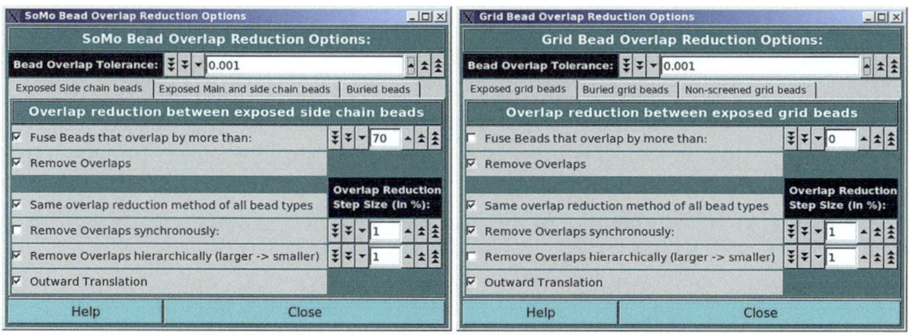

Fig. 10.6 The *SoMo* (*left side*) and *AtoB* (*right side*) overlap reduction options modules

sections, except for the outward translation which is present only in the exposed side chains and exposed grid beads sections (see the US-SOMO Help pages for a complete description of all the features available in these modules).

The transformation process from an atomic-level structure to a bead model is activated by pressing either the "Build SoMo Bead Model," the "Build AtoB (Grid) Bead Model," or the "Build SoMo Overlap Bead Model" buttons. Bead models thus generated are automatically saved in a file, whose name is the PDB filename with "_1" added and the extension ".bead_model." Filenames are by default customized by adding a suffix containing a coding of the method used and its settings (which can be turned off by deselecting the "Add auto-generated suffix" checkbox) and additionally by entering a user-selected suffix in the "Bead Model Suffix." The auto suffix will have the "-so," "-a2b," or "-so_ovlp" extensions if the bead model was generated with the *SoMo*, *AtoB*, or *SoMo* with overlaps methods, respectively, and will contain a series of "codes" for the ASA parameter settings and the bead model generation options (see the US-SOMO main Help pages for a complete description of this feature). If the resulting filename is already present in the operating directory, a pop-up menu will offer several choices, including overwriting. This step can be automatically bypassed by selecting the "Overwrite existing filenames" checkbox. Options are available to adjust the bead model(s) file format by selecting the "Bead Model Output Options" from the "SOMO" pull-down menu (not shown). If both the "Overwrite existing filenames" and the "Automatically Calculate Hydrodynamics" checkboxes are selected, the program will complete the full process of generating a bead model and computing its hydrodynamic parameters unattended. This is especially useful when relatively large structures are examined. By default, the *SMI* procedure will be called if *SoMo* or *AtoB* models without overlaps are generated, while *Zeno* will be used if *SoMo* models with overlaps are produced. If the "Automatically Calculate Hydrodynamics" checkbox is not selected (default option), at the end of the model building phase the progress bar will be at 100 %, and the bead model(s) can be visualized with RasMol by clicking on "Visualize Bead Model" (recommended, comparing the original structure with the bead model could reveal previously unforeseen problems). A warning: if a NMR-style file has been uploaded and several/all models selected for bead model generation, pressing "Visualize Bead Model" will open a RasMol window for each one!

The "Grid Existing Bead Model" function allows reduction of the resolution of a previously generated bead model by applying a grid procedure. This button is not available until a PDB file has been processed with any of the bead modeling primary options (see above) or until a previously generated bead model file has been loaded (see below). If this operation is launched, the "-a2bg" suffix is automatically added to the filename of the new bead model.

The results of ASA screening of the original PDB file are written in a text-format file, which can be opened by pressing the "View ASA Results" button. Likewise, a bead model file can be opened in text mode by pressing the "View Bead Model File" button. Bead models previously generated by US-SOMO or coming from other sources like DAMMIN/DAMMIF (Franke and Svergun 2009) can be further processed here by either uploading a single model ("Load Single Bead Model File") or using the "Batch Mode/Cluster Operation" (see Sect. 10.5).

The "SAXS/SANS Functions" button will open the SAS module allowing operations on the current bead model (not dealt with in this chapter).

10.3.3 Hydrodynamic Calculations Area

The options setting for the two hydrodynamic calculation methods using bead models offered in US-SOMO (see Fig. 10.2) are shown in Fig. 10.7. In the left side of Fig. 10.7, the options for the default García de la Torre-Bloomfield *SMI* inversion method (Rai et al. 2005; Brookes et al. 2010a; García de la Torre and Bloomfield 1981; Spotorno et al. 1997) are shown. By default, all calculations are performed for structures (bead models) whose dimensions are in Å, and in standard conditions, *i.e.*, in water at 20 °C. The top part of the hydrodynamic calculations options module lists these values. Users wishing to compute hydrodynamics under different conditions, or using bead models on another scale, can change the required parameters here.

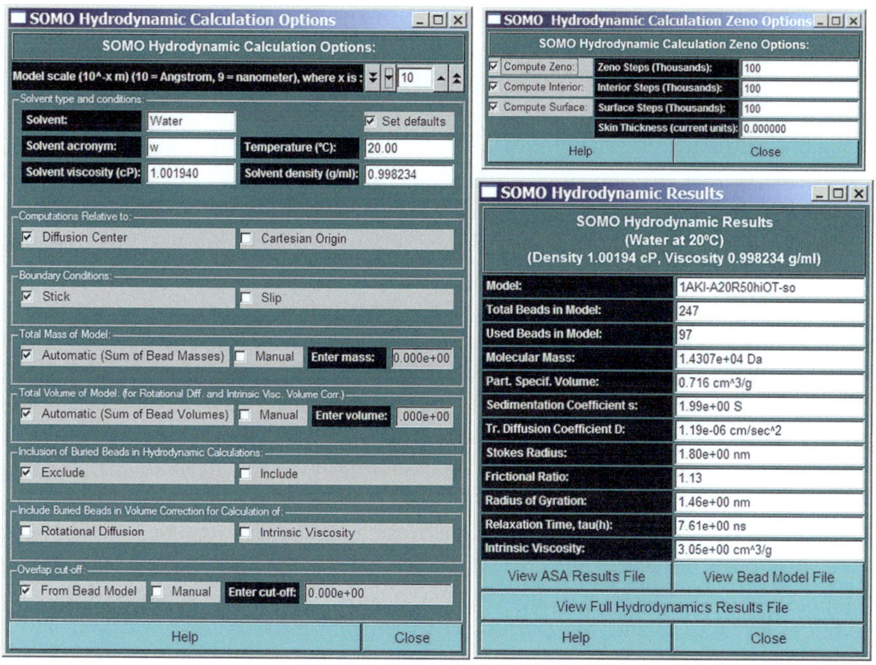

Fig. 10.7 The "Hydrodynamic Calculations Options" modules and "Hydrodynamic Results" panel. *Left side*, options for the standard *SMI* method. *Right side*, *top*, options for the alternative *Zeno* method. *Right side*, *bottom*, "Hydrodynamic Results" pop-up panel

These definitions also apply to the *Zeno* method. By default, all *SMI* computations are carried out relative to the diffusion center of the model, and under the stick boundary conditions (García de la Torre and Bloomfield 1981), but the alternative Cartesian origin, and slip boundary conditions are respectively available. The total mass and total volume of the model (both necessary for the computations of [η], the latter also for the so-called volume correction for the rotational diffusion and [η]; see Spotorno et al. (1997) and references therein) are by default automatically computed from the beads' values. Users can, however, override either of these values by selecting the "Manual" checkbox and entering appropriate values. Entering a manual mass value is especially important when the bead model derives from an incomplete structure and/or including noncoded residues (see the PDB Functions area). By default, the beads labeled as being buried are excluded from the *SMI* hydrodynamic computations, but this can be overridden by selecting the "Include" checkbox. In such a case, it becomes possible to include or exclude (default) the buried beads from the "volume correction" computations for either or both the rotational diffusion and [η]. Finally, the "overlap cutoff," *i.e.,* the level of precision in checking the bead overlaps (see Fig. 10.6), can be set to manual with a different value, to allow for greater overlap tolerance when processing beads generated by other programs (*e.g.*, DAMMIN/DAMMIF).

The *Zeno* computational method involves enclosing an arbitrarily shaped probe object within a sphere and launching random walks from this sphere. The probing trajectories either hit or return to the launch surface ('loss'), whereupon the trajectory is either terminated or reinitiated (Kang et al. 2004). A summary of the ideas behind *Zeno* is given in its dedicated Help page in the US-SOMO manual. In the *Zeno* options module shown in Fig. 10.7, top right side, the first checkbox allows selection of the *Zeno* computation. This will launch a Monte Carlo numerical path integration that generates a large number of random walks in the space outside the body. Sums taken over these random walks yield the electrostatic capacity, the polarizability tensor, the intrinsic conductivity, and, most relevant here, the hydrodynamic radius R_h, the translational diffusion and frictional coefficients D_t and f_t, the intrinsic viscosity [η], and the hydrodynamic volume V_h. The main option of interest here is the number of steps in the "Zeno Steps (Thousands)" field (default, 1000), which controls the accuracy of the calculations at the cost of increasing computational time. The reader is referred to the *Zeno* Help page within US-SOMO for more information on this and the other operations available by selecting the other two checkboxes, as well on the "skin thickness" field.

Once one or more bead model(s) have been generated, or a single existing bead model has been uploaded, the hydrodynamic calculations are started by pressing either "Calculate RB Hydrodynamics SMI" or "Calculate RB Hydrodynamics ZENO" (the "RB" stands for "Rigid Body," meaning that the computations are in the rigid body frame approximation). In a recent examination of the performance of the main available hydrodynamic computations methods/programs (Rocco and

Byron 2015), it was found that while only a slight improvement in accuracy was observed when *SoMo* models without overlaps were processed with *Zeno* in respect to the *SMI* procedure, a significant improvement was present when the primary models with overlaps were employed. Therefore, US-SOMO now offers directly both procedures, but since the *SMI* cannot be used when overlaps are present, the "Calculate RB Hydrodynamics SMI" button will not be available when the "Build SoMo Overlap Bead Model" has been used to generate the model(s). Once the calculations are completed (bottom progress bar at 100 % and "Calculate Hydrodynamics Complete" appears in the progress window), a subset of the results can be visualized by pressing "Show Hydrodynamic Calculations." In the SOMO Hydrodynamic Results pop-up panel (see Fig. 10.7, bottom right side), the conditions under which the calculations were performed are stated first (default, H_2O @ 20 °C). There a series of the most commonly used parameters are reported, among which are the sedimentation coefficient s, D_t, R_h, the frictional ratio f/f_0, the radius of gyration R_g, the harmonic mean of the relaxation times τ_h, and $[\eta]$ (the τ_h field will not be populated if *Zeno* is used). The full list of all the parameters entered/computed is saved in a text-format file that can be opened by pressing the "View Full Hydrodynamic Results File" button in the hydrodynamic results pop-up panel or the "Open Hydrodynamic Calculations File" button in the main panel.

The "Select Parameters to be Saved" button will open another window (see Fig. 10.8) where the user can interactively select among all conditions, results, and parameter values available, to be saved in a comma-separated variable (csv) file for further manipulations either with external spreadsheet programs or by the US-SOMO Model Classifier (see Sect. 10.4). Selecting the "Save parameters to a file" checkbox will enable this feature. "BEST" will open another pop-up window where results from the *BEST* hydrodynamic computation program as implemented within US-SOMO can be analyzed (see Sect. 10.6). "Stop" will halt any operation.

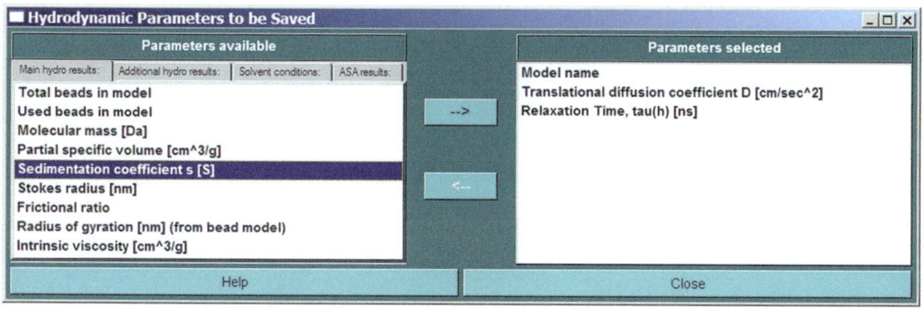

Fig. 10.8 The "Select Parameters to be Saved" module

10.4 The US-SOMO Model Classifier Module

This module presents a tool for selecting a best matching model among a series of models, by comparing their calculated hydrodynamic parameters with user-provided experimental values. Several ranking methods are available in case more than one experimental parameter is known.

In Fig. 10.9, top, the GUI of the Model Classifier is shown. First, the experimental parameters to be used are entered. The selectable parameters are the sedimentation coefficient s [S], the diffusion coefficient D_t [cm^2/s], the Stokes' radius R_h [nm], the frictional ratio f/f_0, the radius of gyration R_g (nm), the harmonic mean of the

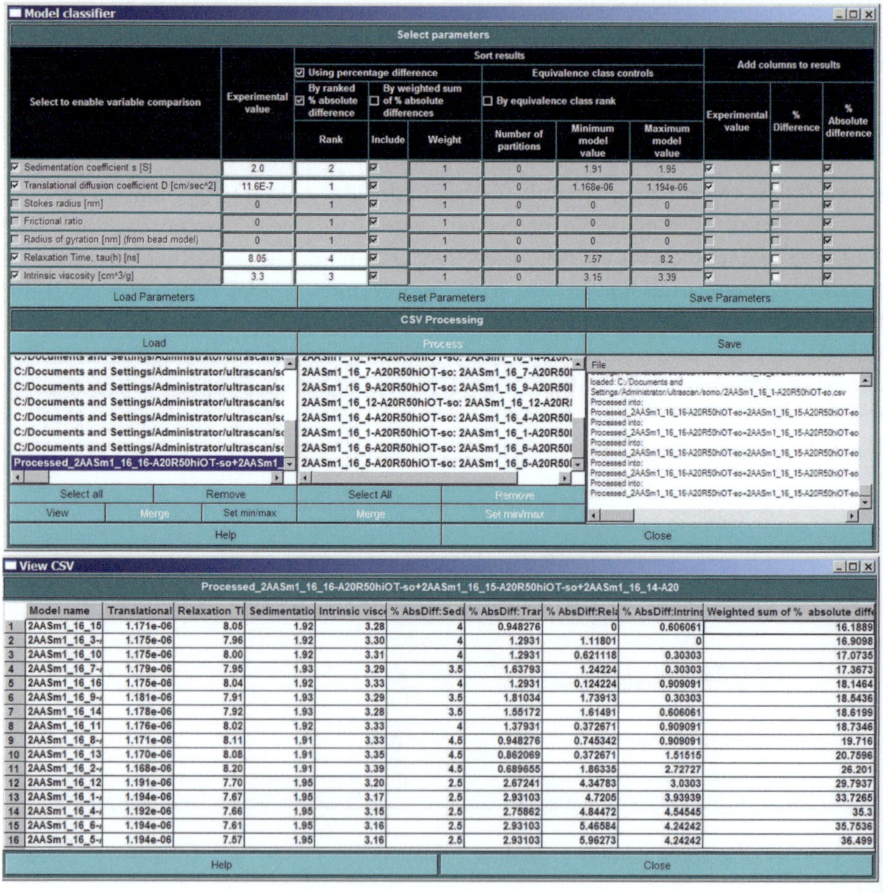

Fig. 10.9 *Top*, the US-SOMO "Model Classifier" interface; shown are the settings and a run using 16 NMR-derived models of RNaseA (2AAS.pdb) whose hydrodynamic parameters were computed and compared with experimental values (Taken from (Brookes et al. 2010a)). *Bottom*, the results of the run are shown through the dedicated "Model Classifier" viewer

relaxation times τ_h [ns], and the intrinsic viscosity [η] [cm^3/g]. The methods used to sort the computed results against the experimental values are set next using several alternative criteria, listed under the "Sort results" group. In the "Using percentage difference" subgroup, they can be ranked by % absolute difference or by the weighted sum of % absolute differences. The first is the simplest procedure, ranking the parameters in a descending order (*i.e.*, 1 = most relevant) in the "Rank" field. The second ranks over multiple parameters without specifically assigning a numerical rank to each parameter. This is accomplished by computing a weighted sum of absolute differences of every included parameter. The user-defined weights do not have to add up to 1, and experimental data with higher confidence should be assigned higher weights.

Alternatively, in the "Equivalence class controls" subgroup, the results can be sorted by equivalence class rank. Equivalence classes partition a range of values. A value that falls into a specific equivalence class is equivalent to all other values within the equivalence class. The range runs from the "Minimum model value" to the "Maximum model value" and is composed of "Number of partitions" equivalence classes. The equivalence class that contains the experimental value is given a distance of zero. Equivalence classes next to the one containing the experimental value are given a distance of 1 and so on. Adding up the distances of each of the selected variables gives the equivalence class rank.

The last three columns under the "Add columns to results" label allow the addition of the experimental values and an additional % difference field to the Model Classifier results (the absolute differences are reported by default if the first ranking method is chosen). The current parameters and the criteria used for the sorting can be saved in a file (extension *.smp) by clicking on the "Save Parameters" button, while "Reset Parameters" will clear all fields. Previously saved parameters can be reloaded by clicking on the "Load Parameters" button.

In the bottom part, the parameters calculated for the models are uploaded. They should be in *.csv files, most easily generated using the "Save parameters to a file" checkbox (and the "Select Parameters to be Saved" module) in the Hydrodynamic Calculations section of the main US-SOMO window (see above). Only the parameters present in the *.csv files, identified through their headers, will then be available in the "Select to enable variable comparison" column. Pressing the "Load" button will open the file system dialog and allow import of the required *.csv files into the left-side window. Files can then be selected by clicking on each filename, which will transfer them to the right-side window, or by pressing the "Select All" button. "Remove" will remove files from the list. "Merge" will join the selected files from the list into one csv file. "Set min/max" will set the "Minimum model value" and "Maximum model value" from the values found in the selected files. The files listed in the right-side window can be then selected for processing by individually clicking on them or by pressing the "Select All" button.

Once files have been selected, the Model Classifier can be launched by pressing the "Process" button, and the progress window at the far right will be updated. At the end, pressing "View" will open a window with all the selected columns, as shown in the bottom part of Fig. 10.9. By pressing the "Save" button, a file system dialog will open to allow the results to be saved in a new *.csv file, which can then be opened with a standard spreadsheet.

10.5 The US-SOMO Batch Mode/Cluster Operation Module

This module (Fig. 10.10, top) was conceived to allow the unattended processing of multiple files for both hydrodynamic, DMD, and SAS calculations. Since some of these operations can be performed only on a remote supercompute cluster, access to

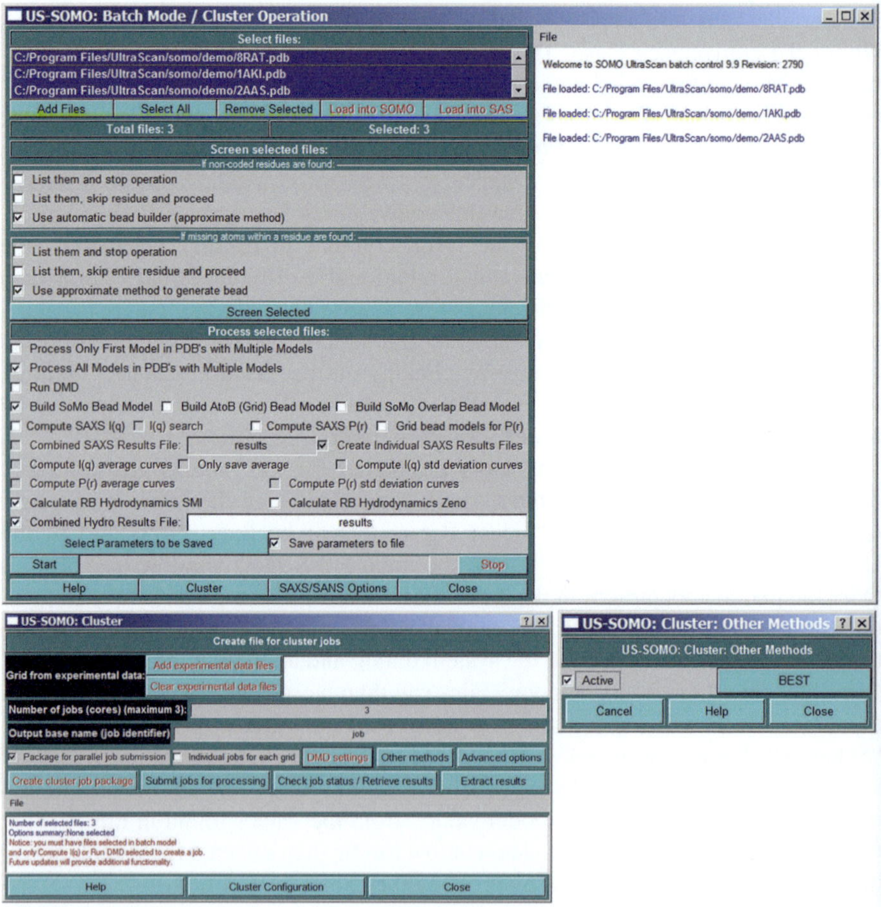

Fig. 10.10 *Top*, the US-SOMO "Batch Mode/Cluster Operation" module. *Bottom*, *left side*, the US-SOMO "Cluster" module GUI. *Bottom*, *right side*, the pop-up "Cluster: Other Methods" pane

the Cluster interface ("Cluster" button; Fig. 10.10, bottom left) is provided within this module.

Operations begin by loading file(s) using the "Add Files" button in the "Select files" section and then either selecting a subset by clicking on individual filenames or all files with the "Select All" button. Files can be removed from the list with the "Remove Selected" button. The "Load into SOMO" and "Load into SAS" buttons become available only if a single file is selected and will just transfer it to either the main US-SOMO or to the SAS modules, respectively. Both PDB and bead model files can be uploaded and selected in this module, but some operations like the "Run DMD" can be performed only on atomic-level structures.

In the "Screen selected files" section, the user can control the level of tolerance for both noncoded residues (first three checkboxes) and for incomplete residues (second three checkboxes; see Sect. 10.3 for a complete discussion of these features). Pressing "Screen Selected" will then verify if the selected files comply with the US-SOMO requirements for processing. This is a relatively quick step and is highly recommended before launching a batch mode operation, since it will be performed anyway when each single file is processed, but the operations will be halted if noncomplying files are then found. A prescreen will allow users to correct the situation and permit fully unattended operations thereafter.

Operations are chosen in the "Process selected files" section. The first two checkboxes control if just the first model or all the models are to be processed when NMR-style files are uploaded. The "Run DMD" checkbox will allow a DMD run to be performed on chosen PDB file(s) (not dealt with in this chapter). The three bead modeling methods available within US-SOMO can be alternatively chosen by selecting either the "Build SoMo Bead Model," the "Build AtoB (Grid) Bead Model," or the "Build SoMo Overlap Bead Model" checkboxes. Next follows a series of checkboxes related to SAS operations, which will be not described here (see the Batch Mode/Cluster Operation Help page for a detailed description of these options). The "Calculate RB Hydrodynamics SMI" or the alternative "Calculate RB Hydrodynamics Zeno" checkboxes (the latter automatically selected if the "Build SoMo Overlap Bead Model" method is checked) allow the hydrodynamic calculations to be performed for bead models, either already present in the uploaded files or after generation from uploaded PDB structures. The "Combined Hydro Results File" checkbox allows saving the hydrodynamic parameter computation results performed on all bead models in a single file, with the averages of all parameters, instead of separate files for each model. A filename for the single results file must be provided in the dedicated space. Otherwise, each file will be named using the general US-SOMO rules and the prefixes present in the main program panel. As with single file operation, selected parameters can be chosen and saved in a *.csv file by accessing the "Select Parameters to be Saved" module (see Fig. 10.8) and selecting the "Save parameters to file" checkbox. The operations are launched by pressing the "Start" button and can be aborted at any stage by pressing the "Stop" button. After launching, the various operations will be reported in the right-side progress window, and the progress bar will become active.

With the exception of DMD, all the other options listed in the Batch Mode/Cluster Operation module can be carried out locally. However, some can be computationally intensive and might require supercomputing in order to be efficiently carried out. For this reason, a cluster interface has been developed, accessible by pressing the "Cluster" button. A complete description of this module is, however, beyond the scope of this chapter, and only a general overview and the *BEST* application (see Sect. 10.6) will be described. See Brookes et al. (2012, 2013b) for more information on cluster usage.

In Fig. 10.10, the main GUI of the Cluster module is shown (bottom left), together with the "Other Methods" pop-up panel that is launched from the "Other methods" button (bottom right). The top part of the module ("Grid from experimental data") deals with SAXS settings not described here. The "Number of jobs (cores) (maximum #)" is adjusted to the number of independent structures considered when the "Package for parallel job submission" checkbox is selected; it can be changed but the value should not be above the maximum # indicated. The "DMD settings" and "Advanced options" buttons will open the DMD settings and the SAXS advanced settings panels, respectively; they will not be discussed in this chapter. Currently, *BEST* is the only option available under "Other Methods."

Once the options have been set, the cluster submission procedure begins with pressing the "Create cluster job package" button. The package is then submitted to the cluster by pressing "Submit jobs for processing," which will open a cluster dialog panel (not shown) where jobs can be seen, clusters can be selected, the status of the operation(s) monitored, and from where the results can be retrieved. The cluster dialog panel can be accessed at any time by pressing the "Check job status/Retrieve results" button. Once the packaged results have been transferred back to the local machine, full datasets can be extracted by pressing the "Extract results" button. All these steps are described at length in the cluster Help section, and cluster access can be defined and configured through a dedicated panel accessed by pressing the "Cluster Configuration" button (not shown).

10.6 The US-SOMO BEST Interfaces

BEST is a software package for the computation of the hydrodynamic properties of (bio)macromolecules that relies on the direct evaluation of the frictional forces acting on surface elements (Aragon 2004) [see also Chap. 12]. *BEST* is made available under US-SOMO as an alternative method to the bead modeling methods we offer for the computation of the hydrodynamic parameters starting from a high-resolution structure. In principle, *BEST* can produce more accurate values with respect to the bead modeling procedures, especially with regard to the rotational diffusion and the intrinsic viscosity, since no "volume correction" is needed. However, some issues such as the proper consideration of the hydration (a recently done comparison between the various hydrodynamic methods (Rocco and Byron 2015) has evidenced that the current *BEST* implementation slightly underestimates, ∼−3 %, the translational frictional properties of proteins), and the requirement to extrapolate values to zero triangle size (see below) need to be considered with care. To this end, we also provide an interface to visualize and statistically analyze the *BEST* results (Fig. 10.11, bottom). Moreover, *BEST* is very computationally intensive, and when many structures are analyzed, for instance when dealing with conformational variability/flexibility, bead modeling can offer a more practical alternative. Due to its requirements, *BEST* is offered only on a supercompute cluster within US-SOMO. To perform the calculations, in *BEST* the smooth atomic surface of the structure needs to be transformed into an ensemble

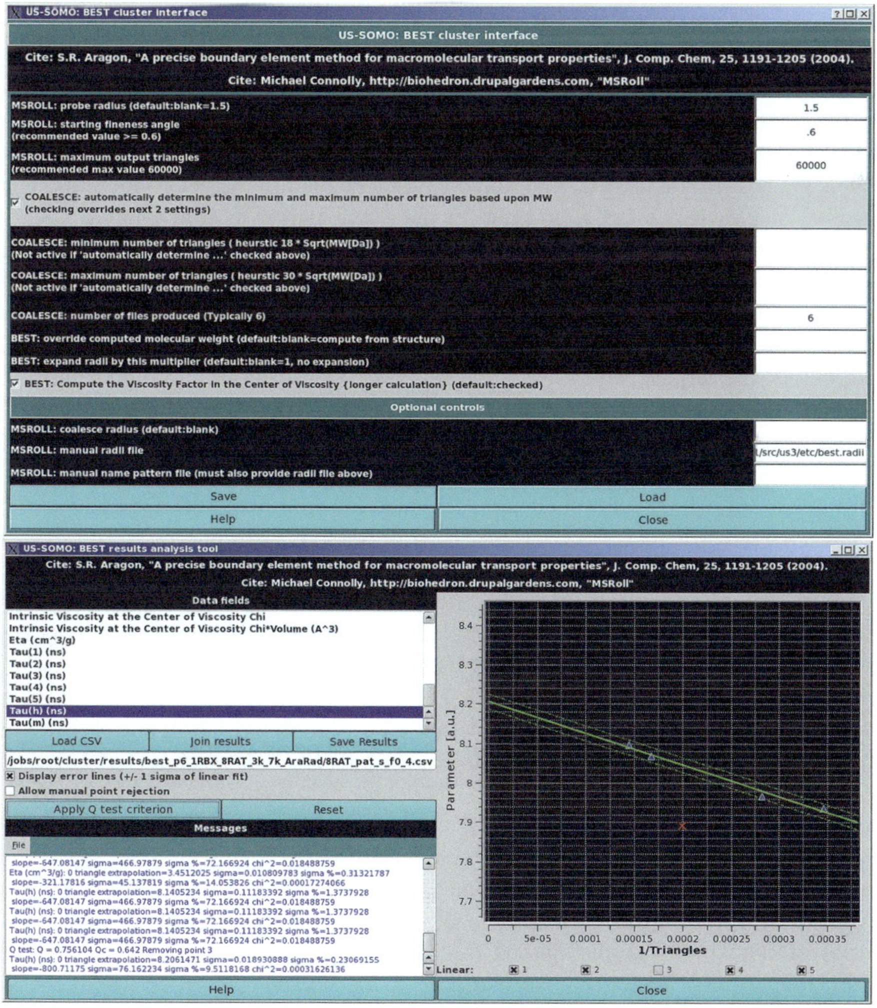

Fig. 10.11 *Top*, the US-SOMO "*BEST* cluster interface" module. *Bottom*, the "*BEST* results analysis tools" module

of triangular elements, allowing the correct evaluation of the surface resistance integrals (Aragon 2004). First, the external program *MSROLL* (Connolly 1993) is used by *BEST* to generate an initial high-resolution triangulated surface of the structure under examination. Then, the *BEST* module *COALESCE* will produce a series of triangulated structures with different resolutions from the initial *MSROLL*-generated surface. The hydrodynamic properties calculated for each triangulated structure are then extrapolated to zero triangle size. The full principles and detailed operation of *BEST* can be found in Aragon (2004) and in Chap. 12, and here we will just describe the various tools and settings we provide.

In the cluster "Other Options" panel (see Sect. 10.5), pressing "BEST" will open the *BEST* settings interface panel. As shown in Fig. 10.11, top, the settings interface allows the user to set the *MSROLL* (Connolly 1993) probe radius (default 1.5 Å), finesse angle, and maximum number of triangles. The options for the *BEST* module *COALESCE* are set next. The first checkbox allows the automatic determination of the optimal maximum and minimum number of triangles based on a heuristic approach involving the structure's molecular weight (see Chap. 12). If this checkbox is left unchecked, these values can be manually entered in the next two fields. The number of files generated, used then for the extrapolation to zero triangle size, is entered in the following field (4 is the minimum suggested value, but 6 can allow for a better checking of the extrapolation). The last two fields in this part of the module allow the user to enter a molecular weight different from that calculated by *BEST* from the structure and to expand (or shrink) the atom radii used by *MSROLL* to compute the surface, which are optimized in *BEST* to take into account a uniform layer of hydration (see Chap. 12). By default, the atomic radii internally used by *BEST* are selected (available in a file called *best.radii*), but any other properly formatted radii file can be uploaded in the "Optional controls" section (a *MSROLL*-formatted radii file can automatically be generated from the values present in the *somo.residue* entries by selecting the "Create MSROLL atomic radii and names files on load residue file" checkbox in the Miscellaneous SOMO Options module; see Fig. 10.5, left panel). Finally, the "BEST: Compute the Viscosity Factor in the Center of Viscosity (longer calculation)" checkbox if unselected will speed up the calculations but at the cost of accuracy (default, checked).

The US-SOMO *BEST* implementation includes assembling all calculated parameters for each model in a csv file. Following retrieval from the cluster and extraction, the *BEST* results can be uploaded in the "BEST results analysis tool" (Fig. 10.11, bottom), accessible by pressing "BEST" from the main panel (see Fig. 10.2).

Upon loading using the "Load CSV" button, the "Data fields" panel will list all the parameters computed by *BEST*, selectable by clicking on each one. The data associated with the selected parameter are plotted *vs.* 1/(number of triangles) in the right-side graphics window, together with a linear regression line and a series of checkboxes corresponding to each data point (see Fig. 10.11, bottom). The "Join results" button allows merging of the separate csv files that the US-SOMO *BEST* implementation will generate, for instance, from NMR-style files, into a single csv file with the data for every parameter grouped together. In this way, averaging can then be easily performed using an external spreadsheet program.

Since *BEST* requires an extrapolation procedure to produce the final value for each parameter, the US-SOMO implementation provides automatic (recommended) or manual (non-recommended) ways to reject outliers from the regression. First, by selecting the "Display error lines (+/– 1 sigma of linear fit)" checkbox, two dotted lines corresponding to $\pm$ 1 standard deviation (SD) will be traced along the regression line. By pressing the "Allow Q test criterion" button, Dixon's *Q*-test (Dixon 1951) is performed and will reject a single outlier if the outlier's computed *Q* value is greater than the critical *Q* value set at a 90 % confidence level. In the example shown in Fig. 10.11, bottom, the third point (marked with a red "X") has been rejected, and the $\pm$ 1 SD lines are retraced after its exclusion (signaled also in the checkboxes below the graph). If more than one point visually appears to be problematic in the regression, it is suggested to rerun the computations including more points. "Reset" will re-include all points in the linear regression. The updated regression data are shown each time in the "Messages" window. All parameters within a single csv file can be independently analyzed. At the end, a new csv file containing all the updated extrapolated values can be saved by pressing the "Save Results" button.

10.7 Conclusions

US-SOMO has now grown into a hub harboring different methods useful in multiresolution modeling. In this chapter, we have dealt only with the hydrodynamic methods which are directly linked to the parameters that AUC can provide. A verification of the accuracy with which the *SoMo* and *AtoB* methods can reproduce experimentally determined hydrodynamic parameters had been presented before (Brookes et al. 2010a, b), with generally more accurate results than alternative bead modeling methodologies. The cost paid is that US-SOMO requires a detailed coding of each residue in order to appropriately convert it into bead(s), somewhat limiting its direct application. However, approximate methods dealing with noncoded residues are provided. In addition, recently two other hydrodynamic computation methods have been implemented within US-SOMO, *Zeno*, which can operate on arbitrarily shaped models (Kang et al. 2004), and *BEST*, using the alternative boundary elements methodology (Aragon 2004) (for the latter, see Chap. 12). A full comparison between all the hydrodynamics methods currently available in US-SOMO, and with *HYDROPRO* (see Chap. 11) using a well-defined set of proteins with carefully verified literature translational diffusion and sedimentation experimental parameters, has been very recently carried out (Rocco and Byron 2015). The results evidenced a slight overestimation on average of D_t and s by the *SoMo* approach ($\sim +2$ %) and a slightly larger underestimation of the same parameters by *BEST* and *HYDROPRO* (~ -3 % and ~ -4 %, respectively). The best results with the standard implementations were obtained using the US-SOMO *AtoB* with a 5 Å grid size ($\sim +1$ %). However, a combination of the *SoMo* bead model generation method, without overlap removal, and the *Zeno* computational tool produced even better results (~0 %) (Rocco and Byron 2015). For this reason, this new combination has been already implemented within US-SOMO. With the future release of a much faster *Zeno* code (J. Douglas, NIST, Gaithersburg, MD, USA, personal communication), this approach could become the method of choice in US-SOMO for the computation of translational frictional properties and [η]. If the computation of rotational diffusion is sought, *BEST* could represent a viable alternative, since it is based on a correct hydrodynamic treatment, even if it is quite computationally intensive, requires an extrapolation to zero plate size, and treats hydration as a uniform layer.

References

Aragon SR (2004) A precise boundary element method for macromolecular transport properties. J Comput Chem 25:1191–1205

Brookes E, Demeler B, Rosano C, Rocco M (2010a) The implementation of SOMO (SOlution MOdeller) in the UltraScan analytical ultracentrifugation data analysis suite: enhanced capabilities allow the reliable hydrodynamic modeling of virtually any kind of biomacro-molecule. Eur Biophys J 39:423–435

Brookes E, Demeler B, Rocco M (2010b) Developments in the US-SOMO bead modeling suite: new features in the direct residue-to-bead method, improved grid routines, and influence of accessible surface area screening. Macromol Biosci 10:746–753

Brookes E, Singh R, Pierce M et al (2012) UltraScan Solution Modeler: integrated hydrodynamic parameter and small angle scattering computation and fitting tools. In: Proceedings of the 1st conference of the extreme science and engineering discovery environment: bridging from the eXtreme to the campus and beyond. ACM, New York. Article #42. doi:10.1145/2335755.2335839

Brookes E, Pérez J, Cardinali B et al (2013a) Fibrinogen species as resolved by HPLC-SAXS data processing within the UltraScan SOlution MOdeler (US SOMO) enhanced SAS module. J Appl Cryst 46:1823–1833

Brookes E, Singh R, Pierce M et al (2013b) US-SOMO cluster methods: year one perspective. XSEDE'13. In: Proceedings of the conference on extreme science and engineering discovery environment: gateway to discovery, ACM, New York. doi:10.1145/24847622484815

Byron O (1997) Construction of hydrodynamic bead models from high-resolution X-ray crystallographic or nuclear magnetic resonance data. Biophys J 72:408–415

Connolly ML (1993) The molecular surface package. J Mol Graph 11:139–141

Demeler B (2005) UltraScan, a comprehensive data analysis software package for analytical ultracentrifugation experiments. In: Scott DJ, Harding SE, Rowe AJ (eds) Modern analytical ultracentrifugation: techniques and methods. Royal Society of Chemistry, Cambridge, pp 210–229

Ding F, Dokholyan NV (2006) Emergence of protein fold families through rational design. PLoS Comput Biol 2:e85

Dixon WJ (1951) Ratios involving extreme values. Ann Math Stat 22:68–78

Dokholyan NV, Buldyrev SV, Stanley HE, Shaknovich EI (1998) Discrete molecular dynamics studies of the folding of a protein-like model. Fold Des 3:577–587

Eswar N, Marti-Renom MA, Webb B, Madhusudhan MS, Eramian D, Shen M, Pieper U, Sali A (2006) Comparative protein structure modeling With MODELLER. Curr Protoc Bioinf Supplement 15:5.6.1–5.6.30, John Wiley & Sons, Inc.

Franke D, Svergun DI (2009) DAMMIF, a program for rapid ab-initio shape determination in small-angle scattering. J Appl Cryst 42:342–346

García de la Torre J, Bloomfield VA (1981) Hydrodynamic properties of complex, rigid, biological macromolecules: theory and applications. Q Rev Biophys 14:81–139

Gerstein M, Chothia C (1996) Packing at the protein-water interface. Proc Natl Acad Sci U S A 93:10167–10172

Halle B, Davidovic M (2003) Biomolecular hydration: from water dynamics to hydrodynamics. Proc Natl Acad Sci U S A 100:12135–12140

Kang EH, Mansfield ML, Douglas JF (2004) Numerical path integration technique for the calculation of transport properties of proteins. Phys Rev E 69:031918/1-11

Kim DE, Chivian D, Baker D (2004) Protein structure prediction and analysis using the Robetta server. Nucleic Acids Res 32(S2):W526–W531 (WEB SERVER ISSUE)

Kuntz ID, Kauzmann W (1974) Hydration of proteins and polypeptides. In: Anfinsen CB, Edsall JT, Richards FM (eds) Advances in protein chemistry, vol 28. Academic, New York, pp 239–345

Lee B, Richards FM (1971) The interpretation of protein structures: estimation of static accessibility. J Mol Biol 55:379–400

Nadassy K, Tomas-Oliveira I, Alberts I et al (2001) Standard atomic volumes in double-stranded DNA and packing in protein-DNA interfaces. Nucleic Acids Res 29:3362–3376

Perkins SJ (1986) Protein volumes and hydration effects: the calculation of partial specific volumes, neutron scattering matchpoints and 280nm absorption coefficients for proteins and glycoproteins from amino acid sequences. Eur J Biochem 157:169–180

Rai N, Nöllmann M, Spotorno B et al (2005) SOMO (SOlution MOdeler): differences between X-ray and NMR-derived bead models suggest a role for side chain flexibility in protein hydrodynamics. Structure 13:723–734

Rocco M, Brookes E (2014) Dynamical aspects of biomacromolecular multi-resolution modelling using the UltraScan Solution Modeler (US-SOMO) suite. In: Howard JA, Sparkles HA (eds) The future of dynamical structural science. NATO science for peace and security series A: chemistry and biology, Springer-Verlag, Berlin/Heidelberg. Chapter 13, pp 185–194. doi:10.1007/978-94-017-8550-1_13

Rocco M, Byron O (2015) Computing translational diffusion and sedimentation coefficients: an evaluation of experimental data and programs. Eur Biophys J 44:417–431

Rocco M, Brookes EH, Byron O (2012) US-SOMO: methods for the construction and hydration of macromolecular hydrodynamic models. In: Roberts GCK (ed) Encyclopedia of biophysics. Springer, Berlin/Heidelberg. doi:10.1007/978-3-642-16712-6

Roy A, Kucukural A, Zhang Y (2010) I-TASSER: a unified platform for automated protein structure and function prediction. Nat Protoc 5:725–738

Sayle RA, Milner-White EJ (1995) RasMol: biomolecular graphics for all. Trends Biochem Sci 20:374–376

Spotorno B, Piccinini L, Tassara G et al (1997) BEAMS (BEAds Modelling System): a set of-computer programs for the generation, the visualization and the computation of the hydrodynamic and conformational properties of bead models of proteins. Eur Biophys J 25:373–384 (Erratum 26:417)

Tsai J, Taylor R, Chothia C, Gerstein M (1999) The packing density in proteins: standard radii and volumes. J Mol Biol 290:253–266

Tsodikov OV, Record MT Jr, Sergeev YV (2002) Novel computer program for fast exact calculation of accessible and molecular surface areas and average surface curvature. J Comput Chem 23:600–609

Voss NR, Gerstein M (2005) Calculation of standard atomic volumes for RNA and comparison with proteins: RNA is packed more tightly. J Mol Biol 346:477–492

Voss NR, Gerstein M (2010) 3V: cavity, channel and cleft volume calculator and extractor. Nucleic Acids Res 38:W555–W562

Vriend G (1990) WHAT IF: a molecular modeling and drug design program. J Mol Graph 8:52–56

Chapter 11
The HYDRO Software Suite for the Prediction of Solution Properties of Rigid and Flexible Macromolecules and Nanoparticles

José García de la Torre

Abstract With basis on the classical concept of bead modeling of polymer hydrodynamics, the HYDRO suite of computer programs allows the calculation of solution properties of macromolecules and nanoparticles of any conformation. Bead or bead/shell models are employed to describe arbitrarily shaped rigid particles, and bead-and-spring models can be used for flexible entities. In addition to general-purpose programs, like HYDRO++, the HYDRO suite contains other programs for calculations starting from specific types of structural information, like atomic or residue coordinates (HYDROPRO, HYDRONMR), or 3D density maps from cryoelectron microscopy (HYDROMIC), or other types constructed by the user (HYDROSUB, HYDROPIX). The programs intended for flexible entities are devised in such a way that they can be applied to a variety of problems, from the simple case of semiflexible wormlike chain to complex structures like those of partially disordered proteins. We provide hints on how the topology and partial flexibility of the structures can be represented by springlike connectors. The HYDRO suite contains also some tools to perform optimization of structural parameters by comparison of calculated and experimental data.

Keywords Hydrodynamic properties • Bead modeling • HYDRO programs • Flexible macromolecules • SIMUFLEX

11.1 Introduction

11.1.1 A Broad Overview

The purely theoretical work of Albert Einstein made it possible to obtain some information on the geometric size of molecules and colloidal/nanoparticles from

J. García de la Torre (✉)
Departament of Physical Chemistry, Regional Campus of Excellence Mare Nostrum, University of Murcia, 30071 Murcia, Spain
e-mail: jgt@um.es

S. Uchiyama et al. (eds.), *Analytical Ultracentrifugation*,
DOI 10.1007/978-4-431-55985-6_11

properties in dilute solution/suspension, and Jean Perrin added further theory to account for particle shape. Later on, Kirkwood and Zimm, among others, provided insights regarding how to include flexibility in the theories concerning the relationship between structure and dilute solution properties. It was later, in the late 1960s, when Bloomfield et al. (1967) made their seminal contribution to account for the details of biomolecular structure that were emerging by that time, so proposing the so-called bead modeling methods. In a previous chapter in this book, Byron presents a clear overview of the antecedents and present status of this field. In this chapter, I shall concentrate in (i) some essential aspects of bead modeling, (ii) how it has been successfully implemented to describe from very rough to very detailed rigid macromolecular and nanostructures in the prediction of hydrodynamic and other solution properties, (iii) how that treatment can be naturally integrated with conformational statistics in order to describe the effects of flexibility, and (iv) further procedures to attack the "inverse problem," i.e., how to extract structural information from solution properties.

11.1.2 The Various Kinds of Bead Modeling

The origins of the bead modeling can be placed in the works of Kirkwood and coworkers about describing either flexible (Kirkwood and Riseman 1948) or rigid polymer chains (Kirkwood and Auer 1951) as a string of beads, i.e., centers of frictional resistance behaving in terms of the Stokes' law for isolated spheres and the Oseen's descriptions of hydrodynamic interaction (HI). The next milestone was set by V. A. Bloomfield et al. (1967), who combined their physical insight with the nascent field of scientific computing. Furthermore, essential theory on hydrodynamic interaction (HI) was developed mainly by Rotne and Prager (1969) and Yamakawa (1970), whose improvements and hinted applicability to rigid particles (Yamakawa and Yamaki 1972; Yamakawa and Tanaka 1972) were the basis for further developments by McCammon and Deutch (1976), Nakajima and Wada (1977), and Bloomfield and García de la Torre (1977a) of advanced bead modeling procedures. In 1981, the last two authors wrote a review of bead modeling describing such further developments in theory and computational procedures (García de la Torre and Bloomfield 1981).

As proposed by Bloomfield et al. in the 1960s, there are two versions of bead modeling. In bead modeling in the strict sense, the purpose is reproducing the size and shape as an array of beads with as few beads as possible. In bead/shell modeling (Bloomfield and Filson 1968), a large number of small beads are used to describe in detail just the surface of the particle, which is where frictional forces really act (Fig. 11.1). Bead and bead/shell models for a variety of macromolecular structures are displayed in Fig. 11.2. For a review and comparison of the modeling strategies, see Carrasco and García de la Torre (1999a). In a recent paper, McCammon and coworkers have presented a detailed appraisal of bead modeling methods (Wang et al. 2013).

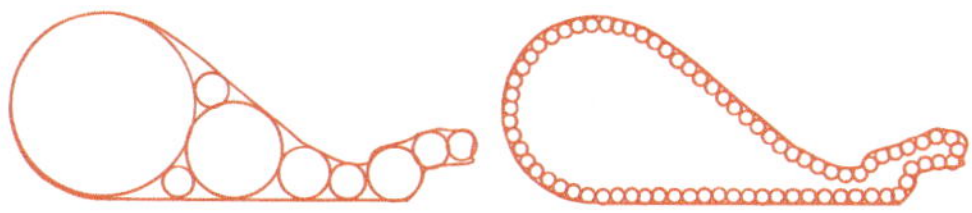

Fig. 11.1 Schemes of (*left*) a bead model in (strict sense) and (*right*) a bead/shell model

In bead modeling in the strict sense, the particle is represented by an array of a moderate number, N, of frictional elements, such that the size and shape of the array match sufficiently that of the actual particle. On the other hand, bead/shell modeling intends a more detailed representation of the shape, describing its surface – which is where hydrodynamic friction takes place – as a shell of a sufficiently large number, N, of small frictional elements ("minibeads").

In bead modeling, the computing time needed for the solution of the hydrodynamic interaction equations is of the order of N^3, so that the importance of N in the model is obvious. By the way, the same happens with the number of elements in finite element modeling (Allison 1998, 1999; Aragón 2004). We recommend the use of a range of decreasing minibead sizes and therefore increasing minibead number, from $N \cong 400$ up to $N \cong 2000$. Thus, extrapolation to zero minibead size is made within the computational procedures, therefore obtaining what would correspond to a smooth surface.

The obvious drawback of computation cost has been addressed in the latest versions of the HYDRO suite having recourse to high-performance computing (HPC) techniques. Thus, a shell calculation with $N \cong 400$–2000 as it is the case of most of the HYDROxxx programs takes typically less than 10 s in an inexpensive personal computer (the term "HYDROxxx" is used here to represent the suite of HYDRO programs available).

11.1.3 Coarse-Grained, Mesoscale, and Multiscale Modeling

Single-valued solution properties, such as hydrodynamic coefficients or the radius of gyration, are obviously related to the low-resolution structure of the solute. Thus, it is certainly justified that the polyhedral head of the T2 virus is represented in Fig. 11.2b by a single bead and its tail a string of beads. As another example, the hydrodynamic properties of an IgG antibody may be influenced by the length of the hinge and the relative size, shape, and disposition of the three subunits but will have little influence of the fine (atomic) details of the protein structure. Thus, a very coarse-grained model bead model (Fig. 11.2c), or a shell model with subunits represented as ellipsoids (Fig. 11.2d), may suffice.

The advent of computational power and availability of high-resolution structures motivated some change of view point in macromolecular dynamics, which evolved to atomic-level descriptions. For some time, single-valued properties characterizing the global dynamics, such as the sedimentation coefficient, were somewhat underestimated, and internal dynamics was approached by atomistic molecular dynamics simulation. Certainly, atomic-resolution structures of globular proteins and nucleic

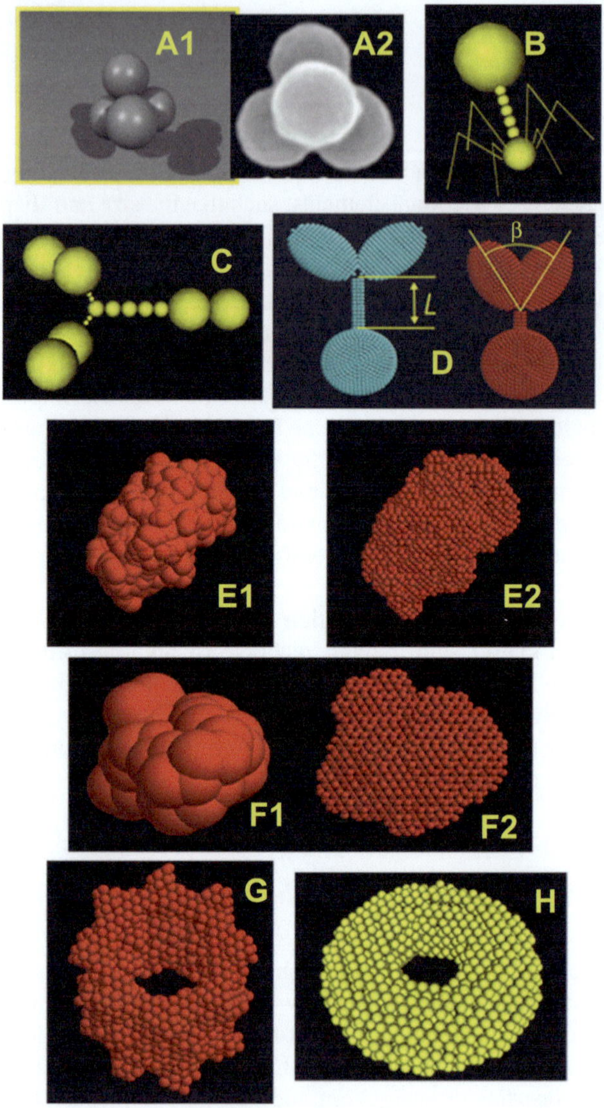

Fig. 11.2 A gallery of bead and bead/shell models. (**a1**) Simplified bead model for a tetrameric structure, (**a2**) as it is indeed found in oligomeric clusters of nanoparticles. (**b**) Bead model for a bacteriophage virus (García de la Torre and Bloomfield 1977b). (**c**) Simplified bead model for a long-hinged IgG antibody (Gregory et al. 1987). (**d**) Mesoscale shell model for antibodies with varying hinge length and subunit arrangement (Amorós et al. 2010). (**e1**) Primary bead model for lysozyme, with one bead per atom, and (**e2**) its derived shell model (García de la Torre et al. 2000a). (**f1**) Primary bead model for BPTI, with one bead per residue, and (**f2**) its derived shell model (Ortega et al. 2011a). (**g**) Shell model for a large protein, chaperonin (CCT), derived from electron microscopy (García de la Torre et al. 2001). (**h**) Shell model for a geometric object: a doughnut-shaped toroid representing a small cyclodextrin molecule (García de la Torre 2001b; Pavlov et al. 2010)

acids are an attractive starting point for predicting solution properties; this is indeed the aim of our HYDROPRO program (García de la Torre et al. 2000a; see Table I in this reference for a review of previous work). Nowadays, it is widely acceptable that such level of structural detail is not capable to describe relevant dynamic events that take place in a long-time scale (or it is just unnecessary in some instances). The present trend is a return to more coarse-grained models, perhaps not as coarse as in the primitive applications. Thus, model elements may represent not atoms but residues, e.g., amino acid or nucleotide residues, or monomers in polymers. This kind of mesoscale representation, with a medium but still very appreciable resolution, saves much computational effort and is still useful in most instances. The book edited by Voth (2009) provides a number of examples. Still, available procedures for atomistic simulation may be useful for predicting properties of the residues or monomers. Thus, the two approaches are employed in what is presently named multiscale modeling, in which the parameters needed for the elements in the coarse-grained model are determined by highly detailed calculation of the entities composing the whole structure. We have employed this approach in rigid bead modeling (e.g., in HYDROSUB) and in the dynamics simulation of flexible chains, as described later in this chapter.

11.1.4 The HYDRO Suite

Following the publication of the first version of the HYDRO program for simple bead models (García de la Torre et al. 1994), a number of other tools, first for rigid particles and then for flexible structures, have been integrated in a suite of computer programs for the prediction of solution properties, including also some tools for the analysis of experimental data (Garcia de la Torre 2014). In this chapter I shall review briefly the set of HYDROxxx programs for rigid particles and then describe the more recently published programs for simulation of flexible structures.

11.2 Rigid Particles

The HYDROxxx programs for the prediction of solution properties of rigid structures are all well documented, with user guides and detailed examples. All of them start from some structural specification and provide as results a number of single-valued solution properties, like diffusion and sedimentation coefficient, the five rotational relaxation times, and the intrinsic viscosity, along with other conformational properties, such as the radius of gyration, the longest distance, and even the particle's covolume, which is needed to evaluate the second virial coefficient (García de la Torre et al. 1999). They also provide the scattering form factor and the distribution of distances, which, accepting the limitations of the models, may be useful as an estimations for light, low-angle X-ray and neutron scattering.

11.2.1 HYDRO++

The original HYDRO program (García de la Torre et al. 1994) was updated more recently (Garcia de la Torre et al. 2007) in the HYDRO++ version, including more accurate descriptions of hydrodynamic properties (Carrasco and García de la Torre 1999b) for an improved calculation of rotational coefficients and intrinsic viscosities (which in previous versions were affected by an ad hoc correction). Furthermore, the accuracy of HYDRO++ has been tested against exact fluid dynamics results for arrays of beads and experimental data of recently constructed clusters of spherical nanoparticles (Fig. 11.2a).

11.2.2 HYDROPRO and HYDRONMR

As mentioned above, HYDROPRO was motivated by the availability of detailed, atomic-level structures of biomacromolecules, coded as PDB files, and followed the aim previously hinted by other authors. The program was conceived to be both accurate and easy to use. Essentially the user has just to supply the PDB file with the atomic coordinates and a few trivial properties of solvent and solute. In the first version of the program (García de la Torre et al. 2000a; García de la Torre 2001a), the programs construct first a primary bead hydrodynamic model (PHM) with one bead per residue (Fig. 11.2e1) with radius a (the present recommended value is 2.9 Å). Then internally the PHM, a bead model with overlapping beads, is replaced by a shell of up to $N_{\text{beads}} \cong 2000$ minibeads (Fig. 11.2e2) The computing time (presently a few seconds in personal computers) is determined by this N_{beads} regardless of the number N_{atoms} of atoms in the model. The same procedure and the a parameter is valid for small oligonucleotides (Fernandes et al. 2002)

The new version of HYDROPRO (Ortega et al. 2011a) maintains the same simple usage, but includes new internal working modes. A novelty is that – in the spirit of present coarse-graining trends – instead of starting with atomic coordinates, in the new modes it just needs the positions of the amino acid residues; the PHM has one bead per residue (Fig. 11.2f1). An obvious advantage is that the program can be still applicable with a lower-resolution structure. Furthermore, this PHM model can be internally processed in two ways (1) being replaced by a shell (Fig. 11.2f2) of up to $N_{\text{beads}} \cong 2000$, as before, and (2) can be used for a bead-model calculation. The latter mode, which deals with a model with important bead overlapping, has been made possible by advances in hydrodynamics of multi-sphere systems (Carrasco and García de la Torre 1999b; Garcia de la Torre et al. 2007, 2010b). In the latter case, the number of elements is $N_{\text{beads}} = N_{\text{residues}}$. Recalling that the computing time is proportional to N^3_{beads}, it is evident that it presents a computational advantage over the two other shell-model modes with $N_{\text{beads}} \cong 2000$ when dealing with cases with less 2000 residues, i.e., smaller than ~200 kD, which is the case for most medium-sized globular proteins.

A thorough check of the performance of the three HYDROPRO models has been made by comparing predicted and experimental values with an extremely large set of data. A summary of the outcome is presented in Fig. 11.3. The percent error in the hydrodynamic (Stokesian) radius, a_T, from diffusion or sedimentation, in the hydrodynamic radius (Einstenian) from intrinsic viscosity, a_I, and in the equivalent radius of gyration, a_G, is indicated for a number of proteins ranging from 56 residues (BPTI) to 10428 (70S ribosome). For the whole set of data, we evaluated the typical percent deviations for each property. As it can be appreciated from the examples in Fig. 11.3, all the procedures give similarly low deviations for the various properties and in the whole range of sizes (for more specific information, see Ortega et al. (2011a)). As a summary, I would enunciate two conclusions: (i) prediction of solution properties can be done with coarse-grained residue-level structures with the same quality as those from atomic-level structure, and (ii) for large protein and macromolecular complexes, the shell-model modes of HYDROPRO provide

Protein	No. res., *Nr*	a_X	Atomic Shell *N*=2000	Residue Shell *N*=2000	Residue Primary *N* = *Nr*
BPTI (4PTI)	56	a_T	-5.8	8.5	6.5
		a_R	-1.8	0.9	0.0
Lysozyme (6LYZ)	129	a_G	-2.5	-4.5	6.4
		a_T	0.9	0.9	0.0
		a_I	-3.1	-3.1	-3.8
		a_R	1.1	1.1	0.3
Chymotrypsinogen (2CGA)	245	a_G	3.3	0.6	9.9
		a_T	-2.1	-2.7	-4.1
		a_I	-4.8	-5.2	-6.2
Beta-Lactoglobulin (1BEB)	324	a_G	-4.2	-3.2	1.9
		a_T	-1.4	1.6	0.3
		a_R	-2.4	0.9	0.1
Oxyhaemoglobin (1HHO)	574	a_T	-0.9	-0.4	-0.9
		a_I	-5.9	-5.4	-5.2
Citrate synthase (1CTS)	874	a_G	0.3	0.0	5.2
		a_T	-3.0	-1.4	-2.1
		a_I	3.1	4.6	4.5
Aldolase (1ADO)	1452	a_T	-0.2	2.6	-0.2
		a_I	-3.6	-2.7	-2.8
Urease (3LA4)	4996	a_T	-3.6	-2.8	-6.9
GroEL (2CGT)	7273	a_G	1.3	3.9	3.0
		a_T	-0.4	3.4	-0.4
IgM (2RCJ)	7514	a_G	-	-4.7	-4.4
		a_T	-	-5.2	-7.1
		a_I	-	-0.7	-0.9
Ribosome 70S (1VSA & 2OW8)	10428	a_G	6.4	4.3	6.4
		a_T	-1.7	-3.4	-0.8

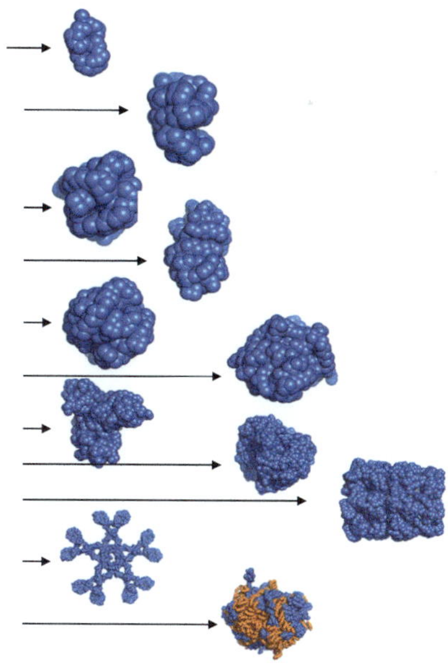

Fig. 11.3 Percent difference between experimental and calculated Stokesian (a_T) and Einstenian hydrodynamic radii (a_I), and radii of gyration (a_G), for a selection of proteins of widely different sizes. Calculation modes are (a) atomic PHM, a=2.9 Å, with shell-model calculation; (b) residues PHM, a=5.0 Å, with shell-model calculation; and (c) residues PHM, a=6.1 Å, with bead-model calculation

predictions with similar accuracy and computational requirements as in the case of small proteins, in contrast with other approaches, like that of Rocco and coworkers (Rai et al. 2005; Brookes et al. 2010), for which computing time grows as the third power of the number of atoms or residues and do not seem feasible for proteins larger than a few hundred kD.

As other HYDROxxx programs, HYDROPRO has been benefited by optimization of the computer code and use of high-performance computing techniques that have dramatically reduced CPU time for any structure, regardless of molecular size, to a few seconds in the simplest personal computer. Furthermore, a user-friendly graphical user interface is available, allowing real-time exploration of changes in structures, parameters, etc., with graphical visualization of bead models and numerical results. Although other approaches, with the same aim as HYDROPRO, are available in the literature and, as HYDROPRO, are of public domain, it can be affirmed that our program is the one most widely used. Since the publication of the year 2000 version, it has received over 600 citations, and the 2011 version, 3 years after its publication, is cited in over 80 references. Recently, some authors are devising quite general tools for structural search by means of ambitious global analysis of NMR, SAXS/SANS, and solution hydrodynamics; HYDROPRO has been the choice for the latter purpose in two significant achievements (Bernadó and Blackledge 2009; Krzeminski et al. 2013).

A few words to mention that a sequel of HYDROPRO is the HYDRONMR program (García de la Torre et al. 2000b), specifically intended for predicting residue-specific NMR T1 and T2 relaxation times. These quantities depend not only (as is the case of the five rotational relaxation times) on the size and shape of the rigid particle, but they are also determined on the location and orientation of the amino acid residue within the protein. Thus, the series of T1/T2 ratios along the sequence of the protein contains a large amount of information. As NMR spectroscopy is somehow far from the reach of this book, the reader is referred to the original publication and to the available computer program from more information.

11.2.3 HYDROMIC

With a purpose similar to HYDROPRO, HYDROMIC (García de la Torre et al. 2001) was conceived to make predictions of solution properties from structures of the (usually large) proteins and macromolecular complexes derived from electron microscopy. Electron-density 3D maps, with a cutoff density, define a 3D shape of the particle. Instead of the PDB atomic coordinates in HYDROPRO, HYDROMIC uses such density map as the primary structural information and, as the first version of HYDROPRO, constructs hydrodynamic bead/shell models for which properties are evaluated (Fig. 11.2g). The use of this tool is not as extended as that of HYDROPRO. Apart from the more limited amount of structural information of this kind, the variety of different formats for the density maps has been a further impediment. HYDROMIC was initially programmed as to work with the "spider"

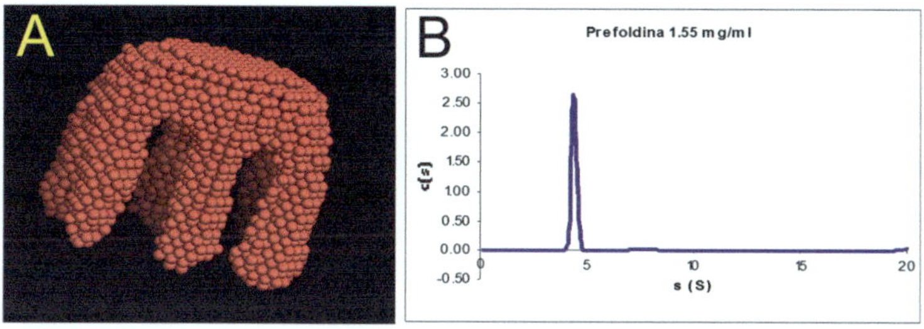

Fig. 11.4 (**a**) Shell model constructed by HYDROMIC when processing the 3D EM map of prefoldin. (**b**) SEDFIT (Schuck 2000) analysis of a sedimentation velocity experiment

format, which was by no means the only one. The latest version included further the common "MRC" format. A public domain tool "em2em" (ImageScience 2014) is available for conversion of other formats.

As a sample of recent application, we advance here results (Ortega et al., to be published) on the experimental and computational characterization of prefoldin, a chaperon protein with a peculiar six-digit hand shape that likely works as an efficient clamp to carry its cargo. With a protein sample and the MRC electron density map, kindly supplied by Prof. J. M. Valpuesta (CSIC, Madrid), we first carried out sedimentation velocity experiments at three concentrations, from 1.55 to 0.19 mg/ml, with both absorbance and interference detection, observing always a sharp peak at $s_{20,w} = 4.5 \pm 0.1$ S, independent of concentration. The HYDROMIC prediction was $s_{20,w} = 4.6$ S (Fig. 11.4).

11.2.4 HYDROSUB

The rationale underlying the HYDROSUB (Garcia de la Torre and Carrasco 2002) method belongs to the concept that has been mentioned above as multiscale coarse-graining modeling (we have also sometimes used the term "crystallohydrodynamics"; Carrasco et al. 2001; Lu et al. 2007). The hydrodynamic model is composed by ellipsoids of revolution or cylindrical subunits, which are internally represented by shell models. It presents the obvious advantage of constructing models with nonspherical elements. In applications, it is particularly suited for large multisubunit complexes whose whole structures are not amenable to direct determination (or is just better handled with a coarse-grained representation), while sufficient information is available separately for the subunits. The size and shape of the ellipsoids or cylinders can be fitted from experimental data (Harding et al. 1997; García de la Torre and Harding 2013). Alternatively, if high-resolution diffraction or NMR information is available for the structure of a subunit, a best-fitting ellipsoid can be found from such structure using either COVOL (Harding et al. 1999) or

```
LOGICAL CONDITION
........
XMAX=80. ; YMAX=80. ; ZMAX=20.
SPACING=1.0
........
R=SQRT((X-40.)**2+(Y-40.)**2)
CONDITION =R.GE.15.AND.R.LE.40.
```

Fig. 11.5 Lines of Fortran code to be inserted in HYDROPIX for specifying a disk with a hole with inner and outer radii 15 and 40, respectively, and thickness 20. Enclosing boxes have opposite corners at (0,0,0) and (80,80,20). The particle's center is at (40,40,10)

ORIEL (Fernandes et al. 2001). One could even "generate" solution properties using HYDROPRO that would be then retrofitted to get subunit dimensions.

Once the size and shape of the subunits are fixed, the structure of the multisubunit complex is defined by a minimum set of a few geometric parameters that determines their arrangements (Fig. 11.2d). This way makes it possible to carry out predictions of solution properties for this kind of structures in terms of such few geometric parameters or deduce them from the properties using tools that we have also devised (*vide infra*).

11.2.5 HYDROPIX

Last but not least, HYDROPIX (García de la Torre 2001b) is a software program that permits the calculation of solution properties for any arbitrarily shaped particle. Instead of being specified by a model of spheres or ellipsoids, the particle is "programmed." A Fortran source code is provided, within which the user has to insert two pieces of code that (1) gives the coordinates of a box that encloses the particle, the program which fills the box with a cubic lattice, and (2) determines whether or not a point within the box (node in the lattice) belongs to the particle (Fig. 11.5).

11.3 Flexible Particles

11.3.1 Introduction

As indicated above, bead models were introduced for the characterization of flexible macromolecules. Beads, joined by suitable (either rigid or partially flexible) connectors, were proposed as models for chain macromolecules. In order to develop theories that would supply analytical expressions for the solution properties, such very coarse-grained models were treated with approximate description of conformational statistics and hydrodynamic interaction. Present textbooks provide pedagogical descriptions of those classical treatments (Rubinstein and Colby 2003; Serdyuk et al. 2007; Hiemenz and Lodge 2007). When computers become accessible, abstract, theoretical work was replaced by computer simulation. A landmark is the proposal by Zimm (1980) of coupling Monte Carlo (MC) simulation with the more rigorous hydrodynamics used for rigid bead models by García de la Torre and Bloomfield (1977a, b). However, by the same time, the availability of computing power was increasing spectacularly, and it was possible to have recourse to computer simulation to solve the problems not accessible to pure theory. A great deal of knowledge on macromolecular dynamics has been achieved by MC methods (Binder 1995). Particularly, the rigid-body Monte Carlo (RBMC) procedure described below has been widely employed to obtain hydrodynamic coefficients and conformational properties of flexible structures ranging from random coils (García de la Torre et al. 1982) to hinged, semiflexible particles (Iniesta et al. 1988) and semiflexible wormlike chains (Amorós et al. 2011).

An alternative for studying, in a rigorous way, every detail of macromolecular hydrodynamics (beyond the limitations of RBMC, *vide infra*) is Brownian dynamics (BD) simulation. In a pioneering paper, Ermak and McCammon (1978) proposed a practical simulation procedure that embodies first principles of Brownian motion with the fluid dynamics concept of hydrodynamic interaction (HI), of which the first applications to macromolecular hydrodynamics appeared in the 1980s (e.g., Allison and McCammon 1984; Diaz et al. 1987). Over the years, improvements of this algorithm and other procedures for DB simulation have been developed; for a recent overview, see Rodríguez Schmidt et al. (2011).

11.3.2 A General Bead-and-Spring Model

Like in the bead models for rigid particles, models for flexible entities are composed by beads, which are the elements at which the frictional forces act. We stress here the relevance of using bead models for rigid particles, rather than other descriptions (e.g., Aragón 2004); a large body of knowledge and developments, from concepts to computer codes, for rigid bead models can be used for flexible ones.

The additional ingredients are those intended for representing the intramolecular interactions and the internal degrees of freedom. These will be expressed in terms of interbead potentials in RBMC or forces in BD. The most basic ones are connectors joining neighbor beads that must remain somehow bonded. Rigid, fixed bond-length constraints present some implementation difficulty, and it is generally preferable to employ “springs,” i.e., bonds of variable stiffness, with the simplest case being the Hookean spring potential:

$$V_{ij}^{(conn)}\left(l_{ij}; H_{ij}, l_{ij,eq}\right) = \frac{1}{2} H_{ij}\left(l_{ij} - l_{ij,eq}\right)^2$$

where l_{ij} is the instantaneous bond length and $l_{ij,eq}$ is its equilibrium value. A value of H_{ij} much greater than $k_B T/l_{ij,eq}^2$ ensures that the bond is very stiff; the r.m.s. fluctuation of l_{ij} is $(k_B T/H_{ij})^{1/2}$; for instance, it is only 10 % if $H_{ij} = 100 k_B T/l_{ij,eq}^2$. A more sophisticated – and more widely applicable potential – is the one accounting for arbitrary spring stiffness and finite extensibility, $l_{ij,max}$,with a potential with three parameters $V_{ij}^{(conn)}(l_{ij}; H_{ij}, l_{ij,eq}, l_{ij,max})$ (del Río Echenique et al. 2009; García de la Torre et al. 2010a).

Angles between two neighbor bonds may be constrained by a bending potential, involving the position of three beads:

$$V_{ijk}^{(ang)}\left(\alpha_{ijk}; H_{ijk}, \alpha_{ijk,eq}\right) = \frac{1}{2} Q_{ijk}\left(\alpha_{ijk} - \alpha_{ijk,eq}\right)^2$$

where α_{ijk} is the angle subtended by the bond vectors i→j and j→k ($\alpha_{ijk} = 0$ if the vectors are aligned), and $\alpha_{ijk,eq}$ is the equilibrium value of this angle. Quasi-rigid bond angles may be determined by a fixed $\alpha_{ijk,eq}$ and a sufficiently high value of $Q_{ijk}/k_B T$. Another case is the bead-and-connector representation of bending flexibility of wormlike chains; then the equilibrium configuration is straight, and bond lengths have all the same l_{eq}, and the force constant is related to the persistence length (Hagerman and Zimm 1981; Allison 1986; Garcia de la Torre 2007), by

$$P = Q\, l_{eg}\ / k_B T$$

while the contour length, L, is related to the number of quasi-rigid bonds in the discrete bead-and-connector representation as $L \cong N\, l_{eq}$.

Another essential interaction refers to interaction between nonbonded pairs of beads, which reflect long-range intramolecular interactions. An essential contribution is that from attractive/repulsive van der Waals interactions, usually referred to as excluded volume (EV) effects. The most easy way to account them for is the hard-sphere potential, $V^{(nonbond)}(r) = 0$ if $r > r_{cutoff}$ and $V^{(nonbond)}(r) = \infty$ if $r < r_{cutoff}$, where r_{cutoff} is a distance that can be taken as the sum of some effective radii of the elements. In the case of BD, where the intervention of forces rather than potentials is required, continuous and differentiable potentials are needed. One obvious candidate is the Lennard-Jones potential:

$$V_{ij}^{(nonbond)}\left(r_{ij};\ \varepsilon_{ij}, \sigma_{ij}\right) = 4\varepsilon_{ij}\left[\left(r_{ij}/\sigma_{ij}\right)^{12} - \left(r_{ij}/\sigma_{ij}\right)^{6}\right]$$

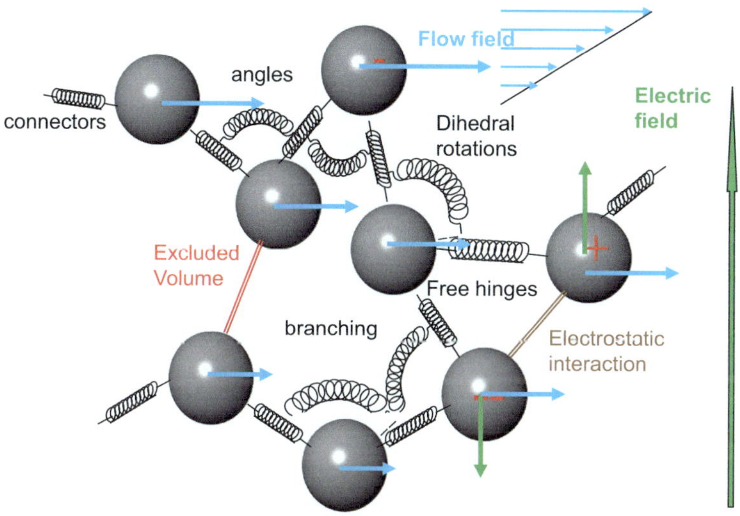

Fig. 11.6 Schematic representation of a bead-and-connector model for flexible particles with arbitrary topology and diverse intra- and extra-molecular interactions

where ε_{ij} and σ_{ij} are the classical Lennard-Jones energetic and geometric parameters, respectively. Many other forms of continuous, differentiable potentials have been proposed in the MC and molecular or Brownian dynamics literature.

Of course, many other intramolecular interactions may be relevant. Such as is the case of electrostatic interactions, usually mediated by the ionic strength of the solvent; a $V^{(\mathrm{electDH})}(r)$ can be included by a Debye-Hückel potential between charged beads. The interaction between the molecule in a flow field or its charged elements and dipolar bonds in an electric field may also be included in the model and simulation procedures. Last but not the least, the method is not restricted to the simplest, linear topology; instead, any other (say circular, branched, etc.) can be considered. A general overview of such a general bead-and-connector model is depicted in Fig. 11.6

11.3.3 MONTEHYDRO and SIMUFLEX

This general mechanic model is implemented in our programs MONTEHYDRO and SIMUFLEX for flexible particles. The user can choose from a menu including a variety of intramolecular interactions and external agents. MONTEHYDRO (García de la Torre et al. 2005) carries out Monte Carlo simulations and computes overall properties in the MCRB scheme, i.e., as averages over conformations considered as instantaneously rigid. This scheme is somehow approximate (Fixman 1986; Rodriguez Schmidt et al. 2012), but the bias that it may introduce seems appreciable only for very long and very flexible chains, and in practical instances it is assumed to be of the same order as the uncertainty of experimental data.

However, the neglect of internal dynamics makes MCRB inadequate for the prediction of more detailed, local-scale aspects of macromolecular dynamics. Brownian dynamics (BD) simulation is the most general and rigorous approach (although more complex to carry out and more computationally intensive). Based on the same mechanical model – now in terms of forces, not potentials – BD has been implemented in our package SIMUFLEX (García de la Torre et al. 2009). It actually consists of BROWFLEX, basically a BD simulation engine, and ANAFLEX, which carries out a variety of analysis of the BD trajectories generated by BROWFLEX, from the simple statistics for the mean square displacement of the center or mass, which provides a rigorous evaluation of the diffusion coefficient, to the reorientational correlation times required for NMR relaxation. The overall properties alternatively evaluated by MONTEHYDRO and SIMUFLEX are in good agreement, but SIMUFLEX allows for a direct simulation of the internal dynamics. A nice application of this possibility is the simulation of single-molecule events, such as the different ways on unfolding of a DNA molecule in an elongational flow (Perkins et al. 1997; del Río et al. 2009), or the effect of strong centrifugal forces in an extremely long-chain molecule (like those very long viral DNAs) which produce the so-called anomalous sedimentation consisting of an unexpected effect of rotor speed on the observed sedimentation coefficient (Zimm 1974; Zimm et al. 1976; Schlagberger and Netz 2008).

11.3.4 Examples: Dendrimers and Intrinsically Disordered Proteins

It seems worth to mention very briefly two recent applications of MONTEHYDRO and SIMUFLEX in two fields of current, intense activity: dendrimers (del Río et al. 2009), as synthetic, polymeric nanomaterials, and intrinsically disordered proteins, a major challenge of present protein biophysics. In both cases, a multiscale approach was followed, avoiding to fit parameters against experimental data; instead, the parameters of the coarse-grained models are extracted from existing structural information or gathered from atomistic simulations, not of the whole molecule but of its constituent entities.

Dendrimers are regularly branched polymers with an absolutely defined topology and molecular size. Branches are small flexible molecular entities having a few chained atoms. One bead and one spring represented each branch. Their effective hydrodynamic radius was evaluated by the RB treatment using molecular dynamics, MD (rather than MC) simulation of a single branch, which was employed also to determine the distribution of end-to-end distance, according to which parameters H_{ij}, $l_{ij,eq}$, and $l_{ij,max}$ for $V_{ij}^{(conn)}$ were fixed. Similarly, MD simulations of a branched trimer were employed to obtain statistic needed to fix the angular parameters

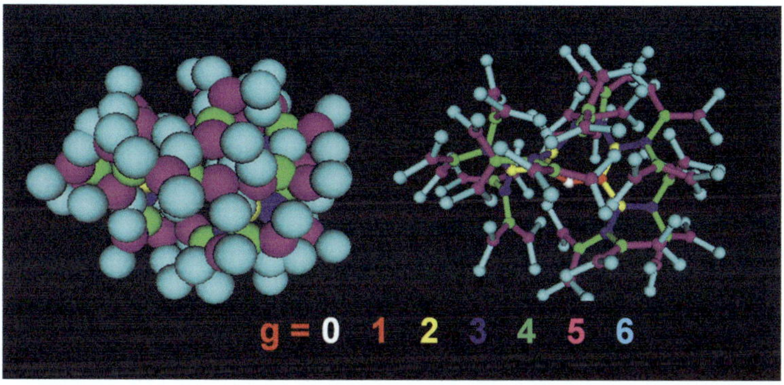

Fig. 11.7 Snapshot of a sixth-generation dendrimer of mono-polybenzylether with 127 branches (*left* is actual model; at *right*, bead sizes reduced to show the connectors)

H_{ijk}, $\alpha_{\text{ijk,eq}}$. In this way, the general mechanic model (Fig. 11.6) for dendrimers is parameterized in an ab initio manner. A snapshot of an instantaneous conformation of a sixth-generation dendrimer is displayed on Fig. 11.7. For four different kinds and various generation numbers, the hydrodynamic radii and radius of gyration were predicted with accuracy of 3–5 %. Furthermore, the BD simulation allows for the simulation of internal dynamics, with greater mobility on going from the inner to the outer branches, which is reflected in NMR relaxation times, and is related to the application of dendrimers in drug encapsulation (García de la Torre, to be published).

We have used the methodology to predict overall properties and internal dynamics of a subclass of intrinsically disordered proteins, in which one can differentiate quasi-rigid domains and flexible linkers or tails. In the coarse-grained model, with one bead per amino acid residues, the linkers and tails are modeled as flexible chains with virtual C^{α}–C^{α} bonds with $l_{\text{eq}} = 3.8$ Å and H sufficiently high so that the virtual bond is quasi-rigid. For coherence with the hydrodynamic representation of amino acid residues in HYDROPRO, the hydrodynamic radius of the residue was as there 6.0 Å. Parameters for the angular potential were taken from a statistics of angles between consecutive C^{α}–C^{α}–C^{α} virtual bonds in the coil regions of proteins (Kleywegt 1997). The excluded volume parameters were taken so that calculations of solution properties of fully disordered (chemically denaturated) proteins were accurately described (García de la Torre, to be published). For the globular, quasi-rigid domains, a special intramolecular potential (the so-called Gõ model, Clementi et al. 2000) is employed for the so-called essential pairs (Sobolev et al. 1999).

Several proteins, displayed in Fig. 11.8, were considered in this study. Excellent agreement with experimental SAXS/SANS results was always found for the radius of gyration, and in spite of the coarse-grained modeling, even the scattering intensities and distribution of distances were quite accurately reproduced. For ZipA, the experimental and calculated sedimentation coefficients were 2.2 and 2.1 S, respectively. For BTK, these values were 3.9 and 3.3, with a more deficient

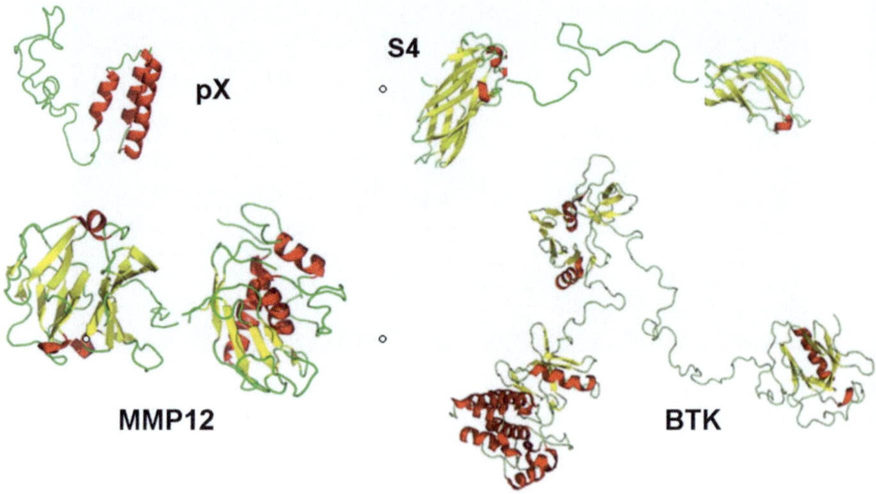

Fig. 11.8 Snapshots of conformation of some intrinsically disordered proteins

agreement that can be understood considering not only the amount and complexity of the modeling and computational methodology but also the complex structure of the protein itself, with 659 residues structured in four globules connected by four linkers.

11.3.5 Wormlike Chains

The Kratky-Porod wormlike chain (WC) is the essential model for many polymers having a continuous stiffness, and its utility encompasses, in addition to the paradigmatic case of double-stranded DNA, most polysaccharides and a number of synthetic polymers. The classical work by Yamakawa and Fuji (1973, 1974) has been for many years the basis for the determination from solution properties of the three essential parameters: the persistence length, P; mass per unit length, $M_L = M/L$; and hydrodynamic diameter, d. However, it is well known that their equations do not cover the whole range of conformations, gauged by the ratios L/d and L/P. Thus, it fails for short, thick rods and for long flexible chain is affected by the well-known preaveraging approximation.

Pursuing the description of conformation and dynamics of semiflexible chains and DNA in particular has been an essential purpose of this author for many years (García de la Torre et al. 1975; García de la Torre and Horta 1976). In an attempt to provide a computational framework that would be able to predict solution properties of WCs for the whole range of the parameters, we recently undertook a computer simulation (Amorós et al. 2011), yielding numerical results and a computer program which have been tested very satisfactorily with experimental data of sedimentation and diffusion coefficients, intrinsic viscosity, and radius of gyration of DNA from 8 to 200,000 base pairs. As we describe later on, this program is the basis for a tool intended for the inverse problem of determining the WC parameters from experimental data.

11.4 Global Fitting for Structural Determination: HYDROFIT

An obvious aim of measuring and calculating solution properties of macromolecules and nanoparticles is to obtain information about its structure in solution. Unfortunately, single-valued properties do not convey sufficient information content as to provide detailed structural information. Still, in some cases, the structural search is reduced to the determination of a set of parameters characterizing an assumed kind of structure. The computer programs for calculation of properties could be used in a trial-and-error manner in the search for such parameters. We have developed tools to make the search easy and systematic. A brief summary is presented here; for details, see Ortega et al. (2011b).

11.4.1 HYDFIT

The HYDROxxx programs admit the calculation, in a single run, for multiple structures. Users can code ancillary programs for producing the data files, but we have also developed tools like MULTIHYDRO or MULTISUB that facilitate that task. Among the output from the HYDROxxx programs is a file intended to be read by the HYDFIT program, which is in charge of finding the best-fitting structure by comparison with a set of experimental properties. The program works optimally with a varied set of properties, including, say, sedimentation or diffusion coefficients, intrinsic viscosity, radius of gyration, longest distance, etc. Internally HYDFIT seeks to minimize a target function Δ^2 that measures the square deviations between calculated and experimental results. In order to treat simultaneously different properties in an equilibrated manner, the analysis is made in terms of equivalent radii (Ortega and García de la Torre 2007).

The value 100Δ is representative as the typical percent error for the whole set of properties.

As a pedagogical example, the HYDFIT user manual describes a hypothetical case in which a double-stranded short DNA oligonucleotide is bent, at a point and with some angle to be determined. The short DNA is modeled as a rigid, bent array of beads. MULTIHYDRO helps in the construction of a series of models in which the position and bent angle are varied. HYDRO++ computes the properties for all

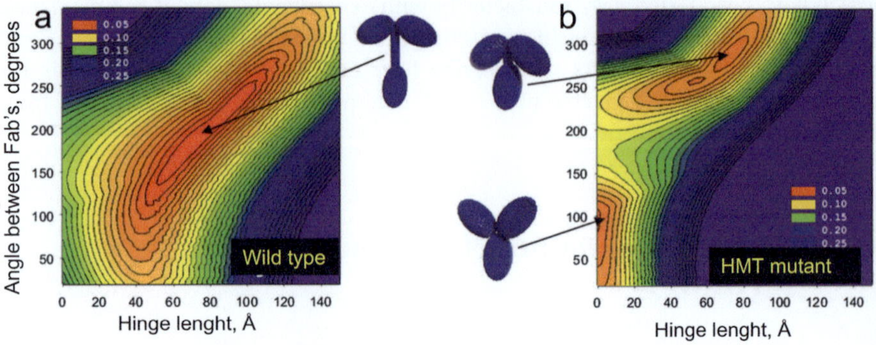

Fig. 11.9 Contour plots of the target function Δ in the HYDROSUB/HYDFIT analysis of two species of antibody IgG3 (*left* wild type, *right* mutant HMT)

the structures in a single run and produces the summary file, which is processed by HYDFIT. The bent position and angle can be unambiguously determined from solution properties.

Another real application has been the determination of the differences between the wild type (WT) of antibody IgG3 and one of its mutants (M15). Coarse-grained, reduced models, in which the essential parameters where the length of the hinge, L_h, and the angle, β, between the Fab subunits (Fig. 11.2d) were submitted to HYDROSUB calculation. This was done for multiple conformations, with varying L_h and β, generated by MULTISUB. Then, HYDFIT processed the data searching for minima of Δ (Fig. 11.9). The analysis reveals a well-defined conformation for the WT, characterized by a remarkably long hinge. For M15 there are some structures that fit the data with nearly the same deviation (note that the detection of these possible ambiguities is another merit of the HYDROFIT approach), but all are characterized by a much shorter hinge.

11.4.2 Multi-HYDFIT

While HYDROFIT is intended for the determination of the structure of a single molecule or sample, Multi-HYDFIT attempts the fit of various properties for a series of samples having in common the same model parameters. Such is the case for a series of samples of a wormlike macromolecule with varying molecular weight, for which various properties may be available. With the same rationale as in HYDROFIT, here again the target function is minimized for the whole series of samples and all the properties in a truly global fit of the WC parameters. A number of examples of using Multi-HYDFIT are provided in the original reference (Amorós et al. 2011; see also the supporting information of this paper). Just to present here

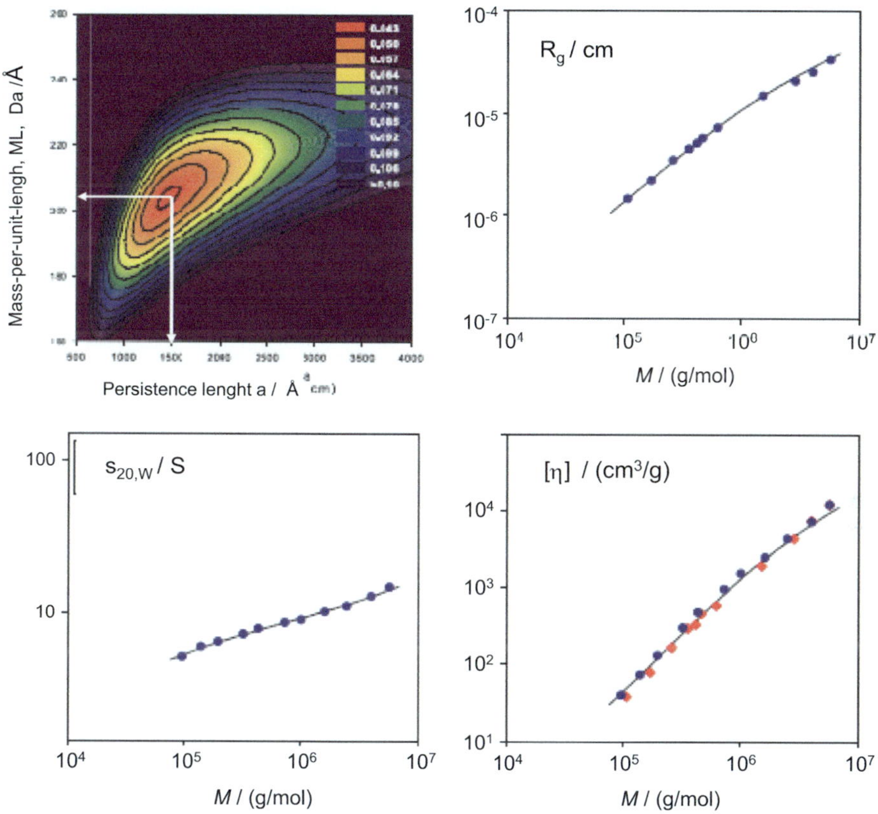

Fig. 11.10 Multi-HYDFIT fit of experimental data and calculated values for three properties of schizophyllan. $P = 1500$ Å, $M_L = 205$ Da/ Å, $d = 23$ Å (Amorós et al. 2011). The contour plot displays Δ vs. P and M_L, showing the position of the best-fitting minimum

one of them, I choose the analysis of properties of a very stiff polysaccharide: schizophyllan. Data for four properties (Yanaki et al. 1980, Kashiwagi et al. 1981) covering two decades in molecular weight were globally fitted by Multi-HYDFIT, collecting data in a single file, and running the program, which took barely 2 min (Fig. 11.10).

11.5 Conclusions

The HYDRO suite – which is the fruit of nearly 40 years of the authors' work – provides a collection of tools that have been consistently developed, in the context of the theory of hydrodynamic interaction of bead models. Utilities are available for both rigid and flexible structure, with the added bonus of some programs for the inverse problem of structural determination from solution properties. Ample documentation, including users' guides and worked examples, are available for all the programs, which can be downloaded freely and anonymously from our web site (Garcia de la Torre 2014).

Acknowledgments This chapter was written while the author was supported by grant CTQ2012-33717 from *Ministerio de Ciencia y Competitividad*, including FEDER funds. The author's group is one of the groups of excellence in the Region of Murcia, supported by *Fundación Séneca*, grant 04531/GERM/06 and grant 19353/PI/14.

References

Allison SA (1986) Brownian dynamics simulation of wormlike chains. Fluorescence depolarization and depolarized light scattering. Macromolecules 19:118–124

Allison SA (1998) The primary electroviscous effect of rigid polyions of arbitrary shape and charge distributions. Macromolecules 31:4464–4474

Allison SA (1999) Low Reynolds number transport properties of axisymmetric particles employing stick and slip boundary conditions. Macromolecules 32:5304–5212. Addition/Correction: (1999) 32:7710–7710

Allison SA, McCammon JA (1984) Transport properties of rigid and flexible macromolecules by Brownian dynamics simulation. Biopolymers 23:167–187

Amorós D, Ortega A, Harding SE, García de la Torre J (2010) Multi-scale calculation and global-fit analysis of hydrodynamic properties of biological macromolecules: determination of the overall conformation of antibody IgG molecules. Eur Biophys J 39:361–370

Amorós D, Ortega A, García de la Torre J (2011) Hydrodynamic properties of wormlike macromolecules: Monte Carlo simulation and global analysis of experimental data. Macromolecules 44:5788–5797

Aragón S (2004) A precise boundary element method for macromolecular transport properties. J Comput Chem 25:1191–1205

Bernadó P, Blackledge M (2009) A self-consistent description of the conformational behaviour of chemically denaturated proteins from NMR and small angle scattering. Biophys J 97:2839–2845

Binder K (ed) (1995) Monte Carlo and molecular dynamics simulations in polymer science. Oxford University Press, New York

Bloomfield VA, Filson DP (1968) Shell model calculations of translational and rotational frictional coefficients. J Polym Sci Part C 25:73–83

Bloomfield VA, Dalton WO, Van Holde KE (1967) Frictional coefficients of multisubunit structures. I. Theory. Biopolymers 5:135–148

Brookes E, Demeler B, Rocco M (2010) Developments in the US-SOMO bead modeling suite: new features in the direct residue-to-bead method, improved grid routines, and influence of accessible surface area screening. Macromol Biosci 10:746–753

Carrasco B, García de la Torre J (1999a) Hydrodynamic properties of rigid particles. Comparison of different modelling and computational procedures. Biophys J 76:3044–3057

Carrasco B, García de la Torre J (1999b) Improved hydrodynamic interaction in macromolecular bead models. J Chem Phys 111:4817–4826

Carrasco B, García de la Torre J, Davis KG, Jones S, Athwal D, Walters C, Burton DR, Harding S (2001) Crystallohydrodynamics for solving the hydration problem for multi-domain proteins: open physiological conformations for human IgG. Biophys J 93:181–196

Clementi C, Nyemeyer H, Onuchic J (2000) Topological and energetic factors: what determines the structural details of the transition state ensemble and "En-route" intermediates for protein folding? An investigation for small globular proteins. J Mol Biol 298:937–953
del Río EG, Rodríguez Schmidt R, Freire JJ, Hernández Cifre JG, García de la Torre J (2009) A multiscale scheme for the simulation of conformational and solution properties of some dendrimer molecules. J Am Chem Soc 131:8549–8556
Diaz FG, Iniesta A, García de la Torre J (1987) Brownian dynamics simulation of rotational correlation functions of simple rigid models. J Chem Phys 87:6021–6027
Ermak DL, McCammon JA (1978) Brownian dynamics with hydrodynamic interactions. J Chem Phys 69:1352–1360
Fernandes MX, Bernadó P, Pons M, García de la Torre J (2001) An analytical solution to the problem of the orientation of rigid particles by planar obstacles. Application to membrane systems and to the calculation of dipolar couplings in protein NMR Spectroscopy. J Am Chem Soc 123:12037–12047
Fernandes MX, Ortega A, López Martínez MC, García de la Torre J (2002) Calculation of hydrodynamic properties of small nucleic acids from their atomic structures. Nucleic Acids Res 30:1782–1788
Fixman M (1986) Translational diffusion of chain polymers. I. Improved variational bounds. J Chem Phys 84:4080–4086
García de la Torre J (2001a) Hydration from hydrodynamics. General considerations and applications of bead modelling to globular proteins. Biophys Chem 63:159–170
García de la Torre J (2001b) Building hydrodynamic bead-shell models for rigid particles of arbitrary shape. Biophys Chem 94:265–274
Garcia de la Torre J (2007) Dynamic electro-optic properties of macromolecules and nanoparticles in solution. A review of computational and simulation methodologies. Coll Surf B 56:4–15
Garcia de la Torre J (2014) http://leonardo.inf.um.es/macromol/. Accessed 15 Apr 2014
García de la Torre J, Bloomfield VA (1977a) Hydrodynamic properties of macromolecular complexes. I. Translation. Biopolymers 16:1747–1763
García de la Torre J, Bloomfield VA (1977b) Hydrodynamic properties of macromolecular complexes. III. Bacterial viruses. Biopolymers 16:1779–1793
García de la Torre J, Bloomfield VA (1981) Hydrodynamic properties of complex, rigid, biological macromolecules. Theory and applications. Q Rev Biophys 14:81–139
Garcia de la Torre J, Carrasco B (2002) Hydrodynamic properties of rigid macromolecules composed of ellipsoidal and cylindrical subunits. Biopolymers 63:163–167
García de la Torre J, Harding SE (2013) Hydrodynamic modelling of protein conformation in solution: ELLIPS and HYDRO. Biophys Rev 5:195–206
García de la Torre J, Horta A (1976) Sedimentation coefficient and X-Ray scattering of double-helical model for deoxyribonucleic acid. J Phys Chem 80:2028–2035
García de la Torre J, Freire JJ, Horta A (1975) A bead and spring model for the stiffness of DNA. Biopolymers 14:1327–1335
García de la Torre J, Jiménez A, Freire JJ (1982) Monte Carlo calculation of hydrodynamic properties of freely jointed, freely rotating and real polymethylene chains. Macromolecules 15:148–154
García de la Torre J, Navarro S, López Martínez MC, Díaz FG, López Cascales JJ (1994) HYDRO: a computer software for the prediction of hydrodynamic properties of macromolecules. Biophys J 67:530–531
García de la Torre J, Carrasco B, Harding SE (1999) Calculation of NMR relaxation, covolume and scattering-related properties of bead models using the SOLPRO computer program. Eur Biophys J 28:119–132
García de la Torre J, Huertas ML, Carrasco B (2000a) Calculation of hydrodynamic properties of globular proteins from their atomic-level structure. Biophys J 78:719–730
García de la Torre J, Huertas ML, Carrasco B (2000b) HYDRONMR: prediction of NMR relaxation of globular proteins from atomic-level structures and hydrodynamic calculations. J Magn Reson 147:138–146

García de la Torre J, Llorca O, Carrascosa JL, Valpuesta JM (2001) HYDROMIC: prediction of hydrodynamic properties of rigid macromolecular structures obtained from electron microscopy. Eur Biophys J 30:457–462
García de la Torre J, Ortega A, Pérez Sánchez HE, Hernández Cifre JG (2005) MULTIHYDRO and MONTEHYDRO: conformational search and Monte Carlo calculation of solution properties of rigid and flexible macromolecular models. Biophys Chem 116:12–128
Garcia de la Torre J, del Rio Echenique G, Ortega A (2007) Calculation of rotational diffusion and intrinsic viscosity of bead models for macromolecules and nanoparticles. J Phys Chem B 111:955–961
García de la Torre J, Hernández Cifre JG, Ortega A, Rodríguez Schmidt R, Fernandes MX, Pérez Sánchez HE, Pamies R (2009) SIMUFLEX: algorithms and tools for simulation of the conformation and dynamics of flexible molecules and nanoparticles in dilute solution. J Chem Theory Comput 5:2606–2618
García de la Torre J, Ortega A, Amorós D, Rodríguez Schmidt R, Hernández Cifre JG (2010a) Methods and tools for the prediction of hydrodynamic coefficients and other solution properties of flexible macromolecules in solution. A tutorial minireview. Macromol Biosci 10:721–730
García de la Torre J, Amorós D, Ortega A (2010b) Intrinsic viscosity of bead models for macromolecules and nanoparticles. Eur Biophys J 39:381–388
Gregory L, Davis KG, Sheth B, Boyd J, Jefferis R, Naves D, Burton K (1987) The solution conformations of the subclasses of human IgG deduced from sedimentation and small angle X-ray scattering studies. Molec Immunol 24:821–829
Hagerman P, Zimm BH (1981) Monte Carlo approach to the analysis of the rotational diffusion of wormlike chains. Biopolymers 10:1481–1502
Harding SE, Horton JC, Coelfen H (1997) The ELLIPS suite of macromolecular conformation algorithms. Eur Biophys J 25:347–359
Harding SE, Horton JC, Jones S, Thornton JM, Winzor D (1999) COVOL: an interactive program for evaluating second virial coefficients from the triaxial shape or dimensions of rigid macromolecules. Biophys J 76:2434–2438
Hiemenz PC, Lodge TP (2007) Polymer chemistry, 2nd edn. CRC Press, Boca Raton
ImageScience (2014) https://www.imagescience.de/em2em.html. Accessed 1 May 2014
Iniesta A, Diaz FG, García de la Torre J (1988) Transport properties of rigid bent-rod macromolecules and semiflexible broken rods in the rigid body approximation. Biophys J 54:269–275
Kashiwagi Y, Norisuye T, Fujita H (1981) Triple helix of schizophyllum commune polysaccharide in dilute solution. 4. Light scattering and viscosity in dilute sodium hydroxide. Macromolecules 14:1220–1225
Kirkwood JG, Auer PL (1951) The visco-elastic properties of solutions of rod-like macromolecules. J Chem Phys 19:281–287
Kirkwood JG, Riseman J (1948) The intrinsic viscosities and diffusion constants of flexible macromolecules in solution. J Chem Phys 16:565–573
Kleywegt GJ (1997) Validation of protein models from C^{α} coordinates alone. J Mol Biol 273:371–376
Krzeminski M, Marsh JA, Neale C, Choy WY, Forman-Kay JD (2013) Characterization of disordered proteins with ENSAMBLE. Bioinformatics 29:398–399
Lu Y, Harding S, Michaelsen T, Longman E, Davis KG, Ortega A, Grossman G, Sandlie I, Garcia de la Torre J (2007) Solution conformation of wild type and mutant IgG3 and IgG4 immunoglobulins using crystallohydrodynamics: possible implications for complement activation. Biophys J 93:3733–3744
McCammon JA, Deutch JM (1976) Frictional properties of nonspherical multisubunit structures. Application to tubules and cylinders. Biopolymers 15:1397–1408
Nakajima H, Wada Y (1977) A general method for the evaluation of diffusion constant, dilute-solution viscoelasticity and the complex dielectric constant of a rigid macromolecule with an arbitrary configuration. Biopolymers 16:875–893

Ortega A, García de la Torre J (2007) Equivalent radii and ratios of radii from solution properties as indicators of macromolecular shape, conformation and flexibility. Biomacromolecules 8:2464–2475

Ortega A, Amorós D, García de la Torre J (2011a) Prediction of hydrodynamic and other solution properties of rigid proteins from atomic- and residue-level models. Biophys J 101:892–898

Ortega A, Amorós D, García de la Torre J (2011b) Global fit and structure optimization of flexible and rigid macromolecules and nanoparticles from analytical ultracentrifugation and other dilute solution properties. Methods 54:115–123

Pavlov GM, Korneeva EV, Smolina NA, Schubert US (2010) Hydrodynamic properties of cyclodextrin molecules in dilute solution. Eur Biophys J 39:371–379

Perkins TT, Smith DE, Chu S (1997) Single polymer dynamics in an elongational flow. Science 276:2016–2021

Rai N, Nollman M, Rocco M (2005) SOMO (Solution Modeller) differences between X-ray and NMR-derived bead models suggest a role for side chain flexibility in protein dynamics. Structure 13:722–734

Rodriguez Schmidt R, Hernández Cifre JG, García de la Torre J (2011) Comparison of Brownian dynamics algorithms with hydrodynamic interaction. J Chem Phys 135:084116

Rodriguez Schmidt R, Hernández Cifre JG, García de la Torre J (2012) Translational diffusion coefficients of macromolecules. Eur J Phys Educ 35:130

Rotne J, Prager S (1969) Variational treatment of hydrodynamic interaction on polymers. J Chem Phys 50:4831–4837

Rubinstein M, Colby RH (2003) Polymer physics. Oxford University Press, Oxford

Schlagberger X, Netz R (2008) Anomalous sedimentation of self-avoiding flexible polymers. Macromolecules 41:1861–1871

Schuck P (2000) Size-distribution analysis of macromolecules by sedimentation velocity ultracentrifugation and Lamm equation modeling. Biophys J 78:1606–1619

Serdyuk IN, Zaccai NR, Zaccai J (2007) Methods in molecular biophysics. Cambridge University Press, Cambridge

Sobolev V, Sorokine A, Prilusky J, Abola EE, Edelman M (1999) Automated analysis of interatomic contacts in proteins. Bioinformatics 15:327–332

Voth GA (ed) (2009) Coarse-graining of condensed phase and biomolecular systems. CRC Press, Boca Ratón

Wang N, Huber G, McCammon JA (2013) Assessing the two-body diffusion tensor calculated by bead models. J Chem Phys 138:204117

Yamakawa H (1970) Transport properties of polymer chains in dilute solution: hydrodynamic interaction. J Chem Phys 53:436–443

Yamakawa H, Fujii M (1973) Translational friction coefficient of wormlike chains. Macromolecules 6:407–415

Yamakawa H, Fujii M (1974) Intrinsic viscosity of wormlike chains. Determination of the shift factor. Macromolecules 7:128–135

Yamakawa H, Tanaka G (1972) Translational diffusion coefficients of rodlike polymers: application of the modified Oseen tensor. J Chem Phys 57:1537–1546

Yamakawa H, Yamaki J (1972) Translational diffusion coefficients of plane-polygonal polymers: application of the modified Oseen tensor. J Chem Phys 57:1542–1546

Yanaki T, Norisuye T, Fujita H (1980) Triple helix of schizophyllum commune polysaccharide in dilute solution. 3. Hydrodynamic properties. Macromolecules 13:1462–1466

Zimm BH (1974) Anomalies in sedimentation. IV. Decrease in sedimentation coefficients of chains at high fields. Biophys Chem 1:279–292

Zimm BH (1980) Chain molecule hydrodynamics by the Monte-Carlo method and the validity of the Kirkwood-Riseman approximation. Macromolecules 13:592–602

Zimm BH, Schumaker VN, Zimm CB (1976) Anomalies in sedimentation. V. Chains at high fields, practical consequences. Biophys Chem 5:265–270

Chapter 12
Accurate Hydrodynamic Modeling with the Boundary Element Method

Sergio R. Aragon

Abstract The integral equations of hydrodynamics are presented for both stick and slip boundary conditions, and results of computations including rigid amino acids are used to obtain a new interpretation of the significance of the hydration parameter used in hydrodynamic modeling. The dynamics of the protein surface perturbs water at that boundary, giving rise to additional viscous energy dissipation which is mimicked by a uniform solvation of 1.1 A thick with stick boundary conditions. BEST (Aragon SR, J Comput Chem 25:1191–12055, 2004) has been used to study 49 different proteins, ranging in molecular weight from 9 to 400 kDa, and we have shown that a model using a 1.1 A thick hydration layer describes all protein transport properties very well. Molecular dynamics (MD) simulation has been used to investigate the origin of a handful of significant discrepancies in some multimeric proteins. A preliminary study of dimeric α-chymotrypsin using approximate implicit water MD is presented. In addition I describe the successful validation of modern protein force fields, ff03 and ff99SB, for the accurate computation of solution structure in explicit water simulation for small proteins using trajectories around 10 ns duration. We have also studied a 150 kDa flexible monoclonal IgG antibody, trastuzumab, with multiple independent trajectories encompassing over 320 ns of simulation. The close agreement within experimental error of the computed and measured properties allows us to conclude that MD does produce structures typical of those in solution and that flexible molecules can be properly described using the method of ensemble averaging over a trajectory.

Keywords Boundary elements • Hydrodynamics • Protein molecular dynamics • Diffusion • Intrinsic viscosity

---

S.R. Aragon (✉)
Department of Chemistry, San Francisco State University, San Francisco, CA, USA
e-mail: aragons@sfsu.edu

S. Uchiyama et al. (eds.), *Analytical Ultracentrifugation*,
DOI 10.1007/978-4-431-55985-6_12

12.1 Introduction

Hydrodynamic modeling plays an important role in the interpretation and study of global molecular motion in liquids. A large number of experimental techniques measure relaxations which include global molecular motions (Aragon 2011). An important method apart from these purely spectroscopic methods is ultracentrifugation. This technique induces the molecule to flow in the presence of a centrifugal field, and its steady-state drift is carefully measured to obtain the sedimentation coefficient (Richards 1980). The sedimentation coefficient is proportional to the average translational diffusion coefficient D and includes a term that contains the specific volume of the molecule in question, but D can be measured directly from the broadening of the sedimentation boundary. Great advances have been made in ultracentrifugation in recent times allowing the deconvolution of mixtures of several molecules (Schuck 2000). As experimental techniques advance in precision, a greater need in accuracy and precision in hydrodynamic modeling arises for the proper interpretation of experimental measurements that depend on hydrodynamic transport properties.

There are three different methodologies to compute hydrodynamic transport properties. The most well-established method is the hydrodynamic interacting bead methodology for the solution of mobility problems. This methodology is discussed at length in Chaps. 10 and 11 of this volume. The second methodology is the boundary element method (BE) – the subject of this chapter (note that Chap. 11 also discusses a graphical interface for BEST within the US-SOMO software). The third methodology, like the BE method, is relatively new – the diffusive Monte Carlo approach (Kang et al. 2004). This last method is not capable of computing tensor values of hydrodynamic transport coefficients, and is most useful for the computation of the average translational diffusion coefficient, but it can handle flexible molecules and provide a decent approximation to the intrinsic viscosity (see also http://web.stevens.edu/zeno/). In this chapter the boundary element method is presented in detail and the differences with the bead methodology are briefly highlighted.

The Stokes creeping flow equations represent the solvent as a mathematical continuum for the case of an incompressible fluid at very low Reynolds number. These differential equations can be solved exactly as a boundary value problem for only a few systems with smooth boundaries: the triaxial ellipsoid (and its degenerate brethren such as a sphere), the toroid, and the dumbbell (Kim and Karilla 1991). To represent a molecule of an arbitrary shape, the early workers (Bloomfield et al. 1967; Garcia de la Torre and Bloomfield 1977a; Teller et al. 1979) used an assembly of beads, at first as a coarse-grained representation. In bead modeling the hydrodynamic interaction of two spheres is given in general as an infinite series expansion in the distance between the spheres. When that distance between spheres exceeds the sum of the diameters, the tensor to first order in the bead size for stick boundary conditions is given by a variational expression first obtained by Rotne and Prager (Rotne and Prager 1969) for the case of equal diameter spheres, which was later generalized to two unequal bead sizes (Garcia de la Torre and Bloomfield 1977b). However, when atomistic resolution is attempted, a problem arises – there does not exist a hydrodynamic interaction tensor for unequal diameter spheres. This has led bead modelers to use a basic approximation: resize spheres to make them of equal diameter if they overlap (Garcia de la Torre et al 2000a) or, even more coarsely, assign a single atomic effective radius (AER) to all the heavy atoms of a molecule in order to avoid this problem (Garcia de la Torre et al 2000b). Other workers have produced variants of the bead methodology, including clever techniques such as the AtoB program to go from an atomistic representation of a protein to a bead representation with control of the degree of coarseness (Byron 1997). Another prominent bead implementation is the SOMO program (Rai et al 2005; Brookes et al. 2010) which is incorporated in the UltraScan sedimentation analysis package. A slightly different bead methodology has also been proposed (Durchschlag and Zipper 2003). The bead methodology is successful to a certain degree – the results are generally not accurate enough to correctly interpret subtle effects of hydration or molecular conformation that other more accurate hydrodynamic treatments are able to handle. In addition, Goldstein has fully explained why the typical implementations of bead hydrodynamics fail to give correct answers for the rotational diffusion of linear bead assemblies and why such programs thus need the "volume correction" (Goldstein 1985). An early implementation of bead methodology (Spotorno et al. 1997) actually included a module to perform correct Goldstein hydrodynamics, but the routine was not included in the later incarnation of what became the SOMO program. This appears to be the case in most bead implementations used at the present time, including all the work from the Garcia de la Torre group. In the boundary element method, the issues of bead overlaps or volume corrections do not arise because the computation focuses exclusively on the hydrodynamic surface represented as interacting triangles instead of beads. As a result, BE calculations are extremely accurate.

An implementation of the BE method was first provided in hydrodynamics in 1975 (Youngren and Acrivos 1975a), even though the basic mathematics was known much earlier (Odqvist 1930). These authors pointed out that the Stokes equations, ordinarily written as partial differential equations with specified boundary conditions, could also be written down exactly as an integral equation for the velocity field outside an arbitrarily shaped body and implemented an algorithm for its solution. In addition, in integral equation form, it is a simple matter to treat stick, slip, or a mixture of the two boundary conditions because they are incorporated into the integral equation (Youngren and Acrivos 1975a; Hu and Zwanzig 1974; Allison 1999). In bead methodology, a rigorous treatment of the slip boundary condition does not exist and only ad hoc approximations have been attempted so far (Venable and Pastor 1988). In the BE method, the starting equation is exact, as was emphasized much later (Wegener 1986), while in the bead methodology, the hydrodynamic interaction tensors are approximate.

The integral equation of hydrodynamics is a Fredholm integral equation of the first kind. Kim and Karilla expounded at length in their modern micro-hydrodynamics treatise about the pitfalls of using this equation due to the fact that it is ill-conditioned (Kim and Karilla 1991). These authors developed a complex methodology in order to overcome this difficulty – the completed double layer boundary integral method, which has not found much favor so far. The integral equation is ill-conditioned because the hydrodynamic interaction matrix that arises when the integral equation is discretized has a zero eigenvalue due to the condition that the Oseen tensor has zero divergence. Such an eigenvalue makes the matrix singular and not invertible. This accounts for the observation of early implementers of the BE method (Allison 1999) that as the number of surface elements increased, the results of the BE method decreased in quality. Essentially, the round-off error in the matrix computation allowed it to be invertible for small sizes but as the matrix size increases, instability arises. However, Aragon published a new implementation of the BE method for stick boundary conditions in which a robust regularization method was incorporated in a program called BEST (Aragon 2004). This allowed the solution of the Stokes equations to unprecedented accuracy, as was amply demonstrated in a recent review (Aragon 2011). In that review it is shown that the BE computations can be as accurate and precise as full analytical solutions for the case when such solutions exist.

This chapter is organized as follows. In Sect. 12.2, the integral equation of hydrodynamics is presented and its solution for either stick or slip boundary conditions via the BE method is discussed. In Sect. 12.3, a thorough discussion is given on the significance of the hydration parameter that is used in all hydrodynamic modeling methods with an eye toward identifying the important contributions of the macromolecule in the determination of this parameter. In Sect. 12.4, a review of the accuracy of the BE method is presented with an emphasis on translational diffusion by a variety of experimental methods and the intrinsic viscosity. Outstanding problems with multimeric proteins are described. In Sect. 12.5, a review of the successful treatment of flexible antibodies in conjunction with molecular dynamics simulations is presented. It should be mentioned that even though this volume is mainly concerned with the sedimentation coefficient, we will discuss translation, rotation, and intrinsic viscosity in our effort to demonstrate that a properly formulated hydrodynamic model must yield accurate results using the same parameters for all transport properties, not just translation.

12.2 The Integral Equations of Stokes Flow

For solute molecules larger than the solvent, consideration of the solvent as a continuum is an excellent approximation, and the governing equations, in the limit of small Reynolds number appropriate for the diffusion process, are known as the Stokes or creeping flow equations (Kim and Karilla 1991). Whereas bead methods aim to directly solve a mobility problem which cannot be formulated exactly, an alternative method is to solve a resistance problem which can be formulated exactly as an integral equation. As is shown below, once one has precise friction tensors, it is straightforward to compute the mobility: the diffusion tensors.

12.2.1 Stick Boundary Conditions

For the case of macromolecules in aqueous solution, "stick" boundary conditions are appropriate [but see Sect. 12.3 below for a discussion of why this is so]. In stick boundary conditions, the velocity vector of the fluid at the body surface is zero, i.e., the solvent moves with the body. In this case, the velocity field of the flow, $v(y)$ at position y in the fluid, can be written exactly as an integral over the particle surface (SP):

$$\mathbf{u}(\mathbf{y}) = \mathbf{u}_0(\mathbf{y}) + \oiint \mathbf{T}(\mathbf{x}, \mathbf{y}).\mathbf{f}(\mathbf{x}) dS_x \tag{12.1}$$

where $\mathbf{u}_o(\mathbf{y})$ is the flow velocity of the fluid if the particle was not there (which can be taken to be zero for diffusive motion) and $\overleftrightarrow{\mathbf{T}}(\mathbf{x}, \mathbf{y})$ is the Oseen hydrodynamic interaction tensor. The surface stress force, $\mathbf{f}(\mathbf{x})$, is the unknown quantity that we must obtain. Once this quantity is known, the transport properties of the macromolecule can be directly computed, as shown below. The Oseen tensor (Oseen 1927; Kim and Karilla 1991) given by

$$\mathbf{T}(\mathbf{x}, \mathbf{y}) = \frac{1}{8\pi\eta\,|\mathbf{x}-\mathbf{y}|}\left[\mathbf{I} + \frac{(\mathbf{x}-\mathbf{y})(\mathbf{x}-\mathbf{y})}{|\mathbf{x}-\mathbf{y}|^2}\right] \tag{12.2}$$

is an exact representation of the hydrodynamic interaction of the infinitesimal surface elements. The solvent viscosity is η. Thus the starting expressions for the calculation, unlike the bead modeling case, are exact; moreover, the equation is applicable to bodies of arbitrary shape.

Since Eq. (12.1) is an integral equation, the solution requires the discretization of the particle surface. The method, however, can be iterated to obtain arbitrary precision. The surface is discretized by replacing it with a collection of N patches that smoothly tile the molecular surface. The details of the solution have been presented previously (Aragon 2004; Aragon 2011). The solution of a linear system of equations containing a superposition of hydrodynamic interactions between the surface patches yields the unknown surface stress force, from which the overall frictional force and torque on the body are computed. Since the velocities and angular velocities are known, the 6×6 friction tensor can be extracted from the total force and torque. The 6×6 friction tensor is composed of 4 3×3 blocks: $\overleftrightarrow{\mathbf{K}}_{tt}$, $\overleftrightarrow{\mathbf{K}}_{tr}$, $\overleftrightarrow{\mathbf{K}}_{rt}$, $\overleftrightarrow{\mathbf{K}}_{rr}$. There are actually only three independent 3×3 friction tensors because the translation-rotation coupling $\overleftrightarrow{\mathbf{K}}_{tr}$ tensor is the transpose of the $\overleftrightarrow{\mathbf{K}}_{rt}$ tensor. This coupling is small unless the body has a screwlike axis of symmetry (Brenner 1967). The 6×6 translation-rotation diffusion tensor is given exactly as the inverse of the 6×6 complete friction tensor whose four 3×3 blocks are the $\overleftrightarrow{\mathbf{K}}$ mentioned above. It is straightforward to show that the 3×3 diagonal blocks of the complete diffusion tensor can be obtained from the friction tensors by an easy 3×3

matrix inversion:

$$\overleftrightarrow{\mathbf{D}}_{\mathrm{tt}} = kT\left[\overleftrightarrow{\mathbf{K}}_{\mathrm{tt}} - \overleftrightarrow{\mathbf{K}}_{\mathrm{tr}}.\overleftrightarrow{\mathbf{K}}_{\mathrm{rr}}^{-1}.\overleftrightarrow{\mathbf{K}}_{\mathrm{rt}}\right]^{-1} \tag{12.3}$$

$$\overleftrightarrow{\mathbf{D}}_{\mathrm{rr}} = kT\left[\overleftrightarrow{\mathbf{K}}_{\mathrm{rr}} - \overleftrightarrow{\mathbf{K}}_{\mathrm{tr}}.\overleftrightarrow{\mathbf{K}}_{\mathrm{tt}}^{-1}.\overleftrightarrow{\mathbf{K}}_{\mathrm{rt}}\right]^{-1} \tag{12.4}$$

Note that the above expressions show that unless the rotation-translation coupling is strictly zero, it is not correct so simply invert the friction tensors to obtain the diffusion tensors – other authors have glossed over this fact (Carrasco and Garcia de la Torre 1999).

BEST computes diffusion tensors in the center of diffusion and the friction tensors in the center of resistance. Details have been presented (Aragon 2004). Furthermore, the more complex expressions for the computation of the intrinsic viscosity are available (Allison 1999; Hahn and Aragon 2006). In the paper by Hahn and Aragon, it is also shown that the center of viscosity is not equivalent to the center of diffusion and that a full matrix inversion is required to calculate the viscosity factor in the center of viscosity. These authors also found that the viscosity factor calculated at the body centroid is an excellent approximation to the true value for globular proteins. In centrosymmetric particles, all of these "centers" coincide.

12.2.2 Slip Boundary Conditions

In the case of slip boundary conditions, the normal component of the velocity of the fluid at the body surface is zero, but the tangential component is unconstrained. Thus, the fluid is said to "slip" past the body surface. This boundary condition has been typically used for small molecules diffusing in organic solvents and is not normally considered for macromolecular diffusion. We consider it here because in order to elucidate why the stick boundary condition is useful for macromolecules such as proteins, it will be convenient to consider the diffusion of the amino acid building blocks in water. In Sect. 12.3 this discussion will lead us to a reinterpretation and full understanding of the hydration parameter that all hydrodynamic modeling methods must use. Note that for two identical surfaces, the stick boundary condition causes greater amount of viscous energy dissipation (more drag) than the slip boundary condition. This effect can be easily observed in the exact computations for a sphere where D_T (stick)/D_T (slip) = 2/3 (Kim and Karilla 1991).

The integral equation for slip boundary conditions is more complex, requiring two integrals, the second of which also contains the unknown velocity of the fluid at the body surface. It is given by (Kim and Karilla 1991; Odqvist 1930)

$$\begin{aligned}\frac{1}{2}\mathbf{u}(\mathbf{y}) &= \frac{1}{2}\mathbf{u}_0(\mathbf{y}) + \oint\!\!\oint \mathbf{T}(\mathbf{x},\mathbf{y}).\mathbf{f}(\mathbf{x})dS_x \\ &-\frac{3}{4\pi}\oint\!\!\oint \frac{(\mathbf{x}-\mathbf{y})\mathbf{n}(\mathbf{x}).(\mathbf{x}-\mathbf{y})(\mathbf{x}-\mathbf{y}).\mathbf{v}(\mathbf{x})}{|\mathbf{x}-\mathbf{y}|^5}dS_x\end{aligned} \tag{12.5}$$

The normal unit vector to the surface at position $\mathbf{x}$ is denoted by $\mathbf{n}(\mathbf{x})$. The extra tensor that appears in this case is more singular than the Oseen tensor but causes no problems inside the integral. This equation can also be solved by discretization (Allison 1999). With slip boundary conditions, the tangential components of the surface stress $\mathbf{f}(\mathbf{x})$ are zero, and we have only N unknown normal components of $\mathbf{f}(\mathbf{x})$ for a surface divided into N triangles. Allison has shown (personal communication) that it is possible to eliminate the unknown velocity at the surface and obtain an equation for the normal components of the surface stress forces. In turn, these components suffice to compute the total force and torque on the body and thus the friction tensors. Using Eqs. (12.3) and (12.4), one can then compute the diffusion tensors as before. Allison's equation requires two matrix inversions to obtain the solution – the full derivation is omitted here – it is a clever modification of his previously published work (Allison 1999). The solution has been programmed in a Fortran program which enables computations with at most $N = 1000$ surface triangles. The regularization of the slip computations is a work in progress and we are limited to the treatment of small molecules at the present time. Nevertheless, the accuracy of this program has been demonstrated in several works (Allison 1999; Sturlaugson et al. 2010).

In the next section, we address the issue of the hydrodynamic hydration thickness, a parameter that is required for computed hydrodynamic transport properties to agree with experiment.

12.3 Hydrodynamic Hydration Reinterpreted

Hydrodynamics, the representation of a solvent as a continuum medium, is surprisingly effective in the description of the transport properties of large and small molecules, provided one uses an appropriate boundary condition at the surface of the solute molecule. For macromolecules such as proteins and nucleic acids dissolved in water, it is well known that the use of "stick" boundary conditions yields an excellent agreement with experimentally measured transport properties, as long as we assume that there is a thin hydration layer of about 1.1 Å thick around the macromolecule. In other words, the experimental diffusion coefficients are smaller than hydrodynamics predicted in the absence of "hydration." We describe the combination of stick boundary conditions and an empirical hydration parameter as a "hydrodynamic model." Such a hydrodynamic model is useful in the prediction of transport properties but we do not claim by its use that the water actually forms a layer that moves with the protein as was believed by early workers (Kuntz and Kauzmann 1974; Squire and Himmel 1979). Yet the modern literature is still filled with claims that hydration water moves rigidly attached to the macromolecule. That naïve interpretation is simply not in agreement with the experimental fact that individual water molecules have a residence time at the surface of a protein in the 10 ps time scale, while rotational correlation times of proteins are in the 10 ns time scale and beyond. Magnetic relaxation dispersion (MRD) measurements of ^{1}H (Venu et al. 1997) and ^{17}O (Denisov and Halle 1996; Halle 1999) demonstrate that water molecules at a protein surface have mean residence times between 10 and 50 ps nearly independent of location on the protein surface. Molecular dynamics simulations come to the same conclusion (Makarov et al. 2000; Luise et al. 2000; Henchman and McCammon 2002). Thus, there must exist a dynamical effect at the protein surface that is mimicked by the presence of an immobile hydration layer.

A big step toward the correct interpretation was provided in a lucid paper (Halle and Davidovic 2003), hereafter denoted as HD. These authors noted that a plausible consequence of the perturbation of the solvent by a solute molecule is that the viscosity of the first solvation layer is different and larger from the bulk viscosity of the solvent. We shall see that an equivalent way of looking at this effect is to realize that there are extra sources of viscous energy dissipation at the protein surface layer. The Stokes creeping flow equations can be solved for rotation and translation of a sphere with such a thin layer of more viscous material around it with stick boundary conditions (Brilliantov and Krapivsky 1991). Let the bulk solvent viscosity be η_o, the viscosity of the thin layer be η_s, the sphere volume V_p, and the volume of the thin layer V_s. The volume of the thin layer can be well expressed by $V_s = A_p\ \lambda$, where A_p is the sphere area and λ the layer thickness. The results of the calculations can be expressed as the ratio of the diffusion coefficients with and without solvent perturbation and are given (without the approximations used by HD) by

$$D_R/D^o{}_R = 1-(1-\alpha)\ \ (1- <\eta_o/\eta_s>) \tag{12.6}$$

$$D_T/D^o{}_T = 1-(\lambda/\ (R+\lambda))\ \ (1- <\eta_o/\eta_s>) \tag{12.7}$$

for rotation (R) and translation (T), respectively, $\alpha = R^3/(R+\lambda)^3 = V_p/(V_p+V_s)$ $= 1/(1+\lambda\ r)$, and $r = A_p/V_p$. Note that these formulas have correct limits: if the layer thickness is zero, $\alpha = 1$, or if $b = 1$, the ratios are unity; if the layer viscosity is much larger than the bulk viscosity, then the last parenthesis is unity, giving a result with a geometrically larger size. In the above equations, I have inserted a spatial average of the viscosities in an anticipation of applying the sphere results to nonspherical solutes. For a water molecule, there is a direct relationship between the viscosity and the rotational correlation time $\tau = 1/6D_R \approx V\eta\,/\,kT$, where V is the water molecule volume. The relationship is missing a correct shape factor and a slip correction factor, but in the ratio $<\ \eta_o/\eta_s\ > = <\ \tau_o\,/\,\tau_s\ >$, such factors cancel out. Thus, as HD noted, the viscosity ratio can be related to a dynamical quantity that can be measured in a protein solution. HD further take into account the distribution of relaxation times in a congested environment and obtain a simple relationship:

$$< \tau_o / \tau_s > \; = 2\, \tau_o / < \tau_s > \tag{12.8}$$

Furthermore, HD argue that the shape effects are effectively the same for the unperturbed and the solvent perturbed case, so that the formulas (12.6) and (12.7) can be applied to globular proteins even though they are not exactly spherical. Now let's apply these relations with appropriate values for two cases that we will find useful to compare: lysozyme and glycine. Using the measured values of the parameters for these cases, we obtain the following ratios for the diffusion coefficients:

$$\begin{aligned}\text{Lysozyme}: V_p &= 16\ \text{nm}^3, A_p = 64\ \text{nm}^2, \lambda = 0.2\ \text{nm}, < \tau_s >= 5.5\ \tau_o \\ D_R/D^o{}_R &= 0.72 \ \text{ and } \ D_T/D^o{}_T = 0.89\end{aligned}$$

$$\begin{aligned}\text{Glycine}: V_p &= 68.3\ \text{Å}^3, A_p = 89.7\ \text{Å}^2, \lambda = 2\ \text{Å}, < \tau_s >= 2.5\ \tau_o \\ D_R/D^o{}_R &= 0.86 \ \text{ and } \ D_T/D^o{}_T = 0.93\end{aligned}$$

For both the protein and the amino acid constituent, the unperturbed diffusion coefficients are slowed down, and in the case of the proteins, the results agree well with experiment – the 30 % slowing in rotation is what is needed to match the measured values for lysozyme. HD and Aragon and Hahn have demonstrated that the result is accurate for proteins in general. For proteins, the 10 % decrease for translation also agrees with experiment. On the other hand, the results for glycine are in complete disagreement with experiment (note that the thin shell formulas make an insignificant error for a small molecule like glycine). For the moment, let's follow HD and compute the effective thickness that is required in order to mimic the dynamic solvent effect in a stick boundary condition calculation with a fixed solvent thickness δ. That is, we apply the standard hydrodynamic model by surrounding the protein with a layer of immobile water of thickness δ and computing the transport properties with stick boundary conditions. In this case, we will obtain a different expression for the ratios of diffusion coefficients with or without such hydration. Equating the relationship containing δ and that for Eqs. (12.6) and (12.7), we can solve for δ, in terms of the parameters of the dynamic perturbation model. The results for rotation and translation are

$$\delta_R \cong \frac{\lambda\,(1-b)\left(1-\frac{\lambda b}{R}\right)}{1+\lambda b/R} \tag{12.9}$$

$$\delta_T = \frac{\lambda\,(1-b)}{1+\lambda b/R} \tag{12.10}$$

In Eqs. (12.9) and (12.10) that I have derived with the aid of Mathematica, $b =< \tau_o/\tau_s >$. Equation (12.9) is an excellent approximation to the exact result and

Eq. (12.10) is exact. These expressions have two immediate consequences. First, the values of the thicknesses required for translation and rotation are predicted to differ, and second, the thicknesses depend on molecular weight (through the variable r) and temperature (through the variable b). If we use the approximate formulas quoted by HD that relate area and volume to protein molecular weight:

$$A_p = 7.43\ \mathrm{Mw}^{0.81}\mathrm{nm}^2,\ V_p = 1.02\ \mathrm{Mw}^{1.03}\mathrm{nm}^3 \tag{12.11}$$

and express $R = 3/\mathrm{r}$, the above formulas yield values of δ_R in the range of 1.18 Å for Mw = 300 kDa to 1.05 Å for Mw = 10 kDa, while for δ_T they yield 1.24 Å for Mw = 300 kDa to 1.18 Å for 10 kDa. Thus, the molecular weight dependence is only slight, and the value of the deltas for either case is almost the same and in remarkable agreement with the value of $\delta = 1.1$ Å found empirically in our work. Given all the approximations used above, the agreement can be taken as excellent.

Thus we arrive at the conclusion that the dynamical coupling of the protein with the solvent does indeed generate extra viscous energy dissipation which is manifest as increased viscosity in a thin layer around the protein. Before we return to the amino acid case, it is worth noting one other issue: the fact that this layer is predicted to be uniform over the entire protein surface. The key observation here is that (Harpaz et al. 1994) about 60 % of the protein surface is composed of atoms in hydrophobic residues, leaving only 40 % for potential preferential interaction sites. In addition, the MRD measurements of water residence times and the MD quoted previously show no significant variation over the entire protein surface.

Now we must discuss the relevance of the calculation of the transport properties of single components of a protein – the amino acids. In Table 12.1 we show the experimental values for the translational diffusion coefficient of three fairly rigid amino acids and the results of various hydrodynamic calculations. As input to the hydrodynamics, the zwitterionic amino acid structures were obtained with Spartan software (wavefunction.com) using Hartree-Fock with a 631 g* basis set. The triangulations were done with MSROLL using the united atom model, but hydrogens were not removed from the pdb files – the hydrogens are so small that this makes very little difference, given that the slip program has a typical 3 % error.

The immediately obvious result is that the stick boundary condition calculation fails in that the values are about 17 % lower than experiment, yet that calculation is the fastest possible result with stick boundary conditions because we have omitted

Table 12.1 Values of D_T for amino acids at 20 °C in water, 10^{-6} cm^2/s

Molecule	Stick unhydrated	Slip unhydrated	Slip 0.8 Å hydration	Experiment
Glycine	8.00	11.5	9.1	9.33[a]
Alanine	7.53	10.5	8.4	8.31[a],8.07[b]
Serine	7.31	10.2	8.2	8.17[a],7.75[b]

Stick unhydrated calculations done with BEST, slip calculations done with nutrn4 (Allison 1999)
[a]Ma et al. (2005)
[b]Germann et al. (2007)

all solvation. A full slip calculation with no hydration is too high, however. Thus we see that a slip boundary condition calculation requires a small layer of "solvation" in order to agree with experiment, as expected since water molecules do show slowed down rotational correlation times at the surface of amino acids (Halle 1999). For these cases we find that a value around 0.8 Å models the perturbation of the solvent adequately. The discrepancies with experiment are on the order of 3 %, which is the precision of the slip program and the experimental data. If we take the HD model at face value, we actually expect, as calculated previously, that the stick translational diffusion coefficient should be faster than experiment by about 7 %, yet we observe an opposite larger deviation. The fact that a slip hydrodynamic calculation better describes the diffusion of a small molecule is not a surprise – many previous workers have shown the appropriateness of such a change in boundary conditions when the solute is not much larger than the solvent (Bauer et al. 1974; Youngren and Acrivos 1975b). Thus we need to take this result seriously, for it is not an artifact of the continuum nature of hydrodynamics when applied to a smaller molecular scale.

Proteins, on the other hand, require a larger hydration layer on top of the slower stick boundary condition. Does a protein have an additional mechanism for viscous energy dissipation that a single amino acid does not have? In Fig. 12.1 we show the root mean square deviation of atoms in a protein as calculated by a molecular dynamics simulation on human serum albumin (1AO6.pdb). The MD simulation was carried out in explicit solvent after adding the missing residues to the crystal structure with Sali's Modeller program (Sali and Blundell 1993; Fiser et al. 2000) and preparing the molecule with the proper disulfide bonds with tleap in AMBER Tools 13 with the ff99SB force field of AMBER 12 (Case et al. 2012) using periodic boundary conditions in pmemd.cuda (Salomon-Ferrer et al. 2013) with a GTX 580 GPU. The simulation was carried out for 2 ns and data points were collected every 2 ps. The reference frame for the simulation was the first structure obtained after energy minimization and constrained heating steps. The figure, typical of any well-folded protein, shows us that the protein atoms on the average execute excursions of about 3 Å in position and that such excursions are already 2 Å by 50 ps, the characteristic residence time of water molecules near the surface. Thus, we see that the protein does have an additional mechanism for viscous energy dissipation that

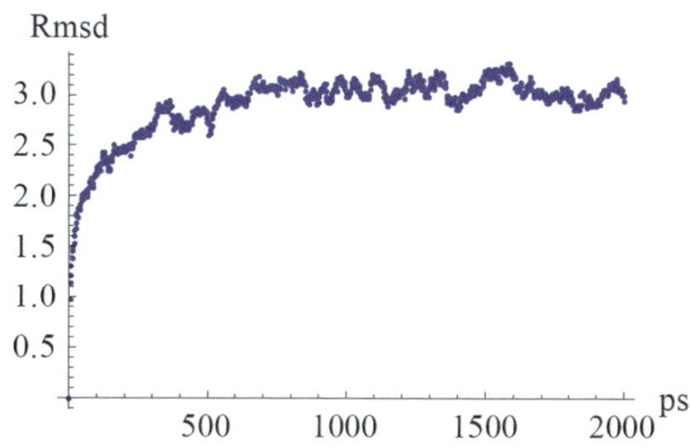

Fig. 12.1 The RMSD (average root mean square deviation in Å) of all atoms in human serum albumin (1AO6) from an arbitrary first frame. The diameter of a water molecule is 3 Å by comparison

is not present in the amino acids – the protein surface is dynamic! The protein is not rigid like our sample amino acids, and thus the protein atoms jostle the water molecules at the protein surface over the diameter of a water molecule. The entire first hydration layer on the protein is definitely disturbed, and this motion generates extra viscous energy dissipation. In other words, the reason we must use the slip boundary condition for a small molecule is that such tend to be nearly rigid and perturb the solvent little.

Our point of view can be clarified by considering the hypothetical case of a protein with no preferential interaction sites which was completely rigid. Then in that case, the interaction of water on its surface would be just the same as on a single rigid amino acid, and we would expect slip boundary conditions to be applicable for the hydrodynamics with a small solvation layer. The energy dissipation in that case would be determined by the geometric features of the protein surface and nothing else. The extent of the perturbation of the solvent would be somewhat larger than that of an amino acid due to more surface roughness. Note that the case of deeply buried water molecules causes no concern because such water molecules do not participate in viscous energy dissipation – only the free water molecules near the surface do. The real protein on the other hand definitely perturbs the water around it much more because of the dynamics of its surface atoms – it is not rigid. We have a layer of water in which extra energy dissipation occurs, thus the viscosity of that layer is larger, and as HD have shown, this slows down the diffusive motions of the macromolecule. Now there are two ways to materially make the calculations reflect this slowing down. One is to use *Slip* boundary conditions and a layer thickness of about 3 Å, considered as fixed hydration, or the alternative is to use *Stick* boundary conditions (which automatically increases the energy dissipation at the surface), and a much smaller thickness of fixed hydration, namely, 1.1 Å. Thus note that in this view, the use of stick boundary conditions and a fixed hydration layer is just a mimic of the effect of a layer in which additional viscous energy dissipation occurs at the protein surface due to its roughness and dynamics. As shown above, just considering how the rotation of water molecules is slowed down near the protein surface, one can calculate the 1.1 Å value that agrees with what BEST determined empirically.

This new view has several immediate consequences. First of all, it assigns the primary cause of the hydrodynamic hydration layer to protein surface atom motion, and thus predicts that the layer, to first order, should be uniform over the protein. The atoms of hydrophobic or hydrophilic residues move about the same. Any preferential interaction site is a perturbation on top of this picture. This perhaps explains why the uniform 1.1 Å parameter used in BEST works so well for monomeric proteins which tend to be very compact and well folded. On the other hand, it also predicts that multimeric proteins that have extra flexible loops or whose component chains can jiggle with respect to one another could need a higher value of the fixed hydration layer to account for a larger amount of solvent being perturbed due to this flexibility. The HD theory predicts that the hydration layer thickness should increase slightly with molecular weight and be temperature dependent. The extra dynamics in some multimeric proteins could add an additional factor. This can be tested with molecular dynamics simulations.

This new hydration interpretation also predicts something important for segmentally flexible proteins such as antibodies. If you have a protein composed of compact segments which move with respect to one another in time scales much longer than the residence time of water molecules on the surface of the protein, then the stick BC plus your standard 1.1 Å fixed layer will work just fine, provided you average the diffusion coefficients over an ensemble of shapes produced during an MD simulation. This was precisely demonstrated in our work with trastuzumab in 2010 where the agreement with experimental transport properties was excellent. This work is reviewed below. This new view also predicts that disordered proteins will require more hydration than the standard 1.1 Å with stick boundary conditions, due to the extra viscous energy dissipation provided by the flexible portions. Previous work (Rai et al. 2005) has also suggested a special role for the dynamics of flexible side chains in affecting the overall hydrodynamics of a protein, however, even though these authors were aware of the HD work, they did not quite make the connection with hydrodynamic boundary conditions that is highlighted here. In addition one can readily see that the amount of water that is involved in the thin 1.1 Å layer is not an accurate representation of the amount of water that is actually perturbed by the protein surface because the stick boundary condition has artificially imposed most of the extra energy dissipation. There is much more water than this layer indicates, and this helps to explain why "hydration water" measured in this way always comes out short (Aragon and Hahn 2006).

This new view also predicts that if you change the solvent, then you should be able to estimate the thickness of the layer by measuring the solvent rotational slowing at the macromolecule surface and that the layer thickness will depend principally on the solvent, and not the macromolecule (which contributes second-order effects). For solvents that are perturbed very little at the boundary with solute, little or no solvation will be needed in a hydrodynamic model and the appropriate boundary condition should be slip. Lastly, we can comment that hydrodynamics will be applicable as long as the solute size is larger than any free volume contained in the solvent, for otherwise, ballistic motion into such volumes can occur and the motion will be faster than predicted by hydrodynamics (Bauer et al. 1974; Sturlaugson et al. 2010).

To summarize: the introduction of a solute molecule into solvent perturbs the structure of the solvent at the solute boundary, thus increasing the solvent viscosity at the interface. The presence of the solute boundary is a wall that disrupts the solvent organization. If that wall has solvent size nooks and crannies, the solvent is perturbed a little more, and if the wall is not rigid, the largest amount of perturbation occurs. For amino acid-sized solutes, only the main smooth wall effect is present and the solvent perturbation is small, requiring the use of solvated slip boundary conditions. For macromolecules, the much larger perturbation is dominated by the solute surface dynamics and a convenient model is a solvated stick boundary condition.

12.4 Studies of Globular Proteins

We have previously reported on extensive calculations on proteins over a very broad range of molecular weight that show excellent agreement with experiment for the three basic transport properties using the same 1.1 Å hydration parameter.

This work has been previously reviewed (Aragon 2011) – here we mention only the salient points and update some of the work.

In the BE method, we need to define the "perturbed boundary" or "hydrated surface" of a protein, so an atomic resolution structure is a required input. In our previous work (Aragon and Hahn 2006; Hahn and Aragon 2006), where 49 proteins ranging from 9 to over 465 kDa where studied with BEST and stick boundary conditions, Connolly's program MSROLL (Connolly 1981, 1983, 1993) was used to roll a probe sphere of solvent size (1.5 Å) around the atomic arrangement defining the molecular surface, after all protein atoms have been "inflated" by 1.1 Å to account for the required hydrodynamic hydration. The atomic radii are only used to define the hydrodynamic surface to be triangulated. The fine triangulations produced by MSROLL are further processed by COALESCE (Aragon 2004), a program that can generate sub-triangulations, preserving the topological properties of the surface. A sample triangulation is shown in Fig. 12.2 (left) for ribonuclease. A sequence of such sub-triangulations with increasing numbers of triangles are analyzed by BEST to produce accurate transport properties via extrapolation to zero triangle size, as shown in Fig. 12.2 (right). The value of the hydration thickness was assigned by simply matching the measured translational diffusion coefficient of a set of four well-characterized small proteins (ribonuclease, myoglobin, lysozyme, and chymotrypsinogen) with the uniform increase in atomic size required for the computation to agree. Thereafter ALL proteins, large or small, were treated with the same value of the hydration parameter for all the transport properties.

The original work of Aragon and Hahn neglected to add missing residues in about 25 % of the proteins studied – the data presented in Table 12.2 for monomeric proteins has been corrected for such omission. It is clear that the agreement with

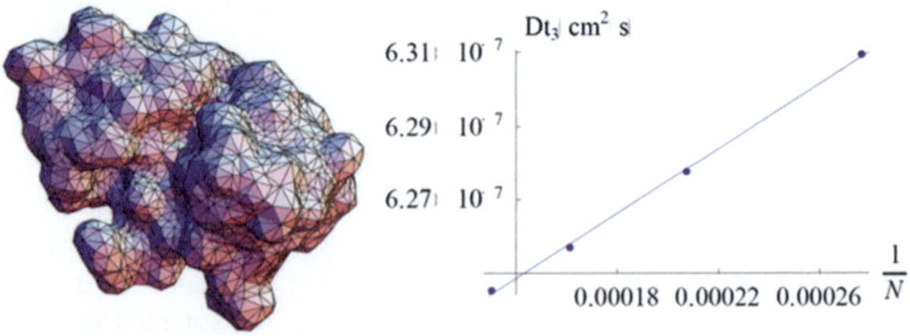

Fig. 12.2 *Left Panel*: Triangular tessellation of ribonuclease with 4952 triangles. *Right Panel*: The least squares fit line for the third eigenvalue of the translational diffusion tensor Dt extrapolation vs. 1/N of human serum albumin (1AO6)

Table 12.2 The intrinsic viscosity and translational diffusion coefficient of monomeric proteins

Protein	s^a	Mass (kDa)	$[\eta]$ (cm^3/g)			$D_t(10^{-7}cm^2/s)$		
			Calc	Exp.	Δ^b	Calc.	Exp.	Δ^b
Cytochrome C (1HRC)	1	12.4	3.07	2.74	13	11.63	11.1–12.1	0
Ribonuclease A (8RAT)	1	13.7	3.32	3.30,3.50	2.4	10.93	10.68	2.3
α-Lactalbumin (1A4V)	1	14.2	3.42	3.01, 3.4	6	10.74	10.57, 10.6	2
Lysozyme (1AKI)	1	14.3	3.14	2.66,3.0	11	11.08	10.6, 11.2	1.6
Myoglobin (1WLA)	1	17.2	3.15	3.25	3	10.10	10.4, 10.5	−3
Soyb.Tryp.Inhib. (1AVU)	1	20.1	3.18	2.8	14	9.88	9.8	1
β-Trypsin (1TPO)	1	23.3	2.99	3.1	4	9.58	9.3	3
Trypsinogen (1TGN)	1	24.0	3.00	2.96	1	9.49	9.68	−2
α-Chymotrypsin (4CHA)	1	25.2	3.25	3.00	8	9.11	9.33*	−3
Chymotrypsinog. A (2CGA)	1	25.7	3.20	2.5,3.13	4	9.16	9.23	−1
Carbonic anhyd. B (2CAB)	1	28.8	3.02	2.76,3.2, 3.7	−5	8.90	8.89	1
Zn-α2-Glycoprotein (1ZAG)	1	32.6	4.79	5.0	−4	7.30	7.96	−9
Pepsin (4PEP)	1	34.5	3.33	3.09,3.35	3	8.10	8.01, 8.71	−3
G-ADPActin (1J6Z)	1	43.0	3.92	3.7	6	7.43	7.15, 7.88	−1
Taka-amylase A (6TAA)	1	52.5	3.15	3.3	−3	7.22	7.37	−2
Human serum alb. (1AO6)	1	66.5	4.26	3.9, 4.2, 4.73	0	5.99	5.9, 6.1, 6.3	−2
Ovotransferrin (1OVT)	1	76.0	4.00	3.8	5	5.87	5.9	1
Lactotransferrin (1LFG)	1	77.1	4.00	4.0	0	5.85	5.6	4

[a]Number of subunits

[b]The percent difference between the calculated and experimental value determined from the average of the experimental values. References for experimental work are available in the original paper (Hahn and Aragon 2006)

experiment is excellent with the exception of 1ZAG whose *Dt* was recalculated from the reported sedimentation coefficient (Burgi and Schmid 1961). It is notable that the molecular weight calculated by these authors is high, possibly indicating the measured value of S is also high. It is also worth noting that the value of a more recent measurement of the sedimentation coefficient of ribonuclease (Moody et al. 2005) yields a value of *Dt* (with 8 % uncertainty) that conflicts with previous measurements quoted in Table 12.2, yielding a value 10 % higher. It is interesting that, as shown in Table 12.5, one cannot interpret this discrepancy as originating from a change of shape of ribonuclease upon going into solution because the molecular dynamics data agrees very well with the hydrodynamics of the crystal structure. The data of Moody et al. does show pH dependence, with data at lower pH tending to agree better with the hydrodynamics.

Table 12.3 The intrinsic viscosity and translational diffusion coefficient of multimeric proteins

Protein	s	Mass kDa	$[\eta]$ (cm^3/g) Calc.	Exp.	Δ[a]	Dt(10^{-7} cm^2/s) Calc.	Exp.	Δ[a]
Superoxide dismu. (2SOD)	2	32.5	3.57	3.3	9	8.10	8.27	−2
b-Lactoglobulin (1BEB)[b]	2	36.7	3.68	3.4–4.2	−3	7.72	7.3	6
a-Chymotrypsin (4CHA)	2	50.4	3.31	4.1,4.25	−21	7.16	7.1,7.4	−1
Concanavalin (1GKB)	2	51.4	3.60	4.1	−12	6.96	6.2	12
Triosephos. isom. (1YPI)	2	53.2	3.59	3.75	−4	6.80	6.76	6
Ricin (2AAI)	2	61.5	3.33	2.96	13	6.61	6.0	10
Oxyhemoglobin A (1HHO)	2	63.2	2.89	2.77	4	7.03	6.78	4
Alkaline phosphat. (1ALK)	2	94.6	3.09	3.4	−7	5.96	5.7	4
Citrate synthase (1CTS)	2	98.0	3.20	3.95	−20	5.82	5.8	0
Inorganic pyrophos. (1FAJ)	6	117.3	3.52	4.0	−12	5.33	5.7	−6
Aldolase (1ADO)	4	157.1	3.84	3.4,4.0,4.04	0	4.66	4.29–4.8	4
Catalase (4BLC)	4	235.7	3.01	3.9	−23	4.42	4.1	8
b-Galactosidase (1BGL)	4	465.8	3.84	3.78	2	3.26	3.13	4

s – number of subunits

[a]The percent difference between calculated and experimental values determined from the average of experimental values

[b]Heavy atoms only. References for experimental work are available in the original paper (Hahn and Aragon 2006)

The multimeric protein data of Aragon and Hahn is shown in Table 12.3. The few significant discrepancies with experiment for proteins found by these authors are worth mentioning in more detail. Whereas the computed transport properties of the 18 monomeric proteins treated as rigid objects generally agreed within experimental error (and the discrepancies were randomly distributed), there was a subset of 4 out of 13 multimeric proteins (α-chymotrypsin, citrate synthase, inorganic pyrophosphates, catalase) that showed large negative systematic deviations in the intrinsic viscosity exceeding –20 %. In addition, the translational diffusion data for the multimeric proteins shows mostly positive systematic deviations from experiment.

Note that the translational diffusion coefficient is a functional of shape divided by a characteristic length, the rotational diffusion tensor components are functionals of shape divided by a volume, but the intrinsic viscosity is exclusively a functional of shape and is thus the most sensitive of the measurements to changes in molecular shape. The results of our protein study indicated that for monomeric proteins, and most multimeric proteins, the crystal structure was a good representation of the average structure in solution. Given that there were only four very deviant cases out of 13 in the multimeric protein set, the most reasonable conclusion is that either the crystal structure and the average solution structure are significantly different for these proteins, or these proteins are significantly more flexible than others in the data set. The technique of molecular dynamics simulation in combination with hydrodynamic computations can be used to address these questions.

As mentioned previously, BEST is capable of computing transport properties to extremely high accuracy and statistical precision. For smooth surfaces, the accuracy is better than 0.02 %, while for amorphous molecular surfaces, the statistical precision is typically around 0.2 %. The ultimate limit in the precision is the accuracy of the atomic coordinates themselves. This limit is comparable also to the impossibility of defining the molecular surface to very high precision because doing so would require more than 20,000 triangles and that requires very large machine memories and much time for the computation. A practical limit for the values of transport coefficients, given these considerations, is about 1 % in translation and 2 % in rotation and intrinsic viscosity. These limits are still much better than most experimental uncertainties. The experimental data set could also be clearly improved. For example, the published *Dt* for Taka-amylase was not extrapolated to infinite dilution even though concentration data are presented in the paper. We have done the extrapolation and presented that improved value in Table 12.2. The 4CHA monomer was recently measured by analytical ultracentrifugation (Ghirlando 2011), and that greatly improved value is now presented in Table 12.2. There is a need for a carefully measured set of transport properties for proteins varying across the entire range of molecular weight – AUC would be an ideal technique to obtain accurate values of *Dt*.

12.5 Combination of Molecular Dynamics Simulation and Hydrodynamic Modeling

In the previous section, we described work in which proteins were assumed to be rigid objects with the crystal structure representing the average solution structure. This picture works very well for most proteins; however, we would like to know what effect the structural fluctuations present in solution have on the measured transport properties of globular proteins and also how to describe proteins that are flexible or have flexible subdomains. The technique of molecular dynamics (MD) simulation is well suited for this task. Modern-day parallel graphical processing units (GPUs) enable us to study even large-sized proteins with an explicit solvent simulation. MD work on the large multimeric proteins is ongoing in this laboratory to test the hypothesis that some have different structure in solution than in the crystal and also investigate their degree of flexibility.

12.5.1 *Implicit Water MD of α-Chymotrypsin*

One of the multimeric proteins that may have a significantly different structure in solution compared to the crystal is α-chymotrypsin. This protein has a dimer-monomer equilibrium that is pH dependent (Schwert and Kaufman 1951) and was treated with AMBER's sander module (Version 9) at constant pH (Mongan et al. 2004) using an implicit solvent model. In an implicit solvent model, water is approximately represented by a continuum fluid with no viscosity, thus the dynamics occur much faster than in a real molecular system, allowing a short simulation to display significant changes. The protocol is similar to that of explicit water simulations described below. In addition, it is possible to choose a typical salt concentration for the system.

Constant pH implicit water simulations done at pH 7, where the protein exists as a monomer in solution, do indeed demonstrate that the initial crystal structure falls apart, and the two pieces separate in time (not shown). At pH 3, however, where the protein is a dimer in solution, the simulation keeps the protein together and deforms its shape, elongating somewhat as the simulation proceeds over 3 ns. The initial and a sample deformed shapes are shown in Fig. 12.3. The trajectory graphs for the translational diffusion coefficient and for the intrinsic viscosity are shown in Fig. 12.4. The graphs clearly show the deformation of the structure as the simulation proceeds as relaxation of the values occurs within the first 1–2 ns of trajectory. The transport properties computed as an average over the last 1 ns of simulation agree much better with experiment (Schwert and Kaufman 1951) than those of the crystal structure. The hydrodynamic analysis is shown in Table 12.4, where data

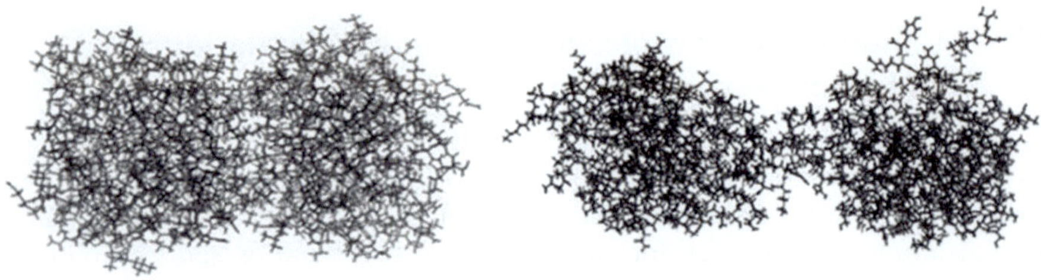

Fig. 12.3 α-chymotrypsin structures. *Left panel*: crystal structure (4cha.pdb). *Right panel*: AMBER 9 typical geometry after 1 ns molecular dynamics with implicit solvent at pH = 3.0 (Taken from Aragon 2011)

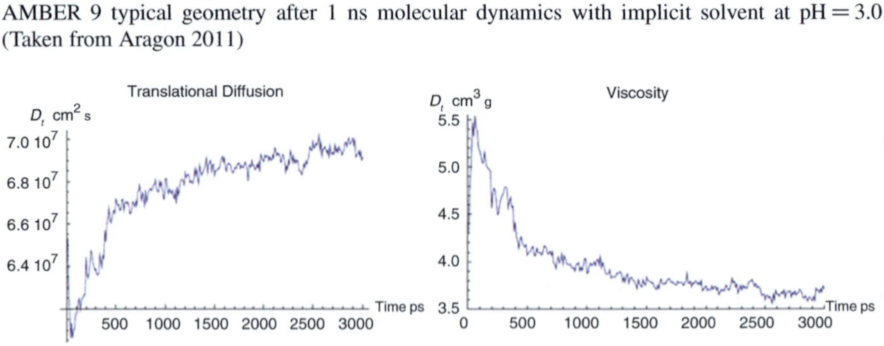

Fig. 12.4 The translational diffusion coefficient (*left panel*) and the intrinsic viscosity (*right panel*) of α-chymotrypsin (4CHA) from an MD trajectory with implicit water at pH 3.0. As the molecule shape deforms from the initial crystal structure, the transport properties evolve and settle down after 2.5 ns (Taken from Aragon 2011)

Table 12.4 The intrinsic viscosity and translational diffusion coefficient of α-chymotrypsin (4CHA) and β-lactoglobulin (1BEB) from implicit water MD

			$[\eta]$ (cm^3/g)			$Dt(10^{-7}\ cm^2/s)$		
Geometry	n	Mass (kDa)	Calc.	Exp.	Δ	Calc.	Exp.	Δ
1beb.pdb	1	36.7	3.7	4.1[a]	−7.5	7.7	7.3[b]	5.5
Sander			**4.0**		−2.4	**7.5**		2.7
4cha.pdb	2	49.7	3.3	4.1[c]	−19	7.2	7.4[c]	−3
Sander			**3.7**		−10	**6.9**		−7

[a]McKenzie and Sawyer (1967)
[b]Ogston (1949)
[c]Schwert and Kaufman (1951)

for an additional monomeric protein, β-lactoglobulin, is shown as a control. Note that the D_t value of the α-chymotrypsin dimer was not corrected for concentration dependence and has a larger than normal uncertainty. The β-lactoglobulin MD results are only slightly improved from the crystal structure results, indicating that the force field is sufficiently accurate to model the system well. This is a result in the right direction but the implicit solvent model is a coarse representation of the aqueous medium.

What can we learn from a more realistic solvent model? Recently, the 4CHA simulations were repeated with explicit solvent MD (data not shown). To perform simulations without direct control of pH with explicit water solvent, the protein was titrated at pH 7 and 3 using the online pH – server module at the University of Virginia. Sufficient counterions were added to make the system neutral, but no extra ions were added to adjust toward an experimental ionic strength. The simulations were carried out to 20 ns with the protocols described below, but the results were not as expected. At pH 7, there was no indication that the protein dimer would separate into monomers. At pH 3, the prominent relaxation features in Fig. 12.4 were not seen – there doesn't appear to be a significant shape change during the simulation. AMBER 14 has just been released with new capabilities to carry out constant pH simulations with explicit water and this work will be repeated once more. On the other hand, the discrepancies could be due to greater than normal surface flexibility in this protein and further MD work should give us a clue.

12.5.2 Explicit Water MD of Small Proteins

Here we report on MD simulations of several small proteins, including some with flexible subdomains in order to validate the method for hydrodynamic computation. In our method, we generate a sequence of snapshots of the protein structure along a simulation trajectory or set of trajectories. Instead of trying to compute the transport properties directly from the trajectory (which requires very long trajectories), we simply compute the transport properties of each snapshot, taken as a rigid structure, and then average over hundreds or thousands of snapshots, depending on how flexible the protein is. We call this the trajectory ensemble average method. We have used the AMBER (Versions 9,10,12) suite of programs (Perlman et al. 1995), and in particular the parallel program pmemd, to perform explicit water simulations with a TIP3P water model in an octahedral box with periodic boundary conditions. A typical simulation protocol consists of four steps: (1) energy minimization of the solvated system at constant volume and fixed protein coordinates to relax close contacts with solvent, (2) energy minimization of the entire system at constant volume with no restraints on the protein atoms, (3) 20 ps of MD simulation at constant volume with temperature increasing from 0 to 300 K with mild restraints on protein atoms, and (4) production run of MD simulation at constant pressure of 1 atm and temperature of 300 K with no restraints.

The first issue we must confront is the accuracy of the force fields that will yield the computed structures during the simulation. We have used three modern force fields that have been developed for accurate modeling of proteins: ff03, ff99SB, and more recently ff2012SB. These are compared in detail by Hornak et al. (2006) who show that ff03 performs slightly better for small systems such as ubiquitin, while ff99SB performs better for larger systems in the prediction of NMR order parameters which are sensitive to detailed local conformational structure. ff2012SB is a further improvement on ff99SB. In order to validate these force fields for whole molecule scale structure probed by hydrodynamics, we performed computations of small monomeric proteins whose crystal structure is a good predictor of solution structure and investigated whether this agreement is maintained during MD simulation. If the force field and simulation process are good, the agreement with experiment will be maintained.

Aragon and Hong have studied several small proteins with explicit water MD simulation (lysozyme, ribonuclease, bpti, human and mouse ubiquitin) using the AMBER pmemd parallel program with a protocol as described previously and an electrostatic cutoff that varied between 15 and 12 Å, depending on the size of the octahedral solvent box (Hong 2009). The solvent contained only as many ions to make the system neutral, but no added salt. The typical buffer used in experiments has a viscosity about 1 % higher than pure water and around 0.1 mM salt which serves to screen electrostatics. In addition, this work included a comparison with implicit water MD (not shown) on the same proteins and found a systematic discrepancy of about 15 % compared to explicit water simulations. The more salient points of the data obtained in this study will be reviewed here. Some of the transport properties of ribonuclease are shown in Fig. 12.5.

Note that, unlike the MD trajectory observed in Fig. 12.4 for α-chymotrypsin, the graph of the transport properties for ribonuclease along the trajectory does not show a relaxation at small times.

The graph fluctuates about the average from the initial points in the trajectory, indicating that the crystal structure is already close to the minimum energy in

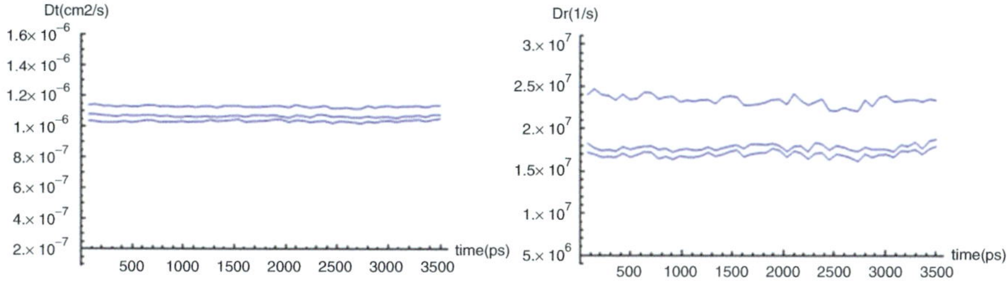

Fig. 12.5 *Left panel*: The translational diffusion tensor eigenvalues along the MD trajectory for ribonuclease (7RSA). Note the small difference between the eigenvalues, justifying the use of the average. *Right panel*: The rotational diffusion tensor eigenvalues along the MD trajectory. Note the symmetric top appearance of the eigenvalues. In both cases, the data shows only small thermal fluctuations characteristic of a globular protein

solution and the structure shows only thermal fluctuations, not a deformation. This result is typical of all the small proteins in this MD study. Average transport properties of lysozyme, ribonuclease, and human ubiquitin are shown in Table 12.5.

The first two molecules belong to the initial parametrization set for the determination of the hydration thickness of proteins from the translational diffusion coefficient, so the discrepancy between experiment and the crystal structure is much less than 1 %. It is noteworthy, however, that the MD simulation value for *Dt* also agrees to this level of precision, indicating that the ff03 force field is an excellent descriptor of the structure in solution. The agreement is less satisfactory for the intrinsic viscosity, but the experimental error in these determinations can vary between 5 and 10 %, making both the crystal structure values and the MD simulation values statistically equivalent.

The human ubiquitin molecule has a 6-residue end chain whose last 4 residues are quite flexible, compared to the fairly rigid structures of the other two proteins. However, despite this flexibility, the crystal structure is quite a good representative of the translational diffusion coefficient. In the crystal structure, the conformation of the chain sticks straight out of the molecule, while in the molecular dynamics structures, it is generally folded inward. The MD average intrinsic viscosity has a substantial difference with that from the crystal, but unfortunately we are not aware of an experimental measurement to make a fruitful comparison. This example shows that the translational diffusion coefficient is not very sensitive to small conformational changes in solution. The effects of shape can be offset by a change in size, leaving the value of *Dt* relatively unchanged. The intrinsic viscosity is sensitive only to shape and is a much better discriminator – the MD trajectory structures of ubiquitin show that only the last 4 residues, comprising about 5 % of the molecule, are actually flexible. In the case of ubiquitin, the table also shows that making the water model more realistic by using a four-point model yields insignificant change in the computed transport properties. Thus, we can conclude that a TIP3P water model yields an excellent descriptor of the conformations in solution even though

Table 12.5 Transport properties for lysozyme, ribonuclease, and ubiquitin

Protein	Data type	Intrinsic viscosity (cm^3/g)	Translational diffusion Dt (10^{-6} cm^2/s)	Rotational diffusion tensor Dr1 ($10^7/s$)	Dr2 ($10^7/s$)	Ave. Dr ($10^7/s$)
Lysozyme (6LYZ) 14.3 kDa	MD TIP3P	3.33 ± 0.01	1.09 ± 0.01	1.79 ± 0.01	2.52 ± 0.01	2.03 ± 0.01
	Crystal	3.21	1.10	1.82	2.57	2.07
	Exp.	298−3.00	1.11 ± 0.05	–	–	167 ± 0.08
			1.12 ± 0.02			2.0
			1.06 ± 0.01			
Ribonuclease (7RSA) 13.7 kDa	MD TIP3P	3.59 ± 0.01	1.08 ± 0.02	1.74 ± 0.1	2.34 ± 0.1	1.94 ± 0.1
	Crystal	3.39	1.10	1.83	2.50	2.05
	Exp.	3.30−3.50	1.068	–	–	2.01
Ubiquitin (1UBQ) 8002E6 kDa	MD TIP3P	3.54 ± 0.1	1.267 ± 0.003	2.75 ± 0.1	3.97 ± 0.1	3.16 ± 0.1
	MD TIP4P	3.57 ± 0.1	1.263 ± 0.002	2.73 ± 0.1	3.90 ± 0.1	3.16 ± 0.1
	Crystal	3.31	1.270	2.80	3.98	3.19
	Exp.		1.30 ± 0.01	3.17 ± 0.1	3.70 ± 0.1	3.34 ± 0.1

References to experimental work are available in Hong (2009)

the diffusion coefficient of water is more than twice the experimental value (Mark and Nilsson 2001). The timing of the dynamics is faster than in a real solution (allowing useful data to be obtained from shorter trajectories), but the range of structures thermally sampled is unaffected.

For the experimental rotational diffusion data for lysozyme in Table 12.5, at first glance it may seem that the two values imply a range of experimental error, but theoretically the value measured by fluorescence (Cross and Fleming 1986), which samples all the eigenvalues of the diffusion tensor, may be different from the depolarized dynamic light scattering value (Dubin et al. 1971). If the rotational diffusion tensor is diagonalized in the same principal axes system as the polarizability of lysozyme, then the birefringence value will not depend on the faster "axial" eigenvalue, called D_{r2} in Table 12.5. Aragon has also implemented a very accurate BE method (POL) for the solution of the electrostatic equations for the determination of classical polarizabilities (Aragon and Hahn 2007). Using the program POL, with the identical triangulation input file used for the hydrodynamics, it can be shown that both the polarizability and rotational diffusion tensor are diagonalized in essentially the same principal axes – despite its irregular shape, lysozyme is optically a symmetric top! Thus, the depolarized light scattering value should be compared to the average of the two smaller eigenvalues shown in Table 12.5 as D_{r1}. The MD value of $D_{r1} = 1.79\ 10^7\ s^{-1}$ is in good agreement with the light scattering experimental value of $1.67\ 10^7\ s^{-1}$ of Dubin et al. The fluorescence value samples all the eigenvalues because the transition moment is unlikely to be oriented along the principal axes of the rotational diffusion tensor. The fluorescence value $D_r = 2.0\ 10^7\ s^{-1}$ (Cross and Fleming 1986) compares very well with the average of the MD (2.03) or crystal structure (2.07) eigenvalues of D_r.

The MD simulations in explicit water appear to provide a very good description of the solution structure of small proteins as measured by hydrodynamic transport properties. Thus, in combination with the data from local structure provided by NMR order parameters, both the whole molecule scale structure and the local structure are well described by the ff03 force field. In the next section, we describe similar results for a large flexible protein.

12.5.3 Explicit Water MD Simulations of Trastuzumab

Brandt and coworkers have carried out explicit water MD simulations of a medium-sized flexible protein, trastuzumab, a monoclonal humanized IgG antibody produced by Genentech which is used in the treatment of breast cancer (Brandt et al. 2010). This study used the ff99SB force field of Simmerling and coworkers (Hornak et al. 2006) for its enhancement of the description of alpha helix secondary structure in proteins. The antibody is a larger flexible system (150 kDa) whose range of motion is very dependent on an accurate representation of the forces between atoms – the flexibility is due to a small hinge length of protein helix in the middle of the molecule. The simulation of trastuzumab required the construction of a model from pieces that could be crystallized because flexibility has impeded the determination of the structure of the entire antibody by X-ray crystallography. The construction procedure relied on an approximate structure for the hinge postulated by Padlan (1994) and the in silico mutation of residues to make the model identical in atomic composition to trastuzumab. This initial construct was subject to energy minimization with the ff99 force field to eliminate construction artifacts, and subsequent 20 ns MD simulation with the TIP3P water model, using a protocol as described above for the small proteins. The final structure produced by that simulation was subsequently used in eight independent 40 ns (TIP3P, 300 K, 1 atm, 2 fs time step, SHAKE) simulations with ff99SB and Glycam04 force fields carried out in parallel in Genentech computer clusters. A snapshot of the trastuzumab structure from one of the independent simulations is shown in Fig. 12.6. The 0.34 μs piecewise trajectory was analyzed by computing the transport properties with the BEST suite, using a 1.1 Å uniform hydration model and compared to experiment. The transport properties were averaged over 3000 structures from the simulation, but separately for each subsimulation. The translational diffusion coefficient of trastuzumab was measured by dynamic laser light scattering and the intrinsic viscosity was measured by a rolling ball viscometer. Both measurements were carefully extrapolated as a function of concentration.

A summary of the results of this study are presented in Table 12.6. The values of the transport properties for each subsimulation are shown, along with the overall

Fig. 12.6 Ribbon structure of trastuzumab taken from one of the multiple MD trajectories (Aragon 2011)

Table 12.6 Summary of hydrodynamic analysis of trastuzumab MD simulation data, experimental hydrodynamic results, and literature values; 20 °C pure water

Trajectory	$Dt(10^{-7}\ cm^2/s)$*	τ_r(ns)**	$[\eta]$ (cm^3/g)
Average[a]	4.08 (±0.07)	173 (±11)	6.24 (±0.3)
Experiment[b]	4.09 (±0.01)		6.37 (±0.2)
Literature[c]		168,180	6.20 (±0.5)

*For computational results: $Dt = \text{Tr}(\mathbf{D}\text{tt})/3$
**For computational results: $\tau_r = (6\text{Dr})^{-1}$, where $\text{Dr} = \text{Tr}(\mathbf{D}\text{rr})/3$
[a]For all MD data, values quoted are the average of eight trajectories and the standard deviation
[b]Uncertainties quoted are the standard error of extrapolations to $c = 0$
[c]See (Brandt 2010) for references to experimental values. Experimental τ_r values are for rabbit IgG and bovine IgG; intrinsic viscosity values are for human IgG1

average. It is immediately apparent that the experimental data and the simulation ensemble averages agree extremely well.

The MD simulation is able to determine the translational diffusion coefficient with a precision of 1.7 %, and it agrees with the experimental measurement to 0.25 %. The rotational correlation time was determined by MD to within 6.3 % and agreed with literature values for other IgG's to better than 5 %. Finally the intrinsic viscosity was determined to within 4.8 % and agreed with the measurement within 2 %, well within the measurement uncertainty of 3 %. The high precision of the experimental measurements and the high precision of the hydrodynamic computations are key components of the extremely good agreement observed in this study. The only other published MD study of a complete antibody in solution used much smaller length trajectories and did not make comparisons with experiment (Chennamsetty et al. 2009a, b). This study demonstrates that the force fields used generate an excellent representation of the solution structure of the antibody. The original paper (Brandt et al. 2010) contains a movie of the complete simulation trajectory in the published supplementary data, along with several figures showing the transport properties along the multiple MD trajectories.

12.6 Conclusions

The high precision implemented via the BE method in BEST has allowed us to generate a general model to numerically treat the transport properties of proteins with a single hydration parameter for all proteins regardless of size or flexibility. The hydration thickness of 1.1 Å is a model parameter that represents the increased viscosity of water around the protein when hydrodynamic stick boundary conditions are used. The dynamical origin of this parameter shows that the layer mimic should be uniform around the protein. The hydration model we have utilized allows for atomic size variation, unlike the approximate models of other authors (Garcia de la Torre et al. 2000b) who have proposed a single atomic equivalent radius (AER) for all heavy atoms. A similar picture is obtained for nucleic acids as shown in a previous review (Aragon 2011), where it appears that DNA may have more hydration water in the grooves. Nevertheless, the recognition that macromolecular surface dynamics may be an important contributor to the increased viscous energy dissipation needs to be more thoroughly investigated.

Our studies of proteins led us to propose that some multimeric proteins have a conformational rearrangement upon going into solution from the crystal. Our preliminary work using MD simulation appears to bear this out, but the possibility of enhanced surface dynamics requiring a larger hydration parameter must still be considered. In order to validate that the structures generated by MD are actually representative of solution structure, we have performed simulations on a number of small proteins, rigid and flexible, and one medium-sized flexible protein. The good agreement we obtain with experiment demonstrates that we have validated both the force fields and the hydrodynamic hydration model for proteins. Our application with the precise hydrodynamics in BEST in combination with the trajectory ensemble average method yields very good agreement with experiment for both small and large proteins, flexible or not.

A large number of proteins have been studied by our group and others and there is broad agreement between experiment and computation, yet there remain several intriguing discrepancies. Some of these discrepancies may be due to older experimental data – it would be quite useful, using modern instrumentation such as AUC to produce a reference set of carefully measured transport properties in the near future.

Acknowledgments The author thanks David Hahn for performing implicit water MD for the α-chymotripsyn and β-lactoglobulin molecules. Financial support for some of the small protein and trastuzumab work was provided by Genentech. This work was funded in part by NIH grant GM52588 to S. Aragon and by the Center for Life Sciences at SFSU.

References

Allison SA (1999) Low Reynolds number transport properties of axisymmetric particles employing stick and slip boundary conditions. Macromolecules 32:5304–5312

Aragon SR (2004) A precise boundary element method for macromolecular transport properties. J Comput Chem 25:1191–1205

Aragon SR (2011) Recent advances in macromolecular hydrodynamic modeling. Methods 54:101–114

Aragon SR, Hahn DK (2006) Precise boundary element computation of protein transport properties: diffusion tensors, specific volume, and hydration. Biophys J 91:1591–1603

Aragon SR, Hahn DK (2007) Polarizability and Kerr constant of proteins by boundary element methods. Colloids Surf B Biointerfaces 56:19–25

Bauer DR, Brauman JI, Pecora R (1974) Molecular reorientation in liquids. Experimental test of hydrodynamic models. J Am Chem Soc 96:6840–6843

Bloomfield VA, Dalton WO, van Holde KE (1967) Frictional coefficients of multisubunit structures. I. Theory. Biopolymers 5:135–148; Ibid (1967) Frictional coefficients of multisubunit structures. II. Application to proteins and viruses. Biopolymers 5:149–159

Brandt JP (2010) Construction, molecular dynamics simulation, and hydrodynamic validation of an all-atom model of a monoclonal antibody. M.S. thesis, SFSU

Brandt JP, Patapoff TA, Aragon SR (2010) Construction, MD simulation, and hydrodynamic validation of an all-atom model of a monoclonal IgG antibody. Biophys J 99:905–913

Brenner H (1967) Coupling between the translational and rotational Brownian motions of rigid particles of arbitrary shape. II. General theory. J Colloid Interface Sci 23:407–436

Brilliantov NV, Krapivsky PL (1991) Stokes laws for ions in solutions with ion-induced inhomogeneity. J Phys Chem 95:6055–6057

Brookes E, Demeler B, Rosano C, Rocco M (2010) The implementation of SOMO (Solution MOdeller) in the UltraScan analytical ultracentrifugation data analysis suite: enhanced capabilities allow the reliable hydrodynamic modeling of virtually any kind of biomacromolecule. Eur Biophys J 39:423–435

Burgi W, Schmid K (1961) Preparation and properties of Zn-α2-glycoprotein of normal human plasma. J Biol Chem 236:1066–1074

Byron O (1997) Construction of hydrodynamic bead models from high- resolution X-ray crystallographic or nuclear magnetic resonance. Biophys J 72:408–415

Carrasco B, Garcia de la Torre J (1999) Hydrodynamic properties of rigid particles: comparison of different modeling and computational procedures. Biophys J 75:3044–3057

Case DA, Darden TA, Cheatham TE III, Simmerling CL, Wang J, Duke DE, Luo R, Walker RC, Zhang W, Merz KM, Roberts B, Hayik S, Roitberg A, Seabra G, Swails J, Goetz AW, Kolossváry I, Wong KF, Paesani F, Vanicek J, Wolf RM, Liu J, Wu X, Brozell SR, Steinbrecher T, Gohlke H, Cai Q, Ye X, Wang J, Hsieh MJ, Cui G, Roe DR, Mathews DH, Seetin MG, Salomon-Ferrer R, Sagui C, Babin V, Luchko T, Gusarov S, Kovalenko A, Kollman PA (2012) AMBER 12. University of California, San Francisco

Chennamsetty N, Helk B, Voynov V, Kayser V, Trout BL (2009a) Aggregation-prone motifs in human immunoglobulin G. J Mol Biol 391:404–413

Chennamsetty N, Voynov V, Helk B, Kayser V, Trout BL (2009b) Design of therapeutic proteins with enhanced stability. Proc Natl Acad Sci U S A 106:1937–11942

Connolly ML (1981) Molecular surface program. QCPE Bull 1:75–83

Connolly ML (1983) Analytical molecular surface calculation. J Appl Crystallogr 16:548–558

Connolly ML (1993) The molecular surface package. J Mol Graph 11:139–141

Cross AJ, Fleming GR (1986) Influence of inhibitor binding on the internal motions of lysozyme. Biophys J 50:507–512

Denisov VP, Halle B (1996) Protein hydration dynamics in aqueous solution. Faraday Discuss 103:227–244

Dubin SB, Clar NA, Benedek GB (1971) Measurement of the rotational diffusion coefficient of lysozyme by depolarized light scattering: configuration of lysozyme in solution. J Chem Phys 54:5158–5164

Durchschlag H, Zipper P (2003) Modeling the hydration of refinement and atomic models of proteins: prediction of structural and hydrodynamic parameters from X-ray diffraction and scattering data. Eur Biophys J 32:487–502

Fiser A, Do RK, Sali A (2000) Modeling of loops in protein structures. Protein Sci 9:1753–1773

Garcia de la Torre J, Bloomfield VA (1977a) Hydrodynamic properties of macromolecular complexes. I. Translation. Biopolymers 16:1747–1763

Garcia de la Torre J, Bloomfield VA (1977b) Hydrodynamic properties of macromolecular complexes. III. Bacterial viruses. Biopolymers 16:1779–1793

Garcia de la Torre J, Huertas ML, Carrasco B (2000a) Calculation of hydrodynamic properties of globular proteins from their atomic-level structure. Biophys J 78:719–730

Garcia de la Torre J, Huertas ML, Carrasco B (2000b) HYDRONMR: prediction of NMR relaxation of globular proteins from atomic-level structures and hydrodynamic calculations. J Magn Reson 147:138–146

Germann MW, Turner T, Allison SA (2007) Translational diffusion constants of the amino acids: measurement by NMR and their use in modeling the transport of peptides. J Phys Chem 111:1452–1455

Ghirlando R (2011) The analysis of macromolecular interactions by sedimentation equilibrium. Methods 54:145–156

Goldstein RF (1985) Macromolecular diffusion constants: a calculational strategy. J Chem Phys 83:2390–2397

Hahn DK, Aragon SR (2006) Intrinsic viscosity of proteins and platonic solids by boundary element methods. J Chem Theory Comput 2:1416–1428

Halle B (1999) Magnetic relaxation dispersion: principles and applications. In: Bellissent-Funel MC (ed) Hydration processes in biology. IOS, Amsterdam, pp 232–249

Halle B, Davidovic M (2003) Biomolecular hydration: from water dynamics to hydrodynamics. PNAS 100:12135–12140

Harpaz Y, Gerstein M, Chothia C (1994) Volume changes on protein folding. Structure 2:641–649

Henchman RH, McCammon JA (2002) Structural and dynamic properties of water around acetylcholinesterase. Protein Sci 11:2080–2090

Hong Q (2009) Molecular dynamics of small proteins in solution. MS. thesis, SFSU

Hornak V, Abel R, Okur A, Strockbine B, Roitberg A, Simmerling C (2006) Comparison of multiple Amber force fields and development of improved protein backbone parameters. Proteins 65:712–725

Hu CM, Zwanzig R (1974) Rotational friction coefficients for spheroids with the slipping boundary condition. J Chem Phys 60:4354–4357

Kang EH, Mansfield ML, Douglas JF (2004) Numerical path integration technique for the calculation of transport properties of proteins. Phys Rev E 69:031918-1–031918-10

Kim S, Karilla SJ (1991) Microhydrodynamics. Butterworth-Heinemann, New York

Kuntz ID Jr, Kauzmann W (1974) Hydration of proteins and polypeptides. Adv Protein Chem 28:239–345

Luise A, Falconi M, Desideri A (2000) Molecular dynamics simulation of solvated Azurin: correlation between surface solvent accessibility and water residence times. Proteins 39:56–57

Ma Y, Zhu C, Ma P, Yu KT (2005) Studies on the diffusion coefficients of amino acids in aqueous solutions. J Chem Eng Data 50:1192–1196

Makarov VA, Andres BK, Smith PE, Pettit BM (2000) Residence times of water molecules in the hydration sites of myoglobin. Biophys J 79:2966–2974

Mark P, Nilsson L (2001) Structure and dynamics of the TIP3P, SPC, and SPC/E water models at 298K. J Phys Chem A 105:9954–9960

McKenzie HA, Sawyer WH (1967) Effect of pH on β-lactoglobulin. Nature 214:1101–1104

Mongan J, Case DA, McCammon JA (2004) Constant pH molecular dynamics in generalized Born implicit solvent. J Comput Chem 25:2038–2048

Moody TP, Kingsbury JS, Durant JA, Wilson TJ, Chase SF, Laue TM (2005) Valence and anion binding of bovine ribonuclease A between pH 6 and 8. Anal Biochem 336:243–252

Odqvist FGK (1930) On the boundary value problems in hydrodynamics of viscous fluids. (German). Math Z 32:329–375

Ogston AG (1949) The Gouy diffusiometer: further calibration. Proc Roy Soc London 196:272–285

Oseen CW (1927) Hydrodynamik. Academiches Verlag, Leipzig

Padlan EA (1994) Anatomy of the antibody molecule. Mol Immunol 3:169–217

Perlman DA, Case DA, Caldwell JW, Ross WS, Cheathamm TE III, Debold FSD, Seibel G, Kollman GP (1995) AMBER, a package of computer programs for applying molecular mechanics, normal mode analysis, molecular dynamics and free energy calculations to simulate the structural and energetic properties of molecules. Comput Phys Commun 91:1–41

Rai N, Nollmann M, Spotorno B, Tassara G, Byron O, Rocco M (2005) SOMO (SOlutionMOdeler): differences between X-Ray and NMR-derived bead models suggest a role for side chain flexibility in protein hydrodynamics. Structure 13:723–734

Richards EG (1980) An introduction to physical properties of large molecules in solution. Cambridge University Press, London

Rotne J, Prager S (1969) Variational treatment of hydrodynamic interaction in polymers. J Chem Phys 50:4831–4837

Sali A, Blundell TL (1993) Comparative protein modelling by satisfaction of spatial restraints. J Mol Biol 234:779–815

Salomon-Ferrer R, Goetz AW, Poole D, Grand S, Walker RC (2013) Routine microsecond molecular dynamics simulations with AMBER – Part II: particle Mesh Ewald. J Chem Theory Comput 9:3878–3888

Schuck P (2000) Size-distribution analysis of macromolecules by sedimentation velocity ultracentrifugation and lamm equation modeling. Biophys J 78:1606–1619

Schwert GW, Kaufman S (1951) The molecular size and shape of the pancreatic proteases. III. α-Chymotrypsin. J Biol Chem 190:807–816

Spotorno B, Piccinini L, Tassara G, Ruggiero C, Nardino M, Molina F, Rocco M (1997) BEAMS (BEAds Modelling System): a set of computer programs for the generation, the visualization and the computation of hydrodynamic and conformational properties of bead models of proteins. Eur Biophys J 24:373–384

Squire PG, Himmel ME (1979) Hydrodynamics and protein hydration. Arch Biochem Biophys 196:165–177

Sturlaugson AL, Fruchey KS, Lynch SR, Aragon SR, Fayer MD (2010) Orientational and translational dynamics of polyether/water solutions. J Phys Chem 111:5350–5358

Teller DC, Swanson E, de Haen C (1979) The translational friction coefficients of proteins. Methods Enzymol 61:103–124

Venable RM, Pastor RW (1988) Frictional models for stochastic simulations of proteins. Biopolymers 27:1001–1014

Venu K, Denisov VP, Halle B (1997) Water ^{1}H magnetic relaxation dispersion in protein solutions. A quantitative assessment of internal hydration, proton exchange, and cross-relaxation. J Am Chem Soc 119:3122–3134

Wegener WA (1986) On an exact starting expression for macromolecular hydrodynamic models. Biopolymers 25:627–637

Youngren GK, Acrivos A (1975a) Stokes flow past a particle of arbitrary shape: a numerical method of solution. J Fluid Mech 69:377–402

Youngren GK, Acrivos A (1975b) Rotational friction coefficients for ellipsoids and chemical molecules with the slip boundary condition. J Chem Phys 63:3846–3848

Part IV
Applications of AUC: Material Science

Chapter 13
Hydrodynamic Analysis of Synthetic Permanently Charged Polyelectrolytes

Christine Wandrey and Hamideh Ahmadloo

Abstract Synthetic permanently charged polyelectrolytes are intensively used as process auxiliary agents in several branches of industry. Particularly important is their application in solid/liquid separation processes, which are parts of water purification, wastewater treatment, and papermaking. The macromolecular and hydrodynamic characteristics of such synthetic polyelectrolytes strongly influence the efficacy and applicability of the polymers. Despite progress in the understanding of the general solution behavior of polyelectrolytes, their comprehensive characterization remains a challenge. This concerns almost all technologically and practically interesting polyelectrolytes, which typically have very high molar masses and are heterogeneous with respect to the molar mass. In some cases, the chain architecture deviates from the linearity. Analytical ultracentrifugation provides advantages due to the absence of interaction with any column material such as used for chromatographic methods. This chapter presents and discusses the advantages and limitations of analytical ultracentrifugation in terms of polyelectrolyte characterization.

Keywords Analytical ultracentrifugation • Densitometry • Dilution viscometry • Molar mass • Polyelectrolytes • Sedimentation velocity

13.1 Introduction

Polyelectrolytes (PEL) are macromolecules containing a large number of ionic or ionizable groups, or both, in their polymer backbone chain (integral type) or in pendent side groups (pendent type). In addition, neutral monomeric units can be present. PEL can be either synthetic, natural, or modified natural macromolecules. Nucleic acids, proteins, and some polysaccharides are examples of natural PEL. Modified natural PEL are chitosan and cellulose-based anionic and cationic PEL. Figure 13.1 presents a general classification of PEL.

C. Wandrey (✉) • H. Ahmadloo
Institut d'Ingénierie Biologique et Institut des Sciences et Ingénierie Chimiques, École Polytechnique Fédérale de Lausanne, EPFL-SV-IBI-LMRP, Station 15, CH-1015 Lausanne, Switzerland
e-mail: christine.wandrey@epfl.ch

S. Uchiyama et al. (eds.), *Analytical Ultracentrifugation*,
DOI 10.1007/978-4-431-55985-6_13

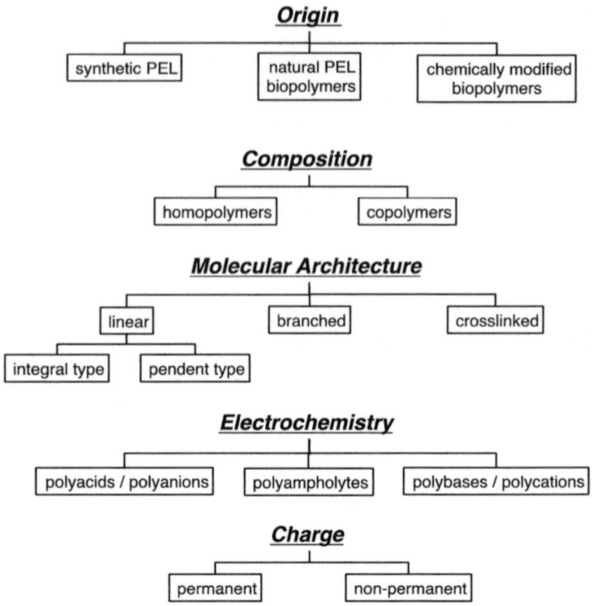

Fig. 13.1 Classification of polyelectrolytes in terms of their origin, composition, molecular architecture, electrochemistry, and charge type

In the case of positive charges on the polymer chain, polycations (polybases) and low molar mass negatively charged conterions are obtained upon dissociation of these cationic PEL. Accordingly, anionic PEL yield polyanions (polyacids) and positively charged counterions. Dissociation of a sufficient number of charge-bearing groups is the prerequisite of the water solubility of PEL.

The presence of ionic or ionizable groups classifies the PEL either as permanently charged or nonpermanently charged PEL. The degree of dissociation of the permanently charged PEL is almost constant over a wide range of the pH. Contrary, the dissociation behavior of nonpermanently charged PEL strongly depends on the pH, i.e., on the extent of protonation or ionization of the ionizable groups. An equivalent number of oppositely charged low molar mass ions, called counterions, always compensate the charges of the polymer chain in order to ensure electroneutrality. Quaternary ammonium groups represent structures of permanently charged PEL. Amino, imino, and carboxyl groups are nonpermanently charged examples. While the charge density of the first group is almost independent of the pH, the charge density of the latter depends on the pH of the medium, on the degree of protonation and ionization. Applying the classification of Fig. 13.1, this chapter will focus on *synthetic permanently charged cationic copolymers of the pendent type*.

The solution behavior of synthetic permanently charged PEL as well as the chain conformation and the hydrodynamic volume are governed by factors such as the chain constitution, the nature of the ions, the molar mass, and the ionic strength. The pH may have an influence on the chemical stability but is not expected to affect the chain dimensions significantly. Different from many natural PEL, secondary

Table 13.1 Theoretical exponents of relationships relating the intrinsic viscosity $[\eta]$, the sedimentation coefficient s_0, and the diffusion coefficient D_0 to the molar mass (Elias 1990)

	Exponent		
Shape	$[\eta]$	s_0	D_0
Rod	2	0.15	−1
Coil[a]	0.5…0.76	0.41…0.5	−0.5… −0.588
Coil[b]	0.5	0.5	−0.5
Coil[c]	<0.5	>0.5	> −0.5
Sphere	0	2/3	−1/3

[a]in a good solvent
[b]in a theta solvent
[c]branched molecule

interactions such as hydrogen bonds or hydrophobic interactions leading to specific structure formation are absent or less important, and isolated macromolecules become present upon sufficient dilution.

Nevertheless, comparing scaling relationships of synthetic neutral polymers and synthetic PEL, differences become obvious from which can be concluded that the shape of PEL molecules in solution refers more to a rodlike/extended coil shape than to a molecule in a good solvent or to a spherical shape. Draining effects can also play a role. Table 13.1 shows the theoretical exponents of the relationships relating the intrinsic viscosity, the sedimentation, and the diffusion coefficient to the molar mass of macromolecules (Elias 1990; Harding 2005). Increasing the ionic strength of the solution reduces the Debye length and, consequently, the chains become more flexible, the electrostatically excluded volume decreases, and the hydrodynamic volume decreases as well. Scaling relationships established for different ionic strengths confirm such behavior (Dautzenberg et al. 1994). Despite generally knowing that low molar mass salt has to be added to aqueous solutions of PEL in the case of permanently charged molecules, quantification and reliable PEL characterization remains a challenge especially when technical products instead of model PEL are the subject of the intended characterization study.

The macromolecular characterization of nonpermanently charged PEL can be simplified by changing the pH in such a way that no dissociation takes place. However, the suppression of dissociation could impede the solubility for not sufficiently hydrophilic polymer backbone chains.

The aim of this chapter is to present and discuss the advantages and limitations of analytical ultracentrifugation in terms of polyelectrolyte characterization. In this context, the main focus will be not on presenting a perfect characterization study but rather on practical aspects and on emphasizing a number of pitfalls. A series of synthetic permanently charged cationic PEL was selected for this purpose. Some attention will also be given to the sample preparation and supporting methods, which are necessary for a comprehensive interpretation of the experimental characterization data.

13.2 Polyelectrolyte Synthesis

Figure 13.2 shows the general chemical structure of the PEL analyzed. The cationic copolymers are composed of the neutral monomer acrylamide (AM) and the cationic monomer acryloyloxyethyltrimethylammonium chloride (AEAC). AM/AEAC copolymers of different molar masses and chain architecture were synthesized by inverse emulsion polymerization according to a technology described previously in detail (Hernandez Barajas et al. 2004; Rasteiro et al. 2010). Due to the similar monomer reactivity of the two monomers in the range of the monomer batch composition applied herein, an almost uniform product in terms of the chemical composition can be expected up to high conversion (Hernandez Barajas and Hunkeler 1997). Contrary, with respect to the molar mass, the copolymers will be nonuniform. The monomer batch formulation yielded copolymers containing 45 wt% of the cationic monomer. Considering the different molar masses of the two different monomer molecules, this composition refers to a molar fraction of 0.23 cationic monomeric chain units, from which a theoretical average charge distance of 1.1 nm can be estimated. At this charge distance, no counterion condensation is expected and all counterions contribute to the ionic strength of the PEL solution (Manning and Zimm 1965; Manning 1965).

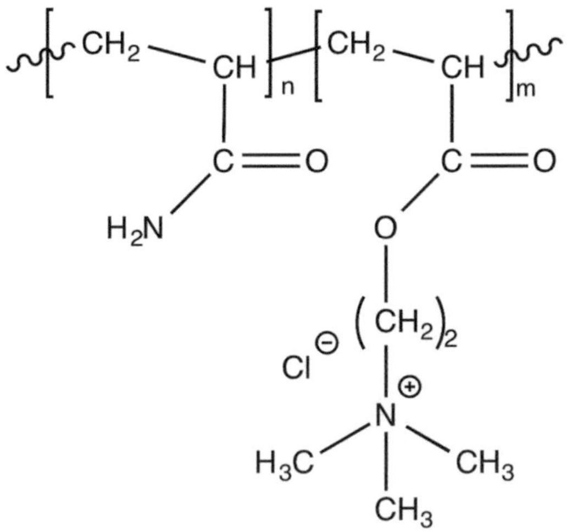

Fig. 13.2 General chemical structure of the cationic copolymers AM/AEAC. The molar masses of the monomers are $MM_{AM} = 71.079$ g/mol and $MM_{AEAC} = 193.674$ g/mol, n and m refer to the molar percentage of the monomers

13.3 Supporting Methods

13.3.1 Solution Preparation

The hydrodynamic characterization of PEL requires the absence of any impurities, the exact knowledge of the PEL concentration, and the PEL to be dissolved molecularly. Depending on the quality of the PEL sample, whether it was provided as a powder, emulsion, or aqueous solution, the purification will differ. For aqueous solutions or dissolved powders, the purification may include ultrafiltration and dialysis in order to remove low molar mass impurities such as residual monomer, initiator, or any other additives. In the case of PEL emulsions, as it was the case for the samples described in this chapter, repeated precipitation and redissolution in acetone, alcohol, and water, respectively, were performed prior to freeze-drying of the purified samples.

In general, PEL are hydrophilic substances, which adsorb water from the atmosphere and do not become completely dry upon freeze-drying. The moisture content depends on the chemical structure and usually varies in the range of 5–15 wt%. Several methods can be used, either alone or in complementary fashion, to determine the concentration of the PEL accurately (Wandrey and Hunkeler 2002). These include:

- Determination of the dry content of aliquots and considering the analyzed moisture content when preparing the PEL solutions
- Determination of the PEL solution concentration by UV/vis spectroscopy if UV absorbing groups are present in the PEL
- Potentiometric titration of the counterions of homopolymers and of copolymers in case their chemical composition is known

The concentration of the aqueous solution of the AM/AEAC copolymers was determined by potentiometric chloride titration after having analyzed the chemical composition by FTIR and titration (Losada et al. 2009).

Figure 13.3 presents an overview of the several steps of the solution preparation of the AM/AEAC copolymers, which include the separate preparation of polymer and salt solutions in water at pH 3.5. Dissolution of the PEL in water and not in the salt solution prevents incomplete dissolution and aggregation. The low pH prevents hydrolysis of the ester bond of the cationic monomer unit during solution storage and characterization. After mixing the aqueous PEL solution of known concentration c_p with the salt solution of concentration c_s at a mixing ratio 1:1, a stock solution containing 1/2 c_p and 1/2 c_s is obtained. Dialysis versus 1/2 c_s is recommended as discussed and justified by Budd (2002).

The selection of an optimum salt type and salt concentration requires special attention. As outlined by Budd (2002), performing the characterization at constant salt concentration c_s and not at a constant chemical potential μ can cause an error

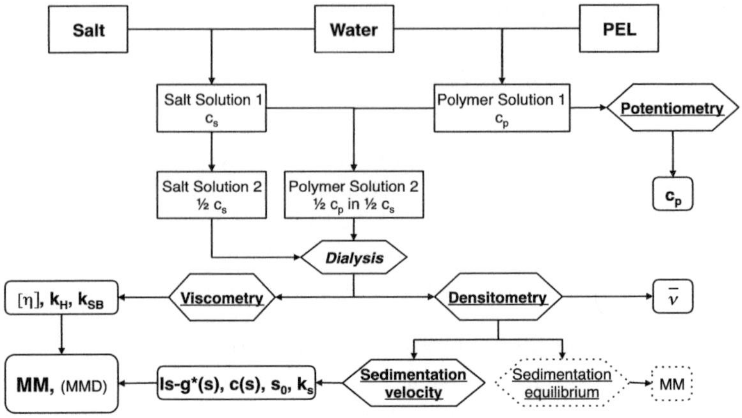

Fig. 13.3 Principal characterization scheme including solution preparation, supporting methods, and analytical ultracentrifugation

in the density increment *(dρ/dc)* of about 20 %, as presented for poly(sodium styrene sulfonate) in 0.5 mol/l NaCl. This error results from the nonequilibrium *Donnan* potential. The error increases with increasing ionic strength. Performing intensive sample dialysis in order to achieve the Donnan equilibrium can have some negative side effects such as change of the initial PEL concentration by solvent permeating through the membrane and adsorption of PEL molecules on the dialysis membrane material. The latter preferably occurs for cationic PEL and is particularly critical when working at very low PEL concentration. Further, sedimentation velocity experiments are sensitive to the added salt concentration. Depending on the rotation velocity and run duration, a radial salt gradient can be formed being more pronounced at higher salt concentration.

Considering all these potential influences, it may be advantageous to identify a minimum salt concentration, which is sufficiently high to prevent the *primary charge effect* during the AUC run, but does not necessarily require intensive dialysis.

A low salt concentration is also particularly favorable for the characterization of cationic PEL by sedimentation velocity. Cationic PEL have in general significantly higher partial specific volume values than anionic PEL (Wandrey et al. 1999). While for polycations values between about 0.7 and 0.9 ml/g have been reported in water, this range was only 0.4–0.6 ml/g for polyanions. These experimental values are in agreement with theoretical calculations (Durchschlag and Zipper 1994; Gianni and Lepori 1996). Consequently higher rotation velocities are needed to sediment cationic PEL, as are the AM/AEAC copolymers presented in this chapter. Their stock solutions prepared according to the scheme in Fig. 13.3 have been dialyzed. Upon dialysis, the PEL and salt stock solutions were used to determine the intrinsic viscosity by dilution viscometry and the partial specific volume by densitometry.

13.3.2 Dilution Viscometry

Dilution viscometry is a relatively fast and simple method to obtain information in terms of the hydrodynamic volume of dissolved polymer molecules. A capillary viscometer is frequently used to determine the intrinsic viscosity $[\eta]$, which has the unit ml/g, so referring to a volume occupied in solution by one gram of polymer. However, $[\eta]$ does not only depend on the molar mass of the polymer but also on the molecule architecture, the shape of the solute, and the solvent quality as shown in Table 13.1. For PEL, scaling exponents between those for neutral polymers in good solvents and rodlike shapes have been reported, depending primarily on the ionic strength/salt concentration of the solvent.

The value $1/[\eta]$ can be taken as an estimate of the overlap concentration, the concentration below which completely separated polymer molecules can be expected. Further, $[\eta]$ can be combined with the limiting sedimentation coefficient s_0 to calculate molar masses relying on Mandelkern-Flory-Tsvetkov (Mandelkern and Flory 1952; Mandelkern et al. 1952; Tsvetkov 1979; Tsvetkov et al. 1970, 1989; Pavlov and Frenkel 1995; Pavlov 1997).

The intrinsic viscosity of the AM/AEAC copolymers was analyzed in 0.1 mol/l and 0.05 mol/l NaCl solution. The $[\eta]$ values, extrapolated according to Schulz and Blaschke (1941) Eq. (13.1), are presented in Table 13.2.

$$\eta_{red} = [\eta] + k_{SB}\,[\eta]\,\eta_{sp} \tag{13.1}$$

with $\frac{\eta_p - \eta_0}{\eta_0} = \eta_{sp}$ and $\eta_{red} = \frac{\eta_{sp}}{c_p}$, η_{p} and η_0 are the viscosity of the PEL solution and the solvent, respectively.

All extrapolations to obtain $[\eta]$ according to Eq. (13.1) yielded linear relationships and confirmed sufficient differences of the three linear copolymers E1, E2, and E4, as it was intended by the polymer synthesis. As expected, the $[\eta]$ values are higher in 0.05 mol/l NaCl solutions than in 0.1 mol/l NaCl solutions (Table 13.2). $c_s = 0.05$ mol/l NaCl was selected for the subsequent characterization. This lower salt concentration was sufficiently high to achieve linear Schulz-Blaschke extrapolations. However, the potential error of not dialyzing upon further sample dilution will be reduced.

Table 13.2 Characterization by dilution viscometry in solvents of different ionic strength, NaCl concentrations 0.05 mol/l and 0.1 mol/l; extrapolation according to Schulz and Blaschke (1941), $T = 20$ °C

	NaCl: 0.05 mol/l		NaCl: 0.1 mol/l	
AM/AEAC copolymer	$[\eta]$ (ml/g)	k_{SB}	$[\eta]$ (ml/g)	k_{SB}
E1	2460	0.13	1703	0.18
E2	1550	0.19	1301	0.26
E4	1250	0.08	802	0.29
E1++	1771	0.25	909	0.54

13.3.3 Densitometry

In contrast to proteins, for which the partial specific volume $\overline{v}$ in many different buffer solutions can be taken from databases, $\overline{v}$ has to be determined experimentally for most synthetic PEL dissolved in a salt solution. Very precise values are obtained using a digital densitometer equipped with an oscillating capillary (Wandrey et al. 1999). The prerequisite for obtaining precise $\overline{v}$ values is an accurately measured concentration. In the case of the PEL studied here, the concentration was accurately measured using potentiometric titration (Wandrey and Hunkeler 2002).

The partial specific volume only depends on the chemical structure of the solute and the solvent quality but not on the molar mass if this is sufficiently high. Thus $\overline{v}$ can even be used to verify copolymer compositions provided that the copolymers are dissolved in the same solvent. Concentration-dependent density measurements yielded according to Eq. (13.2) almost identical values of $\overline{v}$ for E1 (0.700 ml/g), E2 (0.703 ml/g), and E4 (0.703 ml/g), confirming the same chemical composition of these copolymers.

$$\rho = \rho_0 + (1 - \overline{v}\,\rho_0)\,c_p \tag{13.2}$$

ρ and ρ_0 denote the density of the solvent and the copolymer solution, respectively.

Most importantly, the buoyancy term $(1-\overline{v}\rho_0)$ is needed to calculate sedimentation coefficients and molar masses from AUC raw data.

13.4 Analytical Ultracentrifugation

The sedimentation velocity technique of AUC was used to characterize five PEL samples. Due to the absence of light-absorbing groups in all copolymers, only the interference optical system of the Beckman Optima XL-I could be used for the raw data acquisition. The same sample solutions as used for the density measurements were placed in double-sector Epon cells sealed with quartz windows. All characterization results are summarized at the end of Sect. 4.

13.4.1 Synthetic Boundary Experiments

Prior to the sedimentation velocity experiments, synthetic boundary experiments confirmed the complete dissolution of all samples and the absence of molecular aggregates. Figure 13.4 shows typical raw data of a synthetic boundary experiment. The concentration profiles of ten scans acquired at 5000 and 10,000 rpm with a scan

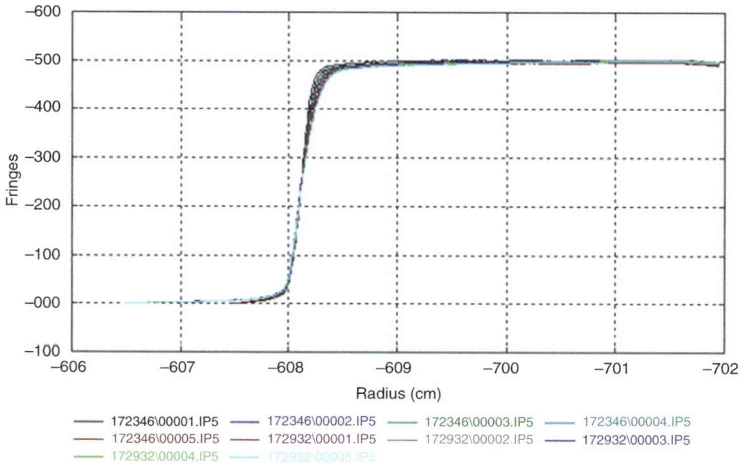

Fig. 13.4 Typical example of a synthetic boundary experiment to check the sample homogeneity using the interference optical system. Overlay of five scans taken at 5000 rpm and five scans taken at 10,000 rpm, delay 1 min, $c_p = 2.18 \times 10^{-3}$ g/ml in 0.05 mol/l NaCl, $T = 20$ °C, (XL-Graph, Philo)

delay of 1 min overlay at the same plateau of 5 fringes. A previously published study on medium to highly branched PEL composed of the same two monomers AM and AEAC, but at different molar ratios, reported the quantification of aggregated/cross-linked portions by such synthetic boundary experiments (Bourdillon et al. 2006).

13.4.2 Sedimentation Velocity Experiments

13.4.2.1 The Limiting Sedimentation Coefficient s_0

Figure 13.5 presents a representative example of sedimentation profiles. For each sample, five concentrations have been analyzed in a concentration range corresponding to initial fringe plateaus of one to eight. The high of the fringe plateau is sensitive to the chemical structure of the PEL, its chain architecture, and the PEL concentration in the sample sector of the centerpiece, as well as the salt concentrations in both sectors. Plotting the initial plateau value vs. the concentration is a possibility to check the correct sample behavior in all cells during the sedimentation velocity run. For example, the evaporation of small amounts of solvent in any cell will become detectable. Figure 13.6 shows almost the same slopes, 2.64×10^3 and 2.66×10^3 for E2 and E4, but 2.44×10^3 for the branched E1++, and 2.15×10^3 for F1, which contains only 35 wt% of the cationic monomer.

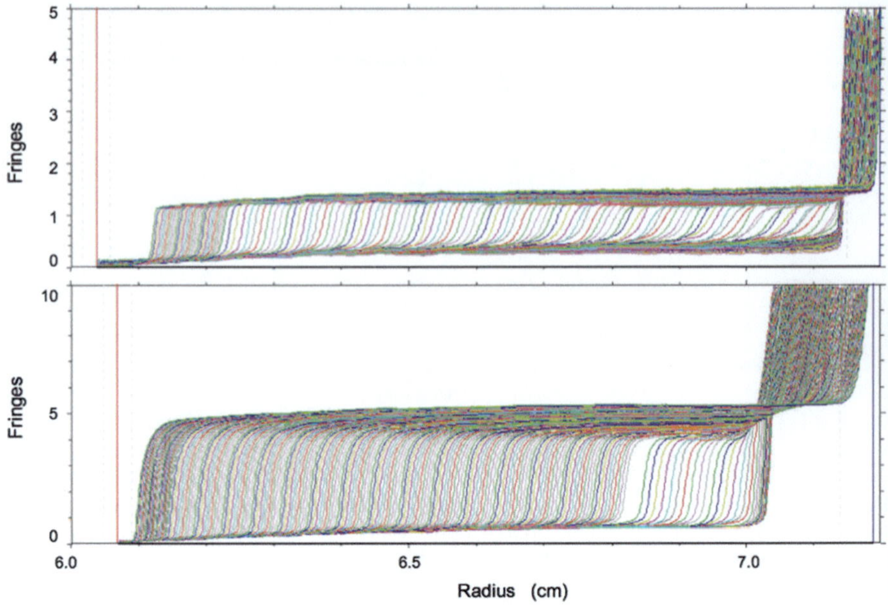

Fig. 13.5 Sedimentation velocity interference raw data showing the fringe signal intensity of different concentrations of E1, $c_p = 4.47 \times 10^{-4}$ g/ml (*top*) and $c_p = 1.80 \times 10^{-3}$ ml/g (*bottom*) in 0.05 mol/l NaCl, 35,000 rpm, $T = 20$ °C, 30 scans/delay 2 min, 150 scans/delay 5 min, 30 scans/delay 10 min, (Sedfit, Schuck)

Sedimentation coefficient distribution curves obtained by fitting the raw data using the *ls–g*(s)* and the *c(s)* model of Sedfit (Schuck 2000; Scott and Schuck 2005) are shown in Fig. 13.7. The curves reveal apparent sedimentation coefficients ranging from approximately 1 to 4 S for concentrations in the range of 5×10^{-4} to 2.5×10^{-3} g/ml.

From plots $1/s_{app}$ vs. c_p, the first- and second-order fits according to

$$\frac{1}{s_{app}} = \frac{1}{s_0}\left(1 + k_s c_p + k_s' c_p^2 + \cdots\right) \tag{13.3}$$

yielded the s_0 from the intercept $1/s_0$ and k_s from the slope k_s/s_0. The values are summarized in Table 13.3. In addition, Fig. 13.8 shows the fits for the samples E1 and E4 for which the highest and lowest molar masses are expected from the intrinsic viscosity values. Assuming increasing molar masses in the order of E1 > E2 > E4 and comparing the results of the first- and second-order fits in Table 13.3, the following becomes obvious. Whereas there is no continuity of $1/s_0$ and the slope k_s/s_0 for the linear extrapolation values, it is for the second-order

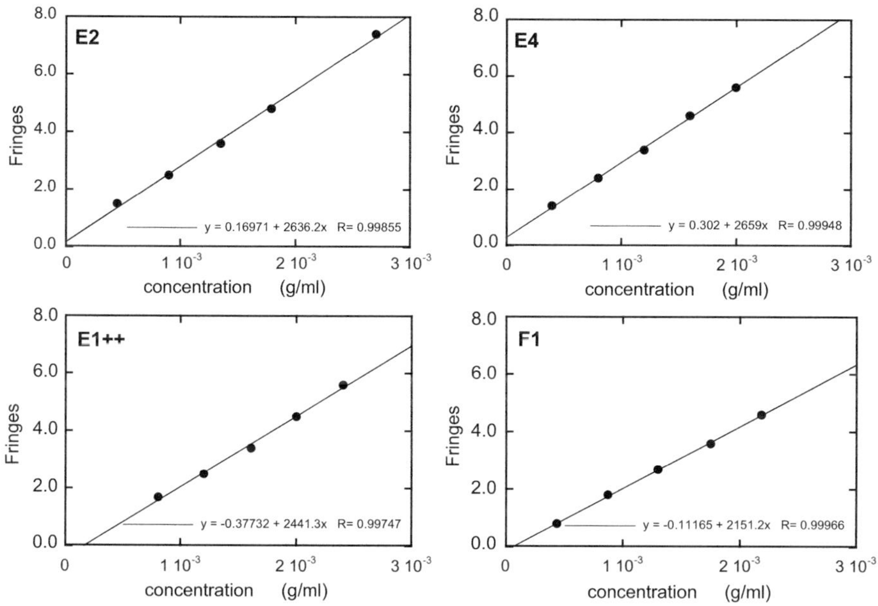

Fig. 13.6 Concentration dependence of the fringe signal of E2, E4, E1++, F1, demonstrating the influence of the chain architecture and the chemical composition. E2 and E4 have the same chemical composition, E1++ is a branched PEL, F1 contains less cationic monomer than the E samples

extrapolation and the regression coefficients (R-values) in Table 13.3 are better also. In general, higher limiting sedimentation coefficients and k_s values result from the nonlinear fits suggesting higher molar masses. Comparing in addition the two graphs in Fig. 13.8, the lower difference between the two fit procedures is confirmed for E4. Considering the overlap concentrations calculable from the intrinsic viscosities as 4.06×10^{-4}, 6.45×10^{-4}, and 8.00×10^{-4} g/ml for E1, E2, and E4, it becomes clear that the experimental concentrations of E1 deviate more from the overlap concentration than the experimental concentrations of E4. These experimental concentrations have been selected as a compromise in terms of the signal to noise ratio of the interference raw data. However, from the range of the sedimentation coefficients shown in Fig. 13.7, it can be hypothesized that shifting the experimental polymer concentrations closer to or even somewhat more below the overlap concentration will still yield sufficiently different apparent sedimentation coefficients as long as the raw data remain reliable. Nevertheless, the nonlinear fit seems to be a good option to obtain s_0 for PEL of high molar masses for which the overlap concentration is very low.

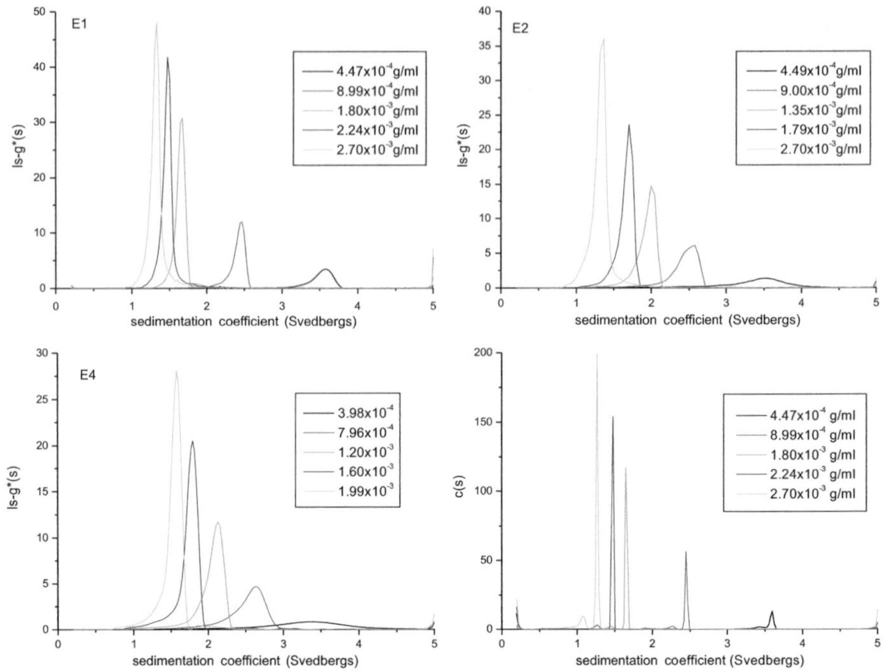

Fig. 13.7 Sedimentation coefficient distribution curves: ls−g*(c) of E1, E2, E4, and c(s) of E1, five concentrations in 0.05 mol/l NaCl, 35,000 rpm, $T = 20$ °C

Table 13.3 Results of first- and second-order fitting according to Eq. (13.3)

PEL	First order $1/s_0 \times 10^{-12}$ (s^{-1})	$k_s/s_0 \times 10^{-15}$ (ml · s/g)	R	Second order $1/s_0 \times 10^{-12}$ (s^{-1})	$k_s/s_0 \times 10^{-15}$ (ml · s/g)	R
E1	2.157	2.064	0.9947	1.551	3.165	0.9997
E2	2.331	2.008	0.9966	1.891	2.739	0.9999
E4	2.204	2.243	0.9992	1.987	2.711	0.9999

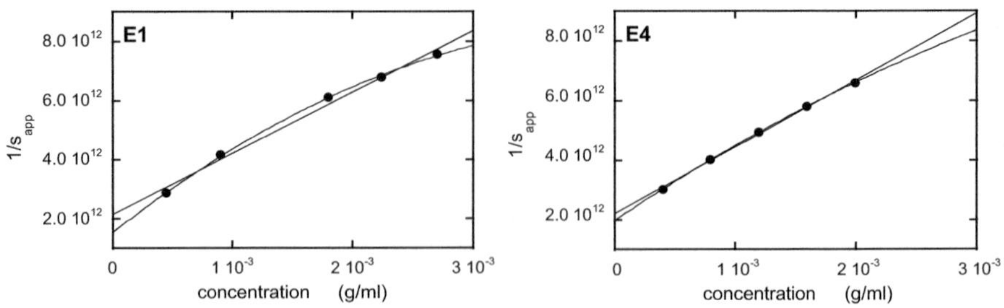

Fig. 13.8 Calculation of s_0 by first- and second-order fits according to Eq. (13.3)

13.4.2.2 The Frictional Ratio f/f_0

Calculating the sedimentation coefficient distribution curves as shown in Fig. 13.7 but using the *c(s)* model instead of the *ls–g*(s)* fitting requires either predefining a frictional ratio or fitting it. Due to the weak size dependence of $D \sim s^{-1/2}$ and the weak dependence of the frictional ratio on macromolecular shape (Scott and Schuck 2005), a weak dependence of s on f/f_0 results. This is demonstrated in Fig. 13.9 for frictional ratios between 2 and 50 and appropriate values in Table 13.4. Fitting the frictional ratio for the experimental data according to *c(s)*, nonrealistic high values were obtained, but the s_{app} values leveled already off for $f/f_0 > 2$, with s_{app} values corresponding to the value obtained from the *ls–g*(s)* fits. Contrary, as expected, a wide range of apparent molar masses M_{app} resulted for the f/f_0 range in Table 13.4. As a conclusion from these findings, $f/f_0 > 2$ was suggested, but could not reliably be quantified by *c(s)* fits alone for the PEL studied here.

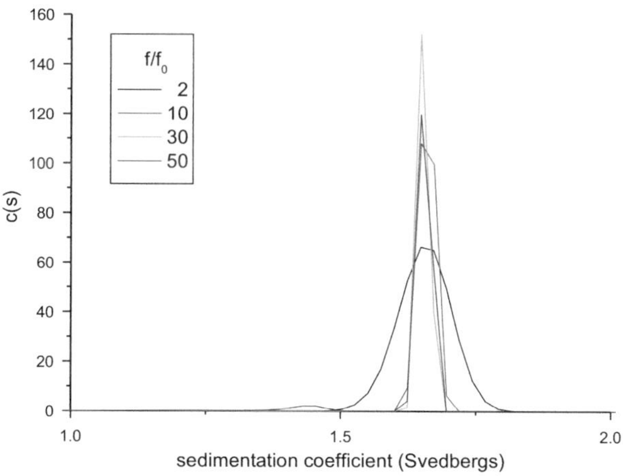

Fig. 13.9 Impact of the frictional ratio f/f_0 on the calculation of s_{app}, E1, $c_p = 1.8 \times 10^{-3}$ g/ml

Table 13.4 Influence of the frictional ratio f/f_0 on the calculation of s_{app} and M_{app} using the *c(s)* model and comparison with *ls–g*(s)*

c(s) model			ls–g*(s) model
f/f_0	s_{app} (S)	M_{app} (g/mol)	s_{app} (S)
2	1.655	2.37×10^4	–
10	1.658	2.64×10^5	–
30	1.652	1.37×10^6	–
50	1.654	2.96×10^6	–
–	–	–	1.639

Selected representative sample: E1, c_p $= 1.80 \times 10^{-3}$ g/ml

Table 13.5 Summary of characterization results. The intrinsic viscosity $[\eta]_{SB}$, the partial specific volume $\overline{\nu}$, and the limiting sedimentation coefficient s_0 were obtained from experiments performed in 0.05 mol/l NaCl solution, $T = 20$ °C

PEL	Cationic (wt%)	$[\eta]_{SB}$ (ml/g)	k_{SB}	$\overline{\nu}$ (ml/g)	s_0 (S)	k_s (ml/g)	$k_s/[\eta]_{SB}$	$M_{s\eta}$ (g/mol)
E1	45	2460	0.13	0.700	6.45	2042	0.83	1.91×10^6
E2	45	1550	0.19	0.703	5.29	1449	0.93	1.34×10^6
E4	45	1250	0.08	0.703	5.03	1364	1.09	9.44×10^5
E1++	45	1771	0.25	0.739	9.00	2618	1.48	3.30×10^6

13.4.2.3 Molar Masses

Molar masses have been calculated combining the sedimentation coefficient and the intrinsic viscosity according to

$$M_{s\eta} = \left(\frac{R}{A_0}\right)^{\frac{3}{2}} [s]^{\frac{3}{2}} \left(\frac{[\eta]}{100}\right)^{\frac{1}{2}} \tag{13.4}$$

$$[s] \equiv \frac{s_0 \eta_0}{(1 - \overline{\nu}\rho_0)} \tag{13.5}$$

with R the gas constant, $A_0 = 3.4 \times 10^{-10}$ (g cm^2/K s^2 mol$^{1/3}$) the hydrodynamic invariant (Tsvetkov et al. 1984; Pavlov 1997), η_0 the viscosity of the solvent, and $(1 - \overline{\nu}\rho_0)$ the buoyancy term. All calculated molar masses are summarized in Table 13.5.

Considering the molar copolymer composition with a molar fraction of the cationic monomer of 0.23, the average contour lengths L of E1, E2, and E3 can be calculated as 5.75 μm, 4.04 μm, and 2.84 μm, respectively. With a chain diameter d of 1 nm, the average axis/aspect ratios L/d of fully extended PEL chain rods would be 5750, 4040, and 2840.

13.4.3 Scaling Relationships

The intrinsic viscosities $[\eta]_{SB}$ obtained according to Schulz and Blaschke (1941), the sedimentation coefficients s_0 from second-order extrapolations according to Eq. (13.3), and the molar masses $M_{s\eta}$ of E1, E2, and E4 have been used to establish the relationships

$$s_0 = 3.49 \times 10^{-15} M^{0.35} \ (s) \tag{13.6}$$

and

$$[\eta] = 1.76 \times 10^{-3} M^{0.98} \left(\frac{ml}{g}\right) \quad (13.7)$$

as shown in Fig. 13.10. Comparing the exponents of both scaling relationships 0.35 and 0.98, with the values indicated in Table 13.1, a shape between coils in a good solvent and rods can be concluded, as expected for PEL. This is in agreement with high values of f/f_0 suggested in Sect. 4.2.2.

The values of the sedimentation coefficient and the molar mass in Table 13.4 are apparent values and do not fulfill Eq. (13.6). They have been used to identify the most probable frictional ratio from which the molecular dimensions of the PEL molecules can be concluded. Calculating the coefficient of Eq. (13.6) for each s-M pair of the four frictional ratios in Table 13.4 and plotting these values vs. f/f_0, as shown in Fig. 13.11, a frictional ratio of almost 5 can be estimated, which corresponds to the coefficient 3.49×10^{-15} in Eq. (13.6) and confirms a strong asymmetry of the PEL coil.

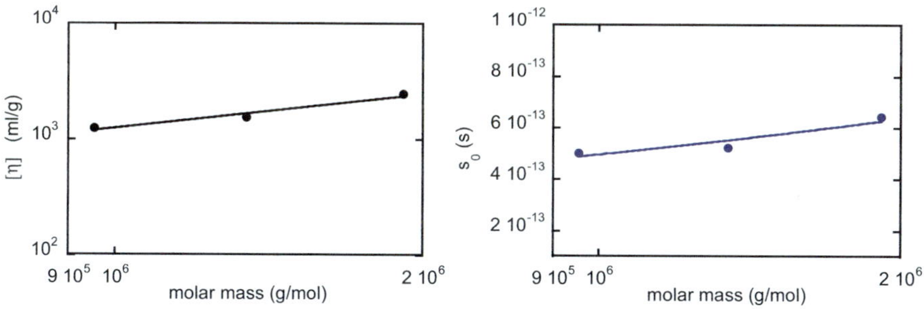

Fig. 13.10 Scaling of the intrinsic viscosity $[\eta]$ and the limiting sedimentation coefficient s_0 vs. the molar mass

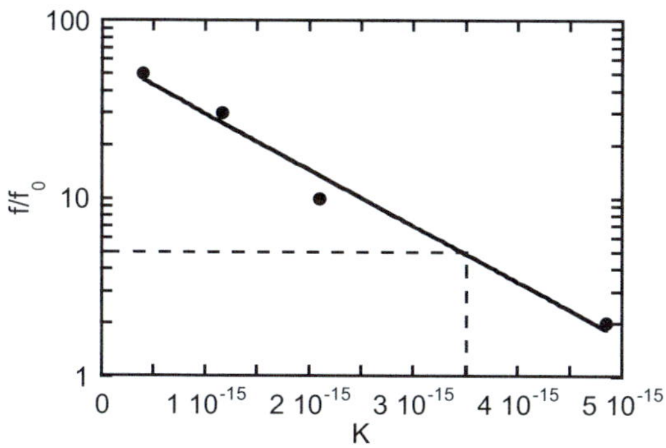

Fig. 13.11 Scaling f/f_0 vs. the coefficient K calculated according to Eq. (13.6) using the s_{app} and M_{app} values of Table 13.4. The dotted lines demonstrate the estimation of f/f_0 for E1

This interpretation is further supported by the values of the Wales-van Holde parameter $k_s/[\eta]$ (Wales and van Holde 1954) in Table 13.5. Creeth and Knight (1965) have shown empirically values of ~1.6 for compact spheres and non-draining but adopt lower values for more extended structures. Rowe (1977, 1992) derived values of ~0.2 for long rod-shaped molecules not considering free-draining effects and assuming a sufficient ionic strength to suppress PEL effects. Considering free-draining effects, Lavrenko et al. (1992) obtained limiting values of 0 for free draining and one for non-free draining. Other authors (Berth et al. 2002) have argued that low values of $k_s/[\eta]$ are due to draining effects rather than related to strong asymmetry.

With $k_s/[\eta]$ values between 0.83 for E1 and 1.09 for E4, an intermediate situation can be concluded, with a more rodlike or draining tendency for the sample of the higher molar mass. To demonstrate the reliability of the Wales-van Holde parameter, the characterization result of a branched copolymer having the same chemical composition as the three linear samples has been included in Table 13.5. Despite the same chemical composition, the partial specific volume deviates as expected for the different chain architecture (Bourdillon et al. 2006). The $k_s/[\eta]$ value of 1.48 is closer to a spherical geometry. For previously characterized medium to highly branched PEL of similar chemical structure, values in the range of 1.3–1.8 have been reported (Bourdillon et al. 2006).

13.5 Conclusions

Using hydrodynamic methods such as analytical ultracentrifugation and dilution viscometry for the characterization of practically interesting high molar mass permanently charged PEL, valuable information can be obtained that is not available from other methods. Despite not always being quantitative and not yet fully supported by the appropriate theory, important conclusions are possible. Compared to proteins and biopolymers such as polysaccharides, a relatively limited number of different chemical structures of permanently charged PEL have been characterized comprehensively by hydrodynamic methods. This may partly explain the remaining uncertainties.

Acknowledgments The authors thank aquaTech Geneva, Switzerland, for having provided the polyelectrolytes and the Swiss National Science Foundation for financial support (grants 200020_119818/1 and 200020_135162/1).

References

Berth G, Cölfen H, Dautzenberg H (2002) Physicochemical and chemical characterization of chitosan in dilute aqueous solution. Progr Colloid Polym Sci 119:50–57

Bourdillon L, Hunkeler D, Wandrey C (2006) The analytical ultracentrifuge for the characterization of polydisperse polyelectrolytes. Progr Colloid Polym Sci 131:141–149

Budd PM (2002) Determination of molar masses of polyelectrolytes. In: Tripathy S, Kumar J, Nalwa HS (eds) Handbook of polyelectrolytes and their applications, vol 2. Amer Sci Publishers, Stevenson Ranch, pp 91–115

Creeth JM, Knight CG (1965) On estimation of shape of macromolecules from sedimentation and viscosity measurements. Biochim Biophys Acta 102(2):549

Dautzenberg H, Jaeger W, Kötz J, Phillip B, Seidel C, Stscherbina D (1994) Polyelectrolytes: formation, characterization, application. Carl Hanser, Munich

Durchschlag H, Zipper P (1994) Calculation of the partial specific volume of organic compounds and polymers. Prog Colloid Polym Sci 94:20–39

Elias H-G (1990) Makromoleküle, Bd 1. Grundlagen: Struktur-Synthesse-Eigenschaften 5. Aufl., Hüthig & Wepf, Basel, Heidelberg/New York

Gianni P, Lepori L (1996) Group contributions to the partial molar volume of ionic organic solutes in aqueous solution. J Solut Chem 25:1–42

Harding SE (2005) Analysis of polysaccharide size, shape and interactions. In: Scott DJ, Harding SE, Rowe AJ (eds) Analytical ultracentrifugation -techniques and methods. RCS Publishing, London, pp 231–252

Hernandez Barajas J, Hunkeler D (1997) Inverse-emulsion polymerization of acrylamide using copolymeric surfactants: Mechanism, kinetics and modeling. Polymer 38:437–447

Hernandez Barajas J, Hunkeler D, Wandrey C (2004) Polyacrylamide copolymeric flocculants with homogeneous branching: heterophase synthesis and characterization. Polym News 29:239–246

Lavrenko PN, Linow KJ, Görnitz E (1992) The concentration dependence of the sedimentation coefficient of some polysaccharides in very dilute solution. In: Harding SE, Rowe AJ, Horton JC (eds) Analytical ultracentrifugation in biochemistry and polymer science. RCS, Cambridge, pp 517–531

Losada R, Käuper P, Wandrey C (2009) Analysis and study of novel terpolymers composed of acrylamide, mono-charged and double-charged cationic monomers. Polym Test 28:688–695

Mandelkern L, Flory PJ (1952) The frictional coefficient for flexible chain molecules in dilute solution. J Chem Phys 20:212–214

Mandelkern L, Krigbaum WR, Scheraga HA, Flory PJ (1952) Sedimentation behavior of flexible chain molecules-polyisobutylene. J Chem Phys 20:1392–1397

Manning GS (1965) Cluster theory of polyelectrolyte solutions. 2. Additivity rules osmotic pressure and Donnan equilibrium. J Chem Phys 43(12):4260–4267

Manning GS, Zimm BH (1965) Cluster theory of polyelectrolyte solutions, 1. Activity coefficients of mobile ions. J Chem Phys 43(12):4250–4259

Pavlov GM (1997) The concentration dependence of sedimentation for polysaccharides. Eur Biophys J 25(5–6):385–397

Pavlov GM, Frenkel S (1995) Sedimentation parameter of linear polymers. Prog Colloid Polym Sci 99:101–108

Philo J, XLGraph, http://www.jphilo.mailway.com/

Rasteiro MG, Garcia FAP, Ferreira PJ, Antunes E, Hunkeler D, Wandrey C (2010) Flocculation by cationic polyelectrolytes: relating efficiency with polyelectrolyte characteristics. J Appl Polym Sci 116:3603–3612

Rowe AJ (1977) Concentration dependence of transport processes- General description applicable to sedimentation, translational diffusion, and viscosity coefficients of macromolecular solutes. Biopolymers 16(12):2595–2611

Rowe AJ (1992) The concentration dependence of sedimentation. In: Harding SE, Rowe AJ, Horton JC (eds) Analytical ultracentrifugation in biochemistry and polymer science. RCS, Cambridge, pp 394–406

Schuck P (2000) Size distribution analysis of macromolecules by sedimentation velocity ultracentrifugation and Lamm equation modeling. Biophys J 78(3):1606–1619

Schuck P, Sedfit, http://www.analyticalultracentrifugation.com/

Schulz GV, Blaschke F (1941) An equation to calculate the viscosity number for very small concentrations [identifying the molecular weight of macromolecular materials, IX]. J Prakt Chem 158:130–135

Scott DJ, Schuck P (2005) A brief introduction to the analytical ultracentrifugation of proteins for beginners. In: Scott DJ, Harding SE, Rowe AJ (eds) Analytical ultracentrifugation -techniques and methods. RCS Publishing, London, pp 1–25

Tsvetkov VN (1979) Structure and properties of rigid-chain polymer molecules in solutions. Vysokomol Soed Ser A 21(11):2606–2623

Tsvetkov VN, Eskin V, Frenkel S (1970) Structure of macromolecules in solution. Butterworths, London

Tsvetkov VN, Lavrenko PN, Bushin SV (1984) Hydrodynamic invariant of polymer molecules. J Polym Sci A Polym Chem 22(11):3447–3486

Tsvetkov VN, Novakovskii VB, Strelina IA et al (1989) Hydrodynamic properties and conformation of molecules of polynaphthoylene benzimidazoles in sulfuric acid. Vysokomol Soed Ser A 31(1):40–44

Wales M, van Holde KE (1954) The concentration dependence of the sedimentation constants of flexible macromolecules. J Polym Sci 14(73):81–86

Wandrey C, Hunkeler D (2002) Study of polyion counterion interaction by electrochemical methods. In: Tripathy S, Kumar J, Nalwa HS (eds) Handbook of polyelectrolytes and their applications, vol 2. Amer Sci Publishers, Stevenson Ranch, pp 147–172

Wandrey C, Bartkowiak A, Hunkeler D (1999) Partial molar and specific volume of polyelectrolytes: comparison of experimental and predicted values in salt-free solutions. Langmuir 15(12):4062–4068

Chapter 14
Different Levels of Self-Sufficiency of the Velocity Sedimentation Method in the Study of Linear Macromolecules

Georges M. Pavlov

Abstract The general principles for the interpretation of the sedimentation velocity for linear macromolecules are discussed in combination with the characteristics obtained by other hydrodynamic methods. The several levels of self-sufficiency of the sedimentation velocity method are demonstrated. They include scaling correlation between velocity sedimentation coefficient s_0 and concentration coefficient $k_s = K_{sks}{s_0}^{bks}$, application of the sedimentation parameter β_s for molar mass estimation, and qualitative conformation zoning of linear macromolecules using the normalized double logarithmic plots ($k_s M_L$vs $[s]/M_L$). The last example of self-sufficiency of the method is the study of the concentration dependence of the sedimentation velocity in combination with continuous c(s) distribution using general scaling law model of Sedfit software. This model allows to establish the relationship $s_0 = K_s M^{bs}$ for flexible polymers and perform further gross conformational analysis and analysis of the molar mass distribution. It is shown that the K_s coefficient is strongly dependent on the polymer concentration.

Keywords Velocity sedimentation • Sedimentation coefficient • Sedfit • Sedimentation parameter • Molar mass • Scaling relations

14.1 Introduction

Apparently, molecular hydrodynamics goes back to Archimedes of Syracuse and his concept of buoyancy which characterizes the difference in density between a particle/body and the surrounding liquid in which this particle moves (contrast of densities). Already in the new times, Stokes has obtained a relationship between

G.M. Pavlov (✉)
Institute of Macromolecular Compounds, Russian Academy of Sciences, Bol'shoi prospect 31, St. Petersburg 199004, Russia

Department of Physics, St. Petersburg State University, 7-9, Universitetskaya nab., St. Petersburg 199034, Russia
e-mail: g.m.pavlov@spbu.ru; georges.pavlov@mail.ru

S. Uchiyama et al. (eds.), *Analytical Ultracentrifugation*,
DOI 10.1007/978-4-431-55985-6_14

the translational friction coefficient of a rigid sphere and its radius. In 1911 Svedberg and Estrup used the precipitation rate of colloidal particles in the Earth's gravitational field (with acceleration of 1 g) to assess their size. They obtained the following relation when the notion of the velocity sedimentation coefficient was not yet introduced:

$$r_h = \left[(9/2)\,(uh)/\,(d-d)\,g'\right]^{1/2} = \Big[\,(9/2)\,(s\eta\overline{\upsilon}/\,(1-\overline{\upsilon}\,\rho_0)\Big]^{1/2} \qquad (14.1)$$

The left side of the relationship (14.1) is the original one, and the right side is its modern paraphrase.

Here r_h is the hydrodynamic radius of the particle, η is the medium viscosity, $\delta \equiv \rho_0$ is the density of the solvent, $d = 1/\overline{\upsilon}$ is the density of the particle, $\overline{\upsilon}$ is the partial specific volume, and $u/g' \equiv s$ is the sedimentation coefficient according to the definition given later by Svedberg.

Conspicuously, in 1913, Dumanskiy, Jabotinsky, and Evseev (see also Introduction by T. Svedberg in Svedberg and Pedersen (1940)) were the first to use a *centrifuge*. They studied the precipitation rate of colloidal particles at $n = 2000$ rpm ($\approx$300 g) to measure particle sizes using the same relationships (14.1) which they have derived from the Stokes law. In 1913–1926, Svedberg and his coworkers have performed an enormous amount of creative and constructive work to design and build the *ultracentrifuge* which allows to achieve the fields of $\approx$ 200 000 g. On May 19, 1927, Svedberg gave his Nobel lecture entitled "The ultracentrifuge" (Svedberg 1927). Thus, a new scientific direction has emerged in molecular biophysics and nascent polymer science – analytical ultracentrifugation. In fact, Svedberg's lecture predicted the widespread use of analytical ultracentrifugation in future studies of various dispersed systems. Further results of the analytical ultracentrifugation obtained before the 1940s were summarized in the book (Svedberg and Pedersen 1940), which so far has not lost its value. The first investigations conducted by Svedberg himself were devoted to studying velocity sedimentation of globular proteins from different organic sources (e.g., Svedberg and Faehraeus 1926; Svedberg and Sjoegren 1928). These data represent an extensive library of velocity sedimentation coefficients of proteins and are given in a special chapter (Pedersen in the monograph Svedberg and Pedersen 1940). In a short while, the investigations of cellulose and its derivatives as well as the synthetic polymers were started (e.g., Stamm (1930); Singer and Gross (1934a, b); Kraemers and Nichols (1940)). In the mid-1950s, the study of DNA properties by AUC also began (see Peacocke and Schachman (1954), Kawade and Watanabe (1956)).

Currently, analytical ultracentrifugation methods are widely used in biophysics, polymer science, nanoscience, colloid science, and other related areas (Munk 1991; Schuster and Laue 1994; Laue and Stafford 1999; Lebowitz et al. 2002; Scott et al. 2005; Maechtle and Boerger 2006). Analytical ultracentrifugation has some experimental variations, and the most important and used among them are sedimentation velocity and sedimentation equilibrium (Schachman 1959; Fujita 1975). In this part, we will consider only the application of the first method.

In 1997 Beckman Instruments, Inc. (Palo Alto, USA) launched a new analytical ultracentrifuge. The progress made by Beckman Instruments in XLA/I model involves various aspects: up-to-date electrical drive, convenient temperature control of the apparatus, the development of optical detection systems, digital camera, and the online digitization of the measuring data. Since the XLI ultracentrifuge is equipped with two kinds of optics (the UV/Vis absorption detector and the Rayleigh interferometer), it became possible to use the new analytical ultracentrifuge to study a wide range of polymer molecules, both synthetic and natural.

The impressive breakthrough was made with the development of a commercial fluorescence detection system (FDS) for the analytical ultracentrifuge which has significantly extended the sensitivity of the boundary registration going up to picomolar concentrations of labeled solutes (MacGregor et al. 2004; Kroe and Laue 2009). The only problem left is incorporation of the corresponding tags into macromolecules; then it would be virtually possible to follow the sedimentation of individual macromolecule. (It will be indeed gold "nuts and bolts.")

After development of the XLA/I ultracentrifuge, the amount of information obtained in the course of one sedimentation velocity experiment (i.e., number of scans) increased by several orders of magnitude. This change necessitated development of software for processing raw data, and these softwares have been worked out by different teams of researchers. Pioneering works in this area were made by Stafford (1992, 2000) and Stafford and Braswell (2004); he elaborated the time derivative method dc/dt which is used in the Sedanal sedimentation velocity module. Several other methods have been developed for direct fitting of the sedimentation boundary; these methods are based both on approximate (Philo 1994; Behlke and Ristau 1997) and numerical (Schuck 1998; Demeler and Saber 1998) solutions of the Lamm equation. In these approaches, the diffusion process is directly taken into account. As a result, the distribution of sedimentation coefficients for the sample may be obtained, which in turn can be converted into the molar mass distribution. These software programs provide significant opportunities for extracting information from the impressive files of raw data obtained in the sedimentation velocity experiments by now.

The methods of molecular hydrodynamics are among the basic methods in polymer science and molecular biophysics (Tanford 1961; Volkenshtein 1963; Tsvetkov et al. 1970; Cantor and Schimmel 1980; Fujita 1990). The information obtained by these methods is essential for understanding the structure and gross conformation of isolated macromolecules. Namely, these methods are useful for estimating the size and shape of the individual macromolecules, for judging how quick the chain is coiled; in other words, they help to understand the degree of rigidity or flexibility of macromolecules and get the information about the intrinsic nonideality of the chain state (i.e., strength of the long-range intrachain interactions).

This contribution considers the ways of the analysis of the results obtained in the study of individual linear synthetic and natural macromolecules using the hydrodynamic methods, especially velocity sedimentation. The main experimental studies of linear chain molecules have been carried out during the second half of the twentieth century, but these studies are currently being continued with the aid of the modern analytical ultracentrifuge which is, however, now called "ProteomeLabTH XL-I Protein Characterization System" (strange nickname for the ultracentrifuge).

14.2 Essential Concepts of Molecular Hydrodynamics and Parameters Characterizing the Conformation of Linear Polymers

The linear chain *macromolecules contract into a coil* due to free or hindered rotation around valence bonds between atoms which form the backbone. However, the self-crossing configurations should be excluded from the consideration, i.e., *intrachain volume effects* should be taken into account. The *intrachain volume effects* lead to an increase in the chain size and also to changes of the chain size distribution function. In a weakly coiled chains, the *effects of intramolecular draining* are present, i.e., the flow of solvent molecules through a loose polymer coil is observed. This phenomenon leads to an increase in friction loss of the macromolecule moving in solution. The effects of the hydrodynamic interactions should be taken into consideration for the polymer coils moving in the solution. This interaction occurs not only between different macromolecules (*inter*coil interaction or concentration effects) but also within individual macromolecule (*intra*molecular interaction). The *intrachain hydrodynamic interaction* means that the moving chain element drags the solvent with it. The solvent is considered as a structureless continuum. The velocity of the adjacent solvent decreases proportionally to the distance from the chain element causing the flow. Other elements of the same chain involved in this velocity field are addicted by the driving fluid (solvent). These intrachain hydrodynamic interactions between the elements of the same chain in a moving liquid are described by the Oseen hydrodynamic tensor (Tanford 1961; Yamakawa 1971; Tsvetkov 1989; Doi 1996; Teraoka 2002; Rubinstein and Colby 2003).

A linear polymer macromolecule is characterized by its molar mass (M), contour length (L), mean-square end-to-end distance ($<h^2>$), or mean-square radius of gyration ($<R_g{}^2>$) (Tanford 1961; Tsvetkov et al. 1970); see Fig. 14.1. In the Gaussian coil limit, i.e., for the chains in the absence of intrachain excluded volume effects, $<h^2>$ and $<R_g{}^2>$ are related in a simple way: $<h^2> = 6 <R_g{}^2>$. It is worth to mention that this relation is an exact result in the absence of excluded volume, but it also holds, in a proper approximation, for good solvents.

Hydrodynamic diameter (d), persistence length (a), or the Kuhn segment length (A = 2a), and mass per unit length ($M_L = M/L$), are the parameters being common for a homologous series of linear polymers which differs only by the contour lengths. For a homologous series of linear polymers, the mean-square end-to-end distance ($<h^2>$) and the mean-square radius of gyration ($<R_g{}^2>$) are related across the whole range of molar masses to the ratio of the contour length (L) to the

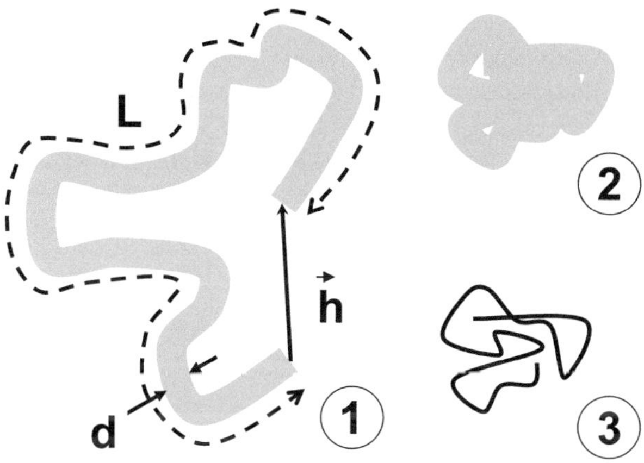

Fig. 14.1 Images of three coils having similar contour lengths ($L_1 = L_2 = L_3$) but different diameters d (1–3) and folded in varying degrees. These coils are characterized by the different statistical segment lengths (A: $A_1 > A_2 = A_3$), different diameters (d: $d_1 = d_2 > d_3$), different thermodynamic qualities of the solvent ($\varepsilon_1 > 0$, $\varepsilon_2 = \varepsilon_3 = 0$), and different intracoil draining effects (drained (1), non-drained (2), and partially drained (3) coils)

persistence length (a) by the following relationships (Kratky and Porod 1949; Benoit and Doty 1953; Landau and Lifschitz 1963):

$$< h^2 >= LA \ [1 - (1 - e^{-x}) / x], \tag{14.2}$$

$$< R_g{}^2 >= a^2 \{(x/3) - [1-(2/x) [1- ((1-e^{-x}) / x)]]\} \tag{14.3}$$

where x = L/a = 2 L/A is the reduced chain length.

The wormlike chain describes the behavior of chain molecules with the L/a values varying from very low to very high ones. For large (L/a) values, the wormlike chain has a Gaussian coil conformation. From this viewpoint, the wormlike chain is a more general model.

14.3 Relationships Between the Experimental Hydrodynamic Values and the Macromolecular Characteristics

The main characteristics obtained by the methods of molecular hydrodynamics are the velocity sedimentation coefficient s_0, the Gralen concentration coefficient k_s, the translational diffusion coefficient D_0, as well as the intrinsic viscosity [η]. All the experimental values should be obtained in the limit of infinite dilution.

Different experimental measurements can be expressed as intrinsic values ($[\eta]$, $[D]$, $[s]$, k_s) which are independent of the solvent properties at a first approximation. Each of these hydrodynamic characteristics is related to common macromolecular characteristics, which are, in the case of linear polymers, M and $< h^2 >$ (Svedberg and Pedersen 1940; Flory 1953; Tanford 1961; Tsvetkov et al. 1970; Yamakawa 1971; Cantor and Schimmel 1980; Pavlov and Frenkel 1986, 1988):

$$[\eta] = \Phi < h^2 >^{3/2} / M, \tag{14.4}$$

$$[D] \equiv D_0 \eta_0 / T = k / \left(P < h^2 >^{1/2} \right), \tag{14.5}$$

$$[s] \equiv s_0 \eta_0 / (\Delta\rho / \Delta c) = M / P < h^2 >^{1/2} N_A \Big), \tag{14.6}$$

$$k_s = B < h^2 >^{3/2} / M, \tag{14.7}$$

where T is the absolute temperature, k is Boltzmann's constant, and N_A is Avogadro's number. The density increment $(\Delta\rho/\Delta c)$ or buoyancy (Archimedes) factor $(1 - \overline{\upsilon}\,\rho_0)$ is used to obtain a partial specific volume ($\overline{\upsilon}$), and ρ_0 is the solvent density. Φ and P are the Flory hydrodynamic parameters; B is also a hydrodynamic dimensionless parameter. The values of $\Phi = \Phi$(L/A, d/A, ε) and $P = P$(L/A, d/A, ε) are functions of the relative contour length (L/A) and relative diameter (d/A), and parameter ε characterizes the thermodynamic quality of the solvent (or the long intrachain interactions). In the case of the absence of the excluded volume effects, when $\varepsilon = 0$, the values of $\Phi = \Phi$(L/A, d/A, $\varepsilon = 0$) and $P = P$(L/A, d/A, $\varepsilon = 0$) are tabulated in the frame of the wormlike cylinder theory (Yamakawa and Fujii 1973, 1974).

In the case of linear polymers of non-globular conformation, the sensitivity of the hydrodynamic characteristics to the changes in the molar mass decreases in the following order: Eq. (14.4) ≈ Eq. (14.7) > Eq. (14.5) > Eq. (14.6).

The comparison of Eqs. (14.5) and (14.6) immediately leads to the second Svedberg relationship, which is an essence of one of the absolute method of the molar mass determination, i.e., sedimentation-diffusion analysis:

$$M_{sD} = (RT / (1 - \overline{\upsilon}\,\rho_0))\,(s_0 / D_0) = R\,[s]\,/aD\Big] \tag{14.8}$$

(The first Svedberg relationship is the relation (14.6).)

14.4 Degree of Solution Dilution

The use of [η], s_0, and D_0 requires extrapolation of the experimentally determined values to zero concentration of polymer *c*. In the vicinity of zero concentration, the following relationships (Tsvetkov et al. 1970; Yamakawa 1971) are applied:

$$\eta = \eta_0 \left(1 + k_{1\eta} c + k_{2\eta} c^2 + \cdots\right) \; or \; \left((\eta/\eta_0) - 1\right)/c = k_{1\eta} + k_{2\eta} c + \cdots \quad (14.9)$$

$$D = D_0 \left(1 + (2A_2 M - (k_s + \overline{\upsilon}))\, c + \cdots\right) \quad (14.10)$$

$$s^{-1} = {s_0}^{-1} \left(1 + (k_s + \overline{\upsilon})\, c + \cdots\right) \quad (14.11)$$

where η and η_0 are the dynamic viscosity of solution and solvent, A_2 is the second thermodynamic virial coefficient, and $k_{1\eta}$, $k_{2\eta}$, k_s are the concentration coefficients. All concentration coefficients in Eqs. (14.9, 14.10, and 14.11) contain molecular information. Two of them are very useful, namely, the *first viscosity concentration coefficient* $k_{1\eta} \equiv [\eta]$, the so-called intrinsic viscosity, and the concentration sedimentation coefficient k_s, the so-called Gralen coefficient.

In order to satisfy the linear regressions (14.9, 14.10, and 14.11) the solutions under study should belong to the zone of *dilute solutions*. This means that the neighboring coils in the solution must be spaced each other by a distance greater than their own size. The *degree of dilution* is usually characterized by dimensionless Debye parameter c[η] (c is polymer concentration expressed in $g \times cm^{-3}$, [η] is intrinsic viscosity expressed in $cm^3 \times g^{-1}$). The product of the intrinsic viscosity and the concentration of the solution represents a good approximation of the volume fraction ϕ of the polymer substance in solution, with $\phi = nv_1/V = mv_1 N_A/VM = c \times 0.36 < h^2 >^{3/2} \times N_A/M = (0.36 N_A/\Phi) c[\eta] \approx c[\eta]$, where $v_1 = 0.36 < h^2 >^{3/2}$ is the volume occupied by the macromolecular coil in solution and m is the overall mass of the polymer dispersed in overall volume of solution V. The condition $c[\eta] = 1$ corresponds to the situation when each coil may contact with others, and the coils start to overlap. The dilute solution regime corresponds to the inequality $c[\eta] \ll 1$; the lower limit of concentration is determined by the sensitivity of the optical detection system used. Thus, the concentration of the solutions of linear macromolecules cannot characterize the degree of dilution of the solution. For example, the polymer solution with a concentration of ≈ 0.4 mg/ml = 0.4×10^{-3} g/cm^3 may be not diluted if the intrinsic viscosity [η] is higher than 2000 cm^3/g (for instance, this is the case for the high molar mass polystyrene Meyerhoff and Appelt (1979)).

Note that since the k_s value also characterizes the specific volume occupied by the macromolecule, the product $k_s c$ will characterize the degree of dilution too. However, for the flexible chain polymers, the condition $k_s c < 1$ will be stronger than the condition $[\eta]c < 1$, as far as $k_s/[\eta] \approx 1.7$. Meanwhile, for the rigid-chain polymers, condition $k_s c < 1$ will overestimate the degree of dilution, because in this case $k_s/[\eta]$ is usually lower than 1.

14.5 Error of k_s Determination

The concentration range should be wide enough to obtain a reliable extrapolation to zero concentration, i.e., to make estimation of both values with good accuracy: intercept (s_0) and slope (k_s/s_0). We shall analyze the errors of determination of the s_0 and k_s values. The mean square errors were determined by the mean square linear approximation of the dependences $s^{-1} = f(c)$ following the relation (14.11). By way of example, let us consider the results obtained in the sedimentation study of chitosan and nitrate chitin solutions (Table 14.1 and Fig. 14.2) (Pavlov and Frenkel 1986). Obviously, the error of k_s value is larger because it includes the error of s_0 value and the error of the slope of the linear dependency (Eq. 14.11). The correlation between the relative errors of s_0 and k_s values is given in Fig. 14.2.

14.6 Scaling Relation Between k_s and s_0

Since the publication of Newman and Eirich (1950) where k_s values were compared with the corresponding values of the intrinsic viscosity, attempts were made to determine the molecular meaning of the k_s coefficient. But in the first place, it was a matter of optimizing the study of the concentration dependence of sedimentation coefficient for a number of polymer-homologous series of fractions/samples, sometimes reaching up to 30–40. For this purpose, the direct comparison between s_0 and k_s represents the most advantageous way.

The comparison of k_s and s_0 allows establishing the cross-scaling relation between two experimental values (Pavlov and Frenkel 1982). Replacing the $<h^2>$ value in the relation $k_s = B <h^2>^{3/2}/M$ with help of the Svedberg relation ($s_0 = (1- \overline{\upsilon}\ \rho_0)M/N_A P_0 \eta_0 <h^2>^{1/2}$) allows to obtain the following relationship:

$$k_s = B\left(\left(1 - \overline{\upsilon}\rho_0\right)/N_A P_0 \eta_0\right)\left(M^2/{s_0}^3\right) \qquad (14.12)$$

Using Kuhn-Mark-Houwink-Sakurada-type scaling relation ($s_0 = K_s M^{bs}$), the following correlation may be obtained:

$$k_s = \left(\left(1 - \overline{\upsilon}\rho_0\right)/N_A P_0 \eta_0\right)\left(B/{K_s}^{2/3}\right) {s_0}^{(2-3bs)/bs} = K_{kss} {s_0}^{\nu} \qquad (14.13)$$

where $\nu \equiv \nu_1 = (2 - 3b_s)/b_s$.

Alternative scaling relations ($k_s \sim {s_0}^{\nu}$) can be obtained: $\nu_2 = (4 - 3b_s)/3b_s$ from the Peterson theory (1964) of straight cylinder sedimentation (see also Fujita et al.

Table 14.1 Some hydrodynamic characteristics, their relative errors, and molar masses of chitosan macromolecules in 0.33 m $CH_3COOH + 0.3$ m NaCl solution at 294K[a]

N	$[\eta]$ cm^{3}g^{-1}	$s_0 \times 10^{13}$ s	$(\Delta s_0/s_0) \times 10^2$	k_s cm^{3}g^{-1}	$(\Delta k_s/k_s) \times 10^2$	$M_{ks} \times 10^{-3}$ g mol^{-1}	$(\Delta M/M_{ks}) \times 10^2$	$M_{sD} \times 10^{-3}$ g mol^{-1}
1	840	2.4	3.8	470	9.8	175	11.2	183
2	790	2.4	0.5	415	1.2	165	8.4	178
3	580	2.35	1.3	420	3.8	155	8.8	148
4	740	2.3	2.2	470	5.5	165	9.4	186
5	740	2.3	0.5	510	2.0	165	8.4	172
6	460	2.2	1.8	450	4.9	150	9.1	113
7	370	2.15	0.5	260	2.3	105	8.5	73
8	360	2.05	2.4	220	13.6	95	11.4	106
9	305	1.9	5.3	270	18.5	95	14.8	82
10	340	1.9	4.2	270	15.9	95	13.2	83
11	440	1.9	2.6	230	13.0	90	11.3	74
12	370	1.8	1.7	240	7.9	80	9.6	80
13	295	1.8	2.8	210	11.4	80	11.0	86
14	310	1.7	2.9	220	12.7	70	11.4	64
15	240	1.65	4.2	200	13.0	60	12.4	55
16	100	1.35	3.7	150	16.7	45	13.1	27
17	170	1.3	3.1	70	28.6	25	17.2	38
18	45	1.0	8.0	60	80.0	16	42.6	14

[a]Chitosan was obtained from the krill that was caught off the coast of Kamchatka. Its degree of deacetylation was 0.80, which corresponds to the molar mass of the repeating unit of $M_0 = 169$. Chitosan fractions were investigated in the following buffer solvent: 0.33 M $CH_3COOH + 0.3$ M NaCl. In this solvent of high ionic strength, the polyelectrolyte effects are suppressed. This solvent was characterized by the density $\rho_0 = 1.015$ g/cm^3 and the viscosity $\eta_0 = 1.05$ mPa × s values at 294 K. Buoyancy factor was determined pycnometrically and was equal to $(1 - \upsilon\rho_0) = 0.38$ with a mean relative error of 4.7×10^{-2}. Velocity sedimentation of these fractions was investigated using the analytical ultracentrifuge MOM 3170 (Hungary) at a rotor speed of 40×10^3 rpm in a double-sector synthetic boundary cell; the height of the centerpiece was 12 mm (Pavlov and Selyunin 1986). The optical system of recording the sedimentation boundary was the Lebedev's polarizing interferometer (Tsvetkov et al. 1970)

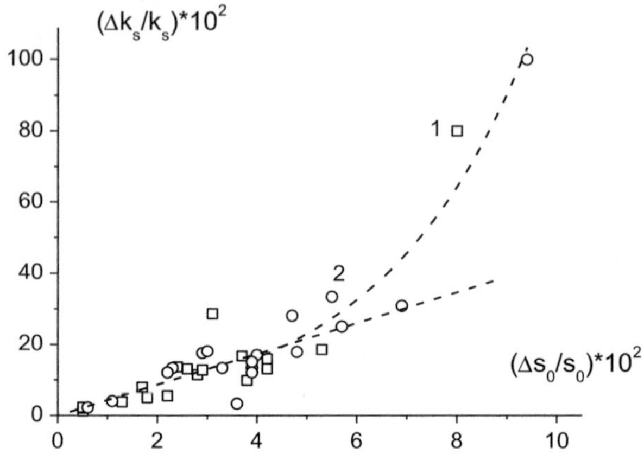

Fig. 14.2 Comparison of the relative errors of velocity sedimentation coefficient s_0 and Gralen coefficient k_s for the chitosan fractions studied in buffer solvent 0.33 M $CH_3COOH + 0.3$ M NaCl (*1*) and chitin nitrate fractions in dimethylformamide (*2*)

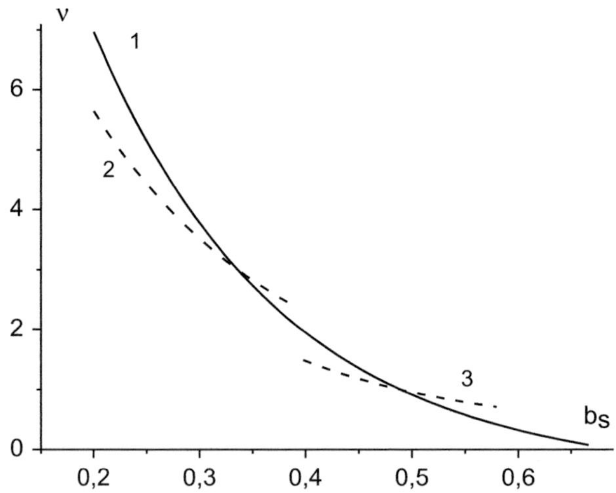

Fig. 14.3 Dependencies of the scaling coefficients ν on b_s: (*1*) $\nu \equiv \nu_1 = (2 - 3b_s)/b_s$ (Pavlov and Frenkel 1982); (*2*) $\nu_2 = (4 - 3b_s)/3b_s$ for the straight cylinder (Peterson 1964); and (*3*) $\nu_3 = (1 - b_s)/b_s$ for the coils under θ-conditions (Imai 1970; Freed 1983; Muthukumar and de Mense 1983)

1966) and $\nu_3 = (1 - b_s)/b_s$ for the random-coil macromolecules under θ-conditions (Imai 1970; Freed 1983; Muthukumar and de Mense 1983). Figure 14.3 shows the dependence of ν on b_s.

The values of ν_1 and ν_2 are close to each other in the range of $b_s < 0.3$ corresponding to rigid-chain macromolecules. At $b_s = 0.5$ (random coils under θ-conditions), we have $\nu_1 = \nu_3$. Consequently, the cross-scaling dependence

$$k_s = K_{kss} s_0^{(2-3bs)/bs} \tag{14.14}$$

has a general meaning and can be used for b_s value estimation for all spectrum of linear macromolecules.

Thus, the study of concentration dependence of sedimentation coefficient for three or four fractions in the possibly wide range of s_0 allows to establish the correlation given in Eq. 14.13 and to obtain a reliable estimation of the concentration coefficients k_s for other fractions of the studied series. Furthermore, the b_s value may be determined (Eq. 14.14); this is a key parameter in transformation of the sedimentation coefficient distribution of a sample into the molar mass distribution, as $ds_0/dM = b_s K_s^{1/bs} s_0^{(bs-1)/bs}$. The b_s value is also necessary for the application of the continuous c(s) model with general scaling law of Sedfit suite to treat the velocity sedimentation data.

It should be noted that the k_s values can vary within wide limits (from a few to several thousand of cm^3/g) for the same homologous series of linear polymer systems. The k_s values are different for linear macromolecules of various conformational status and scale differently with s_0 values (Fig. 14.4). In the case of globular structures, the values of k_s vary slightly ($\approx$(4–10) cm^3/g) and are weakly dependent on the sedimentation coefficient (see Fig. 14.4 and Table 14.2). In the case of more rigid macromolecules, the more of them are draining, the greater are the values of k_s and the steeper is the dependence of k_s on s_0 (Fig. 14.4).

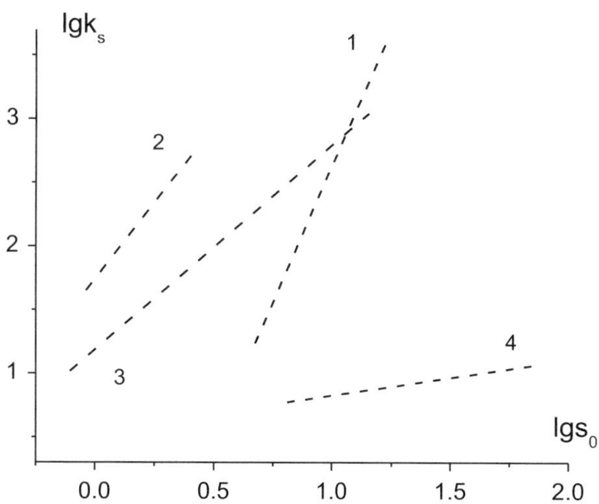

Fig. 14.4 Comparison of the dependencies k_s *vs.* s_0 in double logarithmic scale, characterizing different types of linear chain conformations: (*1*) extra-rigid (*Schizophyllum commune* polysaccharide), $k_s \sim s_0^{4.3\pm0.2}$ (Yanaki et al. 1980); (*2*) rigid (chitosan), $k_s \sim s_0^{2.4\pm0.2}$ (Pavlov and Selyunin 1986); (*3*) flexible (poly(1-vinyl-2-pyrrolidone)), $k_s \sim s_0^{1.61\pm0.07}$ (Pavlov et al. 1990); and (*4*) globular (globular protein), $k_s \sim s_0^{0.25\pm0.07}$ (Creeth and Knight 1965)

Table 14.2 The limits of the scaling indices $\boldsymbol{b_s}$, $\boldsymbol{b_\eta}$, and $\boldsymbol{b_{kss}}$ values for different types of gross polymer conformations

Gross conformations	Types of polymer systems	b_s	b_η	b_{kss}
Globular/rigid sphere	Globular proteins, dendrimers, glycogen	$2/3 = 0.67$	≈ 0	≈ 0
Thick and relatively short macromolecules	Brushlike linear polymer	$0.5 < b_s < 0.67$	$0.5 \geq b_\eta > 0$	$1 > b_{kss} > 0$
Θ-condition	Flexible polymers	0.50	0.50	1
Rigid macromolecules without intrachain volume effect	Rigid macromolecules in the range of very high M	0.50	0.50	1
Thermodynamically good solvent, non-draining coil	Flexible polymers	$0.4 < b_s < 0.5$	$0.85 \geq b_\eta > 0.5$	$2 > b_{kss} > 1$
Rigid macromolecules, crankshaft conformation	Cellulose, cellulose derivatives, aromatic polyamides (moderate M range)	$0.33 < b_s < 0.5$	$1.0 \geq b_\eta > 0.5$	$3 > b_{kss} > 1$
Extra-rigid	Helical structures such as DNA, polysaccharides, schizophyllan, xanthan	$\cong 0.25 < b_s < 0.3$	$1.8 \geq b_\eta \geq 1$	$5 > b_{kss} > 3.7$

14.7 Scaling Relations Between Other Hydrodynamic Values and Molar Mass

To quote from Fujita (1990), "one of the most surprising generalities in the world of polymers is that $[\eta]$ for a series of homologous polymers under a fixed solvent condition follows a simple power law over an extended range of M." We can add that such relationships are also observed for all other hydrodynamic characteristics: s_0, k_s, and D_0 and between them; this was shown above for the $s_0 - k_s$ correlation. In principle, this kind of dependence is characteristic of fractal systems in which any part of the object is similar to the whole (de Gennes 1979; Mandelbrot 1982). The comparison of hydrodynamic characteristics with each other and with molar mass allows to obtain cross (among hydrodynamic characteristics) and canonic (among hydrodynamic characteristics and molecular mass) relationships of Kuhn-Mark-Houwink-Sakurada type or hydrodynamic scaling relationships. In general, they may be given as the following:

$$P_i = K_{ij} P_j^{\,bij} \tag{14.15}$$

where P_i is one of the hydrodynamic characteristics $[\eta]$, D_0, s_0, or k_s and P_j is another hydrodynamic characteristic from this array or molar mass. In the case of homologous series of linear polymers, the scaling indices b_{ij} are intercorrelated (Tsvetkov et al. 1970; Pavlov and Frenkel 1995) in the following ways: $b_{sD} = b_s/b_D$, $b_{D\eta} = b_D/b_\eta$, $b_{s\eta} = b_s/b_\eta$, $|b_D| = (1 + b_\eta)/3$, $|b_D| + b_s = 1$, and $b_{kss} = (2 - 3b_s)/b_s$, where the double underline index refers to cross-correlations and the single underline index refers to correlations of appropriate hydrodynamic characteristic with molar mass. (For instance, b_{sD} is the index of the power function between s_0 and D_0, which is represented as $s_0 = K_{sD} D_0^{\,bsD}$. b_s and b_D are the indices of the power functions between M and s_0 and between M and D_0, respectively: $s_0 = K_s M^{bs}$, $D_0 = K_D M^{bD}$.)

It is important to note that further molecular interpretation of the hydrodynamic characteristics will be the more accurate the wider is the molar mass range of investigated homologous series. At the same time, the b_{ij} value may vary in different ways during the transition *from very high to very low molar masses (contour lengths)* (Bloomfield 1968; Budtov 1992) (Fig. 14.5).

For any kind of linear macromolecules, there is a certain range of contour lengths where the coil maturation happens. Just in this range of the molar masses, the changes in the slopes of KMHS plots may be observed. The change of the slope in the KMHS plots is more easy to observe for $[\eta]$ dependence (b_η value) and less easy to see for the s_0 dependence (b_s value), because $[\eta] \sim \langle h \rangle^{3/2}$ and $s_0 \sim \langle h \rangle^{-1/2}$, i.e., the value of $[\eta]$ is more sensitive to changes in the size and shape of the coil than the s_0 value.

The analyzed polymer can be referred to a certain class according to the type of the change in the respective slope. This is illustrated in Fig. 14.5a. The data obtained for polyvinylpyrrolidone (14.1) demonstrate a decrease in slope from $b_\eta = 0.74$ in

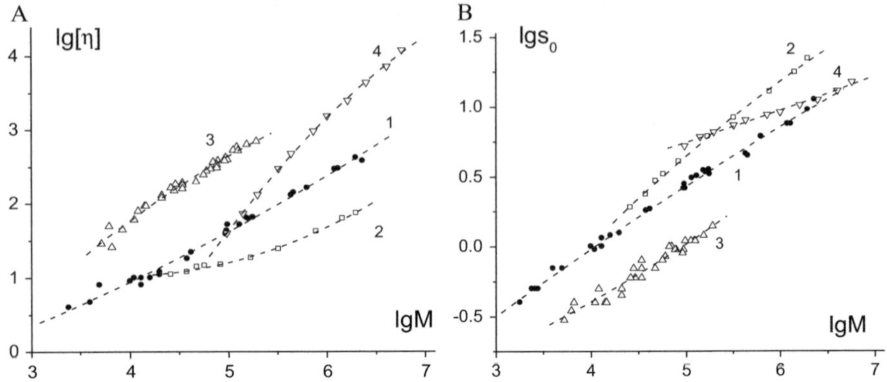

Fig. 14.5 Scaling plots of the hydrodynamic characteristics: (**a**) canonical Kuhn-Mark-Houwink-Sakurada (KMHS) plot $[\eta] = K_\eta M^{b\eta}$ and (**b**) dependencies of velocity sedimentation coefficient on the molar mass in double logarithmic scale or KMHS-type plot $s_0 = K_s M^{bs}$
The comparison of the linear macromolecules of different structures and equilibrium rigidity was made: flexible, poly(1-vinyl-2-pyrrolidone) (PVP) in 0.1 M sodium acetate solutions (*1*) (Pavlov et al. 1990); brushlike, styrene-methyl methacrylate brush copolymer in bromoform (2) (Magarik et al. 1978); rigid, poly(2,2′-p-phenylene-(5-benzimidazole)) isophthalamide (PBIA) in dimethylacetamide + 3%LiCl (*3*) (Pavlov and Selyunin 1986); and extra-rigid, the triple helix of *Schizophyllum commune* polysaccharide (schizophyllan) in water (*4*) (Yanaki et al. 1980)

the high molar mass region to $b_\eta \approx 0.50$ in the low molar mass region; this pattern corresponds to the behavior of flexible macromolecules in thermodynamically good solvents. Thus, in this case, the deviation of b_η value from 0.5 in the high molar mass region is attributed to the intrachain volume effects (intrinsic nonideality). System 2 in Fig. 14.5a represents a thick macromolecule which is characterized by a large transverse dimension. In this case, a decrease in the slope of the $[\eta]$ dependence *vs.* molar mass in the region of small contour lengths of the main chain is also observed. The b_η value can be significantly less than 0.5, which is explained by the comparability of longitudinal and transverse dimensions of the coil when the chain length is short enough. The systems 3 and 4 describe the behavior of the rigid polymer for which is characteristic the approaching of molecular conformation to slightly bending rod with decrease in chain contour length, and b_η becomes higher than 1. In this case, deviation of the b_η value from 0.5 is attributed to the effects of intramolecular draining associated with high equilibrium rigidity of the chains (high values of persistent length). The $[\eta]$ dependence *vs.* molar mass for rigid macromolecules is concave downward in the opposite direction as compared with that for flexible macromolecules which is concave upward (system 1). Figure 14.5b shows the dependence of s_0 *vs.* molar mass for the same systems. In this connection, two circumstances should be noted. First, the changes in the slopes of the curves shown in this figure are much weaker than the slope for the intrinsic viscosity values; second, these changes should show the opposite trend.

Table 14.2 gives the comparison between the b_s, b_η, and b_{kss} indexes for linear macromolecules with different conformational status. Comparison of variation of the b_η and b_{kss} indices with changing conformational type of the macromolecules shows that the direct comparison between the sedimentation coefficient s_0 and the Gralen coefficient k_s is more sensitive to conformational changes of macromolecules than the comparison between $[\eta]$ and M. This fact has led to the establishment of conformational zones of macromolecules considering the normalized plot of $k_s M_L$ vs. $[s]/M_L$. It is one of the examples of self-sufficiency of the sedimentation velocity method in the conformational analysis of linear polymers (Pavlov and Frenkel 1986; Pavlov 1997).

14.8 Another Level of Self-Sufficiency of Velocity Sedimentation Data: Sedimentation Parameter and Hydrodynamic Invariant

Wales and van Holde (1954) were the first to suggest estimating molar masses of polymers from the experimental data obtained in the study of the concentration dependence of sedimentation velocity. As was shown by Wales and van Holde (1954), using the s_0 and k_s values, it is possible to estimate the molar mass of the flexible polymer chains by the following relation:

$$M = 4.85\pi N_A [s]^{3/2} {k_s}^{1/2} \qquad (14.16)$$

A reexamination of this problem from the original theoretical standpoint was carried out later by Rowe (1977). The definitive result for the sedimentation of solvated molecules is given in the following equation:

$$M = N_A (6\pi)^{3/2} [s]^{3/2} \{(3\overline{\upsilon}/4\pi)\,[(k_s/2\overline{\upsilon}) - (\upsilon_s/)]\}^{1/2} \qquad (14.17)$$

where v_s, according to Rowe, is the specific volume occupied by the sedimenting component (solute + entrained solvent + bound solvent). Rowe gave the following approximation for the $v_s/\overline{\upsilon}$ ratio:

$$\upsilon_s/\overline{\upsilon} = k'(k_s/\,[\eta])^{-1} \qquad (14.18)$$

where k' is the Huggins parameter. For the majority of polymer systems, the second term in brackets in Eq. (14.17) is negligible as compared with the first one and is far lower than the error of k_s determination, so Eq. (14.17) can be rearranged as:

$$M \cong 9\pi N_A [s]^{3/2} {k_s}^{1/2} \qquad (14.19)$$

Equations (14.16) and (14.19) differ only in numerical coefficients, and for some comparative purposes (e.g., scaling relation), they can give the same results.

The Wales-van Holde-Rowe Eqs. (14.16 and 14.19) were presented in general form with the introduction of the sedimentation parameter β_s (Pavlov and Frenkel 1986). In fact, this idea has been already implicitly contained in the paper of Wales and van Holde (1954). Actually, the sedimentation coefficients s_0 (first Svedberg equation) and k_s depend differently on molecular mass and size of macromolecules (Eqs. 14.6 and 14.7). The exclusion of $< h^2 >$ from Eqs. (14.6) and (14.7) allows to obtain the following expression:

$$\beta_s \equiv {B_s}^{1/3} {P_0}^{-1} = N_A \, [s] \, {k_s}^{1/3} M^{-2/3} \tag{14.20}$$

where β_s is the sedimentation parameter introduced by Pavlov and Frenkel (1986, 1988, 1995).

Obviously, the sedimentation parameter β_s is an analogue of the hydrodynamic invariant β (Mandelkern and Flory 1952).

The introduction of the sedimentation parameter was supported by the extensive set of experimental data available in the literature for linear uncharged polymers, both synthetic and natural. Analysis of almost entire data library on the velocity sedimentation of macromolecules indicates that the value of parameter β_s is practically invariant both in terms of molar mass and with respect to the structure of the repeating unit of the linear polymer (Pavlov and Frenkel 1995; Pavlov 1997). Table 14.3 lists the average experimental values of the sedimentation parameter β_s, parameter B_s for different polymer classes, and also classical hydrodynamic invariant $\beta \equiv A_0/k$.

The invariance of β_s value makes it possible to use Eq. 14.20 for determining molar mass of the polymer macromolecules using the data obtained only from

Table 14.3 Averaged experimental values of the sedimentation parameter β_s, hydrodynamic invariant β (and A_0), and parameter B_s

Polymer-solvent	${\beta_s}^a 10^{-7}$ mol$^{-1/3}$	$B_s 10^{-23}$ mol^{-1}	$\beta^{**} 10^{-7}$ mol$^{-1/3}$	${A_0}^b 10^{10}$
Flexible-chain polymers under θ-conditions	1.0	1.33	1.08	3.21
Flexible-chain polymers in thermodynamically good solvents.	1.25	2.61	1.08	3.21
Rigid-chain polymers	1.0	1.33	1.24	3.69
β1-4 glucanes and their derivatives	1.0	1.33	1.14	3.39
Globular proteins	1.17	2.14	1.01	3.00

[a]Mean-square errors are ≈ 6 %, $P_0 = 5.11$
[b]$A_0 = k\beta$ – Mandelkern-Flory-Tsvetkov-Klenin hydrodynamic invariant, k is Boltzmann constant
Note that expressed in the same unit the both β and β_s values are close to each other (see Table 14.3, columns 2 and 4)

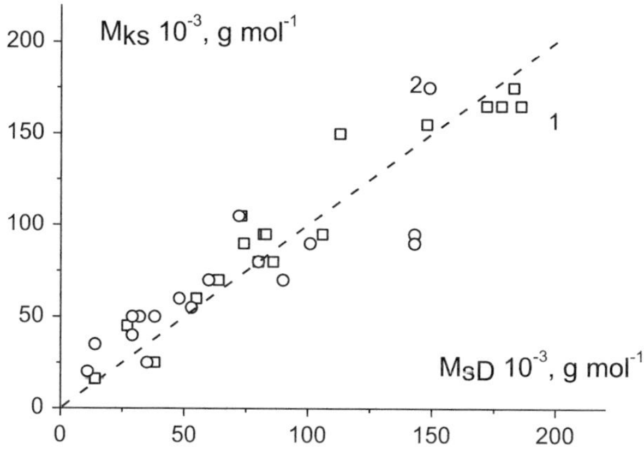

Fig. 14.6 Comparison of the molar mass values for chitosan (1) and chitin nitrate (2) calculated from the data of the sedimentation-diffusional analysis (M_{sD}) and from the s_0 and k_s values (M_{ks}, Eq. 14.21). Chitosan fractions were investigated in a buffer solvent: 0.33 M $CH_3COOH + 0.3$ M NaCl. Chitin nitrate fractions were investigated in dimethylformamide. *Dashed line* is drawn with a slope equal to 1

velocity sedimentation experiments:

$$M_{ks} = (N_A/\beta_s)^{3/2}[s]^{3/2}{k_s}^{1/2} \tag{14.21}$$

Good correlation is observed between independently measured M_{sD} values and those determined from the s_0 and k_s values. Figure 14.6 demonstrates this kind of correlation observed for fractions of chitosan and chitin nitrate.

Knowledge of the molar masses opens up the possibilities for further molecular interpretation of s_0 and k_s values. Useful applications of this concept have been demonstrated in various publications for different polymers (Tarabukina et al. 1991; Pavlov et al. 1995, 2003; Harding et al. 2011a; Morris and Harding 2013).

One of the definitions of Flory-Mandelkern-Tsvetkov-Klenin hydrodynamic invariant β (or $A_0 \equiv k\beta$) (Mandelkern and Flory 1952; Tsvetkov and Klenin 1953) is given by the following relation:

$$\beta \equiv \Phi^{1/3}P^{-1} = N_A\,[s]\,[\eta]^{1/3}M^{-2/3} \tag{14.22}$$

(Traditionally, in this case, the intrinsic viscosity is denominated in 100 cm^3/g.)

The ratios $\Phi^{1/3}/P$ and $B^{1/3}/P$ and thus A_0 and β_s are not strictly constant even theoretically (Table 14.4); they are experimentally found to be invariant (i.e., within the incertitude of the measurements) in the case of homologous polymers with different chain lengths. The theoretical values of the Flory hydrodynamic parameters P_0 and Φ_0 depend on the models and mathematical approximations. The limiting theoretical values of Φ and P for a Gaussian coil ($M \to \infty$), obtained after a preliminary averaging of the hydrodynamic Oseen's tensor, are $\Phi_0 = 2.87 \cdot 10^{23}$

Table 14.4 Theoretical values of Flory hydrodynamic parameters P_0, Φ_0, and hydrodynamic invariant $A_0 \equiv kP_0\Phi_0^{-1/3}$ (g cm^2 s^{-2} K^{-1} mol$^{-1/3}$) for polymer chains in the absence of intrachain excluded volume effects

P_0	Φ_0 10^{-23}mol^{-1}	A_0 10^{10}	References
(Sphere) 6π[a]	(Sphere) 2.5[a]	2.914	Tanford (1961)
7.35[b]	4.3	3.05	Kuhn et al. (1953) and Tsvetkov et al. (1970)
5.11[c]	3.62	4.15	Kirkwood and Riseman (1948)
5.11[c]	2.19	3.51	Hearst and Tagami (1965)
5.11[c]	2.86	3.84	Hearst and Stokmayer (1962) and Yamakawa and Fujii (1973, 1974)
5.99[d]	2.51	3.13	Zimm (1980) and de la Torre et al. (1982, 1984)
5.3[d]	1.9	3.23	Bernal et al. (1991)
6.20[e]	2.36	2.96	Oono (1985)

[a]For sphere $f_0 = 6\pi\eta_0 R$; $[\eta] = 2.5\,(4\pi/3)\,N_A\,(R^3/M)$
[b]Obtained with macroscopic models of polymer chains
[c]Obtained with preaveraging approximation of the tensor of the hydrodynamic interactions
[d]Obtained by Monte Carlo simulation without preaveraging approximation
[e]Obtained with renormalization groups theory

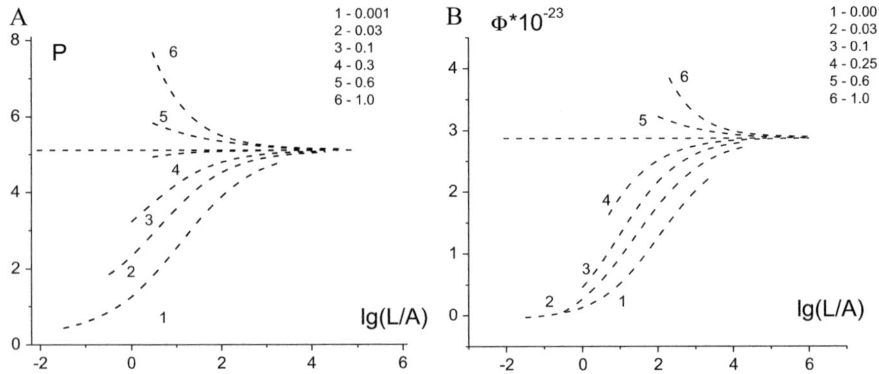

Fig. 14.7 The Flory parameters P (A) and Φ (B) as function of the relative contour length L/A of wormlike cylinder in semilogarithmic scale for different relative thickness d/A of cylinder from $d/A = 0.001$ up to $d/A = 1$ (Yamakawa and Fujii 1973, 1974)

and $P_0 = 5.11$ (Yamakawa 1971; Tsvetkov 1989, Fig. 14.7). They are affected by the preaveraging of the Oseen hydrodynamic tensor, as studied by the Monte Carlo simulation method (Zimm 1980; de la Torre et al. 1982, 1984; Bernal et al. 1991) and by renormalization group calculations (Oono 1985).

Thus, the basis of the similarity of two invariants is the same physical dimensions (cm^3/g); in other words, both $[\eta]$ and k_s are the functions of the specific volume of macromolecules. At the same time, Eqs. 14.20 and 14.22 expose the fundamental difference between β (or A_0) and β_s. This difference consists of the following. The value β is obtained from M and from two experimental values s_0 and $[\eta]$ which, in turn, are obtained in different kinds of experiments. In these experiments, a macromolecule performs different kinds of movement, namely, translational (s_0) and rotational ($[\eta]$). The sedimentation parameter β_s is obtained from M and two experimental values s_0 and k_s obtained in a single series of experiments where the macromolecule is studied under exactly the same translational conditions.

The calculation of the hydrodynamic invariants A_0 and β_s is a check for the mutual coherence of a set of the experimental data. These values should not go beyond a certain range of values indicated in Tables 14.3 and 14.4. Positioning of the A_0 and β_s values in these ranges allows further interpretation of the hydrodynamic characteristics. Otherwise, we need to analyze the causes of mismatch in the primary experimental data or to admit that the sizes of macromolecules are not equivalent in different types of movements.

14.9 Relationship Between Hydrodynamic Values and Conformational Parameters A and d in the Model of Wormlike Chain

In the limit of low molar mass chains ($L/A < 2.3$; $L \gg d$), a model of a weakly bending rod or a cylinder can be used that provides M_L and d through a linear regression (Broersma 1969; Yamakawa and Fujii 1973):

$$[s] = (M_L/3\pi N_A)\,[lnM - ln\,(M_L d) + 0.386) \qquad (14.23)$$

In principle, the slope of $[s] = f(lnM)$ in the range of lowest L/A values (<2) allows estimating linear density of the chain M_L. For this purpose, a sufficient number of experimental data in the corresponding range of molar masses should be available. However, these conditions are not frequently realized. The rare examples are the cases of extra-rigid macromolecules (Yanaki et al. 1980; Sato et al. 1984). The more flexible are the macromolecules, the lower molar mass samples must be studied (Fig. 14.8). This condition imposes significant restrictions on using velocity sedimentation data to obtain the M_L value.

A similar relationship was obtained for the intrinsic viscosity of slightly bending rods:

$$M^2/\,[\eta] = \left(45{M_L}^3/2\pi N_A\right)[lnM - ln\,(M_L d) - 0.697] \qquad (14.24)$$

It should be noted that the intrinsic viscosity data must be obtained in the region of smaller contour lengths (i.e., molar masses) than that used in velocity sedimentation experiments. This condition renders it more difficult to use intrinsic viscosity for determining the M_L value.

The dimensions of long linear chains are determined by two main factors: the short- and long-range interactions. The stronger is the short-range interaction, the stiffer is the chain; therefore, the coil becomes spongy, i.e., permeable for solvent

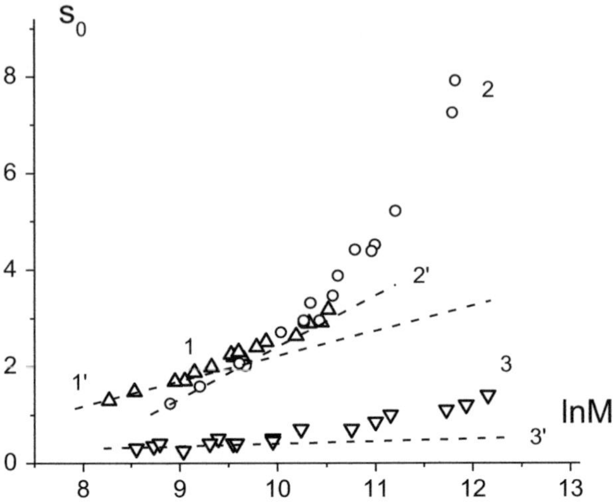

Fig. 14.8 Determination of mass per unit length M_L, the ratio of mass M to contour length L, and the hydrodynamic diameter d of heparin in 0.2 M NaCl (1and 1′) (Pavlov et al. 2003), poly{4-[4-(hexyloxy)phenyl]ethynylphenyl methacrylate} in toluene (2 and 2′) (Pavlov et al. 2012), and poly(2,2′-p-phenylene-(5-benzimidazole)) isophthalamide in dimethylacetamide + 3 % LiCl (3 and 3′) (Pavlov et al. 1985) from the plot of s_0 versus ln M. The linear extrapolation (dashed line 1′, 2′, and 3′) was made using the fractions of smallest M. The slope and intercept allow to determine M_L and d, respectively, within the framework of the weakly bending rod model or the cylinder model (Eq. 14.23), usable in the theoretical limit of $L/A < 2.3$. For heparin chain, the following results were obtained: $M_L = (570 \pm 50)$ g × mol^{-1}nm^{-1} and $d = (0.9 \pm 0.1)$ nm, for poly{4-[4-(hexyloxy)phenyl]ethynylphenyl methacrylate} chain M_L was equal to (1520 ± 60) and d was equal to (2.1 ± 0.1). The data obtained for polyamide benzimidazole macromolecules did not contradict to the theoretical value of $M_L = 190$

molecules. The long-range effects are apparent in very long chains regardless of their stiffness, but it is easier to observe and study the long-range interactions in the flexible polymers. In the case of flexible polymers, the chain is contracted significantly, and there is a high probability of interaction between the remote chain monomer units. The most part of polymer coil becomes impermeable for a solvent; flexible macromolecules manifest the effects of thermodynamic interaction between polymeric material and solvent molecules.

The detailed modeling theories of hydrodynamic characteristics are elaborated for two models of macromolecules *without the volume effects*: wormlike necklace (Hearst and Stokmayer 1962; Hearst 1964) and wormlike cylinder (Yamakawa and Fujii 1973, 1974). These model theories do not take into account the excluded volume effects and attribute the change in coil size solely to the change in persistence length. Note that in the case of chains without volume interactions, the b_η index varies from 1.8 for slightly bending rod (very stiff and/or very short chains) to 0.5 for a Gaussian coil (very long chain independently of the rigidity). Meanwhile, in the case of the chain with volume interactions, the b_η index varies from 0.8 for long chains with strong volume interactions to 0.5 for a Gaussian coil (no volume interactions or very short flexible chains). The procedure taking into account the influence of volume effects on transport characteristics was developed on the basis of linear approximation of the dependence of swelling coefficient of flexible macromolecules (α) on the excluded volume parameter (z),

$$\alpha = \left[< h^2 > / < h^2 >_\theta\right]^{1/2} = 1 + Cz + \cdots, \qquad (14.25)$$

where $< h^2 >$, $< h^2 >_\theta$ are the size of the swelling coil and the size of the coil in θ-conditions, respectively. The numerical coefficient C for short chains depends on the relative contour length (L/A), $C - C(L/A)$, and with increasing chain length, the C value tends to the limit $C = 4/3$ (Yamakawa and Stockmayer 1972). The excluded volume parameter z describes the number of collisions between chain segments per unit volume occupied by the macromolecule. Finally for flexible chains with moderate excluded volume effects, the data on translation friction are treated with the aid of the Cowie-Bywater plot (1965), and the data on intrinsic viscosity are processed using the Burchard-Stockmayer-Fixman plot (1961; 1963). In this case, the change of the coil size is related exclusively to the change in the thermodynamic quality of the solvent.

Up to now, the more complete theory that takes into account the simultaneous influence of draining and excluded volume effects on the translation friction coefficient is Gray-Bloomfield-Hearst (GBH) theory (1967). Unfortunately, this translation friction (sedimentation) coefficient theory has found insufficient dissemination and application, apparently, because the corresponding equation has not been represented in an easy form to use. In the case of the intrinsic viscosity theory (Sharp and Bloomfield 1968), the authors have received only an asymptotic expression for extremely large values of molar mass.

The sedimentation coefficient calculations (Gray et al. 1967) were carried out for distant segment pairs using the parameter ε which characterizes the deviations of the mean-square end-to-end distance from the Gaussian form due to volume swelling effects ($< h^2 > \sim M^{1+\varepsilon}$). For close segment pairs, when volume interactions are small, the Porod-Kratky statistics have been applied. The analytical result for the velocity sedimentation coefficient is given in Eq. (14.10) in Gray et al. (1967, p.1495). This equation may be rearranged for $L/A > 2.3$ in a more compact form as shown in Pavlov et al. (1990):

$$\begin{aligned}[s]P_0 N_A = {} & (3/(1-\varepsilon)(3-\varepsilon)(M_L{}^{(1+\varepsilon)/2}/A^{(1-\varepsilon)/2})M^{(1-\varepsilon)/2} \\ & + (M_L P_0/3\pi)\,[ln\,(A/d) - (d/3A) - \varphi(\varepsilon)] \end{aligned} \qquad (14.26)$$

where P_0 and Φ_0 are Flory hydrodynamic coefficients and $\varphi(\varepsilon) = 1.431 + 2.635\varepsilon + 4.709\varepsilon^2 + \cdots$.

In the case of $\varepsilon = 0$, i.e., in the absence of the intrachain excluded volume effects, Eq. 14.26 is transformed into the expressions obtained in the theories of Hearst and Stockmayer (1962) and Yamakawa and Fujii (1973):

$$[s]\,N_A P_0 = (M_L/A)^{1/2} M^{1/2} + (P_0 M_L/3\pi)\left[ln(A/d\,(A/d)\ -\varphi(0)\right] \qquad (14.27)$$

where φ(0) depends on the model used (φ(0) = 1.431 for the wormlike necklace model (Hearst and Stokmayer 1962) or φ(0) = 1.056 for the wormlike cylinder model (Yamakawa and Fujii 1973), respectively).

Viscometry data in the range of the molar mass values when $L/A > 2.3$ may be treated by changing the variables $[s]P_0N_A = (M^2\Phi_0/[\eta])^{1/3}$ according to Pavlov et al. (1990) and using the corresponding relationship (Eq. 14.26). When $L/A > 2.3$, linear approximations of the s_0 and/or $(M^2/[\eta])^{1/3}$ vs. $M^{(1-\varepsilon)/2}$ may be applied (Eqs. 14.26 and 14.27). The slope of these straight lines allows evaluation of the chain persistence length, and the intercept allows estimating the hydrodynamic diameter of the chain. Some examples are presented in Fig. 14.9, i.e., the treatments of data using the s_0 and $[\eta]$ values for two different systems (polyamide benzimidazole in DMAA + 3 % LiCl at 21° C (Pavlov et al. 1985) and polyisobutylene in n-heptane at 25° C (Abe et al. 1993)).

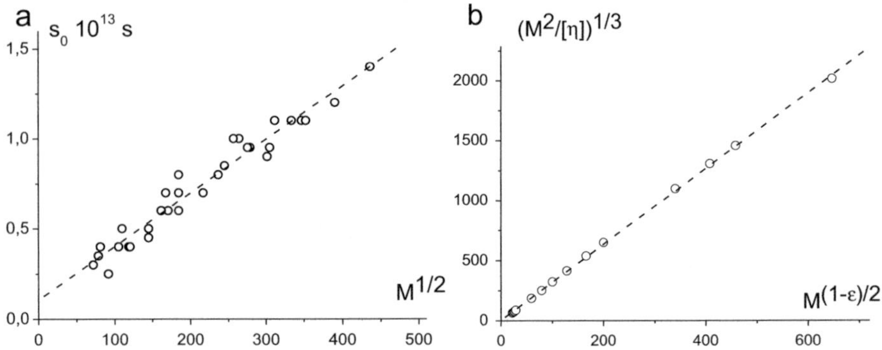

Fig. 14.9 Dependences of s_0 (**a**) and $(M^2/[\eta])^{1/3}$ (**b**) on $M^{(1-\varepsilon)/2}$ used for the estimation of the statistical segment length A (or the persistent length $a = A/2$) and the hydrodynamic diameter d in correspondence with the Gray-Bloomfield-Hearst theory (relationships (25)) for polyamide benzimidazole fractions in DMAA + 3 % LiCl at 21 °C (**a**) and polyisobutylene in n-heptane at 25 °C (**b**) of different sizes. For polyamide benzimidazole macromolecules (Pavlov et al. 1985), the following estimations were obtained: the parameter $\varepsilon = 0$, the values $A_f = (6.9 \pm 0.7)$ nm, and $d_f = (0.4 \pm 0.2)$ nm with $P_0 = 6.0$ (Fig. 14.9a). For polyisobutylene macromolecules (Abe et al. 1993), correspondingly, $\varepsilon = 0.10$, $A_\eta = (1.40 \pm 0.01)$ nm, and $d_\eta = (0.20 \pm 0.07)$ nm with $\Phi_0 = 2.3 \times 10^{23}$ (Fig. 14.9b)

14.10 Multi-Hydfit Program

In the relatively recent works of de la Torre and his colleagues (Ortega and de la Torre 2007, 2013; Amoros et al. 2011), a method of computer processing of the experimental data obtained in independent experiments is developed. The aim of this approach is to obtain the structural and conformational parameters of linear macromolecules. The authors introduce the dimensionless ratio of the experimental values to the corresponding values for the rigid sphere (concept of bead-model hydrodynamics (de la Torre and Bloomfield 1981)). The persistence length, diameter, and mass per unit length can be evaluated using Multi-Hydfit program; this program performs a minimization procedure aimed at finding the best values of a, d, and M_L, satisfying the equations which describe hydrodynamic behavior of persistence cylinders without the excluded volume effects (Yamakawa and Fujii 1973, 1974). The Multi-Hydfit program then "floats" the variable parameters in order to find a minimum of the multi-sample error function (Ortega and Torre 2007), which is calculated using equivalent radii. The equivalent radius is defined as a radius of an equivalent sphere having the same value as the determined characteristic (translation diffusion and/or velocity sedimentation coefficients and intrinsic viscosity). The error function is a dimensionless estimate of the agreement between the experimentally measured characteristic and the theoretical values of a, d, and M_L calculated for the selected hydrodynamic characteristic and for a particular molar mass. The last version of program also includes the option for chains with the excluded volume effects, e.g., for poly(isobutylene) in n-heptane (Amoros et al. 2011). This important addition concerns on a huge class of flexible chain polymers, both synthetic and natural origin.

The final results of Multi-Hydfit program are presented as the maps of conformation-structural parameters like the topographic maps. This is illustrated in Fig. 14.10 which demonstrates the result of handling hydrodynamic data by Hydfit program. The measurements were performed on a series of samples of alternating styrene/diphenylethylene copolymers functionalized with terpyridine in toluene solutions; the molecular mass range was $1.7 < M \times 10^{-3}$ g/mol <25.2 (Pavlov et al. 2009).

14.11 Further Steps of the Analysis of the Hydrodynamics Data of Homologous Series of Linear Macromolecules

It should be noted that as well as for non-computerized procedures, the sought-for estimates are more adequate, the wider is the range of studied molecular masses. It is especially important to have a sufficient number of samples of low molecular masses, when L/A is less than 2, to obtain the reliable and adequate value of M_L. However, the choice between intramolecular effects of excluded volume and/or of intrachain draining in doubtful cases of semiflexible macromolecules is beyond the

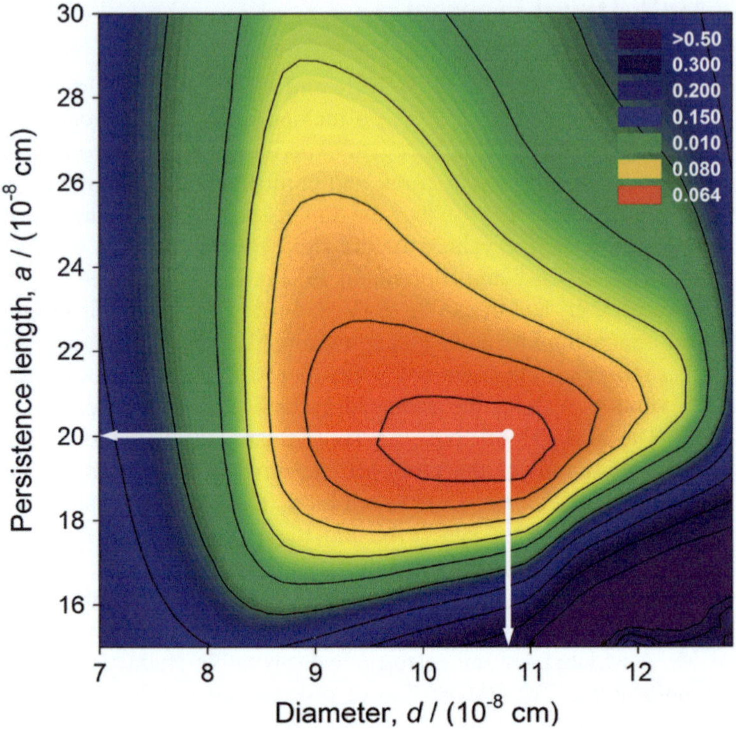

Fig. 14.10 Hydfit contour plot of the percent typical deviation, 100Δ, of the alternating styrene/diphenylethylene copolymers functionalized with terpyridine molecule in toluene with a fixed value of $M_L = 564$ g/(mol nm), showing the minimum of the deviation at the following values of persistence length, $a = 2$ nm and $d = 1.1$ nm (Pavlov et al. 2009)

capabilities of these approaches. One unresolved macromolecular hydrodynamic problem concerns the partition of the influence of the draining and volume effects on the size of the macromolecular chain. The value of parameter ε can/should be divided into two components: one is responsible for the volume effects (ε_v) and the other one for the flow of solvent molecules through the polymer coil (ε_d): $\varepsilon = \varepsilon_v + \varepsilon_d$. In Eq. 14.26, only the ε_v part should be used for estimating statistical segment length and hydrodynamic diameter.

Finally, generalization of the hydrodynamic data, namely, the intrinsic viscosity and velocity sedimentation coefficient values can be made by the double normalization of the canonical KMHS equations by the structural parameter M_{L} and the statistical segment length A taking into account the both parameters (Pavlov et al. 1999; Pavlov 2005, 2007). In this procedure, the chain contour length L is calculated for a unit of the statistical segment length A, i.e., *(L/A)*. The fundamental

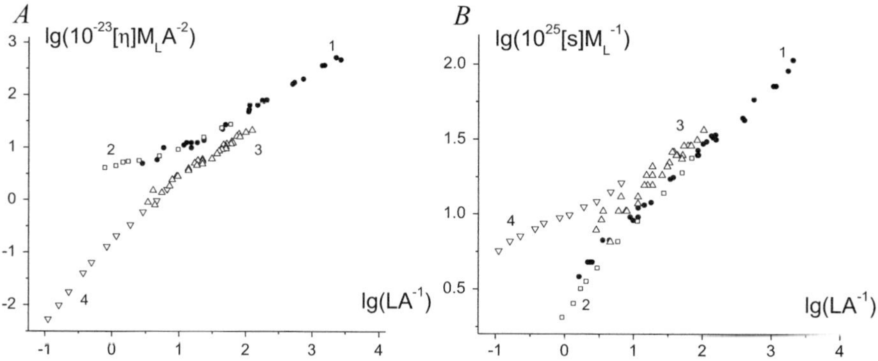

Fig. 14.11 Double normalized plots of $[\eta]M_LA^2$ (**a**) and $[s]M_L{}^{-1}$(**b**) vs. relative contour length LA^{-1} presented in double logarithmic scale. This figure interprets the data given in Fig. 14.5. Comparison of linear macromolecules of different structures, different equilibrium rigidities, and different scaling indices was made; these macromolecules are poly(1-vinyl-2-pyrrolidone) in 0.1 M sodium acetate solutions (*1*), styrene-methyl methacrylate brush copolymer in bromoform (*2*), poly(2,2′-p-phenylene-(5-benzimidazole)) isophthalamide in dimethylacetamide +3 % LiCl (*3*), and the triple helix of *Schizophyllum commune* polysaccharide (schizophyllan) in water (*4*). The following lengths of the statistical segment $A \times 10^8$ cm and the relative hydrodynamic diameters (*d*/*A*) are obtained for comparing macromolecules: 22 and 0.2 (1), 90 and 0.5 (2), 90 and 0.06 (3), 4000 and 0.007 (4)

Flory-Fox (14.4) and Svedberg (14.6) equations can be transformed into the following relationships:

$$[\eta]\, M_L/A^2 = \Phi\,(L/A,\ d/A, \varepsilon) \times (L/A)^{1/2} \tag{14.28}$$

$$[s]\,/M_L = (P\ (L/A,\ d/A, \varepsilon)\, N_A)^{-1} \times (L/A)^{1/2} \tag{14.29}$$

where *L/A* is the relative contour length and *d/A* is the relative chain diameter.

Figure 14.11 shows the data plotted in Fig. 14.5 in new coordinates $[\eta]\ M_L/A^2$ and $[s]/M_L$ *vs.* relative contour length *L/A*. These plots (Fig. 14.11) illustrate the earlier conclusion (Pavlov 2013; Pavlov et al. 2014) that the entire set of the linear macromolecules, regardless of the stiffness realization mechanism (free/hindered rotation around valence bonds, multistrand helix structures, electrostatic and/or thermodynamic interactions, specific interactions that lead to the globule formation), obeys the same laws.

14.12 Examples of Handling of Velocity Sedimentation Data with Sedfit Suite

To estimate the value of sedimentation coefficient for a series of fractions/samples of polymers from the files of raw data obtained with the XLI analytical ultracentrifuge (usually, using the interference optics), the Sedfit suite software may be chosen. This program provides the opportunity to process the velocity sedimentation data (Schuck 2000; Dam and Schuck 2004; Schuck and Zhao 2011). Within the framework of this suite, the raw data can be processed using various programs, and the results can be compared to each other. The treatment of the results obtained in the sedimentation experiments with different linear polymers in the wide range of molar masses will be discussed below.

14.12.1 Model-Less Method for Calculating the Apparent Differential Sedimentation Coefficient Distribution g(s)*

First, the model-less method for calculating the apparent differential sedimentation coefficient distribution $g^*(s)$ denoted by ls$-g^*(s)$ (Schuck and Rossmanith 2000) is employed; this method involves direct linear least-squares boundary modeling with the aid of superposition of sedimentation profiles of ideal *nondiffusing particles* (Figs. 14.12 and 14.13). This model is appropriate for the high molar mass polymer when the diffusion flow is slow and the s/D ratio is more than ≈ 0.3 svedberg/fick (1 svedberg $= 10^{-13}$ s and 1 fick $= 10^{-7}$ cm^2/s). Despite the fact that the average sedimentation coefficient does not differ from the values obtained using other models, the obtained distribution ls-$g^*(s)$ does not reflect the real distribution of sedimentation coefficients (molar masses) because it contains unaccounted diffusion spreading.

In some cases, the additional *ls*$-g^*(s)$ *model with one discrete Lamm equation component* completes the distribution curve in the region of small s values and, respectively, in the range of large D values (Fig. 14.13). This variation of the model not only serves an aesthetic purpose but also is entirely consistent with the general idea that linear synthetic and natural polymers have the essentially continuous molar mass distributions.

The results obtained using different up-to-date software should be compared to those obtained by the classical method based on the definition of sedimentation coefficient: $s \equiv (d\ln r/dt)\omega^{-2}$, where r is the position of sedimentation boundary, ω is the angular velocity, and t is the time of sedimentation (Svedberg and Pedersen 1940). The integral curves can be transformed into differential ones. Two kinds of differentiation are possible. First, the value of dJ/dr is calculated by differentiation of the raw interference curve with the subsequent smoothing. The second way is to compute dJ/dt curves as the difference of the adjacent interference curves when the

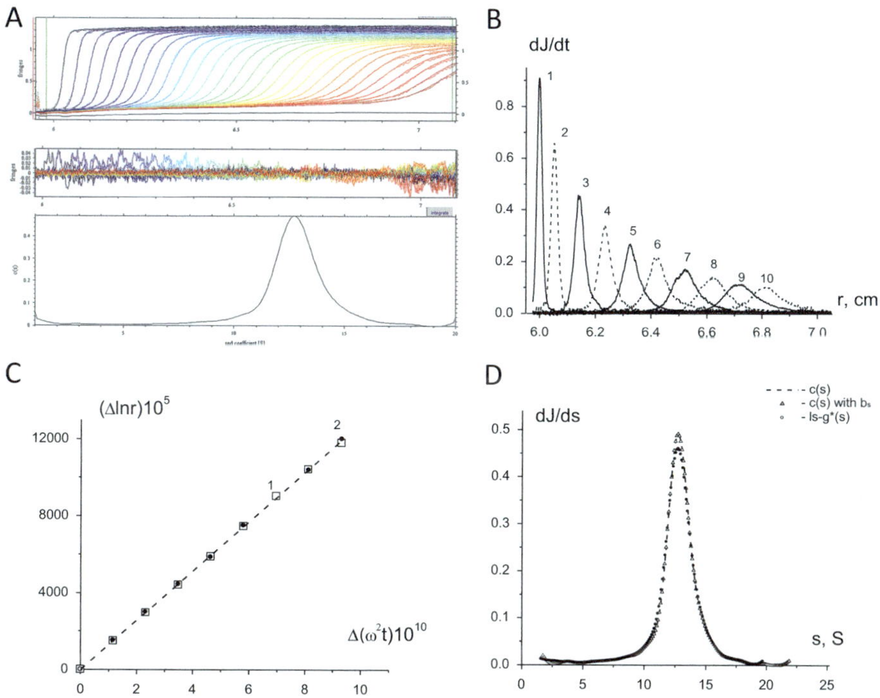

Fig. 14.12 Results obtained in the velocity sedimentation experiments using a pullulan standard (Polymer Standards Service GmbH) with $M_w = 1\ 660$ kg mol^{-1} and a concentration $c = 0.54$ mg cm^{-3} in H_2O at 25 °C, at a rotor speed of 42,000 rpm (Pavlov and Ebel 2006). (**a**) *Top*, superposition of the integral distribution of the pullulan concentration inside the cell collected by interference optics every 240 s; *middle*, corresponding residuals plot obtained by the Sedfit – $c(s)$; *bottom*, distribution obtained with $c(s)$ model. (**b**) Corresponding differential distribution of the sedimentation profiles *dJ/dt*. (**c**) Dependence of Δlnx on Δt; the slope of this dependence yields $s \equiv d\ln r/d(t\omega^{-2}) = 12.8$ S, $1 - dJ/dt$, $2 - dJ/dr$. (**d**) Comparison of the differential distributions of the sedimentation coefficient obtained by different model of Sedfit suite: (1) $c(s)$, $s = 12.78$ S; (2) $ls - g^*(s)$, $s = 12.79$ S; and (3) $c(s)$—bs, $s = 12.64$ S. Thus, the average of the four values is $s = (12.75 \pm 0.04)$ S, and the maximum difference between the obtained values $\Delta s = 0.16$S amounts 1.2 % from the mean value (insignificant deviation)

shift in time is small (Stafford 1994) (Fig. 14.12b, c). The slope of the dependence of the sedimentation boundary shift logarithm ln$r$ on $\omega^2 t$ allows to evaluate the sedimentation coefficient by its definition.

It is necessary to point out that the study of the concentration dependence of the velocity sedimentation is an important task in the hydrodynamic investigation of linear polymers (Eq.14.11 and Fig. 14.14).

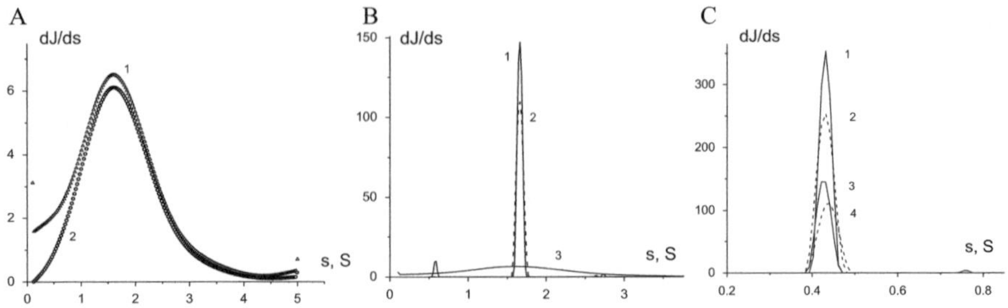

Fig. 14.13 Comparison of differential distributions obtained with different model of Sedfit suite. (**a**) Differential distributions *dJ/ds* of a pullulan sample with $M_w = 11.8$ kg/mol in H_2O at $c = 5.38$ mg/cm^3 obtained with *ls*–*g**(*s*) model (1) and with *ls*–*g**(*s*) with one discrete Lamm Eq. component model (14.2). The last model gives more realistic dependence in the region of very low s values. (**b**) The distributions of sedimentation coefficients obtained for the same sample. The following values were obtained with the different model of Sedfit program: continuous *c*(*s*) distribution – $s = 1.65$ S (2); continuous *c*(*s*)–*bs*, $b_s = 0.458$ – $s = 1.66$ S (1); and *ls* – *g**(*s*) – $s = 1.7$ S (3). In the case of the low molar mass sample which is characterized by the high diffusion coefficient ($D_0 \approx 10 \times 10^{-7}$ cm^2/s), there is a striking difference in the form of *ls*–*g**(*s*) distributions. Here, the diffusion process is not taken into account, which leads to a fictitious broad distribution as compared with two other methods (*c*(*s*) and *c*(*s*)–*bs*). These two methods take into account the dominant diffusion flux during sedimentation. At the same time, the average sedimentation coefficient obtained by *ls*–*g**(*s*) method virtually is the same as those obtained with two other methods. For this sample, the ratio s_0/D_0 is ≈ 0.17 svedberg/fick. (**c**) Distributions of sedimentation coefficients obtained for the fraction of poly(N-methyl-N-vinylacetamide) with $M_{sD} = 5.2$ kg/mol in H_2O at $n = 40{,}000$ rpm. The following values were obtained using the different models included in Sedfit program at the concentration $c = 3.03$ mg/cm^3 (1 and 2): continuous *c*(*s*) distribution – $s = 0.43$S (1); continuous *c*(*s*)–*bs*, $b_s = 0.43$ – $s = 0.43$S (2); at the concentration $c = 1.5$ mg/cm^3 (3 and 4), continuous *c*(*s*) distribution – $s = 0.44$S (3); continuous *c*(*s*)–*bs*, $b_s = 0.43$ – $s = 0.43$S (4). It is noteworthy that, in the case when the sedimentation flow becomes wilted, and the diffusion flow is growing ($D_0 \approx 13 \times 10^{-7}$ cm^2/s), the *ls*–*g**(*s*) model is no longer applicable for evaluating the velocity sedimentation coefficient. For this sample, the ratio s_0/D_0 is ≈ 0.03 svedberg/fick (Pavlov et al. 2010a)

14.12.2 Continuous c(s) Model in Sedfit Suite

Second, the differential distribution of sedimentation coefficient *c*(*s*) is calculated by numerical solution of Lamm equation in which both opposite flows (sedimentation and diffusion) are taken into account (*continuous c(s) model*, Schuck 2000). In order to achieve one-parametric distribution, a scaling law between the diffusion and sedimentation coefficient is invoked in the *c*(*s*) method. The choice of the scaling law is based on the $D_0 = K_{Ds} s_0^{-1/2}$ relation which is correct for roughly globular particles, with the frictional ratio (f/f_{sph}) as a scaling parameter (f is the frictional coefficient of the solute macromolecule; f_{sph} is the frictional coefficient of a rigid sphere with the same "anhydrous" volume (free of solvent) as the macromolecule). Fitting for (f/f_{sph}) in a nonlinear regression will provide the estimate of the weight-average frictional ratio of all macromolecules in solution that actually leads to

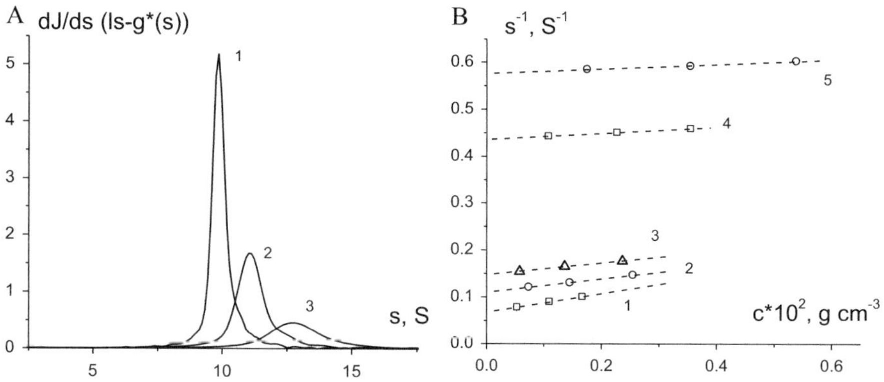

Fig. 14.14 Concentration dependence of velocity sedimentation for pullulan molecules in H_2O, $n = 42{,}000$ rpm, at 25°C. (**a**) Distributions obtained using $ls - g^*(s)$ model for sample with $M_w = 1660$ kg/mol at different concentrations: (1) $c = 1.67$, (2) $c = 1.10$, and (3) $c = 0.54$ mg/cm^3. The area under the distribution curve gives the number of fringes J and corresponds to polymer concentration c. The refractive index increment was calculated from $(\Delta n/\Delta c) = J\lambda/Kcl$, where λ is the wavelength, K is the magnifying coefficient, and l is the optical path length. (**b**) Concentration dependences of the (reciprocal) sedimentation coefficient s^{-1} for pullulan samples of different molar masses and corresponding characteristics s_0 and k_s: (1) $M_w = 1\ 660$ kg/mol, $s_0 = 14.8$ S, $k_s = 290$ cm^3/g; (2) $M_w = 404$, $s_0 = 9.1$, $k_s = 130$; (3) $M_w = 212$, $s_0 = 6.8$, $k_s = 80$; (4) $M_w = 22.8$, $s_0 = 2.3$, $k_s = 15$; and (5) $M_w = 11.8$, $s_0 = 1.7$, $k_s = 8$

determination of the average value of the translational diffusion coefficient D:

$$D_0 = kT(1 - \upsilon\rho_0)^{1/2}/{\eta_0}^{3/2}\left(9\pi 2^{1/2}\right)\left(f/f_{sph}\right)_0^{3/2}(s_0\upsilon)^{1/2} \tag{14.30}$$

The distribution of the sedimentation coefficients for the sample in Sedfit software is denoted as c (s). Usually this kind of designation is used for integral distribution. Indeed, c is a concentration; in the case of interference optics, it is expressed as a number of fringes J, which is unambiguously associated with the concentration; see the relation (14.31). However, the Sedfit program gives the distribution in a differential form. The area under the differential curve is equal to the number of interference fringes (i.e., concentration). Therefore, it is logical and more consistent to denote this distribution by $dc(s)/ds$ or $dJ(s)/ds$. Finally, the differential distribution ($dc(s)/ds$) of the sample is obtained (Figs. 14.12 and 14.13). The area under the $dc(s)/ds$ curve gives the loading concentration of the macromolecules between the minimum and maximum s-value occurring (expressed in number of fringes, J). Note that the concentration dependence of J (fringes number) allows defining an additional parameter – the refractive index increment $\Delta n/\Delta c$

$$\Delta n/\Delta c = J\lambda/Kcl \tag{14.31}$$

where λ is the wavelength, K is the magnifying coefficient, and l is the optical path length. The value of $\Delta n/\Delta c$ is a characteristic of polymer-solvent system and should be virtually constant in a series of polymer homologues until the oligomer region. Obviously, such correlation will be performed in the case of an absorption boundary registration: $OD_{solution} = OD_{solvent} + K_{ab}c$, where K_{ab} is the increment of the polymer absorption in a given solvent ($K_{ab} \equiv dOD/dc$). However, this relationship will hold within the range of much smaller concentration due to quick absorbance saturation.

Needless to say, the series of globular particles is not the best model to describe hydrodynamic behavior of linear macromolecules. Nevertheless, the continuous $c(s)$ distribution model leads to reasonable values of the velocity sedimentation coefficients for many linear polymers.

14.12.3 Continuous c(s) Model with General Scaling Law in Sedfit Suite

Third, recently Sedfit suite has been implemented with *continuous c(s) model with general scaling law* (designated below as $c(s)$–bs). Some information about this add-in program is contained in a publication by Harding et al. (2011b). In this paper, the implementation is called "extended Fujita approach." For direct boundary modeling with distributions of the Lamm equation solutions (Schuck 2000), the measured interference (or absorbance) profiles $a(r, t)$ were modeled as an integral over the differential concentration distribution $c(s)$

$$a\,(r,\ t) = \int c(s)\chi\,(s,\ D(s),\ r,\ t)\ ds + b_{nr}(r) + b_{nt}(t) \tag{14.32}$$

with $b_{nr}\,(r)$ and $b_{nt}(t)$, denoting systematic baseline noise components, and $\chi(s, D(s), r, t)$ denoting the solution of the Lamm equation at unit loading concentration of a species with sedimentation coefficient s and diffusion coefficient D

$$d\chi/dt = r^{-1}d\left[rD(s)\,(d\chi/dr) - s\omega^2 r^2\chi\right]/dr \tag{14.33}$$

where r is the distance from the center of rotation and ω is the rotor angular velocity. Equation 14.33 was solved by finite element methods in a static or moving frame of reference as described in Claverie et al. (1975), Schuck (1998), and Schuck (1998). For each species, the diffusion coefficient $D(s)$ was estimated as a function of the sedimentation coefficient s based on the general scaling law, which may be presented as

$$D_0 = \left(RT/\,(1 - \upsilon\rho_0)\,K_s{}^{1/bs} s_0{}^{(bs-1)/bs} = K_{Ds} \times s^{(bs-1)/bs}\right. \tag{14.34}$$

where the parameters b_s and K_s define the scaling relation between s_0 and M ($s_0 = K_s M^{b_s}$). It should be emphasized that the $D_0 = K_{Ds} {s_0}^{-1/2}$ relation known for globular species is a special case of general Eq. 14.34. To solve the Lamm equation numerically, the range of possible velocity sedimentation coefficient values is typically discretized into 100–200 values. Actually, it is modern paraphrase of the "graphic fractionation" approach (Gralen 1944; Kinell and Ranby 1950). For any reasons, a researcher specifies the power law exponent b_s, and the program fits the best value of K_s^*. The asterisk means that this value is obtained at some known concentration, and these values must be extrapolated to zero concentration to obtain the non-disturbed K_s value. In the finite element method, the solutions of the ideal Lamm equation are obtained using the adaptive grid algorithm which allows fitting the best K_s^* value; finally, each s-value gets the corresponding D-value. As a result, the differential distribution ($dc(s)/ds$) of the sample is also fitted. The relation (14.34) has a general meaning and may be applied to any kind of homologous series of the macromolecular compounds. In principle, the $c(s)–bs$ implementation is more appropriate to treat the velocity sedimentation data on any macromolecules due to the possibly large variation of the b_s value. The K_s value (Eq. 14.15) should not be confused with the Gralen concentration coefficient k_s (Eqs. 14.7).

Note that the distributions of the sedimentation coefficients for the samples of high molar masses obtained using different models ($c(s)$, $c(s)–bs$, and $ls - g^*(s)$) are virtually indistinguishable (Fig. 14.12d). The lower the molar mass, the higher the deviation of the $ls - g^*(s)$ distribution from the real one (Fig. 14.13b). However, up to a certain value of the s/D ratio (>0.2), the average value of sedimentation coefficient s obtained using the $ls - g^*(s)$ model is almost equal to the average values of s obtained with the use of other models.

Another important question arises: How do the fitted K_s^* values sort with those obtained by the straightforward procedure of comparison between the s_0 values and molar masses for a series of samples? The results obtained for the flexible polymers are summarized in Figs. 14.15, 14.16, and 14.17, and the following conclusions can be draw from these data. Uppermost, the K_s^* value strongly depends on concentration. This fact highlights the importance of studying the concentration dependencies of the hydrodynamic characteristics in the appropriate range of concentrations and extrapolating the sought-for characteristics to infinite dilution. This means that the use of a single concentration for estimating samples polydispersity is unlikely to give the reliable results.

The sign of the slope of concentration dependence is determined by the thermodynamic quality of the solvent (compare Figs. 14.15 and 14.17 with Fig. 14.16). The K_s^* values depend in a unified manner on the parameters characterizing the degree of dilution ($c[\eta]$ or ck_s) of samples of different molar masses. The fitted K_s values extrapolated to the infinite dilution are close to those obtained in the traditional way. Here is another proof of self-sufficiency of the velocity sedimentation method coupled with the Sedfit $c(s)–bs$ model in the determination of molecular characteristics of linear polymers.

Explicitly, the use of new software for primary processing of the experimental data should lead to results, which must not contradict with the main conclusions on

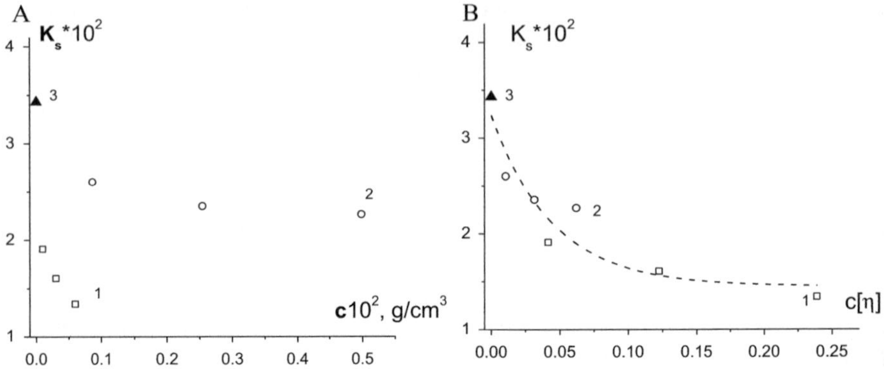

Fig. 14.15 Concentration dependences of fitted K_s^* value parameter (**a**) for polystyrenes of different molar masses in toluene (which is thermodynamically good solvent for polystyrene): (1) $M = 1760$, (2) $M = 17$ kg/mol; dependences of K_s on the $c[\eta]$ parameter (**b**). Triangle point (3) is the K_s value from the independent literature data (Meyerhoff and Appelt 1979). K_s^* values are fitted within the framework of the continuous $c(s)$ model with general scaling law $D_0 = RT/(1-\upsilon\rho_0)K_s^{1/bs}s_0^{(bs-1)/bs} = K_{Ds} \times s^{(bs-1)/bs}$. K_s^* is not to be confused with the concentration Gralen coefficient k_s in Eq. 14.11

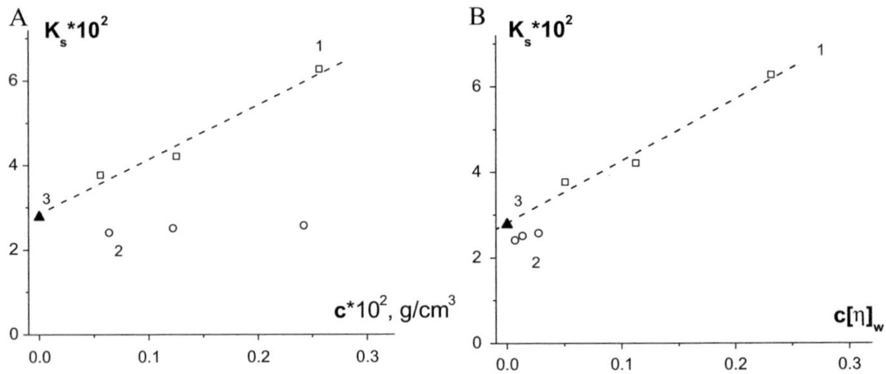

Fig. 14.16 Concentration dependences of fitted K_s^* value (**a**) for two mixtures of polystyrene standards of different molar masses studied in MEK (which is a marginal solvent for polystyrene) and dependences of K_s^* on the dilution parameter $c[\eta]$ (**b**). Full triangle point is the K_s value from the independent literature data (Pavlov et al. 2011). (1) $(1760 + 1060 + 710 + 311 + 194) \times 10^3$ g/mol, 20 % of each in the mixture, $[\eta]_w = \Sigma(0.2[\eta]_i) = 90.2$ cm^3/g; (2) $(95 + 43 + 9 + 4 + 1.8) \times 10^3$ g/mol, 20 % of each in the mixture, $[\eta]_w = \Sigma(0.2[\eta]_i) = 11.6$ cm^3/g, where $[\eta]_i$ is the intrinsic viscosity value of the individual sample

conformational states of linear macromolecules that have been previously received on the basis of the velocity sedimentation coefficients obtained with the fundamental relations. This is true, at least for the flexible linear polymers.

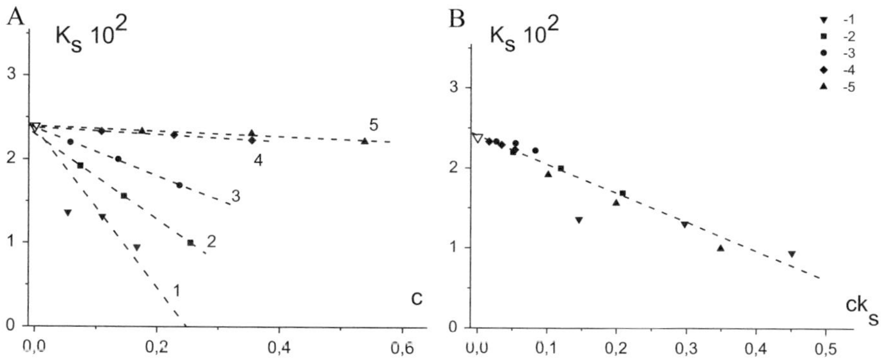

Fig. 14.17 Concentration dependences of fitted $K_s{}^*$ value (**a**) for pullulan of different molar masses (kg/mol) in H_2O (which is thermodynamically good solvent for pullulan): (1) $M = 1660 \times 10^3$ g/mol, (2) $M = 404$, (3) 212, (4) 22.8, (5) 11.8×10^3 g/mol (*left side*) and $K_s{}^*$ dependences on the ck_s parameter characterizing the degree of solution dilution (**b**). Open triangle point is the K_s value from the literature data (Pavlov et al. 1994 and the references therein). $K_s{}^*$ is not to be confused with the concentration Gralen coefficient k_s in Eq. 14.11

14.13 Conclusions

Self-sufficiency of the velocity sedimentation method manifests itself (1) in the qualitative analysis of the conformational status of linear macromolecule from the lgk_sM_L vs. $lg\ [s]/M_L$ plot, (2) in the possibility of an adequate assessment of polymer molar masses and further conformational analysis using the sedimentation parameter β_s, and (3) in the fact that numerical solution of the Lamm equation using the $c(s)$–bs model within Sedfit suite allows to set the scaling relation $s_0 = K_sM^{bs}$ which opens up possibilities for further conformational analysis, as well as molar mass distribution analysis.

In outcome conclusion, we propose the following algorithm for the study of homologous series of macromolecules using XLI coupled with $c(s)$–bs model inside of Sedfit suite.

The proposed algorithm involves the following stages:

0. First of all, for the comprehensive and exhaustive study of any polymer system, the researcher should have a set of samples/fractions with narrow polydispersity and the possibly broadest range of molar masses.
1. Studying of the concentration dependence of the available number of samples/fractions using $c(s)$ or/and $ls - g^*(s)$ models of Sedfit suite.
2. Determining the s_0 and k_s values from the concentration dependencies of s^{-1} and subsequently establishing the $k_s = K_{sks}s_0{}^{bks}$ correlation in order to define the scaling index b_s in the KMHS-type relation $s_0 = K_sM^{bs}$ from the $b_{ks} = (2 - 3b_s)/b_s$ relationship.
3. Using the obtained b_s value, the rehandling of raw set of data must be done using $c(s)$–bs model and $K_s{}^*(c)$ parameter must be fitted. The concentration

dependencies of K_s* must be plotted and extrapolated to zero concentration conditions. Thus, the unperturbed value of K_s will be estimated and the scaling relation $s_0 = K_s M^{bs}$ is known.

4. The molar masses M may be calculated. It is necessary to check the b_s and K_s values using the double logarithmic plot (lgs_0 vs. lgM). The difference of about $\Delta b_s \approx 0.01$ can be considered as insignificant. But if, for example, instead of $b_s = 0.41$, the b_s value turns out to be $b_s = 0.46$, it is necessary to go back to step 3 and repeat the process using the new value $b_s = 0.46$. In other words, the method of successive approximations is necessary to achieve the consistency in the b_s value.
5. With the final set of s_0, k_s, and M values, further molecular interpretation may be reached: assessments of the sedimentation parameter β_s values, Kuhn segment length, the value of hydrodynamic diameter, the degree of macromolecular coiling, etc. Finally, the obtained estimations should be compared with the chemical structure of the repeating unit of the polymer.
6. If successful, this path gives the full set of molecular and conformation characteristics of studied linear polymer and will illustrate self-sufficiency of the sedimentation velocity method in the study of a homologous series of linear polymers.

Finally, it is interesting to analyze possible scope of measuring sedimentation coefficients with the help of the modern analytical ultracentrifuge. The upper limit can be represented by sedimentation of nanoparticles and nano-complexes, and it totals hundreds of thousands of svedberg (e.g., Perevyazko et al. 2010, 2012). Up to now, the lower limit has been apparently achieved in the studies of the cyclodextrins, and it is about one tenth of svedberg (Pavlov et al. 2010b). Thus, one XLI instrument enables us to measure the values of a physical quantity differing by six orders of magnitude (by varying the rotation speed and/or solvent). This is a rare opportunity in the practice of physical experiments.

Acknowledgments GMP is grateful to Dr. O.V. Okatova for her help in preparing this manuscript. This work was partially supported by the St. Petersburg State University research grant (11.38.267.2014).

References

Abe F, Einaga Y, Yamakawa H (1993) Excluded-volume effects on the intrinsic viscosity of oligomers and polymers of styrene and isobutylene. Macromolecules 26:1891

Amoros D, Ortega A, de la Torre GJ (2011) Hydrodynamic properties of wormlike macromolecules: monte carlo simulation and global analysis of experimental data. Macromolecules 44:5788

Behlke J, Ristau O (1997) Molecular mass determination by sedimentation velocity experiments and direct fitting of the concentration profiles. Biophys J 72:428

Benoit H, Doty PM (1953) Light scattering from non-Gaussian chains. J Phys Chem 57:958

Bernal JMG, Tirado MM, Freire JJ, de la Torre JG (1991) Monte Carlo calculation of hydrodynamic properties of linear and cyclic polymers in good solvents. Macromolecules 24:593–598
Bloomfield VA (1968) Hydrodynamic studies of structure of biological macromolecules. Science 161:1212
Broersma S (1969) Translational diffusion constant of a random coil. J Chem Phys 51:233
Budtov VP (1992) Physical chemistry of polymer solutions. Chemistry Edition, St. Petersburg
Burchard W (1961) Uber den Einfluss der losungsmittel auf die Structur linearer Makromolekule. Makromol Chem 50:20
Cantor CR, Schimmel PR (1980) Biophysical chemistry. W.H. Freeman & Company, San Francisco
Claverie JM, Dreux H, Cohen R (1975) Sedimentation of generalized systems of interacting particles. I. Solution of systems of complete Lamm equations. *Biopolymers*. 14:1685
Cowie MG, Bywater S (1965) The use of frictional coefficient to evaluate unperturbed dimensions in dilute polymer solutions. Polymer 6:197
Creeth JM, Knight CG (1965) On the estimation of the shape of macromolecules from sedimentation and viscosity measurements. Biochim Biophys Acta 102:549
Dam J, Schuck P (2004) Calculating sedimentation coefficient distributions by direct modeling of sedimentation velocity concentration profiles. Methods Enzymol 384:185
de Gennes P-G (1979) Scaling concepts in polymer physics. Cornell University Press, Ithaca
de la Torre JG, Bloomfield VA (1981) Hydrodynamic properties of complex, rigid, biological macromolecules: theory and applications. Q Rev Biophys 14:81
de la Torre JG, Jiminez A, Freire J (1982) Monte carlo calculation of hydrodynamic properties of freely jointed, freely rotating, and real polymethylene chains. Macromolecules 15:148
de la Torre JG, Martínez LM, Tirado M, Freire J (1984) Monte carlo study of hydrodynamic properties of flexible linear chains: analysis of several approximate methods. Macromolecules 17:2715
Demeler B, Saber H (1998) Determination of molecular parameters by fitting sedimentation data to finite-element solutions of the Lamm equation. Biophys J 74:444
Doi M (1996) Introduction to polymer physics. Oxford Science Publications/Clarendon Press, New York
Flory PJ (1953) Principles of polymer chemistry. Cornell Univ Press, New York
Freed KF (1983) Concentration dependence of frictional coefficient for polymer chain in solution. J Chem Phys 78:2051
Fujita H (1975) Foundations of ultracentrifugal analysis. Wiley, New York
Fujita H (1990) Polymer solutions. Elsevier, Amsterdam
Fujita H, Teramoto A, Yamashita T, Okita K, Ikeda S (1966) Solution properties of synthetic polypeptides. Sedimentation and viscosity of poly-γ-benzyl-L-glutamate in dimethylformamide. Biopolymers 4:781
Gralen N (1944) Sedimentation and diffusion measurements on cellulose and cellulose derivatives. Almqvista Wiksells, Uppsala
Gray HB, Bloomfield VA, Hearst JE (1967) Sedimentation coefficient of linear and cyclic wormlike coils with excluded-volume effects. J Chem Phys 46:1493
Harding SE, Abdelhameed AS, Morris GA (2011a) On the hydrodynamic analysis of conformation in mixed biopolymer systems. Polym Int 60:2
Harding SE, Schuck P, Abdelhameed AS, Adams G, Koek MS, Morris GA (2011b) Extended Fujita approach to the molecular weight distribution of polysaccharides and other polymeric systems. Method 54:136
Hearst JE (1964) Intrinsic viscosity of stiff-chain macromolecules. J Chem Phys 40:1506
Hearst JE, Stokmayer WH (1962) Sedimentation constants of broken chains and wormlike coils. J Chem Phys 37:1425
Hearst JE, Tagami Y (1965) Shear dependence of the intrinsic viscosity of rigid distributions of segments with cylindrical symmetry. J Chem Phys 42:4149
Imai S (1970) Concentration dependence of the sedimentation coefficient at the theta state. J Chem Phys 52:4212

Kawade Y, Watanabe I (1956) Sedimentation study of sodium deoxypentosenucleate preparations from herring sperm and calf thymus. Biochim Biophys Acta 19:513

Kinell PO, Ranby BG (1950) Ultracentrifugal Sedimentation of Polymolecular Substances. In Advances in Colloid Science 1st Ed., Eds. Mark H, Verwey EJ Vol. 3. New York : Intersci Publ: Inc. p. 161.

Kirkwood J, Riseman J (1948) The intrinsic viscosities and diffusion of flexible macromolecules in solution. J Chem Phys 16:565

Kraemers EO, Nichols JB (1940) Results obtained with svedberg ultracentrifuges. Organic colloids (Except proteins). In: Svedberg T, Pedersen KO (eds) The ultracentrifuge. Oxford University Press, New York, p 416

Kratky O, Porod G (1949) Roentgenuntersuchung gesloester Fadenmolekuele. Rec Trav Chim 68:1106

Kroe RR, Laue TM (2009) NUTS and BOLTS: applications of fluorescence-detected sedimentation. Anal Biochem 390:1

Kuhn H, Kuhn W, Silberberg A (1953) Improved relationships for diffusion and sedimentation constants and for viscosity and streaming birefringence of solutions of polymers. J Polym Sci 16:193

Landau LD, Lifschitz EM (1963) Statistical physics, 3rd edn. Pergamon Press, Oxford

Laue TM, Stafford WF III (1999) Modern application of analytical ultracentrifugation. Annu Rev Biophys Biomol Struct 28:75

Lebowitz J, Lewis MS, Schuck P (2002) Modern analytical ultracentrifugation in protein science: a tutorial review. Protein Sci 11:2067

MacGregor IK, Anderson AL, Laue TM (2004) Fluorescence detection for the XLI analytical ultracentrifuge. Biophys Chem 108:165

Maechtle W, Boerger L (2006) Analytical ultracentrifugation of polymers and nanoparticles. Springer, Berlin

Magarik SY, Pavlov GM, Fomin GA (1978) Hydrodynamic and optical properties of homologous series of styrene-methyl methacrylate graft copolymers. Macromolecules 11:294

Mandelbrot B (1982) The fractal geometry of nature. Freeman and Company, New York

Mandelkern L, Flory PJ (1952) The frictional coefficient for flexible chain molecules in dilute solution. J Chem Phys 20:212

Meyerhoff G, Appelt B (1979) A low shear viscometer with automated recording and application to high molecular weight polystyrene solutions. Macromolecules 12:968

Morris GA, Harding SE (2013) Hydrodynamic modeling of carbohydrate polymers. In: Roberts GCK (ed) Encyclopedia of biophysics. Springer, Berlin/Heidelberg

Munk P (1991) Polymer characterization using the ultracentrifuge. In: Barth HG, Mays JW (eds) Modern methods of polymer characterization. Wiley, New York

Muthukumar M, de Mense M (1983) Concentration dependent frictional coefficient of polymer molecules in dilute solution. J Chem Phys 78:2773

Newman S, Eirich F (1950) Particle shape and the concentration dependence of sedimentation and diffusion. J Colloid Sci 5:541

Oono Y (1985) Statistical physics of polymer solutions: conformation-space renormalization-group approach. Adv Chem Phys 61:301

Ortega A, de la Torre JG (2007) Equivalent radii and ratios of radii from solution properties as indicators of macromolecular conformation, shape, and flexibility. Biomacromolecules 8:2464

Ortega A, de la Torre JG (2013) Hydfit and related packages for linear molecules. In: Roberts GCK (ed) Encyclopedia of biophysics. Springer, Berlin/Heidelberg

Pavlov GM, Frenkel SY (1982) On the concentration dependence of the sedimentation coefficients of macromolecules. Vysokomol Soedin 24:178

Pavlov GM, Selunin SG, Shildiaeva NA, Yakopson SM, Efros LS (1985) Translational friction and intrinsic viscosity of polyamide benzimidazole molecules in solutions. Polym Sci USSR 27:1823

Pavlov GM, Frenkel SY (1986) Determination of the molar mass of linear polymers on the concentration dependence of sedimentation coefficient. Vysokomol Soedin 28:353

Pavlov GM, Selyunin SG (1986) Velocity sedimentation, molar mass and conformational properties of some soluble chitin derivatives. Polym Sci USSR 28:1925
Pavlov G, Frenkel S (1988) The sedimentation parameter of linear polymer molecules in absence of excluded volume effects. Acta Polymer 39:107
Pavlov GM, Panarin EF, Korneeva EV, Kurochkin EV, Baikov VE, Uschakova VN (1990) Hydrodynamic properties of polyvinylpyrrolidone molecules in dilute solutions. Makromolek Chemie 191:2889
Pavlov G.M. Korneeva E.V., Yevlampieva N.P. (1994) Hydrodynamic characteristics and equilibrium rigidity of pullulan molecules. Intern. J. Biolog. Macromolecules 16:318
Pavlov GM, Frenkel SY (1995) Sedimentation parameters of linear polymers. Progr Colloid Polym Sci 99:101
Pavlov GM, Tarabukina EB, Frenkel SY (1995) Self-sufficiency of the velocity sedimentation for the determination of the molecular characteristics of linear polymers. Polymer 36:2043
Pavlov GM (1997) The concentration dependence of sedimentation for polysaccharides. Euro Biophys J 25:385
Pavlov G, Rowe A, Harding S (1997) Conformation zoning of large molecules using the analytical ultracentrifuge. Trends Anal Chem 16:401
Pavlov GM, Harding SE, Rowe AJ (1999) Normalized scaling relations as a natural classification of linear macromolecules according to size. Prog Colloid Polym Sci 113:76
Pavlov G, Finet S, Tatarenko K, Korneeva E, Ebel C (2003) Conformation of heparin studied with macromolecular hydrodynamic methods and X-ray scattering. Euro Biophys J 32:437
Pavlov GM (2005) Normalized Kuhn-Mark-Houwink-Sakurada relationships. Polymer Sci 47A:1129
Pavlov GM, Ebel C (2006) Sedfit utilization for sedimentation coefficient determination of linear noncharged macromolecules. Abstract Book of 15th international symposium AUC, London, p 60
Pavlov GM (2007) Size and average density spectrums of macromolecules obtained from hydrodynamic data. Eur Phys J E 22:171
Pavlov GM, Amoros D, Ott C, Zaitseva II, de la Torre JG, Schubert US (2009) Hydrodynamic analysis of well-defined flexible linear macromolecules of low molar mass. Macromolecules 42:7447
Pavlov GM, Okatova OV, Michailova AV, Ulianova NN, Gavrilova II, Panarin EF (2010a) Conformational parameters of poly(N-methyl-N-vinylacetamide) molecules through the hydrodynamic characteristics studies. Macromol Biosci 10:790
Pavlov GM, Korneeva EV, Smolina NA, Schubert US (2010b) Hydrodynamic properties of cyclodextrin molecules in dilute solutions. Eur Biophys J 39:371
Pavlov GM, Perevyazko I, Okatova OV, Schubert US (2011) Conformation parameters of linear macromolecules by velocity sedimentation and other hydrodynamic methods. Methods 54:124
Pavlov GM, Breul AM, Hager MD, Schubert US (2012) Hydrodynamic and molecular study of poly(4-((4-(hexyloxy)phenyl)ethynyl)phenyl methacrylate) in dilute solutions and conformational peculiarities of brush-like macromolecules. Macromol Chem Phys 213:904
Pavlov GM (2013) Hydrodynamic of macromolecules: conformation zoning of general macromolecules. In: Roberts GCK (ed) Encyclopedia of biophysics. Springer, Berlin/Heidelberg
Pavlov GM, Okatova OV, Gubarev AS, Gavrilova II, Panarin EF (2014) Strong linear polyelectrolytes in solutions of extreme concentrations of one-one valent salt. Hydrodynamic study. Macromolecules 47:2748
Peacocke AR, Schachman HK (1954) Studies on the sedimentation behaviour of thymus deoxypentose nucleic acid with reference to its homogeneity, size and shape. Biochim Biophys Acta 15:198
Pedersen KO (1940) Results obtained with Svedberg Ultracentrifuges. Proteins. In: Svedberg T, Pedersen KO (eds) The Ultracentrifuge, Animal Proteins, Vegetable Proteins, Enzymes and Hormones, and others. Oxford University Press, New York, p 354

Perevyazko I, Vollrath A, Hornig V, Pavlov G, Schubert US (2010) Characterization of poly(methyl methacrylate) nanoparticles prepared by nanoprecipitation using analytical ultracentrifugation, dynamic light scattering, and scanning electron microscopy. J Polym Sci: Polym Chem 48:3924

Perevyazko IY, Bauer M, Pavlov GM, Hoeppener S, Schubert S, Fischer D, Schubert US (2012) Polyelectrolyte complexes of DNA and linear PEI: formation, composition and properties. Langmuir 28:16167

Peterson JM (1964) Hydrodynamic alignment of rodlike macromolecules during ultracentrifugation. J Chem Phys 40:2680

Philo JS (1994) Measuring sedimentation, diffusion, and molecular weights of small molecules by direct fitting of sedimentation velocity concentration profiles. In: Schuster T, Laue T (eds) Modem analytical ultracentrifugation. Birkhauser, Boston, p 156

Rowe AJ (1977) The concentration dependence of transport processes: a general description applicable to the sedimentation, translational diffusion, and viscosity coefficients of macromolecular solutes. Biopolymers 16:2595

Rubinstein M, Colby RH (2003) Polymer physics. Oxford University Press, New York

Sato T, Norisuye T, Fujita H (1984) Double-stranded helix of xanthan. Macromolecules 17:2696

Schachman HK (1959) Ultracentifugation in biochemistry. Academic, New York/London

Schuck P (1998) Sedimentation analysis of noninteracting and self-associating solutes using numerical solutions to the Lamm equation. Biophys J 75:1503

Schuck P (2000) Size-distribution analysis of macromolecules by sedimentation velocity ultracentrifugation and Lamm equation modeling. Biophys J 78:1606

Schuck P, Rossmanith P (2000) Determination of the sedimentation coefficient distribution by least-squares boundary modeling. Biogeosciences 54:328

Schuck P, Zhao H (eds) (2011) Modern analytical ultracentrifugation. Elsevier, Amsterdam

Schuster T, Laue T (eds) (1994) Modern analytical ultracentrifugation. Birkhauser, Boston

Scott DJ, Harding SE, Rowe AJ (2005) Analytical ultracentrifugation: techniques and methods. The Royal Society of Chemistry, Cambridge

Sharp P, Bloomfield V (1968) Intrinsic viscosity of wormlike chain with excluded-volume effects. J Chem Phys 48:2149

Singer R, Gross H (1934a) Ueber der Verhalten von Polystyrolen in der Svedberg'schen Sedimentationsgeschwindigkeits-Zentrifuge. Helv Chim Acta 17:59

Singer R, Gross H (1934b) Ultrazentrifugale Polydispersitaetbestimmungen an hochpolymeren Stoffen. Helv Chim Acta 17:726

Stafford WF (1992) Boundary analysis in sedimentation transport experiments: a procedure for obtaining sedimentation coefficient distributions using the time derivative of the concentration profile. Anal Biochem 203:295

Stafford WF (1994) Boundary analysis in sedimentation velocity experiments. Methods Enzymol 240:478

Stafford WF, Braswell EH (2004) Sedimentation velocity, multi-speed method for analyzing polydisperse solutions. Biophys Chem 108:273

Stamm AJ (1930) The state of dispersion of cellulose in cuprammonium solvent as determined by ultracentrifuge methods. J Am Chem Soc 52:3047

Stockmayer WH, Fixman M (1963) On the estimation of unperturbed dimensions from intrinsic viscosities. J Polymer Sci 1:137

Svedberg T (1927) Nobel lecture: the ultracentrifuge. http://www.nobelprize.org/nobel_prizes/chemistry/laureates/1926/svedberg-lecture.html

Svedberg T, Faehraeus R (1926) A new method for the determination of the molecular weight of the proteins. J Am Chem Soc 48:430

Svedberg T, Pedersen KO (1940) The ultracentrifuge. Oxford University Press, New York

Svedberg T, Sjoegren (1928) The molecular weight of serum albumin and of serum globulin. J Am Chem Soc 50:3318

Tanford C (1961) Physical chemistry of macromolecules. Wiley, New York

Tarabukina EB, Slavina ZN, Frenkel SY (1991) Concentration dependence of the sedimentation coefficient of poly(amidobenzimidazole). Acta Polym 42:266

Teraoka I (2002) Polymer solution: an introduction to physical properties. Wiley, New York
Tsvetkov VN (1989) Rigid-chain polymers. Plenum Press, New York
Tsvetkov VN, Klenin SI (1953) Diffusion of polystyrene fractions in dichlorethan. Dokl Akad Nauk SSSR 88:49
Tsvetkov VN, Eskin VE, Frenkel SY (1970) Structure of macromolecules in solution. Nat Lend Library Sci & Technol, Boston
Volkenshtein MV (1963) Configurational statistics of polymeric chains. Wiley, New York
Wales M, van Holde K (1954) The concentration dependence of the sedimentation constants of flexible macromolecules. J Polymer Sci 14:81
Yamakawa H (1971) Modern theory of polymer solutions. Harper & Row, New York
Yamakawa H, Fujii M (1973) Translational friction of wormlike chains. Macromolecules 6:407
Yamakawa H, Fujii M (1974) Intrinsic viscosity of wormlike chains. Determination of the shift factor. Macromolecules 7:128
Yamakawa H, Stockmayer WH (1972) Statistical mechanics of wormlike chains. Excluded volume effects. J Chem Phys 57:2843
Yanaki T, Norisuye T, Fujita H (1980) Triple helix of schizophyllum commune polysaccharide in dilute solution. Hydrodynamic properties in water. Macromolecules 13:1462
Zimm BH (1980) Chain molecule hydrodynamics by the Monte-Carlo method and the validity of the Kirkwood-Riseman approximation. Macromolecules 13:592

Part V
Applications of AUC: Biological Science

Chapter 15
Applications of Analytical Ultracentrifugation to Membrane Proteins

Karen G. Fleming

Abstract The analytical ultracentrifuge is well suited to address the questions of membrane protein molecular weights and interactions in detergent micelle solutions. This chapter explains the theoretical fundamentals underlying the analysis of the buoyant molecular weights of protein-detergent complexes. Strategies are presented to unambiguously isolate the protein molecular weights and association states for both sedimentation equilibrium and sedimentation velocity experiments. The data behavior and experimental tests for reversibly interacting proteins in micellar solutions are discussed, and the information to be gained from nonreversible distributions is explained. The advantages and disadvantages of both types of experiments as they apply to membrane proteins are discussed. Examples from the literature are presented throughout the chapter to highlight specific applications, strategies, or analysis techniques.

Keywords Membrane protein • Detergent micelles • Buoyant molecular weight • Sedimentation equilibrium • Sedimentation velocity • Density matching • Protein interactions

15.1 Introduction

Membrane proteins form specific complexes within the membrane that are essential for stabilizing their native folds or for creating transient oligomers with specialized functional capabilities. Understanding the fundamental units of membrane protein folds requires knowledge about the oligomeric states that are populated. However, because it is impossible to isolate functional membrane proteins in the absence of membrane-like cosolvents, measuring the masses of membrane protein complexes can be problematic using many traditional methods. For example, Förster resonance energy transfer experiments are often employed to evaluate membrane protein interactions, but these methods contain no information about the masses of any

K.G. Fleming (✉)
T. C. Jenkins Department of Biophysics, Johns Hopkins University, 3400 N. Charles Street, Baltimore, MD 21218, USA
e-mail: Karen.Fleming@jhu.edu

S. Uchiyama et al. (eds.), *Analytical Ultracentrifugation*,
DOI 10.1007/978-4-431-55985-6_15

complexes that are formed and thus require assumptions about stoichiometry in the data analysis. Mass determination is where the analytical ultracentrifuge excels, and thus this technique has a unique role to play in analyzing membrane protein complexes.

Both sedimentation equilibrium and sedimentation velocity have been employed to analyze membrane protein complexes. The choice of which method to use depends on the biological question at hand and the sample requirements. Although equilibrium constants can in principle be extracted from sedimentation velocity data, the equilibrium experiment is a more straightforward technique for measuring this quantity as long as the detergent employed will allow the density matching strategy to be implemented. On the other hand, for an analysis seeking to address the question of how homogeneous a sample is, sedimentation velocity can be more revealing and is certainly faster than sedimentation equilibrium. Sedimentation velocity is also far superior to the traditional method of gel filtration chromatography because the information content is greater and the centrifuge experiments require no matrix, so it eliminates any ambiguities arising from matrix interactions. This chapter will provide an overview of some considerations for setting up the equilibrium experiment and for analyzing the data. In addition to the discussion here, the reader is directed to several tutorials and reviews that emphasize additional aspects of membrane protein analysis using the analytical ultracentrifuge (Fleming 1998, 2008; Ebel 2011).

15.2 Theoretical Considerations for Mass Evaluation Using the Analytical Ultracentrifuge

The experimentally determined quantity in any analytical ultracentrifugation experiment is the buoyant molecular weight, M_B (Casassa and Eisenberg 1964), defined as

$$M_B = M_P\left(1 - \phi'\rho\right) \tag{15.1}$$

where M_B is the product of M_P, the protein molecular weight, and the buoyancy term, $(1 - \phi'\rho)$, which contains terms for the partial specific volume of the complex, ϕ' (ml g^{-1}), and from the solvent density, ρ (g ml^{-1}). The key concept to recognize when strategizing an experimental setup and when analyzing data involving membrane proteins is that all bound components contribute to the experimentally determined buoyant molecular weight. Early work by Reynolds and Tanford showed that the buoyant molecular weight could be approximated as a sum of component buoyant molecular weight terms (Reynolds and Tanford 1976). This allows the protein contribution to be isolated from all others:

$$M_B = M_P\left(1 - \overline{v}_P\rho\right) + n_{CS}M_{CS}\left(1 - \overline{v}_{CS}\rho\right) + n_{H_2O}M_{H_2O}\left(1 - \overline{v}_{H_2O}\rho\right) \tag{15.2}$$

where v_i, n_i, and M_i indicate the partial specific volumes (ml g^{-1}), numbers, and molecular weights of any bound components, the subscript *CS* indicates bound detergent or lipid cosolvents, and H_2O indicates bound water. Note that two different cosolvent terms would be required in cases where both detergent and lipids are components of the membrane protein complex. Another familiar form of this equation is the following:

$$M_B = M_P\left[(1 - \overline{v}_P\rho) + \delta_{CS}\left(1 - \overline{v}_{CS}\rho\right) + \delta_{H_2O}\left(1 - \overline{v}_{H_2O}\rho\right)\right] \tag{15.3}$$

where δ is a dimensionless number that indicates the ratio of grams of a particular component bound per gram of protein.

Because the vast majority of analytical ultracentrifugation experiments are carried out in aqueous solution, the water term, $\delta_{H_2O}\left(1 - \overline{v}_{H_2O}\rho\right)$, is usually a very small value and is traditionally dropped from further consideration yielding the following general expression for the buoyant molecular weight of the membrane protein complex:

$$M_B = M_P\left[(1 - \overline{v}_P\rho) + \delta_{CS}\left(1 - \overline{v}_{CS}\rho\right)\right] \tag{15.4}$$

This expression for the buoyant molecular weight can be used for the analysis of both sedimentation velocity and sedimentation equilibrium experiments.

15.3 Sedimentation Equilibrium of Membrane Proteins

15.3.1 Density Matching Strategy

A particularly attractive strategy for determining the molecular weights of detergent- or lipid-solubilized membrane proteins is to carry out sedimentation equilibrium experiments under conditions where the cosolvent has minimal contribution to the buoyant molecular weight. This can be accomplished by adjusting the experimental conditions such that the buoyancy terms for all the bound cosolvents have a minimal, near-zero value, e.g.:

$$(1 - \overline{v}_{CS}\rho) \rightarrow 0 \tag{15.5}$$

Because the partial specific volume of the cosolvent is an intrinsic property of the molecule, it cannot be manipulated. However the solvent density *can* be altered. In many cases the buffer can be adjusted such that the product of $\overline{v}_{CS}\rho \cong 1$, which is generally referred to as "density matching." In this case, the effective density of the bound detergent is essentially equal to the solvent density. Density matching results in a near-zero value for the buoyancy term in Eqs. 15.2, 15.3, and 15.4 no matter how large the value of δ. This latter point is important because the amount of bound detergent is a quantity that can be large and difficult to independently measure and is frequently unknown.

The best way to adjust the solvent density is using heavy water because this strategy will also maintain a near-zero sum for the water buoyancy term, e.g.:

$$(1 - \overline{\nu}_{H_2O}\rho) \rightarrow 0 \tag{15.6}$$

In this case, Eq. 15.4 reduces to the simplified expression below:

$$M_B = M_P\,[(1 - \overline{\nu}_P\rho)] \tag{15.7}$$

Investigators in many different laboratories have used D_2O or $D_2{}^{18}O$ to achieve density matching. Some examples include the pioneering work of Tanford (Reynolds and Tanford 1976), cytochrome oxidase (Suarez et al. 1984), experiments on the M2 influenza channel and designed transmembrane proteins by DeGrado and coworkers (Gratkowski et al. 2001; Howard et al. 2002; Li et al. 2004) (Kochendoerfer et al. 1999), human plasma paraoxonase (HuPON1) (Josse et al. 2002), and analyses of transmembrane helix-helix interactions by Fleming and coworkers (Fleming 2000, 2002; Fleming and Engelman 2001; Doura et al. 2004; Fleming et al. 2004; Stanley et al. 2006, Burgess et al. 2008). Because the density of heavy water is 1.11 g ml^{-1}, density matching works best when the partial specific volume of the bound detergent is equal to or greater than 0.900 ml g^{-1}. Detergents that meet this criterion are Fos-Choline-12 (DPC) (Gratkowski et al. 2001; Howard et al. 2002; Li et al. 2004), 3-(N,N-dimethylmyristylammonio)propanesulfonate (C14SB) (Fleming et al. 2004), and the polyoxyethylenes $C_{12}E_8$ (Josse et al. 2002). Notable exceptions for density matching are bile salt detergents, *n*-dodecyl-β-D-maltopyranoside (DDM) and decylmaltoside (DM), which is unfortunate because the maltoside detergents are especially useful in structural studies of membrane proteins. Durshlag has assembled a very useful compilation of partial specific volumes for many different molecule types, including detergents (Durshlag 1986).

For completeness it is worth noting that it is also theoretically possible to match the amphiphile density using additives such as glycerol, sucrose, Nycodenz, or other solvent additives (Mayer et al. 1999; Lustig et al. 2000). However if these additives significantly alter the solvent density from that of pure water, then the hydration buoyancy term will be nonzero and need to be considered. Compounding this, preferential binding of the additives may need to be incorporated into the buoyant molecular weight, which will be difficult because one cannot usually independently measure the amount of bound glycerol, sucrose, etc. A third problem is that these compounds themselves may sediment and create a density gradient, which could require a more sophisticated treatment of the solvent density (ρ) term in analysis. For this reason, if it is at all possible, density matching using heavy water is the highly preferable approach.

15.3.2 *Procedure for Density Matching the Detergent in a Buffer of Interest*

When using either a new detergent or a detergent in a new buffer, it is helpful to experimentally determine the density match point for a detergent micelle solution at the same concentration(s) and in the presence of the buffer solutions and temperatures that will be subsequently employed for the membrane protein analysis. Even moving from a Tris-buffered to a phosphate-buffered solution, we have observed that the D_2O percentage required for density matching can shift, which would lead to unnecessary uncertainties in the subsequent experiments with the proteins.

To empirically determine the match point, prepare a series of samples in different H_2O/D_2O (v/v) percentages and containing all components except the membrane protein. If the partial specific volume of the detergent is unknown, good starting points would be samples in 100 % H_2O, 45 % D_20, 55 % H_2O; and a third sample in 90 % D_2O, 10 % H_2O. If the partial specific volume for the detergent micelles is known, the best match point estimate will be obtained by using the D_2O percentage that is predicted to match the detergent calculated as $\%D_2O = 100/\overline{v}_{CS}$ along with at least two additional $\%D_2O$ values that bracket the predicted match point on either side by ± 10 % D_2O. Subject these detergent-only samples to a sedimentation experiment to obtain the concentration gradients at equilibrium. The usage of interference optics for this experiment is preferred to avoid labeling of the micelles with a dye. To test for the possibility of compressibility in these solutions (and hence the need for a density gradient term), we have typically sedimented such samples at 50,000 rpm until equilibrium, which is a speed that is typically higher than that required for the sedimentation equilibrium experiment in the presence of the membrane protein. To mimic the conditions that will be subsequently used in the experiment with the protein, we use six-sector cells for these preliminary matching experiments. If the mixed H_2O/D_2O solutions are close to the density match point, the sedimentation profile at equilibrium will not have the shape of the normal exponential equilibrium data; rather, the distribution will be linear with slope near zero. A buffer density that is slightly too high (too much D_2O) will result in a linear sedimentation profile with a negative slope; and a density lower than the micelles (not enough D_2O) will result in a linear distribution with a positive slope. Figure 15.1 shows that a plot of these slopes as a function of $\%D_2O$ will also be linear in the region near the match point. The percentage of D_2O required to establish the match point will be given by the value of the x-intercept of this plot.

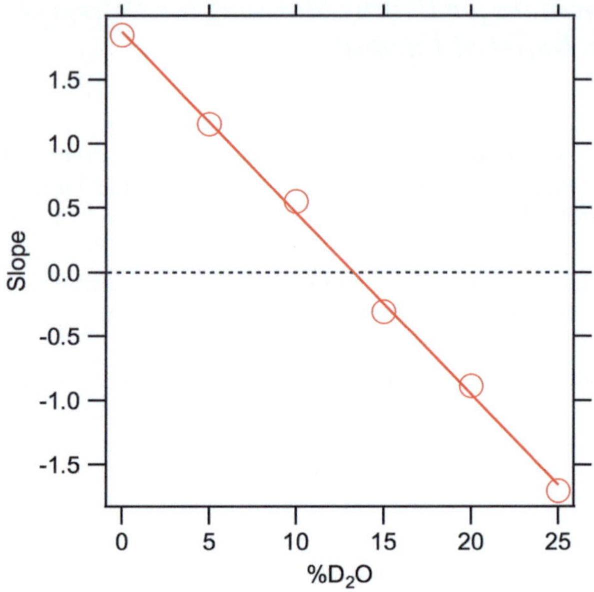

Fig. 15.1 Example of data collected in the density-matching procedure. The y-axis is the slope of the observed linear distribution from six different sedimentation equilibrium experiments (~24 h) of detergent micelles. The x-axis is the percentage of D_2O employed in each condition. The x-intercept is the percentage of D_2O required to match the micelle density under the buffer conditions used. Because buffer densities can be calculated using Sednterp, this experiment can also be used to measure the partial specific volume of the detergent; this value will be the reciprocal of the density value at the x-intercept

15.3.3 Sedimentation Equilibrium Data Collection and Analysis

Under density matching conditions, sedimentation equilibrium data collection and analysis can be carried out in a manner identical to that for soluble proteins (Fleming 2008). Unless the molecular weights are on the order of small peptides, standard six-sector charcoal-filled, epon cells with 1.2 cm optical path and either quartz or sapphire windows can be employed that use typical sample and reference volumes of 110 and 112 μl, respectively. To obtain better data for transmembrane peptides, it may be prudent to use the larger sample volumes (~420 μl) enabled by two-sector cells. This larger volume reduces the throughput, requires longer times to equilibrium (up to 16x), but yields a higher quality data set for analysis. Because the bound detergent does not contribute to the buoyant molecular weight, its contribution can be ignored. Similarly, the density matching procedure also ensures that the bulk, unbound detergent will undergo minimal sedimentation (or flotation) during the experiment, which makes it unlikely for the formation of a density gradient in the solution. Accordingly a constant value for the buffer density can be used and the speeds to be chosen for sedimentation equilibrium would be those that are optimal for that of the membrane protein mass alone. This latter point can be verified experimentally by the experiments used to find the density matching condition, as described in Sect. 15.3.1. Despite the higher viscosity of a buffer containing detergent micelles, the time to reach sedimentation equilibrium is not noticeably longer; however the attainment of the equilibrium condition should always be monitored using a program such as WinMatch. We typically collect an absorbance scan once per hour to enable plotting the approach to equilibrium function. As with soluble proteins, for a completely unknown sample, a rule of thumb often used is to start with a rotor speed with $\sigma = 1$, as defined by Yphantis (1964):

$$\sigma = \frac{M_P\,(1 - \overline{v}_P\rho)\,\omega^2}{RT} \tag{15.8}$$

where ω is the angular velocity in radians per second and the other variables are as defined above. This quantity is easily calculated using the amino acid composition and buffer components (ignoring detergent) using the program Sednterp (Laue et al. 1992). Of course, until a preliminary characterization of the system is complete, the user must make an assumption about the dominant oligomeric form to calculate σ for the early runs.

One additional correction to this term that must be taken into account is the incorporation of deuterons onto the protein through exchangeable hydrogens, a phenomenon that will increase the protein molecular weight. Summing the number of exchangeable protons and assuming a level of deuterium incorporation similar to that of the bulk percentage of D_2O can be used to calculate the increase in molecular weight.

Because detergents can be expensive and do not generally equilibrate well using dialysis, it is usually impractical to bring the reference buffer and sample to dialysis equilibrium. Excellent data can be obtained by carefully hand pipetting the samples. Employing absorbance optics (as compared to interference optics) for the detection of membrane protein distributions by sedimentation equilibrium further avoids the observation of any inequalities in buffer and background detergent concentrations. Any software program that can be used to analyze sedimentation equilibrium data of soluble proteins can also be used to analyze these data. Our group has typically used Nonlin for this purpose.

15.3.4 Analysis of a Reversibly Equilibrating Membrane Protein Interaction

The equilibrium position of an interacting membrane protein complex will depend on the concentration of the protein within its hydrophobic solvent and not on the concentration of the protein with respect to the bulk aqueous volume. Mole fraction (protein to micellar detergent) units have been adopted to describe this equilibrium constant (Fleming 2002). If a protein is equilibrating within a detergent micelle solution, increasing the number of micelles will increase the hydrophobic volume available to the membrane protein, which should favor membrane protein dissociation. Conversely, decreasing the number of micelles reduces the hydrophobic volume and should favor membrane protein association. This assumes that re-equilibration of the membrane protein within the entirety of the hydrophobic solvent is kinetically possible within the time frame of the experiment, which may not always be true. Importantly, what this means for sedimentation equilibrium experiments of suspected interacting membrane protein complexes is that a single experiment conducted in one detergent concentration is insufficient for establishing that the membrane protein is in equilibrium with the entirety of the hydrophobic solvent. Mechanistically, it is not entirely clear how micelles can exchange their protein contents or even what governs the kinetics of this process. Although individual detergent monomers can rapidly and reversibly equilibrate between their (monomeric) soluble forms and their aggregated states characterized by their participation in micelles, it is unlikely that membrane proteins can do the same. Rather it seems more likely that micelles must fuse together and break apart in order for protein contents to mix. To envision a scenario in which a membrane protein can equilibrate throughout the entirety of the micelles present in solution, Tanford suggested the simplification that the hydrophobic solvent available to a membrane protein should be thermodynamically referred to as a distinct phase defined by the amount of detergent in the micellar phase (Tanford 1980).

For a dimeric assembly reaction of a membrane protein within a hydrophobic solvent phase, Fleming showed that the equation to describe equilibrating behavior of membrane proteins within this micellar phase can be written as (Fleming 2002)

$$\Delta G_{app} = \Delta G_x^o + RTln[micellar\ Det]_w \quad (15.9)$$

where ΔG_{app} is the apparent free energy change (kcal mol^{-1}) obtained directly from the sedimentation equilibrium experiment and equals

$$\Delta G_{app} = -RTln\ K_{app} \quad (15.10)$$

where K_{app} equals the experimentally observed association equilibrium constant in bulk M^{-1} units with respect to the total aqueous volume, ΔG_x^o is the mole fraction standard state free energy change for the membrane protein interaction, R is the universal gas constant, T is the temperature in degrees Kelvin, and $[micellar\ Det]_w$ is the concentration of detergent in the micelle phase expressed in molar units referenced to the aqueous scale. This is calculated as

$$[micellar\ Det]_w = [Total\ Det]_w - cmc \quad (15.11)$$

where $[Total\ Det]_w$ is the total detergent concentration expressed as mole detergent per unit of total aqueous volume and cmc is the critical micelle concentration in

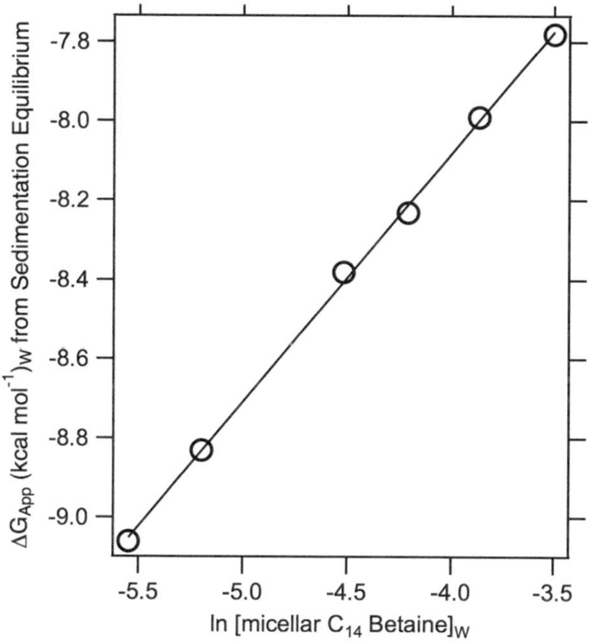

Fig. 15.2 Example of how the apparent free energy change (in kcal M^{-1})$_w$ units varies with the total concentration of micellar detergent. Each ΔG_{app} value was determined by global fitting of nine data sets in a sedimentation equilibrium experiment. Each sedimentation equilibrium experiment filled a six-sector cell with three protein concentrations (OD = 0.9, 0.6, and 0.3 using the bulk aqueous scale) in a background of a constant detergent concentration using the detergent solutions in the reference channels. Three speeds were collected for a total of nine data sets generated for each experiment. Using an eight-hole rotor, data for seven different detergent concentrations can be easily collected. The intercept of the linear fit to these data equals −5.8 kcal mol^{-1} with a slope equal to 0.6 (close to 0.592 kcal mol^{-1}, the value of *RT* at 25°C), demonstrating that the dimerization reaction is behaving as an ideal dilute solution under these conditions

aqueous molar units for that detergent. In the ideal case where ideal mixing is occurring within the hydrophobic solvent, a plot of ΔG_{app} (in bulk molar units) as a function of $ln[micellar\ Det]_w$ should have slope and intercept values equal to *RT* and ΔG^o_x, respectively. Figure 15.2 shows an example of such data collected for the dimerization reaction of the transmembrane domain of glycophorin A in C14SB detergent micelle solutions (Fleming et al. 2004). These data highlight the fact that the apparent equilibrium constant will depend strongly on the concentration of detergent micelles, as it should.

Also shown in Fig. 15.3 is the fact that the mole fraction equilibrium position depends on the detergent identity (le Maire et al. 2000) because it can be clearly observed that the standard dimerization midpoint of glycophorin A transmembrane domains is thermodynamically favored in C_8E_5 as compared to C14SB. A later sedimentation equilibrium study by Dorwart and colleagues demonstrated that

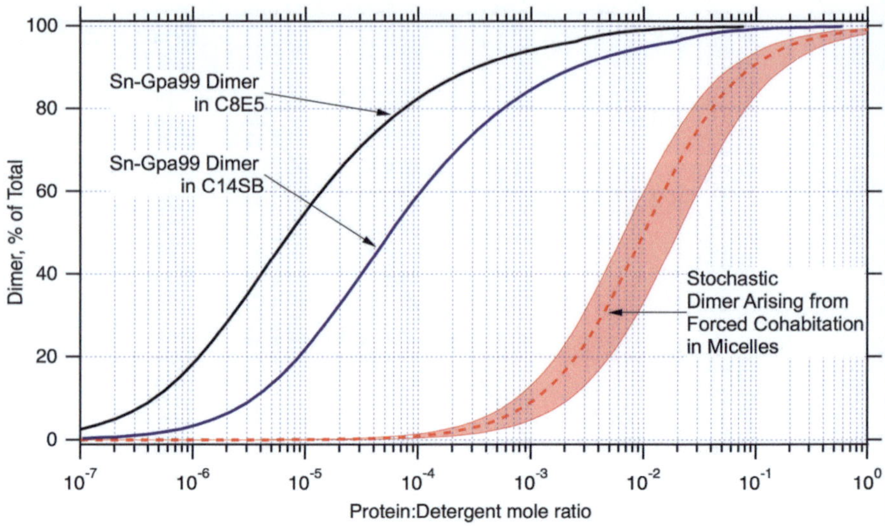

Fig. 15.3 Mole fraction distributions of membrane protein dimers within the micellar phase. The two left curves show the mole fraction distributions of the glycophorin A dimers measured in two different micelles. A comparison of these shows the energetic effect that a specific micelle environment can have on the equilibrium position of an interacting system. The dashed curve on the right represents the stochastic distribution of a dimeric molecular weight arising from "forced cohabitation" of protein dimers caused by limiting detergent micelles. The distributions are shown for micelle aggregation numbers ranging from 50 to 150. This type of forced cohabitation can happen irrespective of whether sedimentation equilibrium or sedimentation velocity experiments are used to evaluate the particle distribution

detergents can even alter the subunit stoichiometry as evidenced by observing either a tetrameric or pentameric assembly of a bacterial mechanosensitive channel depending on the solubilizing detergent (Dorwart et al. 2010). It is worth emphasizing that membrane proteins that assemble into higher-order oligomers like the bacterial channel can also be analyzed for reversibility using the standard formalism by deriving the appropriate equations.

15.3.5 Forced Cohabitation of Membrane Proteins

Even when the contribution of the detergent to the molecular weight can be density matched, "forced cohabitation" of membrane proteins can lead to erroneous results for the resultant, observed molecular weight of a membrane protein. It is within this concentration range that the Tanford "single thermodynamic phase" approximation breaks down and the particulate nature of detergent micelles becomes important in sedimentation analysis. One can think of forced cohabitation as a kind of crowding that arises where multiple membrane proteins occupy a single micelle (or nanodisk or lipid vesicle) even when there are no attractive forces between the proteins. In this situation, the proteins are mechanically linked together by the fact that they occupy the same detergent micelle (or nanodisk or lipid vesicle) and thus will sediment at an oligomeric molecular weight corresponding to the number of proteins trapped in the micelle together.

As with the equilibrating system discussed in Sect. 15.3.4, the key variable to consider in this situation is the detergent to protein molar ratio and its subsequent effect on the micelle (or particle) to protein molar ratio. Forced cohabitation arises when this ratio is too small and the cosolvent is limiting because there are not enough micelles for proteins to distribute themselves. Kobus developed a mathematical description of this phenomenon with respect to protein distributions within micelles (Kobus and Fleming 2005). This statistical approach predicts the probability of non-preferential, random protein occupancy within a cosolvent particle using the formalism of placing balls (proteins) in bins (micelles). For a particular partition function, P, that takes into account physically possible occupancy values an investigator can choose to enumerate, the probability that multiple proteins will occupy the same micelle is described by a Poisson distribution. Although this formalism is simplistic in nature, it predicts that dimer molecular weights of membrane proteins are 50 % populated when the micelle to protein molar ratio is equal. As shown in the "stochastic dimer" distribution in Fig. 15.3, increasing the micelle to protein molar ratio to 10:1 reduces this stochastic dimer population to 10 %.

15.3.6 Non-equilibrating Systems

Both the equilibrating situation in Sect. 15.3.4 and the forced cohabitation in Sect. 15.3.5 assume that a membrane protein can re-equilibrate within the entire micellar phase. If this situation is untrue, then the history of how the membrane protein became incorporated into the micelle (or nanodisk) will dictate the apparent oligomeric states observed in a sedimentation experiment. Testing for equilibration is therefore an essential part of measuring the molecular weight distributions. Nevertheless, sedimentation equilibrium experiments are still extremely useful in non-equilibrating cases because unlike any of the spectroscopic methods, the results will always be a molecular weight or molecular weight distribution. Moreover, in the absence of an interaction, sedimentation equilibrium yields a positive data result, the monomeric molecular weight. It should further be noted that forced cohabitation is not limited to sedimentation equilibrium experiments and will be evident in a velocity experiment where the stochastic accumulation of more monomers produces a distribution of non-equilibrium complexes.

15.4 Sedimentation Velocity Data Collection and Analysis

More recently, several investigators have used sedimentation velocity to analyze membrane protein samples. An excellent overview of this method is presented by Ebel (2011). A distinct advantage of sedimentation velocity experiments is that they take much less time to conduct than sedimentation equilibrium, which is important for proteins with limited stabilities. A disadvantage of sedimentation velocity (which is not restricted to membrane proteins) is that molecular interpretation of the data requires a greater number of assumptions and can therefore have higher uncertainties. The main issue is that the molecular weight information cannot be unambiguously distinguished from the shape, hydration, and bound cosolvent contributions. However, two approaches can be used to obtain a reasonable interpretation of the data that can in many cases distinguish between possible models.

15.5 Approaches to Disentangle Protein and Detergent Contributions to the Sedimentation Coefficient

The first approach is to make the assumption that the ratio of the experimental (f_{exp}) and theoretical anhydrous frictional coefficient (f_{min}) is limited to a small range of values. As argued by Ebel, (f_{exp}/f_{min}) must be greater than 1, and its value should reasonably be between 1.2 and 1.3 for a compact globular complex with average hydration properties (Ebel 2011). Ebel further proposes that the binding of detergent molecules to a membrane protein should, in many cases, render the particle more globular in shape, which has the effect of minimizing the anisotropic shape effect on the sedimentation coefficient. Defining the variables using the following equations, it can easily be observed that

$$R_S = \left(\frac{f_{\text{exp}}}{f_{\text{min}}}\right) R_{\text{min}} \tag{15.12}$$

because

$$\frac{R_S}{R_{\text{min}}} = \frac{f_{\text{exp}}}{f_{\text{min}}} \tag{15.13}$$

where the frictional coefficients and molecular radii have their usual mathematical descriptions:

$$f_{\text{min}} = 6\pi\eta R_{\text{min}} \tag{15.14}$$

$$R_{\min} = \sqrt[3]{\frac{3M_P\left(\overline{v}_P + \delta_{CS}\overline{v}_{CS}\right)}{4\pi N_A}} \tag{15.15}$$

$$f_{\exp} = 6\pi\eta R_S \tag{15.16}$$

and

$$R_S = \sqrt[3]{\frac{3M_P\left(\overline{v}_P + \delta_{CS}\overline{v}_{CS} + \delta_{H_2O}\overline{v}_{H_2O}\right)}{4\pi N_A}} \tag{15.17}$$

Substituting Eqs. 15.12, 15.14, and 15.16 into the Svedberg equation, the following relationship can be obtained:

$$\begin{aligned} s &= \frac{M_B}{N_A f_{\exp}} = \frac{M_B}{N_A 6\pi\eta R_S} = \frac{M_P\left[(1-\overline{v}_P\rho) + \delta_{CS}\left(1-\overline{v}_{CS}\rho\right)\right]}{N_A 6\pi\eta\left(\frac{f_{\exp}}{f_{\min}}\right)R_{\min}} \\ &= \frac{M_P\left[(1-\overline{v}_P\rho) + \delta_{CS}\left(1-\overline{v}_{CS}\rho\right)\right]}{N_A 6\pi\eta\left(\frac{f_{\exp}}{f_{\min}}\right)\left[\frac{3M_P(\overline{v}_P+\delta_{CS}\overline{v}_{CS})}{4\pi N_A}\right]^{1/3}} \end{aligned} \tag{15.18}$$

Of course, Eq. 15.18 will be valid as long as the buffer contains no additional additives such as sucrose or glycerol, which would require the hydration term be reintroduced in the numerator. The hydration term in the denominator is implicit in the frictional coefficient ratio and is always relevant.

While this equation looks daunting, a second strategy utilized in analyzing protein detergent complexes is to use both absorbance and interference optics to visualize the sedimenting particle. Because many detergents have low to no absorbance at 280 nm, using both optical detection systems in the same experiment allows the protein and detergent signals to be deconvoluted from each other: the absorbance optics system will detect the protein concentration distribution, and the interference optics will detect the protein-detergent particle distribution. Deconvolution of the interference signal into the protein and detergent components is accomplished using three pieces of information: (1) the protein concentration obtained from the absorbance signal and Beer's law; (2) the refractive increment of the protein, a known value $\left(\frac{\partial n}{\partial c}\right)_P = 0.187$ (Maezawa et al. 1983; Hayashi et al. 1989); and (3) the refractive increment of the detergent, which is measured in a prior, detergent-only experiment in the ultracentrifuge and can be determined from ΔJ, the fringe shift per unit weight concentration, c, of detergent as follows:

$$\Delta J = \left(\frac{\frac{\partial n}{\partial c}}{\lambda}\right)\left(\frac{c}{1000}\right) \tag{15.19}$$

where λ is the laser wavelength in the standard analytical ultracentrifuge interference optics (675 nm). Ebel has compiled $\frac{\partial n}{\partial c}$ values for a number of detergents (Ebel 2011).

Knowing the protein concentration and using the refractive increments for each species, the interference signal for a sedimenting protein detergent complex is a mass-weighted sum of the protein and detergent contributions. This results in determination of the value for δ_{CS}. Keeping in mind that $\overline{v}_{CS}$ can be independently determined in prior, detergent alone experiments similar to the type described in Sect. 15.3.2, the ratio of the frictional coefficients $\left(\frac{f_{\exp}}{f_{\min}}\right)$ is the only remaining unknown, and it can be reasonably approximated. Furthermore, if these sedimentation velocity experiments are conducted under "density matching" conditions, the numerator in Eq. 15.18 reduces to only the protein term. The final term not previously discussed is the solvent viscosity, which can be measured independently.

15.6 Sedimentation Velocity Characterization of Surfactants

Implicit in the discussion above is the analysis of surfactant solutions in the absence of proteins. Because many surfactants have partial specific volumes close to unity, they do not significantly redistribute under conditions of sedimentation equilibrium. Sedimentation velocity is therefore extremely useful in investigating the hydrodynamic properties of surfactants. In particular, partial specific volume values determined using sedimentation velocity agree well with independent values measured by densitometry (Salvay and Ebel 2006). In addition to this and the refractive increment values required for any of the analyses above, additional information that can be gained about surfactants from sedimentation velocity experiments include the critical micelle concentrations, aggregation numbers, and even global shape information (Ebel 2011) (Polidori et al. 2006; Breyton et al. 2009) (Sharma et al. 2008; Bazzacco et al. 2009). The usage of the interference optics has a distinct advantage in these experiments because the distributions are directly observed without the requirement for the dye-labeling of micelles, which could alter the cmc or aggregation number or both.

15.7 Examples of Membrane Protein Complexes Analyzed in the Ultracentrifuge

Several analyses of the role of sedimentation equilibrium applied to the evaluation of transmembrane helix-helix interactions were mentioned in Sect. 15.3.1. While the sedimentation equilibrium protocol has been feasible since the 1970s, the computer-aided data collection and the dual detection systems on the modern XL-I instruments have stimulated sedimentation velocity experiments more recently. In particular,

Ebel and coworkers have developed protocols to measure the sedimentation profiles of the Ca^{++}-ATPase (Salvay et al. 2007), the ExbB protein (Salvay et al. 2007), and the acridine resistance protein B, which forms homo-trimers (Ebel 2011). As expected for an interaction system, Josse observed a decrease in the average sedimentation coefficient of HuPON1 upon increasing the detergent concentration, indicating dissociation of the enzyme within the micellar phase (Josse et al. 2002). Moreover, even when the detergent has a density less than the buffer – and will thus float in the centrifugal field – Nury et al. have shown that the membrane protein dispersity and molecular weights can be established with confidence (Nury et al. 2008).

The authors of these studies point out that these sedimentation velocity experiments take relatively little time and provided an assessment of homogeneity unmatched by gel filtration chromatography. In addition, matrix interactions are not a concern in the centrifuge experiments because they are conducted in solution. An interesting feature of these studies was the usage of DDM; because it cannot easily be density matched, this detergent has not been widely used in sedimentation equilibrium experiments, yet these velocity studies show that membrane proteins can readily be characterized in these micelles. Despite the fact that DDM will sediment under these conditions, causing a change in the detergent concentration, Ebel has argued that this is not a big problem because the protein-detergent complex sediments faster due to the higher protein density (Ebel 2011). Thus, the protein detergent complex does not experience the changing detergent concentration. Because these experiments also reveal the amount of detergent bound per protein, they offer a unique view into the structure of the membrane protein detergent complex.

Acknowledgments This work was funded by grants from the NSF (MCB 1412108) and the NIH (R01 GM079440).

References

Bazzacco P, Sharma KS, Durand G, Giusti F, Ebel C, Popot JL, Pucci B (2009) Trapping and stabilization of integral membrane proteins by hydrophobically grafted glucose-based telomers. Biomacromolecules 10:3317–3326

Breyton C, Gabel F, Abla M, Pierre Y, Lebaupain F, Durand G, Popot JL, Ebel C, Pucci B (2009) Micellar and biochemical properties of (hemi)fluorinated surfactants are controlled by the size of the polar head. Biophys J 97:1077–1086

Burgess NK, Stanley AM, Fleming KG (2008) Determination of membrane protein molecular weights and association equilibrium constants using sedimentation equilibrium and sedimentation velocity. Methods Cell Biol 84:181–211

Casassa EF, Eisenberg H (1964) Thermodynamic analysis of multicomponent systems. Adv Protein Chem 19:287–395

Dorwart MR, Wray R, Brautigam CA, Jiang Y, Blount P (2010) S. aureus MscL is a pentamer in vivo but of variable stoichiometries in vitro: implications for detergent-solubilized membrane proteins. PLoS Biol 8:e1000555

Doura AK, Kobus FJ, Dubrovsky L, Hibbard E, Fleming KG (2004) Sequence context modulates the stability of a GxxxG mediated transmembrane helix-helix dimer. J Mol Biol 341:991–998
Durshlag H (1986) Specific volumes of biological macromolecules and some other molecules of biological interest. Springer, Berlin
Ebel C (2011) Sedimentation velocity to characterize surfactants and solubilized membrane proteins. Methods 54:56–66
Fleming KG (1998) Measuring transmembrane α-helix energies using analytical ultracentrifugation. Springer, New York
Fleming KG (2000) Probing the stability of helical membrane proteins. Methods Enzymol 323:63–77
Fleming KG (2002) Standardizing the free energy change of transmembrane helix-helix interactions. J Mol Biol 323:563–571
Fleming KG (2008) Determination of membrane protein molecular weight using sedimentation equilibrium analytical ultracentrifugation. Curr Protoc Protein Sci Chapter 7: Unit 7 12 1–7 12 13
Fleming KG, Engelman DM (2001) Specificity in transmembrane helix-helix interactions defines a hierarchy of stability for sequence variants. Proc Natl Acad Sci U S A 98:14340–14344
Fleming KG, Ren CC, Doura AK, Kobus FJ, Eisley ME, Stanley AM (2004) Thermodynamics of glycophorin A transmembrane helix-helix association in C14 betaine micelles. Biophys Chem 108:43–49
Gratkowski H, Lear JD, DeGrado WF (2001) Polar side chains drive the association of model transmembrane peptides. Proc Natl Acad Sci U S A 98:880–885
Hayashi Y, Matsui H, Takagi T (1989) Membrane protein molecular weight determined by low-angle laser light-scattering photometry coupled with high-performance gel chromatography. Methods Enzymol 172:514–528
Howard KP, Lear JD, DeGrado WF (2002) Sequence determinants of the energetics of folding of a transmembrane four-helix-bundle protein. Proc Natl Acad Sci U S A 99:8568–8572
Josse D, Ebel C, Stroebel D, Fontaine A, Borges F, Echalier A, Baud D, Renault F, Le Maire M, Chabrieres E, Masson P (2002) Oligomeric states of the detergent-solubilized human serum paraoxonase (PON1). J Biol Chem 277:33386–33397
Kobus FJ, Fleming KG (2005) The GxxxG-containing transmembrane domain of the CCK4 oncogene does not encode preferential self-interactions. Biochemistry 44:1464–1470
Kochendoerfer GG, Salom D, Lear JD, Wilk-Orescan R, Kent SB, DeGrado WF (1999) Total chemical synthesis of the integral membrane protein influenza A virus M2: role of its C-terminal domain in tetramer assembly. Biochemistry 38:11905–11913
Laue TM, Shah BD, Ridgeway TM, Pelletier SL (1992) Computer-aided interpretation of analytical sedimentation data for proteins. Royal Society of Chemistry, Cambridge
le Maire M, Champeil P, Møller JV (2000) Interaction of membrane proteins and lipids with solubilizing detergents. Biochim Biophys Acta 1508:86–111
Li R, Gorelik R, Nanda V, Law PB, Lear JD, DeGrado WF, Bennett JS (2004) Dimerization of the transmembrane domain of Integrin alphaIIb subunit in cell membranes. J Biol Chem 279:26666–26673
Lustig A, Engel A, Tsiotis G, Landau EM, Baschong W (2000) Molecular weight determination of membrane proteins by sedimentation equilibrium at the sucrose or Nycodenz-adjusted density of the hydrated detergent micelle. Biochim Biophys Acta 1464:199–206
Maezawa S, Hayashi Y, Nakae T, Ishii J, Kameyama K, Takagi T (1983) Determination of molecular weight of membrane proteins by the use of low-angle laser light scattering combined with high-performance gel chromatography in the presence of a non-ionic surfactant. Biochim Biophys Acta 747:291–297
Mayer G, Ludwig B, Muller HW, van den Broek JA, Friesen RHE, Schubert D (1999) Studying membrane proteins in detergent solution by analytical ultracentrifugation: different methods for density matching. Anal Ultracentrifugation V 113:176–181

Nury H, Manon F, Arnou B, le Maire M, Pebay-Peyroula E, Ebel C (2008) Mitochondrial bovine ADP/ATP carrier in detergent is predominantly monomeric but also forms multimeric species. Biogeosciences 47:12319–12331

Polidori A, Presset M, Lebaupain F, Ameduri B, Popot JL, Breyton C, Pucci B (2006) Fluorinated and hemifluorinated surfactants derived from maltose: synthesis and application to handling membrane proteins in aqueous solution. Bioorg Med Chem Lett 16:5827–5831

Reynolds JA, Tanford C (1976) Determination of molecular weight of the protein moiety in protein-detergent complexes without direct knowledge of detergent binding. Proc Natl Acad Sci U S A 73:4467–4470

Salvay AG, Ebel C (2006) Analytical ultracentrifuge for the characterization of detergent in solution. Prog Coll Pol Sci S 131:74–82

Salvay AG, Santamaria M, le Maire M, Ebel C (2007) Analytical ultracentrifugation sedimentation velocity for the characterization of detergent-solubilized membrane proteins Ca++ ATPase and ExbB. J Biol Phys 33:399–419

Sharma KS, Durand G, Giusti F, Olivier B, Fabiano AS, Bazzacco P, Dahmane T, Ebel C, Popot JL, Pucci B (2008) Glucose-based amphiphilic telomers designed to keep membrane proteins soluble in aqueous solutions: synthesis and physicochemical characterization. Langmuir 24:13581–13590

Stanley AM, Chauwang P, Hendrickson TL, Fleming KG (2006) Energetics of Outer Membrane Phospholipase A (OMPLA) dimerization. J Mol Biol 358:120–131

Suarez MD, Revzin A, Narlock R, Kempner ES, Thompson DA, Ferguson-Miller S (1984) The functional and physical form of mammalian cytochrome c oxidase determined by gel filtration, radiation inactivation, and sedimentation equilibrium analysis. J Biol Chem 259:13791–13799

Tanford C (1980) The hydrophobic effect Chapter VI: Micelles Krieger Publishing Company, Malabar, Florida

Yphantis DA (1964) Equilibrium ultracentrifugation of dilute solutions. Biochemistry 3:297–317

Chapter 16
Protein-Ligand Interactions

Shane E. Gordon and Matthew A. Perugini

Abstract This chapter will review the use of analytical ultracentrifugation for measuring protein-ligand interactions. This class of interactions is important in biology, biomedicine and industry and includes a diverse range of ligands from small molecules, such as drugs and enzyme substrates, through to large macromolecular complexes, e.g. lipid emulsions. Given its experimental diversity and sample flexibility, the analytical ultracentrifuge is a premier platform for detecting and quantitating protein-ligand interactions. This will be discussed herein by reviewing recent studies focused on protein-drug, enzyme-substrate and apolipoprotein-lipid interactions.

Keywords Analytical ultracentrifuge • Apolipoprotein E • Biotin protein ligase • Dihydrodipicolinate synthase • Drug • Flotation • Interaction • Lipid • Lipoprotein • Sedimentation

16.1 Introduction

The interaction of proteins with ligands underpins biomolecular communication in living systems in health and disease. Examples include protein-peptide, protein-DNA, protein-RNA, protein-drug, protein-metabolite and protein-lipid interactions. The detection, identification and quantification of such interactions are of interest to the fields of biology, nanotechnology and biomedicine. Of equal importance is the application to the biotechnology and pharmaceutical industries. Indeed, the quantitation of protein-drug interactions is critical for identifying compound 'hits' in the early stages of drug discovery, as well as potent 'lead compounds' required for later stage preclinical and clinical trials (Hughes et al. 2011). Similarly, the rapidly advancing field of industrial enzymology relies on the detection and quantification of protein-ligand interactions essential for the manufacturing of commercially

S.E. Gordon • M.A. Perugini (✉)
Department of Biochemistry and Genetics, La Trobe Institute for Molecular Science, La Trobe University, Melbourne, VIC 3086, Australia
e-mail: M.Perugini@latrobe.edu.au

S. Uchiyama et al. (eds.), *Analytical Ultracentrifugation*,
DOI 10.1007/978-4-431-55985-6_16

Table 16.1 Comparison of common techniques used to measure protein-ligand interactions

Property	AUC	ITC	MST	SPR
Affinity (K_D)	++	+	+	+
	(pM−mM)	(pM−μM)	(pM−μM)	(pM−μM)
Binding kinetics (k_{on}/k_{off})	+	−	−	++
	(k_{off} <10^{-2} s^{-1})			(k_{off} < 10^{-2} s^{-1} k_{on} < 10^6 $M^{-1}s^{-1}$)
Stoichiometry	+	++	+	+
Enthalpy/entropy	+	++	+	+
	(indirectly)		(indirectly)	(indirectly)
Sample heterogeneity	++	−	−	−
Size and shape	++	−	−	−

Note: − denotes 'not obtainable', + represents 'measurable (but with limitations)', and ++ indicates 'measurable (with no significant limitations)'

important products (Lee et al. 2012). Accordingly, there is a need in both industry and academia to measure protein-ligand interactions with precision and accuracy.

Several techniques are commonly employed to quantitate protein-ligand interactions in industrial or academic laboratories. These include isothermal titration microcalorimetry (ITC) (Ladbury and Doyle 2004), surface plasmon resonance (SPR) (Kodoyianni 2011), microscale thermophoresis (MST) (Wienken et al. 2010) and analytical ultracentrifugation (AUC) (Stafford 2009). These techniques offer both complementary and synergistic outputs in terms of measurement of binding thermodynamics, kinetics and stoichiometry (Table 16.1). Although time consuming in terms of sample preparation, data acquisition and data analysis, the analytical ultracentrifuge provides the unique advantage of determining the degree of heterogeneity in an interacting system, as well as measurement of size and shape (Table 16.1). It also provides significant throughput given the advent of the 8-hole rotor, which can accommodate the simultaneous measurement of up to 21 different samples when using three-channel centrepiece cells. AUC is thus a robust, flexible and efficient method for quantitating protein-ligand interactions. To demonstrate this, the chapter will focus on the use of AUC for detecting and quantifying the interactions of proteins with (i) small molecules and (ii) large macromolecular lipid complexes.

16.2 Protein-Small Molecule Interactions

The first class of ligands to be discussed in this chapter is the 'small molecules'. These are defined as any biological or synthetic compound with a molecular mass less than 500 Da. Common examples observed in both industrial and academic settings are pharmaceutical compounds (i.e. drugs) and enzyme substrates. With a view to highlighting the applicability of AUC for quantitating interactions between proteins and these common small molecules, we will firstly describe the interaction of (i) drug-like compounds with biotin protein ligase and (ii) the substrate, pyruvate, with dihydrodipicolinate synthase.

16.2.1 Biotin Protein Ligase

Protein biotinylation, the post-translational and covalent attachment of the cofactor biotin (244 Da), is a key event required for maturation of several metabolic enzymes. In *Escherichia coli*, for example, biotinylation is crucial for the first and committed step in fatty acid biosynthesis, catalysed by the biotin-dependent enzyme acetyl-CoA carboxylase, where biotin operates as a carboxyl carrier (Pendini et al. 2008). This biotinylation is mediated by the ATP-dependent enzyme biotin protein ligase (BPL).

The reaction mechanism of BPL biotinylation is two-step (Bagautdinov et al. 2005). In the first step, the reaction intermediate, biotinyl-5′-AMP, is synthesised from biotin and ATP substrate. In the second step, deadenylation of biotinyl-5′-AMP is coupled to its transfer to the biotin carboxyl carrier protein (BCCP). This process is summarised in the reaction scheme:

$$\text{Biotin} \xrightleftharpoons{\text{ATP} \quad \text{PP}_i} \text{Biotinyl-5'-AMP} \xrightleftharpoons{\text{BCCP}} \text{BCCP:Biotin} + \text{AMP}$$

where PP_i represents the reaction by-product, pyrophosphate.

BPL dimerisation is tightly regulated by the availability of enzyme substrates: biotin and biotin domain substrates (e.g. BCCP) (Fig. 16.1) (for review, see Pendini et al. 2008). In the presence of both binding partners, biotin-bound BPL (holoBPL) forms a transient catalytic complex with BCCP, during which biotin transfer from BPL to BCCP is coupled to the liberation of AMP. Contrastingly, in the absence of the binding partner BCCP, BPL self-associates to form a homodimer (Pendini et al. 2013). The BPL homodimer specifically recognises and complexes with the biotin operator sequence, effecting repression of biotin biosynthesis (Pendini et al. 2013). Thus, the homodimer is the DNA-competent form of BPL. In *E. coli* and *Staphylococcus aureus*, homo- and hetero-dimerisation are mutually exclusive phenomena, competing for equivalent protein-protein interface surfaces (Weaver et al. 2001; Pendini et al. 2013). In this fashion, partitioning between transcriptional repressor and ligase functions of BPL is manifested through changes in oligomeric state. Intriguingly, X-ray crystal structures indicate that this rule does not hold true for the constitutively dimeric BPL from *Pyrococcus horikoshii*, indicating that other regulatory mechanisms may operate in this microbe (Bagautdinov et al. 2005).

Several small molecule inhibitors have recently been described with activity against *S. aureus* BPL (Soares da Costa et al. 2012). Derived by analogy to the enzyme substrate biotin, biotin acetylene was highlighted as a promising pre-drug

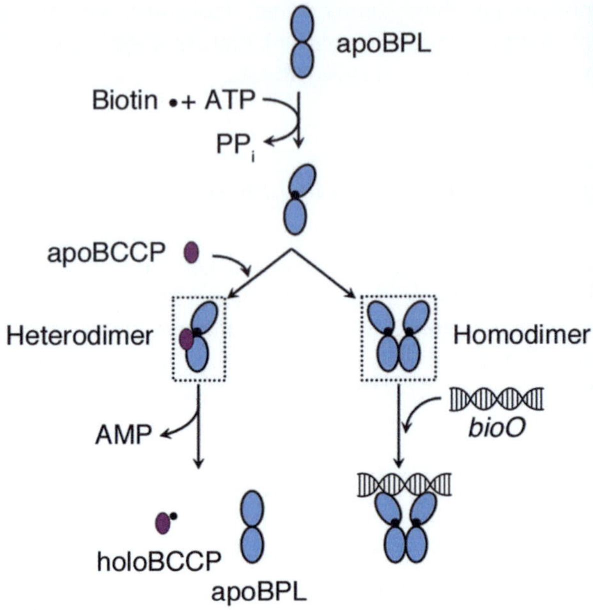

Fig. 16.1 Ligand-dependent reaction cycle of biotin protein ligase (BPL) (adapted from Zhao and Beckett (2008))

compound owing to its selectivity for bacterial over human isoforms of BPL. Recently, Soares da Costa et al. (2014) employed AUC to show that self-association of *S. aureus* BPL was substantially enhanced by the substrate derivative, biotin acetylene.

Data analysis was achieved by fitting sedimentation velocity data to the continuous size distribution model (Schuck 2000; Schuck et al. 2002). The distribution is determined using finite element solutions of the Lamm equation combined with size distribution analysis techniques by regularisation (Schuck 2000; Schuck et al. 2002). In brief, the experimentally observed sedimentation profiles are described as a superposition of the contributions of each subpopulation $c(M)$ of particles with masses between M and $M + dM$ by the integral equation:

$$a(r,t) \cong \int c(M) L(M,D,r,t)\, dM \tag{16.1}$$

where a(r,t) represents the experimentally observed signal at radius r and time t and L(M,D,r,t) denotes the sedimentation profile of an ideally sedimenting monodisperse species of molar mass M and diffusion coefficient D, calculated as the solution to the Lamm equation (Lamm 1929).

Analogously, the size distribution can be calculated as a sedimentation coefficient (s) distribution, $c(s)$, according to

$$a(r,t) \cong \int c(s) L(s,D,r,t)\, ds \tag{16.2}$$

The distributions for *S. aureus* BPL were calculated using maximum entropy regularisation, selecting the most parsimonious distribution within a predefined confidence level (P) for the quality of the fit. The diffusion coefficients (*D*) for all species were determined using the Stokes-Einstein and the Svedberg equations employing a frictional ratio (f/f_0) representing the ratio of the frictional coefficient to that of an anhydrous sphere, which was fitted as a variable. The size distributions were solved on a radial grid of 300 radial values between the meniscus and bottom, a confidence level of $P = 0.95$ and a resolution (N) of 200 sedimentation coefficients between 0 and 8 S, respectively (Soares da Costa et al. 2014).

The resulting *c*(*s*) distribution for *S. aureus* BPL reveals two species in the absence of ligand with standardised sedimentation coefficient ($s_{20,w}$) values of 2.5 S (consistent with the BPL monomer) and 3.8 S (consistent with the BPL dimer) (Fig. 16.2). The dimerisation of *S. aureus* BPL is concentration dependent, since increasing the initial concentration of *S. aureus* BPL shifted the *c*(*s*) distribution in favour of the larger 3.8 S component, whilst lower concentrations favoured the 2.5

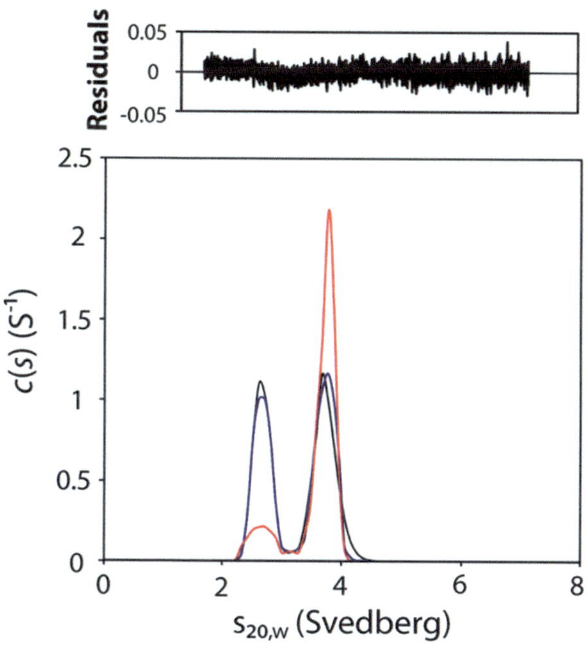

Fig. 16.2 Effects of substrate and substrate-analogue on the sedimentation coefficient distribution of *S. aureus* BPL. Sedimentation velocity data for *S. aureus* BPL (initial loading concentration of 26 μM) in the absence (*black line*) or presence of 100 μM biotin substrate (blue line) and 100 μM biotin acetylene (*red line*) were analysed by the continuous sedimentation coefficient distribution [*c*(*s*)] model using SEDFIT (Schuck 2000). *c*(*s*) (ordinate) is plotted as a function of the sedimentation coefficient, $s_{20,w}$ (abscissa), corrected for solvent temperature and viscosity. Residuals of best fit for a single *c*(*s*) distribution are depicted overhead (adapted from Soares da Costa et al. (2014))

S component. The fact that two distinct species are observed in the $c(s)$ distributions also indicates that the 2.5 S monomer and 3.8 S dimer reside in a slow equilibrium, given that the concentrations of both species do not change significantly during the timecourse of the experiment (Schuck 2000; Schuck et al. 2002). Strikingly, addition of moderate concentrations (100 μM) of biotin acetylene substantially shifted the $c(s)$ distribution in favour of the larger (3.8 S) component (Fig. 16.2). Quantification of this effect was subsequently achieved by performing sedimentation equilibrium experiments to determine the dimer-monomer dissociation constant ($K_D{}^{2\rightarrow 1}$) according to Eq. 16.3:

$$c_{tot} = c_{mon} e^{\left[\frac{\omega^2}{2RT} M_{mon}(1-\bar{v}\rho)\left(r^2-r_0^2\right)\right]} + \left(\frac{1}{K_D^{2\rightarrow 1}}\right)(c_{mon})^2 e^{\left[\frac{\omega^2}{RT} 2M_{mon}(1-\bar{v}\rho)\left(r^2-r_0^2\right)\right]} \tag{16.3}$$

where c_{tot} is the total concentration of the protein; c_{mon}, the concentration of the monomer at radius r_0; $(c_{mon})^2$, the concentration of dimer at radius r_0; ω, the angular velocity; R, the gas constant; T, the absolute temperature; M_{mon}, monomeric molecular weight; $\bar{v}$, partial specific volume of the solute; ρ, solvent density; and $K_D{}^{2\rightarrow 1}$, the dimerisation dissociation constant.

Performing global nonlinear least squares fit at multiple rotor speeds and *S. aureus* BPL enzyme concentrations revealed that the dimer-monomer dissociation constant ($K_D{}^{2\rightarrow 1}$) was 5.8-fold tighter in the presence of biotin acetylene ($K_D{}^{2\rightarrow 1} = 5$ μM) when expressed relative to the unliganded enzyme ($K_D{}^{2\rightarrow 1} = 29$ μM). This result suggests that biotin acetylene inhibition may be imparted by inducing self-association of *S. aureus* BPL into the DNA-competent and catalytically inactive homodimer. This represents an elegant example of a small molecule interaction that induces the self-association of the protein target to a higher-order 'inactive' form. A similar example will now be discussed, but where the small molecule ligand promotes self-association to a higher-order 'active' state.

16.2.2 Dihydrodipicolinate Synthase

The diaminopimelate (DAP) biosynthesis pathway – found exclusively in plants, bacteria and lower fungi – represents a promising avenue for the discovery of novel antibiotics (Dogovski et al. 2009, 2012). The penultimate product of this multistep pathway, *meso*-diaminopimelate (*meso*-DAP), constitutes a crucial cross-linking ingredient in the synthesis of the cell wall of Gram-negative bacteria such as *E. coli*. Similarly, the terminal product of the pathway, (*S*)-lysine, fulfils the equivalent role in the cell wall of Gram-positive bacteria, such as *S. aureus*, and is also a crucial ingredient in biosynthesis of housekeeping proteins and proteinaceous virulence factors (Soares da Costa et al. 2015). The essentiality of the DAP biosynthesis

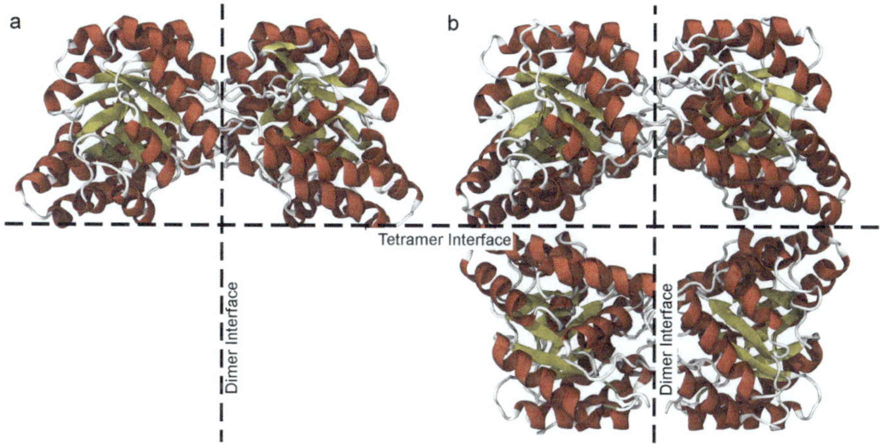

Fig. 16.3 Alternative quaternary architectures of DHDPS. Oligomerisation occurs when DHDPS monomers assemble to form homo-oligomers. Self-association occurs at the dimer interface (vertical broken lines) and/or the tetramer interface (horizontal broken line), forming either (**a**) homodimers or (**b**) homotetramers. Depicted in cartoon representation are native forms of (**a**) dimeric DHDPS from *S. aureus* (PDB ID 3DAQ) and (**b**) tetrameric DHDPS from *E coli* (PDB ID 1YXC) (Figures were generated using VMD (Humphrey et al. 1996))

pathway to these processes (Kobayashi et al. 2003; Dogovski et al. 2013) identifies enzymes in this pathway as attractive antimicrobial targets.

Dihydrodipicolinate synthase (DHDPS) (Fig. 16.3) catalyses the first and committed step in the DAP pathway, namely, the condensation of pyruvate and (*S*)-aspartate semi-aldehyde [(*S*)-ASA], to form the heterocyclic compound hydroxytetrahydrodipicolinic acid (HTPA) (Blickling et al. 1997). All characterised forms of DHDPS conform to an ordered ping-pong bi-substrate kinetic mechanism (Dobson et al. 2004). Under this scheme, binding of the first enzyme substrate, pyruvate, must precede recruitment of the second substrate (*S*)-ASA (Dobson et al. 2004; Muscroft-Taylor et al. 2010) as described by the reaction below:

$$\mathrm{E} \overset{\mathrm{PYR}}{\rightleftharpoons} \mathrm{E{:}PYR} \rightleftharpoons \mathrm{E'{:}H^+} \overset{\mathrm{H^+}}{\rightleftharpoons} \mathrm{E'} \overset{(S)\text{-}\mathrm{ASA}}{\rightleftharpoons} \mathrm{E'{:}}(S)\text{-}\mathrm{ASA} \overset{\mathrm{HTPA}}{\rightleftharpoons} \mathrm{E}$$

where PYR represents the first substrate, pyruvate, and E represents the DHDPS enzyme.

The reaction mechanism has been well characterised using a combination of X-ray crystallography and nuclear magnetic resonance spectroscopy (Blickling et al. 1997; Devenish et al. 2010). For the purposes of this section, a comprehensive discussion of the mechanistic details of catalysis is not warranted. The interested reader is referred to other works in this regard (Blickling et al. 1997; Hutton et al. 2007).

The solution form of DHDPS exists primarily as a homotetramer (Atkinson et al. 2009; Blagova et al. 2006; Pearce et al. 2006) (Fig. 16.3b). More recently, native dimeric variants of the enzyme have been described from bacteria, including the human pathogens, *Pseudomonas aeruginosa* (Kaur et al. 2011) and *S. aureus* (Burgess et al. 2008) (Fig. 16.3a). In all quaternary architectures, two DHDPS monomers self-associate to form a so-called tight-dimer at the *dimer interface* (Fig. 16.3) that captures a substantial proportion (10–13 %) of the monomeric solvent-inaccessible surface area (SISA) (Voss et al. 2010). In tetrameric enzymes, two tight dimers associate at the *tetramer interface* forming the canonical bacterial homotetramer (Fig. 16.3b).

Examination of X-ray structures of apo and substrate-bound DHDPS reveals differences in articulating surfaces between monomers. In tetrameric DHDPS from *Bacillus anthracis*, Voss et al. (2010) identified a 5.2% increase in SISA at the *tetramer interface* in the pyruvate-bound enzyme relative to the apo form, whilst comparatively minor changes are observed at the *dimer interface* (ΔSISA < 0.1 %). Furthermore, pyruvate enhances the thermal stability of tetrameric *B. anthracis* DHDPS, which mutational studies show is the most active oligomeric form of the enzyme (Voss et al. 2010). Likewise, pyruvate stabilises the active dimeric form of *S. aureus* ortholog (Burgess et al. 2008). These studies suggest that substrate-mediated stabilisation of the active quaternary state regulates catalytic function of bacterial DHDPS.

To further investigate this substrate-mediated phenomenon, sedimentation studies were performed in the analytical ultracentrifuge using the fluorescence detection system (MacGregor et al. 2004; Kroe and Laue 2009). The sedimentation velocity profile for apo *S. aureus* DHDPS (40,000 rpm) revealed a primarily single component with $s_{20,w}$ of 3.3 S, consistent with the monomer (Fig. 16.4a). However, the addition of pyruvate promoted self-association of *S. aureus* DHDPS to form a 4.0 S dimer (Fig. 16.4a). The effect of pyruvate on the dimerisation of *S. aureus* DHDPS was then assessed using sedimentation equilibrium experiments at multiple enzyme concentrations and multiple rotor speeds. As shown in Fig. 16.4b, the presence of pyruvate imparted a greater upward curvature to the concentration distribution of *S. aureus* DHDPS, consistent with a greater equilibrium proportion of the dimeric species. Quantification of this phenomenon was subsequently achieved by nonlinear regression analyses of the sedimentation equilibrium data using various models (i.e. monomer-dimer, monomer-trimer, monomer-tetramer and dimer-tetramer). These analyses showed that the best fit was obtained to a monomer-dimer equilibrium model consistent with sedimentation velocity studies, whereas fits to monomer-trimer, monomer-tetramer and dimer-tetramer models were disqualified given poorer fit statistics [i.e. significantly higher root mean square deviation (RMSD) and global reduced χ^2 values]. The $K_D{}^{2\rightarrow1}$ resulting from the best fit using Eq. 16.3 was calculated to be 33 nM for the apo enzyme and 1.6 nM for the pyruvate-bound form. This indicated that the substrate stabilises the *S. aureus* dimer by 20-fold relative to the apo enzyme. This was the first case demonstrating a role for pyruvate in stabilising the quaternary structure of DHDPS.

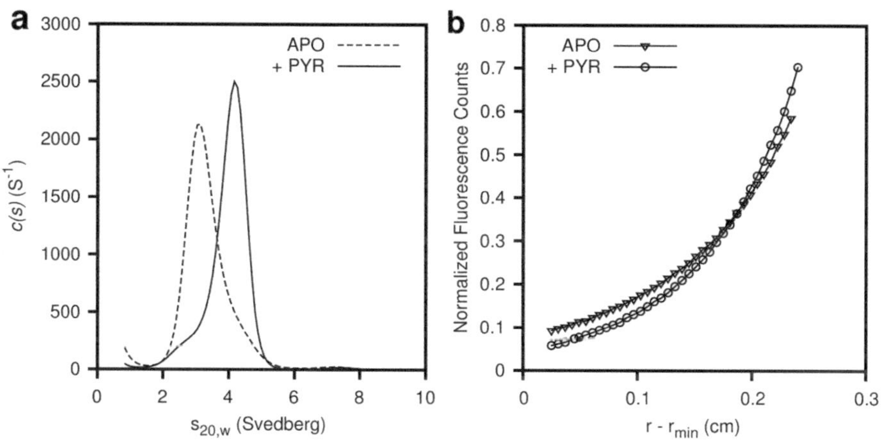

Fig. 16.4 Sedimentation velocity and equilibrium profiles of fluorescently labelled *S. aureus* DHDPS in the presence and absence of substrate. (**a**) Sedimentation coefficient distribution [$c(s)$] model of *S. aureus* DHDPS (initial loading concentration of 25 nM) with (*unbroken line*) and without (*broken line*) 2.0 mM pyruvate substrate. (**b**) Sedimentation equilibrium analyses. Radial distance ($r-r_{min}$) (abscissa) is plotted against the normalised fluorescence count (ordinate). Readings were taken from samples loaded at an initial *S. aureus* DHDPS concentration of 25 nM with (*open circles*) or without (*open triangles*) 2.0 mM pyruvate substrate using fluorescence-detected AUC (refer to Chap. 4). A rotor speed of 10,000 rpm was employed. Global nonlinear least-squares fits to a monomer-dimer equilibrium model are indicated by lines (adapted from Burgess et al. (2008))

Substrate-mediated stabilisation of *S. aureus* DHDPS dimerisation provided an important precedent; however it remained to be seen whether a similar phenomenon applied to tetrameric DHDPS enzymes. Voss and co-workers (2010) therefore assessed the effects of pyruvate on the tetramerisation of *B. anthracis* DHDPS, applying similar experimental strategies described above for *S. aureus* DHDPS (Fig. 16.5). For the apo form of *B. anthracis* DHDPS, $c(s)$ distribution analysis showed two components with $s_{20,w}$ values of 4.0 S and 6.5 S (Fig. 16.5c) typical of dimeric and tetrameric DHDPS forms, respectively (Perugini et al. 2005). Although not baseline resolved, the two distinct peaks resulting from $c(s)$ distribution analysis indicate that the dimer and tetramer reside in a slow equilibrium. This is supported by global sedimentation velocity and sedimentation equilibrium analysis that calculated the off-rate to be $10^{-5.3}$ s^{-1} (Voss et al. 2010). However, in the presence of pyruvate, the equilibrium shifts in favour of the tetrameric 6.5 S species. This effect can also be visualised using the highly complementary and model-independent enhanced van Holde-Weischet (vHW) method (Demeler and van Holde 2004; van Holde and Weischet 1978) (Fig. 16.5d). Finally, the stabilisation effect of pyruvate on the *tetramer interface* was quantified by sedimentation equilibrium experiments. The addition of pyruvate decreased the $K_D{}^{4\rightarrow2}$ from 1.9 μM for the apo form to 0.66 μM in the presence of the substrate, resulting in a 2.9-fold enhancement of tetramerisation.

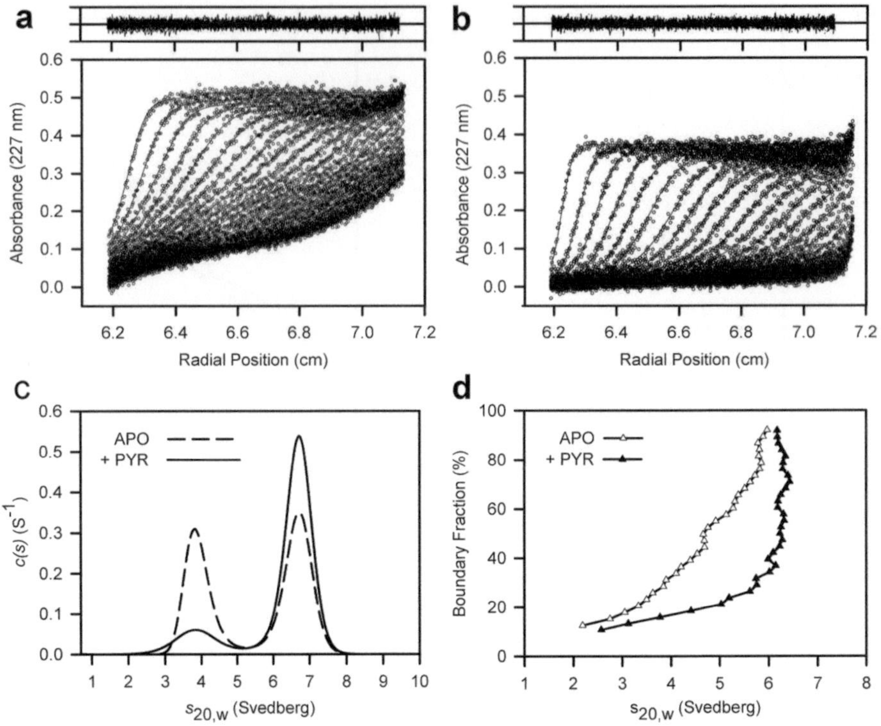

Fig. 16.5 Solution properties of *B. anthracis* DHDPS (initial loading concentration of 1.6 μM) in the presence and absence of 0.6 mM pyruvate substrate. Raw sedimentation velocity absorbance profiles for (**a**) apo and (**b**) pyruvate-bound *B. anthracis* DHDPS are plotted as a function of radial position (*open circles*). Scans were taken at a wavelength of 227 nm, allowing 600 s between successive scans. Best fits to a continuous size distribution [$c(s)$] model are overlaid as lines. Residuals of best fits are depicted above their respective plots. (**c**) Continuous size distributions [$c(s)$] of *B. anthracis* DHDPS with (*unbroken line*) and without (*broken line*) pyruvate substrate derived from data shown in panels (**a**) and (**b**). (**d**) Van Holde-Weischet analysis of apo (*open triangles*) and pyruvate-bound (*closed triangles*) *B. anthracis* DHDPS using the data shown in panels (**a**) and (**b**) (adapted from Voss et al. (2010))

Together, the results of AUC studies of *S. aureus* DHDPS and *B. anthracis* DHDPS provided essential quantitative information that aided in defining the structural basis of substrate-mediated self-association to more active higher oligomeric forms. This mechanism is currently being investigated for microbial intervention strategies to afford the discovery of new antimicrobial agents (Burgess et al. 2008; Dogovski et al. 2012, 2009; Voss et al. 2010).

16.3 Protein-Lipid Interactions

Protein-lipid interactions play an important role in membrane stabilisation, signalling, metabolism and the transport of lipid to cells and tissues. This section will focus on the latter example, mediated by the interactions of apolipoproteins with lipids to form large macromolecular complexes (lipoproteins) that circulate in plasma and cerebrospinal fluid.

16.3.1 Lipoprotein Biology and Apolipoprotein E

Lipoprotein particles are comprised of an outer monolayer of polar biomolecules, namely, apolipoproteins, phospholipid and cholesterol, encompassing a neutral core of triacylglycerol and cholesterol esters (Fig. 16.6) (Gotto et al. 1986; Pownall and Gotto Jr 1999). There are five major classes of lipoproteins that are differentiated on the basis of size, density, lipid composition and apolipoprotein content (Table 16.2). One of the more important apolipoproteins found circulating on chylomicrons, very low-density lipoproteins (VLDL), intermediate-density lipoproteins (IDL) and high-density lipoproteins (HDL), is apolipoprotein E (apoE). ApoE plays a critical role in mediating the uptake of extracellular lipid into cells via interactions with cell surface receptors, including the low-density lipoprotein (LDL) receptor (Hatters et al. 2006). Polymorphisms in the apoE gene give rise to three common isoforms that differ by single amino acid substitutions, namely, apoE2 (^{112}C^{158}C), apoE3 (^{112}C^{158}R) and apoE4 (^{112}R^{158}R) (Hatters et al. 2006; Perugini et al. 2000). The apoE3 isoform is associated with normolipidemia or a healthy state, whereas the apoE2 and apoE4 isoforms are linked to type III hyperlipoproteinemia and cardiovascular disease, respectively (Hatters et al. 2006). Interestingly, apoE4 is also associated with an increased risk of developing late-onset Alzheimer's disease (Corder et al. 1993; Rhinn et al. 2013). Characterizing and comparing the interactions of apoE isoforms with lipids is therefore of interest to biomedical research initiatives aimed at defining the role of apoE in common age-related diseases.

16.3.2 Models of Lipoprotein Particles

Traditionally, apoE-lipid interactions have been characterised using (i) native lipoproteins isolated from plasma (Fig. 16.6), (ii) synthetic lipid vesicles (Fig. 16.7a), (iii) phospholipid discoidal particles or discs (Fig. 16.7b), (iv) short-chain phospholipid micelles (Fig. 16.7c) or (v) synthetic lipid emulsions (Fig. 16.7d) (Derksen and Small 1989; Funahashi et al. 1989; Hatters et al. 2006; Perugini et al. 2000; Weisgraber 1994). To demonstrate the power of the analytical ultracentrifuge in measuring large ligand complexes, the remainder of this chapter will focus on

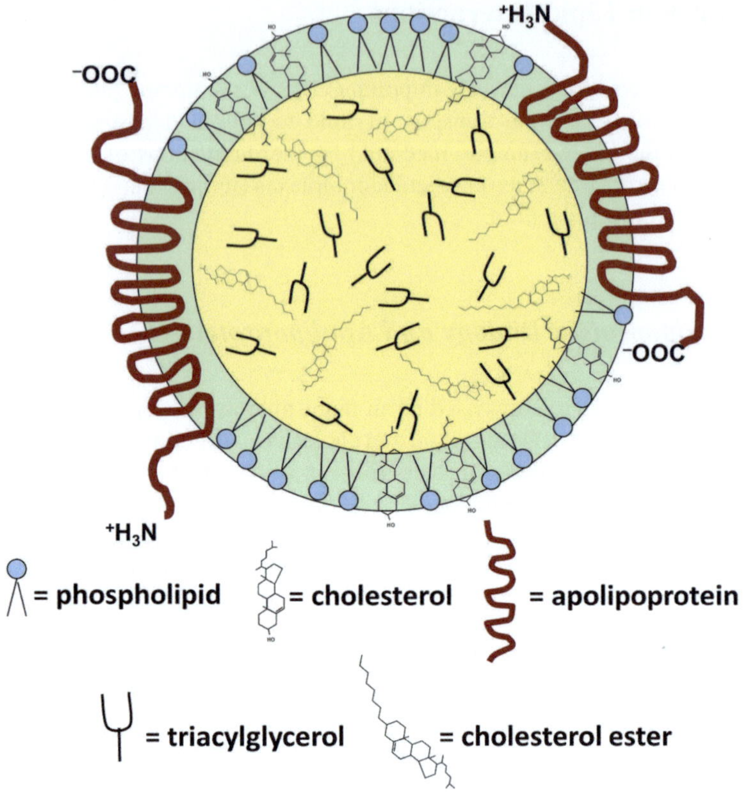

Fig. 16.6 Two-dimensional representation of a plasma lipoprotein particle. Five major classes of plasma lipoproteins exist, namely, chylomicrons, VLDL, IDL, LDL and HDL, differing in lipid and protein composition (Table 16.2). Generally, the outer monolayer (*green*) contains phospholipids, free cholesterol and apolipoproteins (refer to key above). The core of the lipoprotein (*yellow*) is primarily comprised of triacylglycerol and cholesterol esters (refer to key above). Diagram is not to scale

Table 16.2 Summary of lipoprotein classes, sizes and composition[a]

Lipoprotein class	Particle diameter (nm)	Density (g/ml)	Major lipids	Major apolipoproteins
Chylomicrons	80–1200	0.93	80–95 % TAG[b]	A, B, C
VLDL	30–80	0.93–1.006	45–65 % TAG[b]	B, C, E
			25 % cholesterol	
IDL	25–35	1.006–1.019	45 % cholesterol	B, E
LDL	18–25	1.019–1.063	70 % cholesterol	B
HDL	5–12	1.063–1.210	25 % cholesterol	A, C, E

[a]Adapted from Gotto et al. (1986) and Pownall and Gotto (1999)
[b]TAG, abbreviation for triacylglycerol

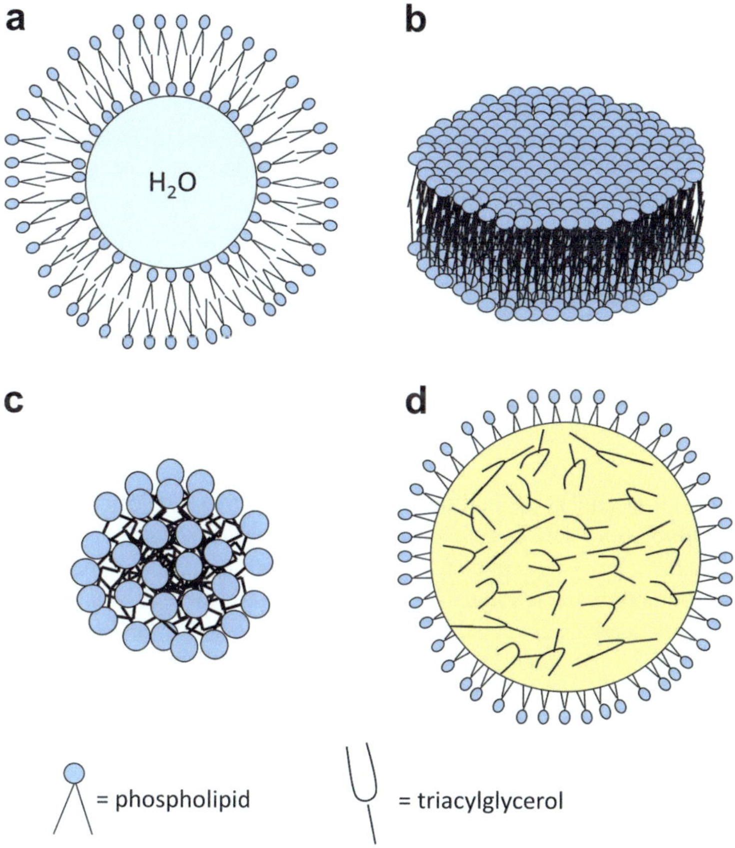

Fig. 16.7 Synthetic lipid particle models. *Top*: Diagrammatical representation of (**a**) a lipid bilayer vesicle, (**b**) a phospholipid disc, (**c**) a short-chain phospholipid micelle and (**d**) a phospholipid-triacylglycerol emulsion. The particles shown are not proportional in size. *Bottom*: Diagrammatical representation of a phospholipid and triacylglycerol used in the four structures above

studies employing AUC analyses of lipid emulsions in the absence and presence of apoE isoforms.

16.3.3 Flotation Velocity Analysis of Synthetic Lipid Emulsions

Compared to bilayers, lipid emulsions better emulate the structure and composition of native lipoproteins and can be synthesised from simple mixtures of polar and nonpolar lipids (MacPhee et al. 1997; Perugini et al. 2002). Emulsions consist of a hydrophobic core of triacylglycerol (and/or cholesterol ester) stabilised by a surface monolayer of phospholipid (and/or cholesterol) (Fig. 16.7d). Phospholipid vesicles are not included within the above definition given their bilayer structure and internal aqueous core (Fig. 16.7a). Likewise, phospholipid discs (Fig. 16.7b) and phospholipid micelles (Fig. 16.7c) are also excluded from this definition since they lack a core compartment comprised of neutral lipids.

Conveniently, lipid emulsions can also be fractionated efficiently by size using preparative sucrose gradient ultracentrifugation (MacPhee et al. 1997; Perugini et al. 2002). This yields emulsion particles similar in size to native lipoprotein particles (Table 16.2). However, given their low density, the fractionated emulsion particles comprised of phospholipid and triacylglycerol float (rather than sediment) in the gravitational field of the analytical ultracentrifuge (Fig. 16.8). Nevertheless, the size distribution of each fraction can be calculated as a flotation coefficient distribution, $c(s_f)$, according to Eq. 16.4:

$$a\left(r,t\right) \cong \int c\left(s_f\right) L\left(s_f, D, r, t\right) ds_f \tag{16.4}$$

One feature of boundary modelling with Eq. 16.4 is that it allows interconversion of the flotation coefficient distribution to a molar mass distribution via the Stokes-Einstein and Svedberg equations upon consideration of the size-dependent particle density of the polydisperse solutes (Perugini et al. 2002; Schuck et al. 2002). For the functional dependence between the density and molar mass of the fractionated emulsion particles, one can assume a spherical monolayer of phospholipid surrounding a core of triacylglycerol, which is supported by transmission electron microscopy (Fig. 16.9a). Consequently, the relationship between the densities of the particles as a function of particle radius ($\propto M^{1/3}$) using values for M and $\bar{v}$ reported by MacPhee et al. (1997) can be calculated from the ratio of phospholipid to triacylglycerol by assuming each phospholipid molecule occupies an area of 60 Å^2 in the monolayer surface and that each triacylglycerol molecule occupies a volume of 1610 Å^3 in the particle core (Drew et al. 1990; Miller and Small 1983). Linear regression least-squares analysis of these data yields the relationship:

$$\bar{v} = 1.042 + 0.0191 * \left[\left(M/10^8\right)^{1/3}\right] \tag{16.5}$$

Subsequently, the molar mass distribution of each emulsion fraction comprised of dimyristoylphosphatidylcholine (DMPC) and triolein (TO) can be calculated with incorporation of Eq. 16.5.

To prevent an ill-conditioned analysis when performing continuous size distribution calculations with many species, a regularisation technique can be included that selects the most parsimonious distribution of species that fits the data within a predetermined confidence limit. However, in contrast to the aforementioned sedimentation coefficient distributions of proteins, for which maximum entropy regularisation seemed advantageous because of its potential to produce sharp peaks for discrete mixtures (Perugini et al. 2000; Schuck 2000; Schuck et al. 2002), Tikhonov-Phillips regularisation with second derivative functionality is more useful for large lipid emulsion particles (Perugini et al. 2002). Furthermore, the continuous

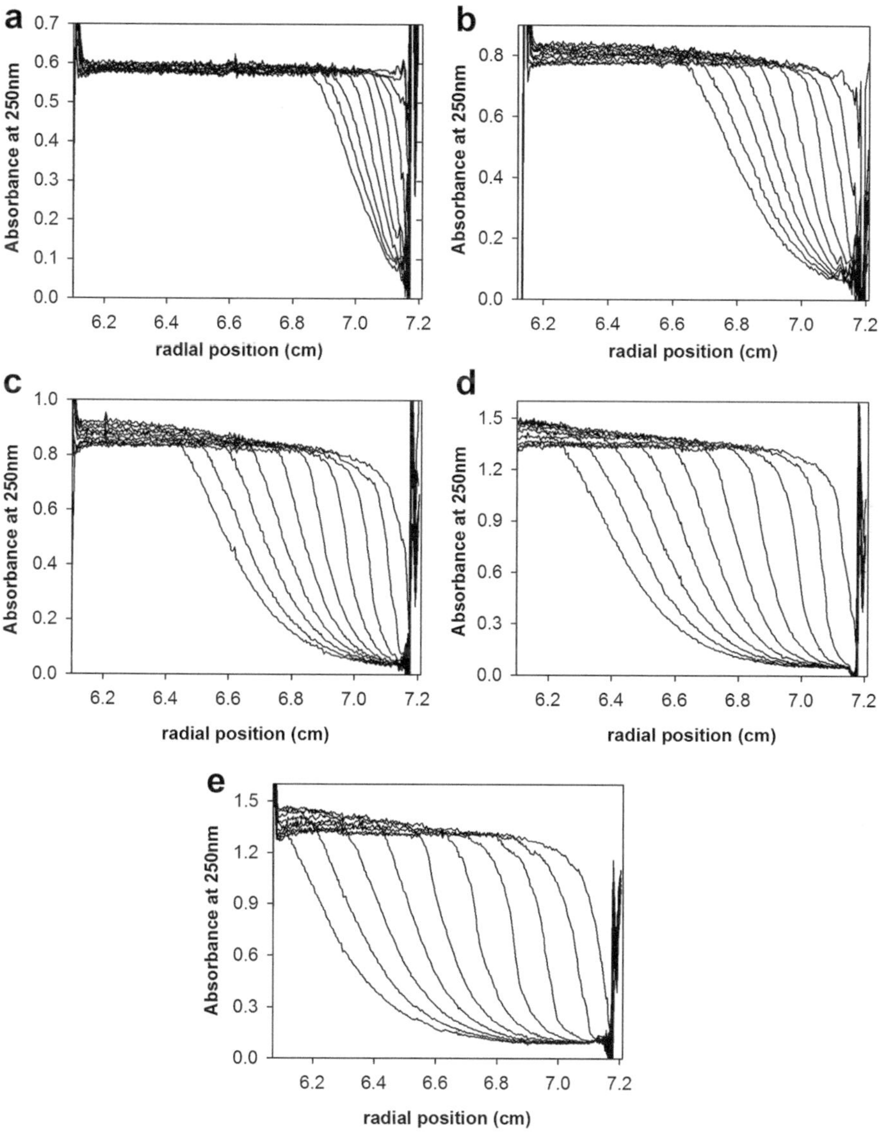

Fig. 16.8 Flotation velocity of DMPC/TO synthetic lipid emulsion fractions. Absorbance at 250 nm is plotted as a function of radial position for emulsion (**a**) Fraction 2, (**b**) Fraction 3, (**c**) Fraction 4, (**d**) Fraction 5 and (**e**) Fraction 6. Flotation velocity experiments were conducted at 20 °C and at a rotor speed of 5000 rpm in 50 mM sodium phosphate, pH 7.4. Data was collected at time intervals of 360 s with a step size of 0.003 cm, without averaging. For a simpler representation, every second scan is shown

size distributions of emulsions are observed to have higher robustness against small oscillations when using Tikhonov-Phillips regularisation, as compared to maximum

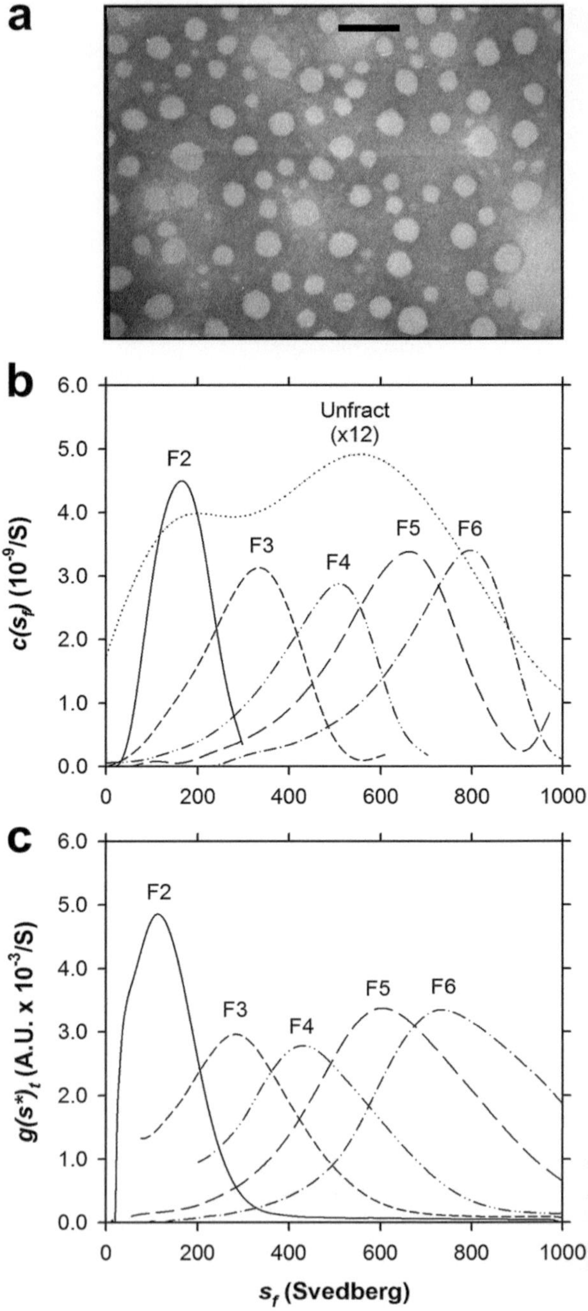

Fig. 16.9 Transmission electron microscopy and flotation velocity size distribution analyses of fractionated DMPC/TO emulsions. (**a**) Electron micrograph of unfractionated lipid emulsions comprised of dimyristoylphosphatidylcholine and triolein negatively stained with 1 % (w/v) sodium phosphotungstic acid, pH 7.4, on a formvar carbon-coated copper grid. The bar marker in panel a (*top*) corresponds to 200 nm. (**b**) Continuous flotation distribution, $c(s_f)$, is plotted as a function of flotation coefficient, s_f, for synthetic lipid emulsion fractions (F) 2, 3, 4, 5 and 6. Unfractionated (Unfract) lipid emulsion distribution has been scaled up a factor of 12. The analysis was performed as documented in Perugini et al. (2002). (**c**) $G(s^*)_t$ (Stafford 1992) is plotted as a function of flotation coefficient for fractions (F) 2, 3, 4, 5 and 6. The same data sets shown in Fig. 16.8 were employed for the distribution analyses in both panels b and c. The RMSD for all fits < 0.01

entropy regularisation (Perugini et al. 2002). Employing Tikhonov-Phillips regularisation with a confidence limit (P) of 0.95, a fixed anhydrous frictional ratio (f/f_0) of

Table 16.3 Hydrodynamic properties of DMPC/TO size-fractionated lipid emulsions[a]

Fraction #	TO:DMPC Molar ratio	s_f (S)	M (× 10^8 Da)	$\overline{v}$ (ml/g^{-1})	R_S (nm)
2	2.05	166	0.94	1.061	34
3	2.62	333	2.3	1.067	46
4	3.18	506	4.00	1.072	55
5	3.86	664	5.60	1.076	62
6	4.29	795	7.00	1.079	67

[a]Lipid emulsions comprised of DMPC and TO were fractionated and characterised by flotation velocity analysis (Figs. 16.8 and 16.9b). The symbols used are s_f, flotation coefficient taken from the ordinate maximum of the best-fit $c(s_f)$ distribution (Fig. 16.9b); M, molar mass taken from the ordinate maximum of the best-fit $c(M)$ distribution, calculated according to Eq. 16.1; $\overline{v}$, partial specific volume calculated from M (as above) according to Eq. 16.5; R_s, particle radii, calculated assuming a spherical particle and using experimentally determined values for M and $\overline{v}$

1.0, and the $\overline{v}$ relationship governed by Eq. 16.5 results in the $c(s_f)$ distributions for fractionated DMPC/TO emulsions shown in Fig. 16.9b. The hydrodynamic properties calculated from the $c(s_f)$ distribution of the smallest fraction (i.e. Fraction 2) through to the largest fraction (i.e. Fraction 6) are summarised in Table 16.3.

The data shown in Fig. 16.8 can also be fitted to obtain an apparent flotation coefficient distribution $g^*(s_f)$ using the direct boundary model for a distribution of non-diffusing particles $ls{-}g^*(s_f)$ (Schuck and Rossmanith 2000). In this method, $ls{-}g^*(s_f)$ is calculated using the same concepts and framework as the $c(s_f)$ distribution analysis, but by replacing the Lamm equation solution $L(s_f,D(s_f),r,t)$ in Eq. 16.4 with the theoretical profiles of non-diffusing species, i.e. step functions $U(s_f,r,t)$, according to Eqs. 16.6 and 16.7 as follows:

$$\mathrm{a}\,(\mathrm{r},\mathrm{t}) \cong \int \mathrm{g} * (\mathrm{s_f})\,\mathrm{U}\,(\mathrm{s_f},\mathrm{r},\mathrm{t})\,\mathrm{ds_f} \tag{16.6}$$

else

$$U\left(s_f,r,t\right) = e^{-2\omega^2 s_f t} \times \begin{cases} 0 & \text{for} \quad r<r^*(t)=r_b e^{\omega^2 s_f t} \\ \end{cases} \tag{16.7}$$

where $U(s_f,r,t)$ describes the ideal behaviour of initially uniformly distributed particles with flotation coefficient s_f (but without diffusion during flotation or radial dilution in the sector-shaped solution column), r_b denotes the bottom position of the solution column, ω denotes the rotor angular velocity and $r^*(t)$ denotes the boundary position (Schuck and Rossmanith 2000; Stafford 1992).

Similar to the $c(s_f)$ distribution shown in Fig. 16.9b, the $g^*(s_f)$ distribution calculated using Tikhonov-Phillips regularisation at a confidence level of $P = 0.95$ results in a similar but broader distribution for each fraction (Fig. 16.9c). The broader $g^*(s_f)$ distributions are attributed to the effects of diffusion, since this method does not allow direct correction for diffusion-based broadening (Schuck and Rossmanith 2000; Stafford 1992).

16.4 AUC Analyses of ApoE-Emulsion Interactions

To compare the interactions of apoE isoforms with size-fractionated lipid emulsions, Perugini et al. (2002) conducted flotation velocity experiments in the analytical ultracentrifuge. The premise for these experiments was studies conducted using native lipoproteins isolated from plasma that demonstrated apoE3 preferentially binds smaller HDL particles, whilst apoE4 prefers to bind larger VLDL particles (Weisgraber 1990). Perugini et al. (2002) hypothesised that these binding preferences were simply due to the size of the lipid particles. Hence, the hypothesis was examined using Fraction 2 as a small particle and Fraction 6 as a large particle (Table 16.3).

Flotation velocity experiments using Fraction 2 enabled the calculation of the buoyant molar mass [$c(M_b)$] distribution in the absence and presence of 1.0 μM apoE3 or apoE4 (Fig. 16.10). Relative to the control, the $c(M_b)$ distribution of Fraction 2 in the presence of 1.0 μM apoE3 was shifted to smaller absolute values, attributed to the binding of apoE3 to these small emulsions (Fig. 16.10a). The difference in the buoyant molar masses at the ordinate maximum of $c(M_b)$ for Fraction 2, with and without apoE3, corresponds to the binding of approximately 118 apoE3 molecules per emulsion particle and a phospholipid to apoE ratio of 360. In contrast, the $c(M_b)$ distribution of Fraction 2 in the presence of 1.0 μM apoE4 was similar to the control. Detailed inspection of this data revealed that the presence of apoE4 results in a minor shift in the $c(M_b)$ distribution to higher buoyant molar masses. This was most significant for small particles with $-M_b < 0.3 \times 10^7$ Da and in the region of the ordinate maximum at $-M_b \approx 0.7 \times 10^7$ Da. The slight shift in $c(M_b)$ is suggested by Perugini et al. (2002) to be due to apoE4-mediated coalescence or cross-linking of the emulsion particles in a similar manner to that reported for lipoprotein lipase (MacPhee et al. 1997). Monte-Carlo analysis was also performed to demonstrate that the observed shifts in the $c(M_b)$ distributions of Fraction 2 in the absence and presence of apoE3 or apoE4 were statistically significant and cannot be attributed to noise affecting the data analysis (Fig. 16.10b).

Perugini et al. (2002) also performed similar flotation velocity analyses of Fraction 6, which revealed the opposite trend with apoE4 preferentially binding this larger emulsion particle compared to apoE3. To validate these results, a direct binding assay was designed and optimised to directly determine the amount of free and bound protein in the analytical ultracentrifuge (Fig. 16.11). This allows the construction of binding isotherms by plotting the ratio of bound protein (P_b) to total phosphatidylcholine (PC) (Fig. 16.12a). The binding data can also be linearised by plotting free protein (P_f) versus PC multiplied by P_f / P_b to yield the affinity

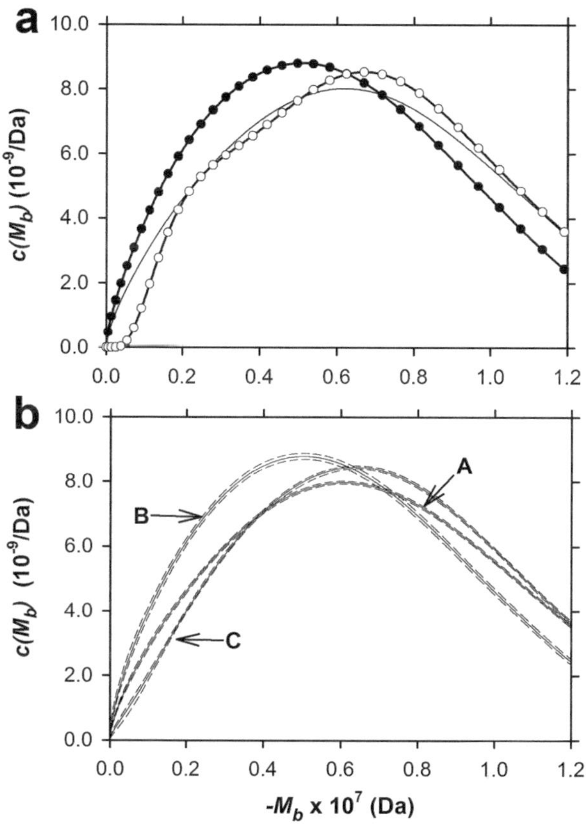

Fig. 16.10 Buoyant molar mass distribution of DMPC/TO Fraction 2 in the presence and absence of apoE3 and apoE4. Experiments were conducted in 50 mM sodium phosphate, pH 7.4, at 20 °C, rotor speed 6000 rpm, and data was collected at time intervals of 360 s with a step size of 0.003 cm. (**a**) Fraction 2 alone (*solid line, no symbols*), Fraction 2 + 1.0 μM apoE3 (*solid symbols + line*) and Fraction 2 + 1.0μM apoE4 (*open symbols + line*). The RMSD for all best-fit $c(M_b)$ distributions presented < 0.01. Total lipid concentration in fraction 2 = 414 μM, i.e., [DMPC, dimyristoylphosphatidylcholine] = 152 μM + [TO, triolein] = 262 μM. (**b**) Results of Monte-Carlo statistical analysis distributions, calculated from 1000 synthetic data sets to a confidence level of $P = 0.95$. Lower (0.025) and upper (0.975) quantiles are depicted as dashed lines, enclosing the mean distribution (*solid lines*) for Fraction 2 alone (labelled A), Fraction 2 + 1.0 μM apoE3 (labelled B) and Fraction 2 + 1.0 μM apoE4 (labelled C) (adapted from Perugini et al. (2002))

(K_D) from the y-intercept and binding capacity (B_{max}) from the slope according to Eq. 16.8:

$$P_f = PC\left(P_f/P_b\right)B_{max} - K_D \tag{16.8}$$

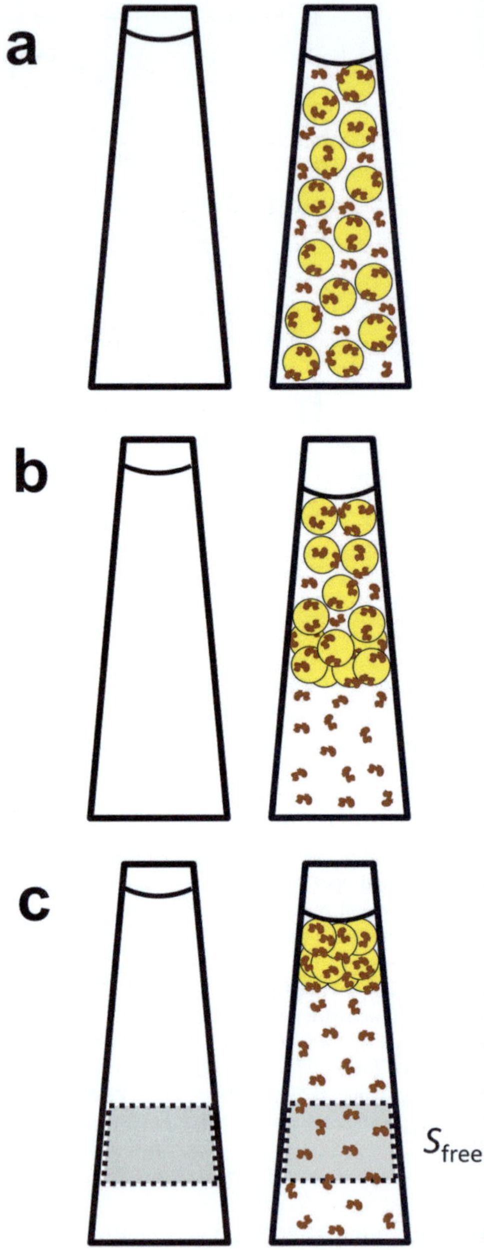

Fig. 16.11 Direct binding assay for measuring protein-lipid emulsion interactions. The yellow circles and brown structures represent lipid emulsion particles and protein, respectively. (**a**) At $t = 0$, showing even distribution of lipid emulsion particles with bound and free protein. (**b**) At t = mid-experiment, showing lipid emulsion particles with bound protein mid-floatation with free protein remaining in the infranatant at high radial positions. (**c**) At t = end of experiment, showing lipid emulsion particles with bound protein at the meniscus with free protein in the infranatant. The signal (absorbance or fluorescence) of the free protein (S_{free}) can easily be measured at high radial positions. With the total signal known, the amount of bound protein can be calculated by subtracting the free from total

The resulting linearised plots according to Eq. 16.8 are shown in Fig. 16.12b, from which the K_{D} and $B_{\max}$ presented in Table 16.4 were calculated. The results of this analysis confirmed that apoE3 and apoE4, although possessing almost identical binding affinity for large emulsion particles, are significantly distinct in their binding

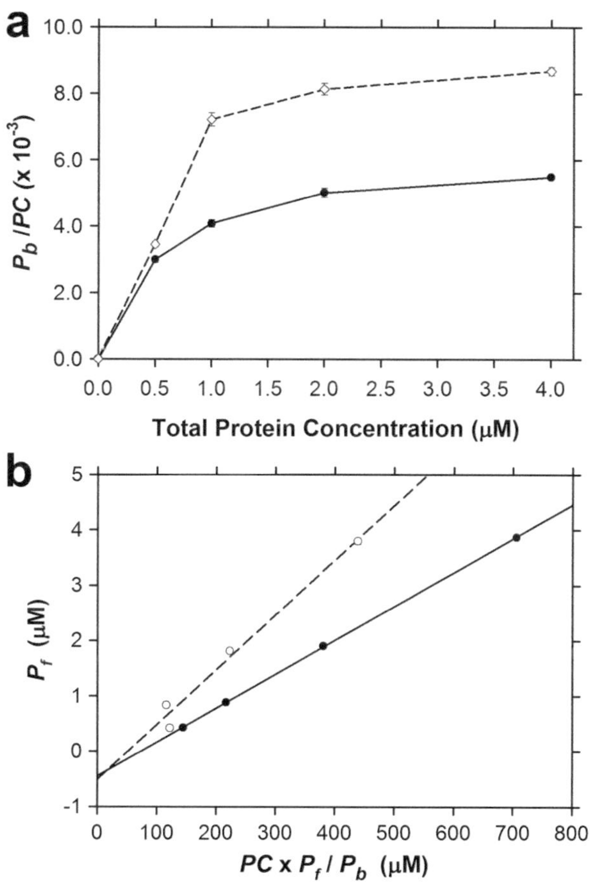

Fig. 16.12 Binding of apoE3 and apoE4 to large DMPC/TO lipid emulsion particles. Binding data was obtained by AUC using the direct binding assay described in Fig. 16.11. (**a**) Binding profile of apoE3 (*solid symbols* + *line*) and apoE4 (*open symbols* + *dashed line*) to Fraction 6 (23 μM DMPC + 102 μM TO). (**b**) Linearised plots of the data from panel a according to Eq. 16.8. The solid and dashed lines represent the least-squares fit to the data for apoE3 ($R = 0.999$) and apoE4 ($R = 0.990$), respectively, where the y-intercept and slope equate to the K_D and B_{max}, respectively. The error bars are calculated from the standard deviation of the absorbance signal averaged over a 0.1 cm radial range at the top of the cell (adapted from Perugini et al. (2002))

Table 16.4 Parameters for the binding of apoE3 and apoE4 to DMPC/TO emulsion Fraction 6[a]

	K_D	B_{max}		
Isoform	(μM)	ApoE/particle	PL/apoE	Amino acids/PL
ApoE3	0.44	1010	163	1.83
ApoE4	0.51	1630	101	2.96

[a] K_D and B_{max} values were calculated by linear least-squares analysis of the data shown in Fig. 16.12b, according to Eq. 16.8

capacities. Moreover, the binding footprint for the apoE4 isoform was calculated to correspond to 2.96 amino acids per phospholipid, compared to 1.83 amino acids per phospholipid for apoE3 (Table 16.4). The authors attribute the difference in B_{max} to variations in the lipid-bound conformations of the apoE isoforms (Perugini et al. 2002). Together, the flotation velocity analyses and direct binding assay results demonstrate the power of the analytical ultracentrifuge in assessing and quantitating the interactions of proteins with large macromolecular ligand complexes.

16.5 Conclusions

AUC is a premier platform for measuring protein-ligand interactions. This technique offers significant advantages over other common biophysical methods (such as ITC, MST and SPR) given the flexibility in accommodating a broad range of (i) *ligand sizes* (from small molecules of $\leq$500 Da right through to large macromolecular ligand complexes of at least 10^8 Da), (ii) *binding affinities*, (iii) *interaction kinetics* and (iv) stoichiometry. The platform also provides flexibility in experimental design from the more conventional sedimentation velocity and sedimentation equilibrium analyses of self-associating systems (such as BPL and DHDPS) through to flotation velocity and direct binding assay studies of fractionated lipid emulsions in the absence and presence of disease-associated apolipoproteins.

References

Atkinson SC, Dobson RC, Newman JM, Gorman MA, Dogovski C, Parker MW, Perugini MA (2009) Crystallization and preliminary X-ray analysis of dihydrodipicolinate synthase from *Clostridium botulinum* in the presence of its substrate pyruvate. Acta Crystallogr Sect F: Struct Biol Cryst Commun 65(Pt 3):253–255

Bagautdinov B, Kuroishi C, Sugahara M, Kunishima N (2005) Crystal structures of biotin protein ligase from *Pyrococcus horikoshii* OT3 and its complexes: structural basis of biotin activation. J Mol Biol 353(2):322–333

Blagova E, Levdikov V, Milioti N, Fogg MJ, Kalliomaa AK, Brannigan JA, Wilson KS, Wilkinson AJ (2006) Crystal structure of dihydrodipicolinate synthase (BA3935) from *Bacillus anthracis* at 1.94 Å resolution. Proteins 62(1):297–301

Blickling S, Renner C, Laber B, Pohlenz H-D, Holak TA, Huber R (1997) Reaction mechanism of *Escherichia coli* dihydrodipicolinate synthase investigated by X-ray crystallography and NMR spectroscopy. Biochemistry 36(1):24–33

Burgess BR, Dobson RC, Bailey MF, Atkinson SC, Griffin MD, Jameson GB, Parker MW, Gerrard JA, Perugini MA (2008) Structure and evolution of a novel dimeric enzyme from a clinically important bacterial pathogen. J Biol Chem 283(41):27598–27603

Corder EH, Saunders AM, Strittmatter WJ, Schmechel DE, Gaskell PC, Small GW, Roses AD, Haines JL, Pericak-Vance MA (1993) Gene dose of apolipoprotein E type 4 allele and the risk of Alzheimer's disease in late onset families. Science 261(5123):921–923

Demeler B, van Holde KE (2004) Sedimentation velocity analysis of highly heterogeneous systems. Anal Biochem 335(2):279–288

Derksen A, Small DM (1989) Interaction of ApoA-1 and ApoE-3 with triglyceride-phospholipid emulsions containing increasing cholesterol concentrations. Model of triglyceride-rich nascent and remnant lipoproteins. Biochemistry 28(2):900–906

Devenish SR, Blunt JW, Gerrard JA (2010) NMR studies uncover alternate substrates for dihydrodipicolinate synthase and suggest that dihydrodipicolinate reductase is also a dehydratase. J Med Chem 53(12):4808–4812

Dobson RC, Griffin MD, Roberts SJ, Gerrard JA (2004) Dihydrodipicolinate synthase (DHDPS) from *Escherichia coli* displays partial mixed inhibition with respect to its first substrate, pyruvate. Biochimie 86(4–5):311–315

Dogovski C, Atkinson S, Dommaraju S, Dobson RCJ, Hor L, Hutton CA, Gerrard JA, Perugini MA (2009) Lysine biosynthesis in bacteria: an unchartered pathway for novel antibiotic design. In:

Doelle HW, Rokem S (eds) Encyclopedia of Life Support Systems (EOLSS), vol 11, Developed under the Auspices of the UNESCO. EOLSS Publishers, Paris, pp 116–136

Dogovski C, Atkinson SC, Dommaraju SR, Downton M, Hor L, Moore S, Paxman JJ, Peverelli MG, Qiu TW, Reumann M, Siddiqui T, Taylor NL, Wagner J, Wubben JM, Perugini MA (2012) Enzymology of bacterial lysine biosynthesis. In: Ekinci D (ed) Biochemistry. InTech Open Access Publisher, Rijeka, pp 225–262

Dogovski C, Gorman MA, Ketaren NE, Praszkier J, Zammit LM, Mertens HD, Bryant G, Yang J, Griffin MD, Pearce FG, Gerrard JA, Jameson GB, Parker MW, Robins-Browne RM, Perugini MA (2013) From knock-out phenotype to three-dimensional structure of a promising antibiotic target from *Streptococcus pneumoniae*. PLoS One 8(12):e83419

Drew J, Liodakis A, Chan R, Du H, Sadek M, Brownlee R, Sawyer WH (1990) Preparation of lipid emulsions by pressure extrusion. Biochem Int 22(6):983–992

Funahashi T, Yokoyama S, Yamamoto A (1989) Association of apolipoprotein E with the low density lipoprotein receptor: demonstration of its co-operativity on lipid microemulsion particles. J Biochem 105(4):582–587

Gotto AM Jr, Pownall HJ, Havel RJ (1986) Introduction to the plasma lipoproteins. Methods Enzymol 128:3–41

Hatters DM, Peters-Libeu CA, Weisgraber KH (2006) Apolipoprotein E structure: insights into function. Trends Biochem Sci 31(8):445–454

Hughes JP, Rees S, Kalindjian SB, Philpott KL (2011) Principles of early drug discovery. Br J Pharmacol 162(6):1239–1249

Humphrey W, Dalke A, Schulten K (1996) VMD: visual molecular dynamics. J Mol Graph 14(1):33–38, 27–38

Hutton CA, Perugini MA, Gerrard JA (2007) Inhibition of lysine biosynthesis: an evolving antibiotic strategy. Mol Biosyst 3(7):458–465

Kaur N, Gautam A, Kumar S, Singh A, Singh N, Sharma S, Sharma R, Tewari R, Singh TP (2011) Biochemical studies and crystal structure determination of dihydrodipicolinate synthase from *Pseudomonas aeruginosa*. Int J Biol Macromol 48(5):779–787

Kobayashi K, Ehrlich SD, Albertini A, Amati G, Andersen KK, Arnaud M, Asai K, Ashikaga S, Aymerich S, Bessieres P, Boland F, Brignell SC, Bron S, Bunai K, Chapuis J, Christiansen LC, Danchin A, Debarbouille M, Dervyn E, Deuerling E, Devine K, Devine SK, Dreesen O, Errington J, Fillinger S, Foster SJ, Fujita Y, Galizzi A, Gardan R, Eschevins C, Fukushima T, Haga K, Harwood CR, Hecker M, Hosoya D, Hullo MF, Kakeshita H, Karamata D, Kasahara Y, Kawamura F, Koga K, Koski P, Kuwana R, Imamura D, Ishimaru M, Ishikawa S, Ishio I, Le Coq D, Masson A, Mauel C, Meima R, Mellado RP, Moir A, Moriya S, Nagakawa E, Nanamiya H, Nakai S, Nygaard P, Ogura M, Ohanan T, O'Reilly M, O'Rourke M, Pragai Z, Pooley HM, Rapoport G, Rawlins JP, Rivas LA, Rivolta C, Sadaie A, Sadaie Y, Sarvas M, Sato T, Saxild HH, Scanlan E, Schumann W, Seegers JF, Sekiguchi J, Sekowska A, Seror SJ, Simon M, Stragier P, Studer R, Takamatsu H, Tanaka T, Takeuchi M, Thomaides HB, Vagner V, van Dijl JM, Watabe K, Wipat A, Yamamoto H, Yamamoto M, Yamamoto Y, Yamane K, Yata K, Yoshida K, Yoshikawa H, Zuber U, Ogasawara N (2003) Essential *Bacillus subtilis* genes. Proc Natl Acad Sci U S A 100(8):4678–4683

Kodoyianni V (2011) Label-free analysis of biomolecular interactions using SPR imaging. BioTechniques 50(1):32–40

Kroe RR, Laue TM (2009) NUTS and BOLTS: applications of fluorescence-detected sedimentation. Anal Biochem 390(1):1–13

Ladbury JE, Doyle ML (2004) Biocalorimetry 2: applications of calorimetry in the biological sciences, 2nd edn. Wiley, Chichester

Lamm O (1929) Die differentialgleichung der ultrazentrifugierung. Ark Mat Astron Fys 21B(2): 1–4

Lee JW, Na D, Park JM, Lee J, Choi S, Lee SY (2012) Systems metabolic engineering of microorganisms for natural and non-natural chemicals. Nat Chem Biol 8(6):536–546

MacGregor IK, Anderson AL, Laue TM (2004) Fluorescence detection for the XLI analytical ultracentrifuge. Biophys Chem 108(1–3):165–185

MacPhee CE, Chan RY, Sawyer WH, Stafford WF, Howlett GJ (1997) Interaction of lipoprotein lipase with homogeneous lipid emulsions. J Lipid Res 38(8):1649–1659

Miller KW, Small DM (1983) Triolein-cholesteryl oleate-cholesterol-lecithin emulsions: structural models of triglyceride-rich lipoproteins. Biochemistry 22(2):443–451

Muscroft-Taylor AC, Soares da Costa TP, Gerrard JA (2010) New insights into the mechanism of dihydrodipicolinate synthase using isothermal titration calorimetry. Biochimie 92(3):254–262

Pearce FG, Perugini MA, McKerchar HJ, Gerrard JA (2006) Dihydrodipicolinate synthase from *Thermotoga maritima*. Biochem J 400(2):359–366

Pendini NR, Bailey LM, Booker GW, Wilce MC, Wallace JC, Polyak SW (2008) Microbial biotin protein ligases aid in understanding holocarboxylase synthetase deficiency. Biochim Biophys Acta 1784(7–8):973–982

Pendini NR, Yap MY, Traore DA, Polyak SW, Cowieson NP, Abell A, Booker GW, Wallace JC, Wilce JA, Wilce MC (2013) Structural characterization of *Staphylococcus aureus* biotin protein ligase and interaction partners: an antibiotic target. Protein Sci 22(6):762–773

Perugini MA, Schuck P, Howlett GJ (2000) Self-association of human apolipoprotein E3 and E4 in the presence and absence of phospholipid. J Biol Chem 275(47):36758–36765

Perugini MA, Schuck P, Howlett GJ (2002) Differences in the binding capacity of human apolipoprotein E3 and E4 to size-fractionated lipid emulsions. Eur J Biochem 269(23):5939–5949

Perugini MA, Griffin MD, Smith BJ, Webb LE, Davis AJ, Handman E, Gerrard JA (2005) Insight into the self-association of key enzymes from pathogenic species. Eur Biophys J 34(5):469–476

Pownall H, Gotto A Jr (1999) Structure and dynamics of human plasma lipoproteins. In: Betteridge D, Illingworth D, Shepherd J (eds) Lipiproteins in health and disease. Oxford University Press Inc., New York, pp 3–15

Rhinn H, Fujita R, Qiang L, Cheng R, Lee JH, Abeliovich A (2013) Integrative genomics identifies APOE $\varepsilon 4$ effectors in Alzheimer's disease. Nature 500(7460):45–50

Schuck P (2000) Size-distribution analysis of macromolecules by sedimentation velocity ultracentrifugation and Lamm equation modeling. Biophys J 78(3):1606–1619

Schuck P, Rossmanith P (2000) Determination of the sedimentation coefficient distribution by least-squares boundary modeling. Biopolymers 54(5):328–341

Schuck P, Perugini MA, Gonzales NR, Howlett GJ, Schubert D (2002) Size-distribution analysis of proteins by analytical ultracentrifugation: strategies and application to model systems. Biophys J 82(2):1096–1111

Soares da Costa TP, Christensen JB, Desbois S, Gordon SE, Gupta R, Hogan CJ, Nelson TG, Downton MT, Ghardi CHTK, Abbott BM, Wagner J, Panjikar S, Perugini MA (2015) Quaternary structure analyses of an essential oligomeric enzyme. Methods Enzymol 562:205–223

Soares da Costa TP, Tieu W, Yap MY, Zvarec O, Bell JM, Turnidge JD, Wallace JC, Booker GW, Wilce MCJ, Abell AD, Polyak SW (2012) Biotin analogues with antibacterial activity are potent inhibitors of biotin protein ligase. ACS Med Chem Lett 3(6):509–514

Soares da Costa TP, Yap MY, Perugini MA, Wallace JC, Abell AD, Wilce MC, Polyak SW, Booker GW (2014) Dual roles of F123 in protein homodimerization and inhibitor binding to biotin protein ligase from *Staphylococcus aureus*. Mol Microbiol 91(1):110–120

Stafford WF 3rd (1992) Boundary analysis in sedimentation transport experiments: a procedure for obtaining sedimentation coefficient distributions using the time derivative of the concentration profile. Anal Biochem 203(2):295–301

Stafford WF 3rd (2009) Protein-protein and ligand-protein interactions studied by analytical ultracentrifugation. Methods Mol Biol 490:83–113

van Holde K, Weischet WO (1978) Boundary analysis of sedimentation-velocity experiments with monodisperse and paucidisperse solutes. Biopolymers 17(6):1387–1403

Voss JE, Scally SW, Taylor NL, Atkinson SC, Griffin MD, Hutton CA, Parker MW, Alderton MR, Gerrard JA, Dobson RC, Dogovski C, Perugini MA (2010) Substrate-mediated stabilization of a tetrameric drug target reveals Achilles heel in anthrax. J Biol Chem 285(8):5188–5195

Weaver LH, Kwon K, Beckett D, Matthews BW (2001) Competing protein:protein interactions are proposed to control the biological switch of the *E coli* biotin repressor. Protein Sci 10(12):2618–2622

Weisgraber KH (1990) Apolipoprotein E distribution among human plasma lipoproteins: role of the cysteine-arginine interchange at residue 112. J Lipid Res 31(8):1503–1511

Weisgraber KH (1994) Apolipoprotein E: structure-function relationships. Adv Protein Chem 45:249–302

Wienken CJ, Baaske P, Rothbauer U, Braun D, Duhr S (2010) Protein-binding assays in biological liquids using microscale thermophoresis. Nat Commun 1:100

Zhao H, Beckett D (2008) Kinetic partitioning between alternative protein-protein interactions controls a transcriptional switch. J Mol Biol 380(1):223–236

Chapter 17
AUC in the High Concentration of Salts/Cosolvent

Christine Ebel

Abstract Macromolecule stability, conformations, and equilibrium are modulated by solvent composition. This chapter presents the formalism allowing to study in analytical ultracentrifugation (AUC) the sedimentation of macromolecules in a solvent with a high salt or other cosolvent concentration. It allows to relate the buoyant molar mass, measured in AUC at low concentration of macromolecule, to the macromolecule molar mass, through thermodynamic preferential binding parameters. These parameters allow to derive values for the numbers of water and cosolvent interacting with the macromolecule, in the framework of models. When modifying the solvent composition, i.e., cosolvent activity, equilibrium constants characterizing conversion between macromolecular species changes are directly related to the differences in the preferential binding parameters, which can be interpreted, in the framework of models, in terms of differences in bound water and cosolvent.

Keywords Analytical ultracentrifugation • Cosolvent • Preferential binding • Solvation • Buoyant molar mass • Wyman linkage

17.1 Introduction

Macromolecule stability, conformations, and equilibrium are modulated by solvent composition. This chapter presents the formalism allowing to study in analytical ultracentrifugation (AUC) the sedimentation of macromolecules in a solvent with a high salt or other cosolvent concentration. The cosolvent may be present in quite high concentration. Due to the various interactions in solution, the solvent

---

C. Ebel (✉)
Institut de Biologie Structurale (IBS), Univ. Grenoble Alpes, CEA, CNRS, Grenoble F-38044, France
e-mail: Christine.ebel@ibs.fr

S. Uchiyama et al. (eds.), *Analytical Ultracentrifugation*,
DOI 10.1007/978-4-431-55985-6_17

composition is affected by the macromolecule, the macromolecule inducing solvent perturbation. As a consequence, the macromolecule behaves differently when compared to its behavior in a diluted solvent. After some brief practical considerations (part 2) and definitions (part 3), we will describe how the fact that the solvent is complex will affect the sedimentation of a macromolecule (part 4). This theoretical presentation is derived from the works of H. Eisenberg (Casassa and Eisenberg 1964; Eisenberg 1976, 2000). We will show that the equations describing the sedimentation of the macromolecule in complex media very conveniently allow to consider separately macromolecule-macromolecule interactions – a topic that will not be developed – and macromolecule-solvent interactions. From the latter, the relation between the buoyant and molar mass of the macromolecule is only slightly modified in a complex solvent, when compared to the case of a diluted solvent, as shown in part 5. We will present the preferential solvent binding parameters, i.e., the thermodynamic parameters that determine the density increments – thus buoyant molar mass in the ultracentrifuge – and the molecular models of solvation that describe them. The following part (part 5), from the Wyman linkage between protein equilibrium and interactions with solvent, will show the insights brought from the study of macromolecule equilibrium at different solvent compositions. The thermodynamic derivations and most examples presented here are based on the works of Eisenberg (Casassa and Eisenberg 1964; Eisenberg 1976, 2000, 2003), Schellman (1994), Timasheff (1993, 1998, 2002), and Record (Courtenay et al. 2000), among others, and were presented in a slightly more extended way in a previous review (Ebel 2007).

17.2 Experimental Considerations

17.2.1 The Sample

Dialysis or gel filtration deals to samples at conditions of constant chemical potential of the solvent components, i.e., the bulk solvent composition is that of the solvent prepared independently. It is the best way to prepare samples and reference buffers. Dilution with the reference buffer will not change the bulk solvent composition. (For a sample prepared in a different way, e.g., by dissolution, experiments may however in general be interpreted with estimates of the solvent composition: composition changes between the solvent used for the solubilizing and the bulk solvent may in some cases be negligible – e.g., if the concentration of macromolecule, μM, is low when compared to that of the solvent salt (0.1 M) or solvent composition may be evaluated by, e.g., measuring the sample density, given the macromolecule concentration is low.) We will consider here (in practice and in principle) a buffer allowing macromolecule charge screening.

17.2.2 Cell Preparation for Analytical Ultracentrifugation

Some points should be emphasized when working with complex solvents. First, solvent should be well defined, with the composition of the bulk solvent and reference solvent the same. The sample is thus preferentially prepared using dialysis or/and chromatography. The sample density should be known at 0.001 g/mL and its viscosity at 0.01 cp. These values can be obtained from tabulated data, using, e.g., the program sednterp (http://sednterp.unh.edu/, available online, created with support from Biomolecular Interactions Technology Center, USA). They can also be measured. We use, e.g., a density meter DMA5000 and a viscosity meter AMVn (Anton Paar, Austria).

The two channels of the centerpieces should be strictly filled at the same level, when measuring the sedimentation with interference optics. Indeed solvent redistribution occurs in the AUC. We will not consider here the consequence of the important cosolvent distribution (Schuck 2004a) – to be considered in, e.g., CsCl – or solvent compressibility of organic media (Schuck 2004b). However even minor salt or buffer component redistribution will alter interference sedimentation profiles, given their concentration exceeds largely that of the macromolecule (0.15 M KCl is 11 mg/mL). One practical way to obtain the same filling is to use boundary centerpieces – with capillaries scribed on the face of the centerpiece between the sectors – which slightly overfill the reference channel with the solvent and centrifuge at low speed (3000 revs/min), allowing the excess solvent to redistribute in the sample channel, stop the centrifuge, and homogenize the channels. Buffer mismatch corrections that can be made, e.g., in the SEDFIT program, are often poorly satisfying. In general, interference data will be more tricky to analyze when compared to optics that probes specifically the macromolecule, e.g., absorbance.

17.2.3 Programs for Analysis of Ultracentrifugation Data

AUC data analysis in high salt or in a complex solvent can be in the general case analyzed as usually. The most popular suites of program are SEDFIT/SEDPHAT, created by P. Schuck (USA), free and available at www.analyticalultracentrifugation.com; SEDANAL, created by W. Stafford (USA), free and available at http://www.sedanal.org; and UltraScan, created by B. Demeler (USA), free and available at http://www.ultrascan.uthscsa.edu.

17.3 Definition of the Components and Symbols Used

17.3.1 Definition of the Components

We consider a three-component system: the main solvent component, water (component 1), one type of macromolecule (component 2), and one additional small solute, the cosolvent (component 3). The numbering is defined by convention. When components 2 and 3 are electrolytes, they have to be defined as electroneutral combination of species (Eisenberg 1976). As an example, nucleic acid molar mass is calculated with the number of counterions required to have electroneutrality. The behavior of the macromolecule is affected because of its interactions with the solvent (water and cosolvent). Because our reference is the anhydrous macromolecule (component 2), which is followed in centrifugation, and that is to be characterized, solvent redistribution corresponds to non-ideality.

17.3.2 Symbols

In the ultracentrifuge, r is the radial distance, t the time of centrifugation, and ω the angular velocity. Temperature T is constant. R is the gas constant and P the pressure. Π is the osmotic pressure, ρ the solution density, and $\rho°$ and $\eta°$ the solvent density and viscosity. The sedimentation coefficient, s; diffusion coefficient, D; and buoyant molar mass, M_b, characterize the macromolecule. K describes the equilibrium between macromolecules.

The components, i, are characterized by their concentrations: c_i in g/mL, C_i, in mol/L, molalities m_i in mol/kg of main solvent (here water), or w_i in weight ratio (g/g of main solvent, here water) ($m_1 = 1$; $w_1 = 1$).μ_i is their chemical potential, M_i their molar mass (g/mol), and $\overline{v}_i$ their partial specific volume (mL/g). A_2 is the second virial coefficient for the macromolecule (mL mol g^{-2}).ϕ' is an apparent partial specific volume characterizing the macromolecule in a given solvent.

Subscripts μ and m indicate constancy of the chemical potential and molalities, respectively, of the solvent components (components *1* and *3*).

N_i and B_i are numbers, in mole of solvent components per mole of macromolecule and g of solvent components per g of macromolecule, respectively, modeling the solvent domain perturbed by the macromolecule (Eisenberg 1976; Schellman 2003; Ebel et al. 2000).

17.4 Sedimentation in a Complex Solvent

17.4.1 Sedimentation Equilibrium in a Complex Solvent

In sedimentation equilibrium, the solution containing the macromolecule is centrifuged at a given angular velocity for a sufficient time (typically 24 h in standard protocols) to obtain the equilibrium condition, in which the concentration profile does not evolve with time. The equilibrium condition corresponds to an invariance of the total potential – composed of chemical, gravitational, and electrostatic terms – throughout the system for each species. Considering that electroneutrality condition is satisfied locally (which is obtained in practice) and in view of the definition as electroneutral species of the components, the electrostatic contribution is eliminated. The equilibrium condition can then be written (Eisenberg 1976, 2000):

$$\frac{d\ln c_2}{dr^2} = \frac{\omega^2}{2}\left(\frac{\partial\rho}{\partial c_2}\right)_\mu\left(\frac{\partial\Pi}{\partial c_2}\right)^{-1} \tag{17.1}$$

Note the variables depend on the position r in the ultracentrifuge. This equation is general for any component among any number of components of any molecular weight, without restriction to thermodynamic non-ideality or to incompressibility (pressure effects, e.g., are manifested in the dependence of the variables). The last term at the right is the osmotic pressure derivative. It can be developed (except for polyelectrolyte in salt-free solutions (Eisenberg 1976, 2000)) in a virial series:

$$\left(\frac{\partial\Pi}{\partial c_2}\right) = RT\left(\frac{1}{M_2} + 2A_2c_2 + \ldots\right) \tag{17.2}$$

In Eq. (17.1), $(\partial\rho/\partial c_2)_\mu$ is the density increment at constant chemical potential of the solvent components, which will be presented below, and of temperature. In the case of dilute non-interacting macromolecule:

$$\left(\frac{\partial\rho}{\partial c_2}\right)_\mu\left(\frac{\partial\Pi}{\partial c_2}\right)^{-1} = \frac{1}{RT}\left(\frac{\partial\rho}{\partial c_2}\right)_\mu M_2 = \frac{1}{RT}M_b \tag{17.3}$$

Equation (17.1) then reduces to:

$$\frac{d\ln c_2}{dr^2} = \frac{\omega^2}{2RT}\left(\frac{\partial\rho}{\partial c_2}\right)_\mu M_2 = \frac{\omega^2}{2RT}M_b \tag{17.4}$$

17.4.2 Sedimentation Velocity in a Complex Solvent

For a three-component system, the sedimentation and diffusion coefficients s and D are complex functions, because there may be interactions between the flows of the solutes 2 and 3, via hydrodynamic and thermodynamic coupling. The general expression of s however simplifies considerably under the assumption that component 3 is at equilibrium. In that case, with m_2 the molality of the macromolecule (mol per kg of component 1, i.e., water), it is:

$$\frac{s}{D} = \left(\frac{\partial \rho}{\partial m_2}\right)_{\mathrm{P},\,\mu_3} \left(\frac{\partial \Pi}{\partial m_2}\right)^{-1}. \tag{17.5}$$

Neglecting terms that contribute to less than 1 %, the Svedberg equation can be written, with c_2 in g/mL (Eisenberg 1976, 2000):

$$\frac{s}{D} = \left(\frac{\partial \rho}{\partial c_2}\right)_{\mu} \left(\frac{\partial \Pi}{\partial c_2}\right)^{-1}. \tag{17.6}$$

Equation (17.6) reduces, in the case of dilute non-interacting macromolecule, using Eqs. 17.2 and 17.3, to:

$$\frac{s}{D} = \frac{1}{RT}\left(\frac{\partial \rho}{\partial c_2}\right)_{\mu} M_2 = \frac{1}{RT} M_b. \tag{17.7}$$

17.4.3 Brief Conclusion Concerning AUC in a Complex Solvent

Table 17.1 compares the sedimentation of a non-ideal macromolecule in a complex solvent to that of a dilute ideal macromolecule in a diluted solvent, assimilated to a one-component solvent, with the intermediate cases. The equations very conveniently allow to consider separately macromolecule-macromolecule interactions and macromolecule-solvent interactions. The concentration dependences of s, D, and of the osmotic pressure derivative, which are out of the scope of this article, are related to weak macromolecule-macromolecule interactions (which can be mediated by macromolecule-solvent interactions!) and used as a predictor in protein crystallization (Saluja et al. 2010; Solovyova et al. 2001).

Table 17.1 Terms related to non-ideality in analytical ultracentrifugation

Concentrated macromolecule in a complex solvent	$\left(\frac{\partial \rho}{\partial c_2}\right)_\mu \left(\frac{\partial \Pi}{\partial c_2}\right)^{-1}$
Diluted homogeneous macromolecule in a complex solvent	$\left(\frac{\partial \rho}{\partial c_2}\right)_\mu \frac{M_2}{RT}$
Concentrated macromolecule in a diluted solvent	$(1 - \rho^\circ \overline{v}_2)\left(\frac{\partial \Pi}{\partial c_2}\right)^{-1}$
Diluted homogeneous macromolecule in a diluted solvent	$\frac{(1-\rho^\circ \overline{v}_2)M_2}{RT}$

17.5 Considering a Diluted Macromolecule in a Complex Solvent

We consider here non-ideality related to solvent composition, or how solvent composition changes the buoyant molar mass measured in AUC. The buoyant term is modified by the interaction of the macromolecule with the solvent component, water, or cosolvent. As a result of these interactions, there may be either cosolvent excess in the vicinity of the macromolecule when compared to the bulk solvent composition, or water excess (positive preferential hydration), as schematized in Fig. 17.1a and b, respectively. We will first emphasize the difference between the density increments measured, either at constant molalities or at constant chemical potential, of the solvent components, before presenting the preferential solvent binding parameters.

17.5.1 The Density Increments ($\partial \rho / \partial c_2$)

17.5.1.1 Partial Specific Volumes Are Derived from ($\partial \rho / \partial c_2$)

This condition, referred by the subscript "*m*" for constant molalities of the solvent components, is used to determine experimentally the partial specific volume of component 2,$\overline{v}_2$. By definition, it is related to the increase of the volume, *V*, of the solution when adding component 2, at constant composition of other components (Eisenberg 1976):

$$\overline{v}_2 = \frac{\left(\frac{\partial V}{\partial m_2}\right)_m}{M_2} \tag{17.8}$$

The partial volume of a component describes not only the changes in the volume of the solution related to the molecular occupation of the component but also the changes in the volume occupied by other components in the solution caused by its presence. For example, typically, the partial specific volume of electrolyte increases when increasing salt concentration. This is true for solvent salts (Monnin

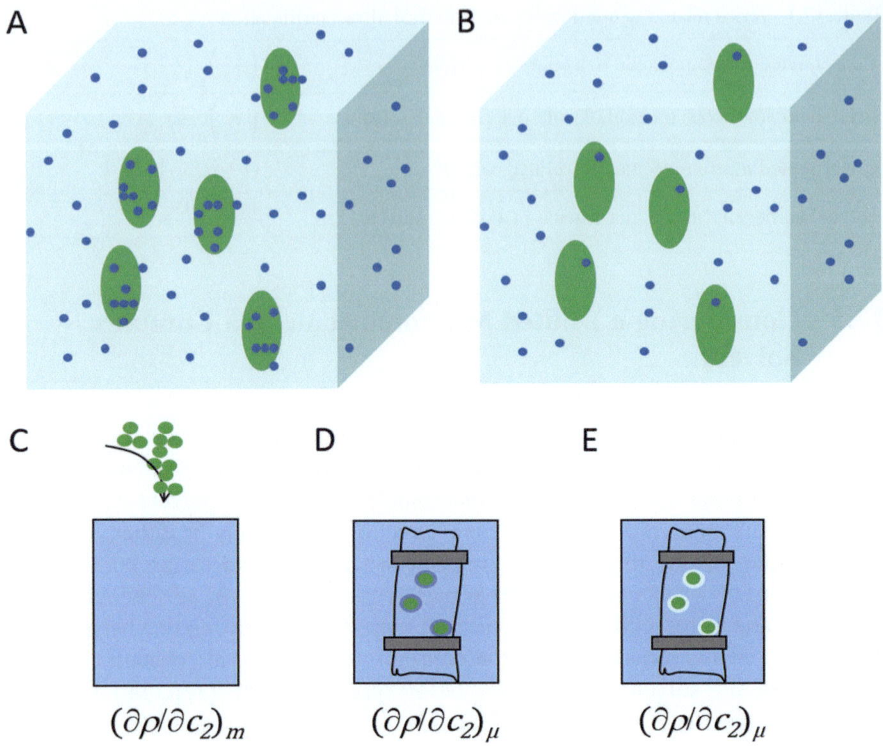

Fig. 17.1 Solvent redistribution related to macromolecule in a two-component solvent. (**a–b**): Schematic representations of solvent redistributions in a two-component solvent. Protein is in *green*, water in *light blue*, and cosolvent in *dark blue*. (**c**): Principle of the measurement of density increment at constant molality of the solvent components, anhydrous macromolecule is added to the solvent and density of the solution compared to that of the solvent. (**d–e**): for measuring density increments at constant chemical potentials of solvent components, the density of the sample after dialysis is compared to that of the bulk solvent. (**a**) and (**d**) correspond to a positive value of the preferential cosolvent binding parameter and a negative value of the preferential hydration parameter; (**b**) and (**e**) correspond to a negative value of the preferential cosolvent binding parameter and a positive value of the preferential hydration parameter

1989), for nucleic acids (Eisenberg 1990), or for very acidic proteins (Ebel et al. 2002). It can be explained considering the effect of water electrostriction is larger in dilute solution, which can be understood considering that, at high salt, most water molecules are already under the influence of a salt ion, or by the formation of ion pairing.

If ρ is the density of the solution with the macromolecule at concentration c_2, and ρ° that of the solvent:

$$(\partial\rho/\partial c_2)_m = \frac{(\rho - \rho^\circ)_m}{c_2} = \left(1 - \rho^\circ \overline{v}_2\right). \tag{17.9}$$

The experimental device that corresponds to this condition is to dissolve component 2 as an anhydrous powder, in the solvent (Fig. 17.1c). Another protocol uses a stock solution of macromolecule dissolved in pure water (Cohen and Eisenberg 1968; Kernel et al. 1999). The overall composition in solvent component is thus defined. These measurements are difficult. They require a precise determination of the macromolecule concentration and of the densities.

17.5.1.2 The Density Increment at Constant Chemical Potential of Solvent Components $(\partial \rho/\partial c_2)_\mu$

If ρ is the density of the dialysate containing the macromolecule at concentration c_2, and $\rho°$ that of the solvent:

$$(\partial\rho/\partial c_2)_\mu = \frac{(\rho - \rho^\circ)_\mu}{c_2}. \tag{17.10}$$

Here, not only the presence of the macromolecule but also of solvent molecules in interaction with the macromolecule affects the density of the solution. It can be measured experimentally by density measurements performed after dialysis (Fig. 17.1d, e).

The density increment at constant chemical potential of solvent components is the buoyant factor determining the sedimentation of the macromolecule in a complex solvent: it is because this is the solvated macromolecule that sediments. In consequence, the density increment at constant chemical potential of solvent components can be measured by analytical ultracentrifugation, given the molar mass of the macromolecule is known. However, because the interest is often in the determination of macromolecule molar mass, e.g., stoichiometry, by analogy with the two-component system, the apparent partial specific volume, ϕ', is often used (see, e.g., Reynolds and Tanford 1976):

$$(\partial\rho/\partial c_2)_\mu = \left(1 - \rho^\circ\phi'\right). \tag{17.11}$$

ϕ' is not a thermodynamic parameter, but an operational parameter. It can be used only in the solvent conditions used for its determination. It is different from the partial specific volume of the "complex," which would consider the solvated macromolecule as the "complex," since the density increment at the left of Eq. 17.8 refers to the anhydrous macromolecule concentration!

17.5.1.3 Illustration of the Differences Between the Two Density Increments

The study of solubilized membrane proteins – a specific case of multicomponent systems – illustrates how it is irrelevant to consider the ϕ' parameters as a partial specific volume: for a specific membrane protein solubilized by detergent and lipid, with partial specific volumes of 0.74, 0.814, and 0.946 mL/g, respectively, the density increments at constant chemical potential were measured in the range 0.96–1.14 and the ϕ' value in the range -0.14–0.04 mL/g (Butler et al. 2004).

As a second example, ion dissociation of polyelectrolyte corresponds to a specific type of solvent interactions. We measured by density for a polyacrylate-based amphipathic polymer a partial specific volume of 0.809 mL/g (Eq. 17.10); we measure, by sedimentation velocity, an apparent specific volume ϕ' of 0.866 mL/g (Eq. 17.11). This considerable difference can be explained by the dissociation of the sodium ions (see below Sect. 5.2.3) (Gohon et al. 2004).

17.5.1.4 About the Measurement of Partial Specific Volumes in Solvent of Different Densities in AUC

Measurement of sedimentation in solvent of various densities is in principle a way to determine both the molar mass and partial specific volume. Because of the hydration of the macromolecule surface – a general phenomena – varying the density using, e.g., sucrose cannot be used for this purpose. Using H_2O/D_2O mixtures to change the solvent density requires taking into account H-D exchange and, in the case of multicomponent system, considering its complexity (e.g., with Eqs. 17.16, 17.17, 17.18, and 17.19 below) (e.g., for a study with a polyelectrolyte, see (Gohon et al. 2004); with membrane proteins, see Le Roy et al. 2013, 2015; Dach et al. 2012). Experiments in $H_2O/H_2{}^{18}O$ were shown to offer an elegant alternative, the solvent part isotopic labeling not being changed and macromolecule molar mass not changed to determine partial specific volume (Brown et al. 2011) and could be used in principle to determine apparent specific volume ϕ'.

17.5.2 The Preferential Solvent Binding Parameters

17.5.2.1 Units and Relations Between These Parameters

These four thermodynamic parameters are related by the solvent composition, or the molar mass of the components, as detailed below. Note that when the preferential cosolvent binding parameter is positive, the preferential hydration parameter is negative and vice versa. Using one of the unit (g/g or mol/mol), and of the parameters (preferential hydration or preferential cosolvent binding), is a matter of convenience. When working with a polyelectrolyte in the presence of salt, or a ligand affecting equilibrium, preferential cosolvent binding in mol/mol may be preferentially used (Timasheff 1993; Ebel et al. 1999); when considering the hydration effects, preferential water binding parameters in g/g may be preferred (Eisenberg 2000; Ebel et al. 2000).

The preferential solvent binding parameters at constant chemical potentials is related to the change in the concentration of solvent components related to the presence of the macromolecule (Casassa and Eisenberg 1964, 1976, 2000). The preferential cosolvent binding parameter $(\partial w_3/\partial w_2)_\mu$, and the related preferential hydration parameter $(\partial w_1/\partial w_2)_\mu$, expressed in g/g, are linked together by the bulk solvent composition. If w_3 is the g/g ratio between the cosolvent and water in the solvent (Eisenberg 2000; Timasheff 1993):

$$\left(\frac{\partial w_1}{\partial w_2}\right)_\mu = -\left(\frac{1}{w_3}\right)\left(\frac{\partial w_3}{\partial w_2}\right)_\mu . \tag{17.12}$$

When expressed in mol/mol units, with (m_3/m_1) as the molar ratio between the cosolvent and water in the solvent:

$$\left(\frac{\partial m_1}{\partial m_2}\right)_\mu = -\left(\frac{m_1}{m_3}\right)\left(\frac{\partial m_3}{\partial m_2}\right)_\mu . \tag{17.13}$$

The relation between the preferential cosolvent parameters expressed in g/g and mol/mol units uses the molar masses M_1, M_3, and M_2, of the main solvent (water), cosolvent, and macromolecule, respectively:

$$\left(\frac{\partial w_3}{\partial w_2}\right)_\mu = \left(\frac{M_3}{M_2}\right)\left(\frac{\partial m_3}{\partial m_2}\right)_\mu ; \tag{17.14}$$

$$\left(\frac{\partial w_1}{\partial w_2}\right)_\mu = \left(\frac{M_1}{M_2}\right)\left(\frac{\partial m_1}{\partial m_2}\right)_\mu . \tag{17.15}$$

17.5.2.2 Preferential Binding Parameters Determine the Density Increment in the Ultracentrifuge

The preferential binding parameters can be experimentally determined, since they determine the buoyant properties of the macromolecule. Preferential binding parameters are thus derived from density increments measured at constant chemical potential of the solvent components (Casassa and Eisenberg 1964; Eisenberg 1976, 2000):

$$\left(\frac{\partial \rho}{\partial c_2}\right)_\mu = (1-\rho^\circ\overline{v}_2) + \left(\frac{\partial w_3}{\partial w_2}\right)_\mu (1-\rho^\circ\overline{v}_3) ; \tag{17.16}$$

$$\left(\frac{\partial \rho}{\partial c_2}\right)_\mu = (1-\rho^\circ\overline{v}_2) + \left(\frac{\partial w_1}{\partial w_2}\right)_\mu (1-\rho^\circ\overline{v}_1) . \tag{17.17}$$

Note that, from Eq. 17.10, $(1 - \rho^\circ \overline{v}_2)$ is $(\partial\rho/\partial c_2)_m$. The combined use of density measurements performed at constant molality and at constant potential of the solvent components was often used by S. N. Timasheff (see, e.g., Arakawa and Timasheff 1985). When using molar units for the concentrations, partial volumes, and preferential binding parameters, Eqs. 17.16 and 17.17 are written as:

$$\left(\frac{\partial\rho}{\partial C_2}\right)_\mu = (M_2 - \rho^\circ\overline{V}_2) + \left(\frac{\partial m_3}{\partial m_2}\right)_\mu (M_3 - \rho^\circ\overline{V}_3) \quad (17.18)$$

$$\left(\frac{\partial\rho}{\partial C_2}\right)_\mu = (M_2 - \rho^\circ\overline{V}_2) + \left(\frac{\partial m_1}{\partial m_2}\right)_\mu (M_1 - \rho^\circ\overline{V}_1) \quad (17.19)$$

17.5.2.3 Molecular Description from the Preferential Binding Parameters

In dialysis experiments, the preferential binding parameter represents the difference of the cosolvent concentration, in g/mL or molality, in the dialysis bag, and in the bulk, per g or mole macromolecule. It corresponds to an equivalent number of grams or moles of solute that would have to be added (or removed) with respect to one gram or mole of the macromolecule to maintain the constancy of the chemical potentials of water and cosolvent. The preferential hydration parameter corresponds to an equivalent number of grams or moles of water that would have to be added (or removed) with respect to one gram or mole of the macromolecule in the (same) dialysis experiment. These parameters can have positive or negative values.

In a structural approach, we can consider the model of a domain close to the macromolecule containing at minimum the solvent perturbed by the macromolecule (Fig. 17.2), comprising N_1 mole, or B_1 gram, of water and N_3 mole, or B_3 gram, of cosolvent. The part of the domain with the same composition as the bulk (i.e., exceeding the perturbed solvent) will not contribute to the preferential binding parameter (see the equations below). The preferential solvent binding parameters can be expressed as (Eisenberg 2000; Timasheff 1993):

$$\left(\frac{\partial w_3}{\partial w_2}\right)_\mu = B_3 - B_1\, w_3 \quad (17.20)$$

$$\left(\frac{\partial w_1}{\partial w_2}\right)_\mu = B_1 - B_3\frac{1}{w_3} \quad (17.21)$$

$$\left(\frac{\partial m_3}{\partial m_2}\right)_\mu = N_3 - N_1\,\frac{m_3}{m_1} \quad (17.22)$$

$$\left(\frac{\partial m_1}{\partial m_2}\right)_\mu = N_1 - N_3\,\frac{m_1}{m_3} \quad (17.23)$$

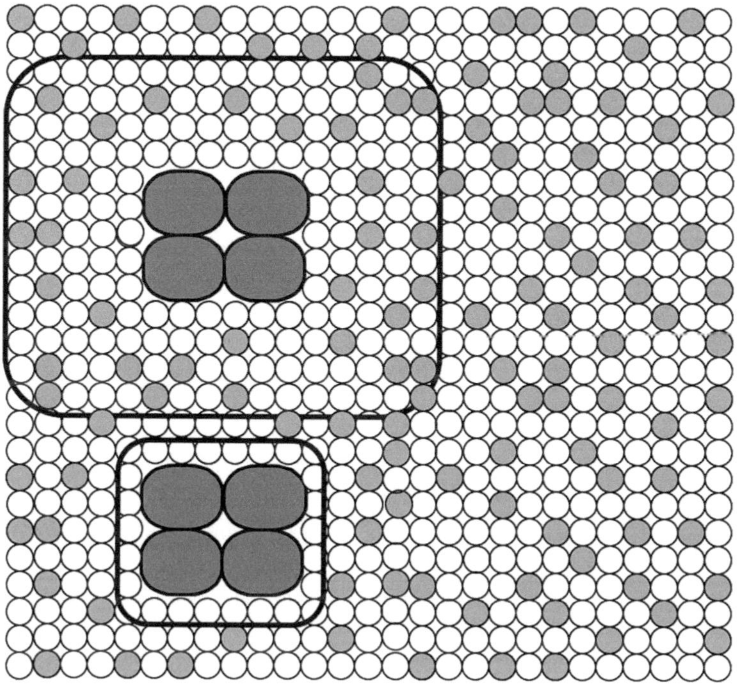

Fig. 17.2 Description of the preferential interactions parameters in the two-domain approach. The domain should contain the solvent of perturbed composition when compared to the bulk. The figure describes the case of preferential hydration

The dissociation of the counterions of a polyelectrolyte induces solvent rearrangements that are equivalent to negative salt binding. For complete counterion dissociation, from a negatively charged polyelectrolyte in a solution containing a salt, the values of $N_{3,\,Donnan}$ corresponding to the Donnan effect are par mole of charge (Eisenberg 1976):

$N_{3,\,Donnan} = -1/2$ for a salt X^+Y^-
$N_{3,\,Donnan} = -1/3$ for a salt $2X^+Y^{2-}$
$N_{3,\,Donnan} = -1/6$ for a salt $X^{2+}2Y^-$
$N_{3,\,Donnan} = -1/4$ for a salt $X^{2+}Y^{2-}$

17.5.2.4 Models for Describing Solvation

It is important to notice that, in the two-domain approach, B_1, B_3, N_1, and N_3 are not thermodynamic parameters. B_1 and B_3, (or N_1 and N_3) are not unique and are defined in the frameworks of models, established usually from measurements done at different cosolvent concentrations. Within these models, we can cite the invariant particle model, in which a solvated particle of invariant composition is defined in a large range of cosolvent composition. This description can be considered when, e.g., the preferential solvent binding parameters are linearly related to the appropriate ratio between the two solvent components (Eq. 17.22). Such a description was fine for describing the solvation of rabbit muscle aldolase in a large range of glucose, sucrose, raffinose, or α-cyclodextrin concentrations: the data were consistent with B_3, values being essentially zero, and B_1, values being constant for each sugar, in the range of 0.18 g/g to 0.3 g/g, depending on the sugar size (Ebel et al. 2000). As a different very simple framework for the interpretation of preferential cosolvent binding parameters, hydration deduced from amino acids composition was postulated as a fixed values for N_1 in order to calculate the value of N_3. In this framework, N_1 corresponds to sites for water which are not exchangeable for the cosolvent (Timasheff 1993; Arakawa and Timasheff 1985). Schellman described the model of a macromolecule with independent binding sites exchanging water and ions and characterized by a unique exchange equilibrium constant K_b defined in the molar fraction activity scale (Schellman 1990, 1994; Schellman and Gassner 1996). In our study of halophilic proteins, which are only stable at multimolar salt concentration, we analyzed salt binding in the framework of this model (Ebel et al. 2002). Record proposed to use a partition coefficient, and estimates of the local domain as a monolayer of hydration water, to derive the number of osmolyte molecules in the local domain. For bovine serum albumin, N_3=58 cosolvent molecules are expected to be in the local domain for a random distribution and 0 for total exclusion. The experimental value for glycerol, N_3=48, is close to the former, while that for the strongly excluded betaine, N_3=8, is close to the latter (Courtenay et al. 2000).

17.6 Link Between Protein Equilibrium and Interactions with Solvent Components

17.6.1 Preferential Binding Parameters are True Thermodynamic Parameters

The preferential binding parameter expresses the mutual perturbation of the chemical potentials of components 2 and 3. The preferential cosolvent binding parameter expressed in molar ratio is the ratio of the partial derivatives of the chemical potentials μ_2 and μ_3 with molality m_3 of the cosolvent (component 3) at constant pressure, temperature, and molality of the macromolecule (Timasheff 1993):

$$\left(\frac{\partial m_3}{\partial m_2}\right)_{T,\,P,\mu_3} = -\frac{\left(\frac{\partial \mu_2}{\partial m_3}\right)_{TPm_2}}{\left(\frac{\partial \mu_3}{\partial m_3}\right)_{TPm_2}} = -\left(\frac{\partial \mu_2}{\partial \mu_3}\right)_{T,\,P,m_2}. \tag{17.24}$$

On the other hand (see, e.g., Ebel 2007):

$$\left(\frac{\partial m_3}{\partial m_2}\right)_{\mu} \cong \left(\frac{\partial m_3}{\partial m_2}\right)_{T,\,P,\mu_3} \tag{17.25}$$

Preferential interactions of the solvent components with the macromolecule determine the change of the chemical potential of the macromolecule when the cosolvent concentration changes.

17.6.2 The Wyman Linkage Relationship

The equilibrium between macromolecular species will be modified when changing solvent composition, if there is a change in the preferential solvent binding parameter upon reaction. Considering the equilibrium between reactants *R* and products *P* (Timasheff 1993):

$$\nu_{R_1}R_1 + \nu_{R_2}R_2 <=> \nu_{P_1}P_1 + \nu_{P_2}P_2 \tag{17.26}$$

If K_{obs} describes the equilibrium constant between the macromolecular species:

$$K_{obs} = \left[\frac{m_{P_1}^{\nu_{P_1}} \times m_{P_2}^{\nu_{P_2}}}{m_{R_1}^{\nu_{R_1}} \times m_{R_2}^{\nu_{R_2}}}\right] \tag{17.27}$$

The Wyman linkage (Wyman 1964) relationship gives:

$$\frac{\partial lnK_{obs}}{\partial lna_3} = \nu_{P_1}\left(\frac{\partial m_3}{\partial m_{P_1}}\right)_{P\mu_3} + \nu_{P_2}\left(\frac{\partial m_3}{\partial m_{P_2}}\right)_{P\mu_3} - \nu_{R_1}\left(\frac{\partial m_3}{\partial m_{R_1}}\right)_{P\mu_3} + \nu_{R_2}\left(\frac{\partial m_3}{\partial m_{R_2}}\right)_{P\mu_3} \tag{17.28}$$

In a condensed simplified form:

$$\frac{\partial lnK_{obs}}{\partial lna_3} = \Delta\left(\frac{\partial m_3}{\partial m_{P_1}}\right)_{P\mu_3} \tag{17.29}$$

In terms of associated water and cosolvent, in the two-domain model, considering and with the reasonable approximation Eq. (17.25):

$$\frac{\partial lnK_{obs}}{\partial lna_3} = \Delta N_3 - \Delta N_1\,\frac{m_3}{m_1} \tag{17.30}$$

When increasing the cosolvent concentration in the solvent, the equilibrium constant will be larger if the preferential cosolvent binding is larger for the products than for the reactants. This is intuitively expected considering reaction of mass action with explicit interaction with solvent: when increasing the cosolvent, there is displacement of the equilibrium toward the macromolecular species interacting more strongly with the cosolvent (less efficiently with water).

17.6.3 Implications

A large variety of macromolecular equilibrium (schematized on Fig. 17.3) were described to be modulated by weak interactions with solvent. "Weak" because these equilibriums are modulated by molar concentrations of cosolvent (e.g., denaturation in 6 M urea or precipitation in 2 M ammonium sulfate). The relative stabilization of the folded and unfolded states is related to the global specific and nonspecific solvent interactions of each of them (Timasheff 1993, 1998, 2002). For example, salts have been classified in the Hofmeister series according to their salting-out properties, which favor folded or precipitated protein states (for a translation of the original manuscript, see Kunz et al. 2004). While the origin of these effects is rather complex – for a recent review, see Salis and Ninham 2014 – salting-out effects were related to the propensity of the salts to induce preferential hydration at high cosolvent concentration in a nonspecific way, independently of the details of the chemical nature of the protein surface. The preferential hydration is related to the effect of the salt on the surface tension of the solution and would favor the macromolecular state with the lowest exposed surface (i.e., the folded rather than the unfolded form) (Timasheff 1993). As an example of such linkage in the framework of membrane protein studies, we analyzed the dimer-monomer equilibrium dissociation constants, measured at different detergent concentrations. The dissociation constants are larger when the detergent concentration increases. This is related to the fact that the dissociated monomers bind more detergent that the dimer (Josse et al. 2002).

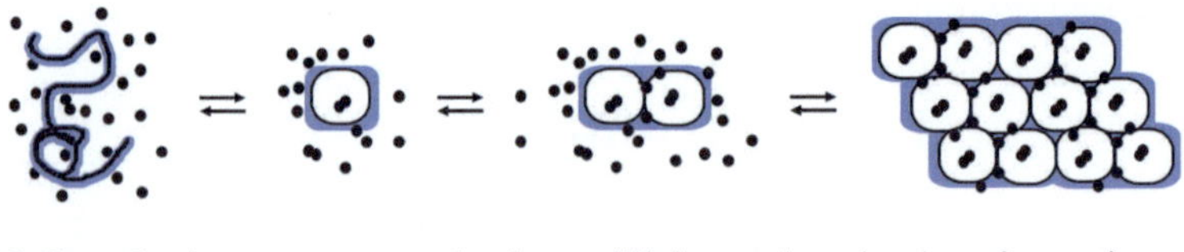

Fig. 17.3 Interplay between macromolecular equilibrium and weak solvent interactions. A, folding; B, multimerization; and C, precipitation/ crystallization are processes that are modulated by interactions with solvent components. When increasing the cosolvent concentration, equilibrium will be displaced toward the macromolecule conformation stabilized by the cosolvent

17.7 Conclusion

The general formalism allowing to describe the sedimentation of macromolecule in complex solvent has been described. It allows to relate the buoyant molar mass, measured in AUC at low concentration of macromolecule, to the macromolecule molar mass, through thermodynamic preferential binding parameters. These parameters allow to derive values for the numbers of water and cosolvent interacting with the macromolecule, in the framework of models. When modifying the solvent composition, i.e., cosolvent activity, equilibrium constants characterizing conversion between macromolecular species changes are directly related to the differences in the preferential binding parameters, which can be interpreted, in the framework of models, in terms of differences in bound water and cosolvent.

Acknowledgments This work used the AUC platform of the Grenoble Instruct center (ISBG; UMS 3518 CNRS-CEA-UJF-EMBL) with support from FRISBI (ANR-10-INSB-05-02) and GRAL (ANR-10-LABX-49-01) within the Grenoble Partnership for Structural Biology (PSB).

References

Arakawa T, Timasheff SN (1985) The stabilization of proteins by osmolytes. Biophys J 47(3):411–414

Brown PH, Balbo A, Zhao H, Ebel C, Schuck P (2011) Density contrast sedimentation velocity for the determination of protein partial-specific volumes. PLoS One 6(10):e26221. doi:10.1371/journal.pone.0026221PONE-D-11-13392 [pii]

Butler PJ, Ubarretxena-Belandia I, Warne T, Tate CG (2004) The Escherichia coli multidrug transporter EmrE is a dimer in the detergent-solubilised state. J Mol Biol 340(4):797–808

Casassa EF, Eisenberg H (1964) Thermodynamic analysis of multicomponent solutions. Adv Protein Chem 19:287–395. doi:10.1016/s0065-3233(08)60191-6

Cohen G, Eisenberg H (1968) Deoxyribonucleate solutions: sedimentation in a density gradient, partial specific volumes, density and refractive index increments, and preferential interactions. Biopolymers 6(8):1077–1100

Courtenay ES, Capp MW, Anderson CF, Record MT Jr (2000) Vapor pressure osmometry studies of osmolyte-protein interactions: implications for the action of osmoprotectants in vivo and for the interpretation of "osmotic stress" experiments in vitro. Biochemistry 39(15):4455–4471

Dach I, Olesen C, Signor L, Nissen P, le Maire M, Moller JV, Ebel C (2012) Active detergent-solubilized H+, K+−ATPase is a Monomer. J Biol Chem 287(50):41963–41978. doi:10.1074/jbc.M112.398768

Ebel C (2007) Solvent mediated protein-protein interactions. In: Schuck P (ed) Protein interactions – biophysical approaches for the study of complex reversible systems, vol 5, Protein reviews. Springer, Berlin, pp 255–287

Ebel C, Faou P, Zaccai G (1999) Protein-solvent and weak protein-protein interactions in halophilic malate dehydrogenase. J Cryst Growth 196:395–402

Ebel C, Eisenberg H, Ghirlando R (2000) Probing protein-sugar interactions. Biophys J 78(1):385–393

Ebel C, Costenaro L, Pascu M, Faou P, Kernel B, Proust-De Martin F, Zaccai G (2002) Solvent interactions of halophilic malate dehydrogenase. Biochemistry 41(44):13234–13244

Eisenberg H (1976) Biological macromolecules and polyelectrolytes in solution, Monographs on physical biochemistry. Clarendon, Oxford
Eisenberg H (1990) Solution properties of DNA: sedimentation, scattering of light, X-rays and neutrons, and viscometry. In: Saenger W (ed) Landolt-Börnstein: numerical data and functional relationships in science and technology, new series biophysics-nucleic acids, vol VII/1C. Springer, Berlin, pp 257–276
Eisenberg H (2000) Analytical ultracentrifugation in a Gibbsian perspective. Biophys Chem 88(1–3):1–9
Eisenberg H (2003) Adair was right in his time. Eur Biophys J Biophys Lett 32:406–411
Gohon Y, Pavlov G, Timmins P, Tribet C, Popot J-L, Ebel C (2004) Partial specific volume and solvent interactions of amphipol A8-35. Anal Biochem 334(2):318–334
Josse D, Ebel C, Stroebel D, Fontaine A, Borges F, Echalier A, Baud D, Renault F, Le Maire M, Chabrieres E, Masson P (2002) Oligomeric states of the detergent-solubilized human serum paraoxonase (PON1). J Biol Chem 277(36):33386–33397
Kernel B, Zaccai G, Ebel C (1999) Determination of partial molal volumes, and salt and water binding of highly charged biological macromolecules (tRNA, halophilic protein) in multimolar salt solutions. Progr Colloid Polym Sci 113:168–175
Kunz W, Henle J, Ninham BW (2004) 'Zur Lehre von der Wirkung der Salze' (about the science of the effect of salts): Franz Hofmeister's historical papers. Curr Opin Colloid Interface Sci 9(1–2):19–37. doi:10.1016/j.cocis.2004.05.005
Le Roy A, Nury H, Wiseman B, Sarwan J, Jault JM, Ebel C (2013) Sedimentation velocity analytical ultracentrifugation in hydrogenated and deuterated solvents for the characterization of membrane proteins. Methods Mol Biol 1033:219–251. doi:10.1007/978-1-62703-487-6_15
Le Roy A, Wang K, Schaack B, Schuck P, Breyton C, Ebel C (2015) AUC and small-angle scattering for membrane proteins. Methods Enzymol 562:257–286. doi:10.1016/bs.mie.2015.06.010
Monnin C (1989) An ion interaction-model for the volumetric properties of natural-waters – density of the solution and partial molal volumes of electrolytes to high-concentrations at 25-degrees-C. Geochim Cosmochim Acta 53(6):1177–1188. doi:10.1016/0016-7037(89)90055-0
Reynolds JA, Tanford C (1976) Determination of molecular weight of the protein moiety in protein-detergent complexes without direct knowledge of detergent binding. Proc Natl Acad Sci U S A 73(12):4467–4470
Salis A, Ninham BW (2014) Models and mechanisms of Hofmeister effects in electrolyte solutions, and colloid and protein systems revisited. Chem Soc Rev 43(21):7358–7377. doi:10.1039/c4cs00144c
Saluja A, Fesinmeyer RM, Hogan S, Brems DN, Gokarn YR (2010) Diffusion and sedimentation interaction parameters for measuring the second virial coefficient and their utility as predictors of protein aggregation. Biophys J 99(8):2657–2665. doi:S0006-3495(10)00989-6 [pii]10.1016/j.bpj.2010.08.020
Schellman JA (1990) A simple model for solvation in mixed solvents. Applications to the stabilization and destabilization of macromolecular structures. Biophys Chem 37(1–3):121–140
Schellman JA (1994) The thermodynamics of solvent exchange. Biopolymers 34(8):1015–1026
Schellman JA (2003) Protein stability in mixed solvents: a balance of contact interaction and excluded volume. Biophys J 85(1):108–125. doi:10.1016/S0006-3495(03)74459-2
Schellman JA, Gassner NC (1996) The enthalpy of transfer of unfolded proteins into solutions of urea and guanidinium chloride. Biophys Chem 59(3):259–275
Schuck P (2004a) A model for sedimentation in inhomogeneous media. I. Dynamic density gradients from sedimenting co-solutes. Biophys Chem 108(1–3):187–200. doi:10.1016/j.bpc.2003.10.016
Schuck P (2004b) A model for sedimentation in inhomogeneous media. II. Compressibility of aqueous and organic solvents. Biophys Chem 108(1–3):201–214. doi:10.1016/j.bpc.2003.10.017
Solovyova A, Schuck P, Costenaro L, Ebel C (2001) Non-ideality by sedimentation velocity of halophilic malate dehydrogenase in complex solvents. Biophys J 81(4):1868–1880

Timasheff SN (1993) The control of protein stability and association by weak interactions with water: how do solvents affect these processes? Annu Rev Biophys Biomol Struct 22:67–97
Timasheff SN (1998) Control of protein stability and reactions by weakly interacting cosolvents: the simplicity of the complicated. Adv Protein Chem 51:355–432
Timasheff SN (2002) Protein-solvent preferential interactions, protein hydration, and the modulation of biochemical reactions by solvent components. Proc Natl Acad Sci U S A 99(15):9721–9726. doi:10.1073/pnas.122225399
Wyman J (1964) Linked functions and reciprocal effects in hemoglobin: a second look. Adv Protein Chem 19:224–284

Chapter 18
Aspects of the Analytical Ultracentrifuge Determination of the Molar Mass Distribution of Polysaccharides

Stephen E. Harding, Gary G. Adams, Richard B. Gillis, Fahad M. Almutairi, and Gordon A. Morris

Abstract Molar mass or 'molecular weight' is one of the most fundamental parameters describing a macromolecule. Because of their polydisperse nature, polysaccharides are usually described by distributions of molar mass. SEC-MALS (size exclusion chromatography coupled to multi-angle light scattering) is often a convenient method of choice, but there are many instances where it is unsuitable. Modern AUC (analytical ultracentrifuge) methods provide a valuable alternative – now easier to use than before – and, after briefly reviewing some older procedures, we highlight two recently published and complementary methods, namely, the 'Extended Fujita' approach for the analysis of sedimentation velocity data and SEDFIT-MSTAR for the analysis of sedimentation equilibrium data. Nonideality needs to be considered and can be dealt with in a standard way. These methods can also indicate if associative phenomena are present, which can then be quantified using more complex AUC algorithms.

Keywords Polydispersity • Nonideality • Sedimentation velocity • Sedimentation equilibrium

---

S.E. Harding (✉) • G.G. Adams • R.B. Gillis
National Centre for Macromolecular Hydrodynamics, School of Biosciences, University of Nottingham, College Road, Sutton Bonington LE12 5RD, UK
e-mail: steve.harding@nottingham.ac.uk

F.M. Almutairi
Biochemistry Department, Faculty of Science, University of Tabuk, P.O. Box 741-Tabuk 71491, Saudi Arabia

G.A. Morris
Department of Chemical Sciences, School of Applied Sciences, University of Huddersfield, Queensgate, Huddersfield HD1 3DH, UK

S. Uchiyama et al. (eds.), *Analytical Ultracentrifugation*,
DOI 10.1007/978-4-431-55985-6_18

18.1 Introduction

One of the most fundamental parameters characterising a macromolecule is its molar mass or 'molecular weight', M, expressed in g/mol or Daltons (Da), or relative molecular mass M_r (relative to the mass of a hydrogen atom or 1/12th of a carbon atom). All three are numerically equivalent. Protein scientists tend to prefer 'molecular weight' or relative molecular mass, polymer scientists molar mass. Polysaccharides seem to be at the halfway house. They are in most cases biological or natural polymers like proteins, although they are polydisperse like synthetic polymers – so nomenclature for these substances tends to be interchangeable between molecular weight and molar mass: in this article we will keep to the latter.

Polysaccharides in common with most polymeric systems are polydisperse and hence will possess a distribution of molar mass (Fujita 1962). This distribution can be described by a distribution plot $f(M)$ versus M, in terms of averages or 'moments', commonly M_n, the number average; M_w, the weight average; and M_z, the z-average, with $M_n < M_w < M_z$, and ratios of the averages M_w/M_n or M_z/M_w, commonly referred to as 'polydispersity indices'.

In common with all other macromolecules, a complication is that due to the effects of nonideality (through co-exclusion and polyelectrolyte behaviour), what is actually measured in an analytical ultracentrifugal determination of molar mass at a finite concentration, c, is the apparent molar mass, M_{app}, with $M_{app} < M$, the true or 'ideal' molar mass, a difference which becomes vanishingly small as the concentration $c \to 0$. So either working at a low enough concentration such that $M_{app} \sim M$ or by measurement at several values of c and performing an appropriate extrapolation to $c \to 0$, the correct M can be obtained. Extrapolation of a distribution $f(M_{app})$ vs. M_{app} is not possible, but comparison/superposability of $f(M_{app})$ vs. M_{app} plots at sufficiently low concentration should give an idea of the reliability of the distribution obtained.

In this short review, we highlight two recently published and complementary methods, namely, the 'Extended Fujita' approach for the analysis of sedimentation velocity data and SEDFIT-MSTAR for the analysis of sedimentation equilibrium data. Nonideality can be dealt with in a standard way, and these methods can also indicate if associative phenomena are present, which can then be quantified using more complex analytical ultracentrifuge (AUC) algorithms. These methods also complement well approaches based on other techniques for obtaining molar mass averages and distributions based on, for example, light scattering, particularly when the latter is found inapplicable for particular systems.

18.2 SEC-MALS

The very first determination of the molar mass distribution $f(M)$ vs. M of a polysaccharide by SEC-MALS (formerly 'SEC-MALLS') – size exclusion chromatography coupled to laser light scattering – was by Horton et al. (Horton et al.

1991a) for a sodium alginate. The rapidity and convenience of the method was clearly illustrated, and the effects of nonideality – and how to correct for them – were demonstrated. Encouragingly, results for the weight average molar mass, M_w, for the whole distribution were in agreement with the results from the independent method of sedimentation equilibrium in the ultracentrifuge (Horton et al. 1991a, b). Since that demonstration nearly a quarter of a century ago, it has become the method of choice for polysaccharide molar mass characterisation. The columns – besides separating materials of different sizes – also acted as online filters helping circumvent the age-old problem of light scattering on solutions of macromolecules, namely, the crippling effects of the presence of trace amounts of dust and other supramolecular contaminants. The method could also be applied to glycoconjugates, and the first demonstration of its application to mucin glycoproteins (which are over 80% glycosylated) was by Jumel et al. 5 years later (Jumel et al. 1996).

Despite the great power and convenience of SEC-MALS, there are limitations, and the need for orthogonal approaches soon became clear. For the large molar mass polysaccharides such as xanthan (M_w ~3–4 × 10^6 g/mol), the columns gave poor separation, and for others such as the highly cationic chitosans, non-inertness or interactions with the columns gave anomalous results. The separation problem can be circumvented by the use of field-flow fractionation systems although problems of non-inertness through anomalous interactions with the membranes can also lead to erroneous results.

18.3 Sedimentation in the Analytical Ultracentrifuge

Sedimentation methods offer a complementary approach to SEC-MALS with their inherent fractionation ability without the need for columns or membranes. Of the two sedimentation methods – sedimentation velocity and sedimentation equilibrium – the former has a greater resolving power of components, but the velocity of sedimentation depends on the shape and friction properties of the macromolecular species as well as the molar mass. Allowance for shape/friction has to be made in order to interpret the data in terms of molar mass. Sedimentation equilibrium by contrast has an absolute basis (no shape/friction contribution), but it is not as resolving due to the lower rotational speeds used. Furthermore, because of the higher concentrations normally required to get an interpretable signal, nonideality is a more serious problem, as shown in Table 2 of Harding et al. (1992) – which compares relative nonidealities for a wide range of polysaccharides. Although obtaining apparent weight average $M_{w,app}$ and apparent z-average $M_{z,app}$ molar masses was possible (and employing a zero concentration extrapolation procedure to eliminate nonideality), obtaining a distribution was more difficult. We showed this was possible for nonideal paucidisperse systems of up to three components, successfully applied to mucins (Harding 1985) (Fig. 18.1a), but nonideal quasi-continuous distributions that characterise polysaccharides proved impossible. A laborious 'long way round' method involving a combination with size exclusion

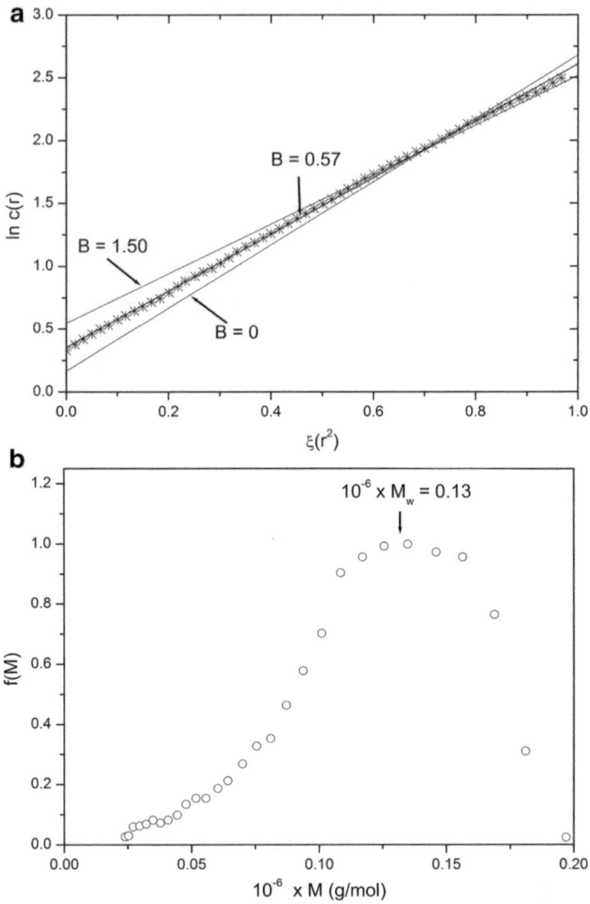

Fig. 18.1 Early attempts at obtaining molar mass distributions of polysaccharides and glycoconjugates from sedimentation equilibrium. (**a**) Modelling log concentration versus distance squared plots for a mucin glycoprotein from a cystic fibrosis patient in terms of a three-component nonideal system, with proportions of the components corresponding to imaging from transmission electron microscopy. $c(r)$ is the concentration in fringe displacement units and $\xi(r^2) = (r^2 - r_m{}^2)/(r^2 - r_b{}^2)$ where r_m is the radial position of the meniscus and r_b the cell base. A three-component system (with relative proportions based on transmission electron microscopy) fits the data for a second (cumulative) thermodynamic virial coefficient $B = 0.15 \times 10^{-4}$ ml.mol.g^{-2} (from Harding (1985)). (**b**) Off-line calibration of preparative SEC columns by sedimentation equilibrium leading to an absolute (and nonideality corrected) molar mass distribution for an alginate (manucol DM in phosphate-chloride buffer, I = 0.3, pH = 6.5). The weight average molar mass obtained for unfractionated material is shown (from Ball et al. 1988)

chromatography or 'gel permeation chromatography' with the M_w of fractions being obtained by sedimentation equilibrium was nonetheless possible, and the log M_w versus elution volume V_e data for the fractions was used to provide an absolute calibration of the SEC/GPC column – successfully applied to alginates (Ball et al. 1988) (Fig. 18.1b), dextran (Ball et al. 1990) and pectin (Harding et al. 1991).

18.4 Sedimentation Velocity: 'Extended Fujita' Method

In a centrifugal field, macromolecular solute molecules (of density, $\rho >$ density of the solvent ρ_o) will sediment towards the cell base; therefore, the region near the meniscus will be depleted of solute, and there will be a region nearer the cell base where the solute concentration is uniform and a transitional region is created (the 'boundary region') where the solute concentration varies with the distance from the axis of rotation (nb. when $\rho < \rho_o$ the movement – flotation – will be in the opposite direction but the same principles apply). It is the rate of movement of the concentration distribution with time which allows the calculation of sedimentation coefficients and the distribution of sedimentation coefficients (see, e.g. Harding et al. 1992; Schuck 2000; Dam and Schuck 2004). The progression of the concentration distribution with time is recorded by an optical system. Since polysaccharides are not usually absorbing in the visible or (near) ultraviolet region, the refractometric or Rayleigh interference optical system is the most useful, using a laser light source. Double-sector cells are employed with solution and reference solvent (dialysate) in each channel. A series of parallel Rayleigh interference fringes are captured on a CCD camera. These register the concentration distribution at regular time intervals throughout the experiment. The change in the distribution with time yields both the weight average sedimentation coefficient (s) measured in seconds (s) or Svedberg units (S) in which 1 S = 10^{-13} s and the distribution of sedimentation distribution $g(s)$.

1. To facilitate comparisons, the s value (a measure of the size and shape of the polysaccharide) is usually corrected to standard conditions (density and viscosity of water at 20.0 °C), to give $s_{20,w}$, and this is done using a database algorithm known as SEDNTERP (Laue et al. 1992).
2. To correct for nonideality, the s (or $s_{20,w}$) value is extrapolated to zero concentration to give $s^o{}_{20,w}$, using, for example, the Gralén relation (Gralén 1944):

$$\frac{1}{s_{20,w}} = \frac{1}{s^o{}_{20,w}}(1 + k_s c) \tag{18.1}$$

where k_s (ml/g) is the concentration dependence regression coefficient. For more severely concentration-dependent systems, other relations such as the equation of Rowe (1977, 1992) can be used. Alternatively, sufficiently low loading concentrations can usually be employed (it is possible to make measurements below 0.1 mg/ml), such that $s_{20,w} \sim s^o{}_{20,w}$ becomes a reasonable approximation. Concentrations should in any case always be lower than the critical overlap concentration, $c^* \sim \chi[\eta]$, where $[\eta]$ is the intrinsic viscosity and χ ~0.3–0.6 (see Harding 1997).

3. Besides nonideality – which needs to be accounted for as described above – the distribution $g(s)$ vs. $s_{20,w}$ (nb. for simplicity in what follows, we use just s for $s_{20,w}$) will be affected by diffusion broadening (although polysaccharides are usually much slower diffusing compared to proteins). Schuck (2000) and

Dam and Schuck (2004) have described a procedure for making an approximate correction based on the assumption that all the species can be represented by an average frictional ratio. The diffusion-corrected distribution is known as a $c(s)$ vs. s plot.

4. Plots of $g(s)$ and $c(s)$ by themselves can provide a useful measure of heterogeneity – e.g. for quasi-continuous distributions of molecular weight – for all polysaccharides and also when there are discrete components present, such as mixed polysaccharide systems like amylose and amylopectin in starch.
5. Plots of $g(s)$ vs. s (or $c(s)$ vs. s) can be converted into molar mass distributions $f(M)$ vs. M provided that the conformation/conformation type (sphere, rod, coil, etc.) of the polysaccharide is known or can be reasonably assumed. The procedure is known as the *Extended Fujita* method (Harding et al. 2011). Fujita (1962) had originally published a method specifically for random coil polymers. The 'Extended' Fujita method extends its application to all conformation types, and the method has recently been incorporated into the highly popular SEDFIT platform of algorithms to estimate the molar mass distribution of heterogeneous systems including polysaccharides and mucins (Harding et al. 2011; Gillis et al. 2013a).

One limitation is that this *Extended Fujita* method does need calibrating for each particular conformational system. The conformation coefficient b and constant κ_s in the transformations:

$$M = (s/\kappa_s)^{1/b} \tag{18.2}$$

and

$$f(M) = \mathrm{d}s/\mathrm{d}M.\, g(s) \tag{18.3}$$

where

$$\mathrm{d}s/\mathrm{d}M = b.{\kappa_s}^{1/b}.s^{(b-1)/b} \tag{18.4}$$

are needed; if the conformation is known, then this will define b: random coils, $b \sim 0.4$–0.5; spheres, $b \sim 0.67$; and rod-shaped molecules, $b \sim 0.2$. Knowledge of both the weight average sedimentation coefficient and corresponding weight average molar mass from a sedimentation equilibrium experiment (or SEC-MALS experiment if it is applicable) can then be used to define κ_s, using Eq. (18.2).

If b is also unknown, then a number of pairs of s-M values are required to define both b and κ_s. Figure 18.2 gives an example of a determination for alginate at a concentration of 0.03 mg/ml. Working at low concentration also offers the additional benefit that complications through hypersharpening (larger molecular weight species being slowed down by having to sediment through solutions of the lower molecular weight species).

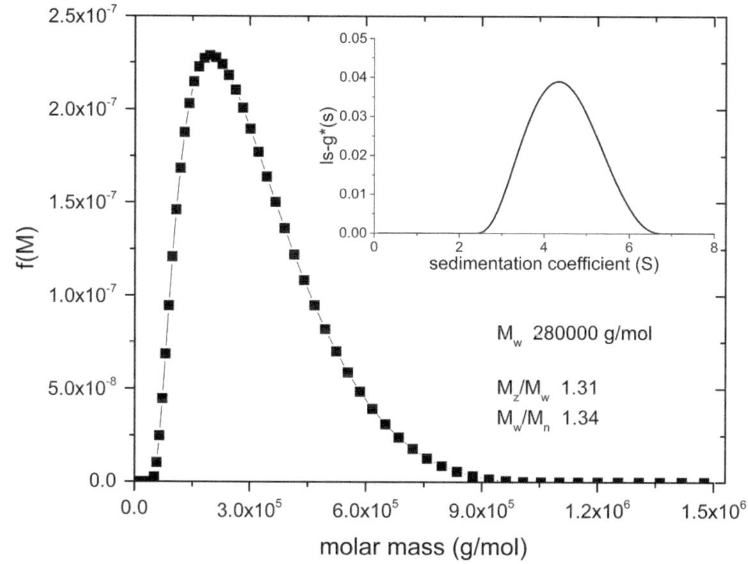

Fig. 18.2 Extended Fujita method estimate of the molar mass distribution $f(M)$ versus M for alginate at a loading concentration of 0.03 mg/ml in 0.3 M NaCl. Transformation from a g(s) vs. s plot (inset) using a value for b ~0.33 (Harding et al. 2011) and $\kappa_s = 0.0685$, the latter calculated from $M_w = 280{,}000$ g/mol (from SEC-MALS) and $s = s_{20,w}$ (at 0.03 mg/ml) = 4.3S. Estimates for M_z/M_w and M_w/M_n are also given

18.5 Sedimentation Equilibrium (SE): SEDFIT-MSTAR

In contrast to sedimentation velocity, sedimentation equilibrium requires lower angular velocities depending on the size of the macromolecule (van Holde 1985). As the solute sediments towards the cell base, the concentration therefore increases at base; this sets up a diffusion gradient, which opposes that of sedimentation. After a certain amount of time, the two processes reach dynamic equilibrium leading to a steady state pattern of solute concentration increasing towards the cell base. As there is no net movement of solute at equilibrium, the final pattern is not affected by frictional/conformation properties and is an absolute function of molar mass and polydispersity. For thermodynamically nonideal and polydisperse systems such as polysaccharides, solute distributions at sedimentation equilibrium can be analysed using the MSTAR algorithm, employing the M^* function, and built into a succession of programmes in FORTRAN (Creeth and Harding 1982; Harding et al. 1992) and PC BASIC (Cölfen and Harding 1997). This algorithm has been recently incorporated into the SEDFIT platform of algorithms, as SEDFIT-MSTAR. Full details are given in Schuck et al. (2014); we just give a very short summary here. Essentially, it yields an estimate for the apparent weight average molar mass for the whole distribution, $M_{w,app}$, using:

1. The M^* function of Creeth and Harding (1982) defined by the integral transformation

$$M^*(r) = \frac{c(r) - c_m}{kc_m\left(r^2 - r_m^2\right) + 2k\int_{r_m}^{r}(c(r) - c_m)\,rdr} \tag{18.5}$$

for sector-shaped solution columns with r the radial position in the ultracentrifuge cell and the meniscus concentration $c_m = c(r = r_m)$. k is defined by:

$$k = (1 - \overline{v}\rho)\,\omega^2/2RT \tag{18.6}$$

with $\overline{v}$ being the partial specific volume; ρ, the solvent density (Fujita 1962); ω, the rotor angular velocity; R, the gas constant; and T, the absolute temperature. $M^*(r)$ has several useful properties, the most important being the M^* extrapolated to the cell base $(r = r_b) = M_{w,app}$, the apparent weight average molar mass for the whole distribution

$$M_{w,app}(r = r_b) = M_{w,app} \tag{18.7}$$

2. The hinge point method: the 'hinge point' in the radial distribution is the radial position at which the local concentration $c(r)$ is equal to the initial cell loading concentration, c^o (which can be evaluated from the conservation of mass equation). SEDFIT-MSTAR algorithm evaluates the local or 'point' apparent weight average molar masses as a function of radial position, $M_{w,app}(r)$, and at the hinge point $(r = r_{\text{hinge}})$

$$M_{w,app}\left(r_{\text{hinge}}\right) = M_{w,app} \tag{18.8}$$

the apparent weight average molar mass for the whole distribution.

SEDFIT-MSTAR provides the facility for obtaining the hinge point by evaluating the initial loading concentration c^o from the conservation of mass equation. SEDFIT-MSTAR also offers a 'smart smoothing' procedure for providing an accurate estimate for the meniscus concentration c_m and baseline correction and also yields an *estimate* for the overall molar mass distribution. An example of the output for a carrageenan which had also been characterised by SEC-MALS is given in Fig. 18.3, and a comprehensive set of other examples for synthetic and real data systems are given in Schuck et al. (2014).

Low concentrations (close to the lowest concentration limit) should be employed to minimise nonideality effects. For standard 12 mm path length cells, a concentration of ~0.5 mg/ml is required to give a sufficient fringe increment between cell meniscus and base (this is considerably higher than the lowest limit for

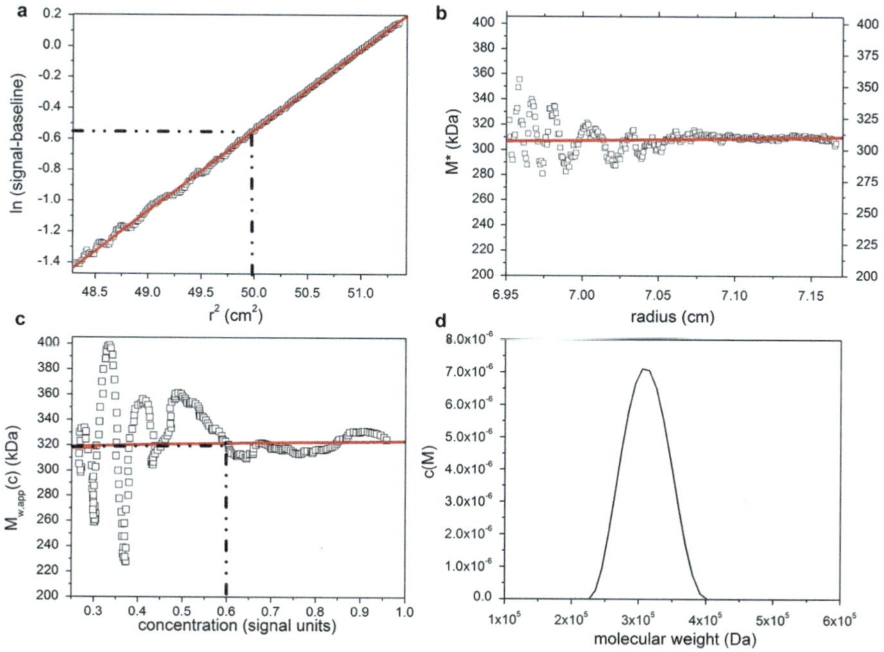

Fig. 18.3 SEDFIT-MSTAR output for analysis on a solution of λ-carrageenan at a loading concentration of 0.3 mg/ml (**a**) log concentration ln*c(r)* versus r^2 plot, where r is the radial distance from the centre (**b**) M^* versus r plot (*open squares*): the value of M^* extrapolated to the cell base $= M_{w,app}$, the apparent weight average molar mass for the whole distribution. Retrieved value for $M_{w,app} = 310{,}000$ g/mol (**c**) point or local apparent weight average molar mass at radial position r plotted against the local concentration $c(r)$ for different radial positions (**d**) of molar mass distribution, $c(M)$ vs. M plot. The *dot-dashed* lines show the position of the hinge point (in panel (**a**)) and the corresponding estimation of $M_{w,app}$ value (panel (**c**)), which retrieves a value for $M_{w,app} \sim 320{,}000$ g/mol. Both retrieved values for $M_{w,app}$ are in agreement with SEC-MALS (From Schuck et al. 2014)

sedimentation velocity experiments). The availability of 20 mm path length cells available from Nanolytics Ltd. (Potsdam, Germany) makes it possible to go as a low as ~0.3 mg/ml, and this is usually sufficient.

However, if working at these low loading concentrations the approximation $M_w \sim M_{w,app}$ is still not valid, the conventional way of dealing with this situation is to perform a series of measurements at different loading concentration and extrapolate back to zero concentration where these effects tend to vanish. The form of the extrapolation can be linear or non-linear. For obtaining $M_{w,app}$ using procedures that do not involve an integration, there is a simple relation relating $M_{w,app}$ and M_w at dilution solution:

$$M_{w,app} = M_w.\{1/(1 + 2BM_w c)\} \tag{18.9}$$

where B is the second thermodynamic virial coefficient (ml.mol.g^{-2}). $M_{w,app}$ values evaluated according to Eq. (18.8) at the hinge point conform to this relation, and a simple linear extrapolation of $1/M_{w,app}$ plotted versus loading concentration c yields the reciprocal of the true M_w from the intercept at $c = 0$. At higher concentrations, the extrapolation may not be linear and an extra virial term in c^2 may be required. Furthermore, for evaluations involving an integral transformation such as Eq. (18.5) to obtain the whole cell distribution M_w, there may also be a speed-dependent enhancement of the nonideality effects (Fujita 1962; Harding 1992) leading to a larger effective value for B and also departure from a linear form of the extrapolation, becoming:

$$M_{w,app} \sim M_w - 2Bc.M_w^2 \left(1 + \lambda^2 M_z^2/12\right) + \cdots \qquad (18.10)$$

where λ the 'speed dependence parameter' $= k.\ (r_b^2 - r_m^2)/2$ with k defined by Eq. (18.6).

So although $M_{w,app}$ from Eq. (18.7) can generally be obtained to a higher precision than from the point average $M_{w,app}$ evaluated from Eq. (18.8) at the hinge point – and without assumptions over conservation of mass – the nonideality effect will be greater. SEDFIT-MSTAR therefore incorporates both methods of $M_{w,app}$ evaluation (Schuck et al. 2014). Once M_w, the whole distribution weight average molar mass, has been obtained, it can also be used with the corresponding value of the sedimentation coefficient to help calibrate the Extended Fujita method, as noted above. Although SEDFIT-MSTAR gives, after allowance for nonideality, an accurate estimate of the weight average molar mass for the whole distribution, M_w, and how $M_w(r)$ varies with radial displacement r and concentration $c(r)$, it should be stressed that the estimated molar mass distribution $c(M)$ vs. M is only low resolution as clearly shown by Schuck et al. (2014): for resolution of discrete types of polydispersity (as opposed to quasi-continuous forms), the Extended Fujita method should be used because of the much better resolving power of sedimentation velocity. Alternatively, the routine MULTISIG can be applied to sedimentation equilibrium data, based on the principle of fitting the concentration distribution $c(r)$ vs. r to a 17-component exponential fit. Good resolution of components is possible, and Gillis et al. (2013b) give an example for chicory root inulin. The method assumes however thermodynamic ideality, so low concentrations must be employed: for very nonideal systems (because of the non-exponential way, the nonideality contribution enters into the fundamental equation for sedimentation equilibrium).

Consideration of the g(s) vs. s, $c(s)$ vs. s or $f(M)$ vs. M distributions from sedimentation velocity or the $M_{w,app}$ or $M_{w,app}(r)$ versus $c(r)$ profiles from sedimentation equilibrium (or both) may suggest self-associative or other forms of interactive behaviour. If this is the case, more sophisticated methods of analysis may then be justified to explore the interaction in further detail using MULTISIG (Gillis et al. 2013b). An outstanding example of this was the discovery of discrete protein-like associative phenomena in aminocelluloses (Heinze et al. 2011). Further detailed analysis of the variation in point number, weight, and z-average molar mass ($M_n(r)$,

$M_w(r)$, and $M_z(r)$, respectively) with radial position and local concentration $c(r)$ in the ultracentrifuge cell in sedimentation equilibrium showed that at least for one class of aminocellulose, the association was fully reversible – a fully reversible tetramerisation (Nikolajski et al. 2014).

18.6 Concluding Remarks

It is hoped that this short review has given an idea of what is possible in terms of polysaccharide molar mass and molar mass distribution determination. So long as adequate consideration is given to issues such as thermodynamic nonideality (particularly for sedimentation equilibrium determinations) and hypersharpening (sedimentation velocity determinations), analytical ultracentrifugation provides an accurate method for characterising polysaccharides. A more extensive review of the methodology is given in Harding et al. (2015) – including the characterisation of polysaccharide conformation and interactions.

References

Ball A, Harding SE, Mitchell JR (1988) Combined low-speed sedimentation equilibrium/gel permeation chromatography approach to molar mass distribution analysis. Int J Biol Macromol 10:259–264

Ball A, Harding SE, Simpkin NJ (1990) On the molar mass distribution of dextran T-500. Gums Stabilis Food Indus 5:447–450

Cölfen H, Harding SE (1997) MSTARA and MSTARI: interactive PC algorithms for simple, model independent evaluation of sedimentation equilibrium data. Eur Biophys J 25:333–346

Creeth JM, Harding SE (1982) Some observations on a new type of point average molecular weight. J Biochem Biophys Methods 7:25–34

Dam J, Schuck P (2004) Calculating sedimentation coefficient distributions by direct modeling of sedimentation velocity concentration profiles. Methods Enzymol 384:185–212

Fujita H (1962) Mathematical theory of sedimentation analysis. Academic, New York

Gillis RB, Adams GG, Wolf B, Berry M, Besong TMD, Corfield A, Kök SM, Sidebottom R, Lafond D, Rowe AJ, Harding SE (2013a) Molar mass distribution analysis by ultracentrifugation: adaptation of a new approach for mucins. Carbohydr Polym 93:178–183

Gillis RB, Adams GG, Heinze T, Nikolajski M, Harding SE, Rowe AJ (2013b) MultiSig: a new high-precision approach to the analysis of complex biomolecular systems. Eur Biophys J 42:777–786

Gralén N (1944) Sedimentation and diffusion measurements on cellulose and cellulose derivatives. PhD dissertation, University of Uppsala, Sweden

Harding SE (1985) The representation of equilibrium solute distributions for non-ideal polydisperse systems in the analytical ultracentrifuge. Biophys J 47:247–250

Harding SE (1992) Sedimentation analysis of polysaccharides. In: Harding SE, Rowe AJ, Horton JC (eds) Analytical ultracentrifugation in biochemistry and polymer science. Royal Society of Chemistry, Cambridge, pp 495–516

Harding SE (1997) The intrinsic viscosity of biological macromolecules. Progress in measurement, interpretation and application to structure in dilute solution. Prog Biophys Mol Biol 68:207–262

Harding SE, Berth G, Ball A, Mitchell JR, Garcia de la Torre J (1991) The molar mass distribution and conformation of citrus pectins in solution studied by hydrodynamics. Carbohydr Polym 16:1–15

Harding SE, Horton JC, Morgan PJ (1992) MSTAR: A FORTRAN algorithm for the model independent molecular weight analysis of macromolecules using low speed or high speed sedimentation equilibrium. In: Harding SE, Rowe AJ, Horton JC (eds) Analytical ultracentrifugation in biochemistry and polymer science. Royal Society of Chemistry, Cambridge, pp 275–294

Harding SE, Schuck P, Abdelhammed AS, Adams G, Kök MS, Morris GA (2011) Extended Fujita approach to the molecular weight distribution of polysaccharides and other polymer systems. Methods 54:136–144

Harding SE, Adams GG, Almutairi F, Alzaharani Q, Erten TM, Kök MS and Gillis RB (2015) Ultracentrifuge methods for the analysis of polysaccharides, glycoconjugates and lignins. Meth Enzymol 562:391–439

Heinze T, Nikolajski M, Daus S, Besong TMD, Michaelis N, Berlin P, Morris GA, Rowe, AJ, Harding SE (2011) Protein-like oligomerisation of carbohydrates. Angew Chem Int Ed 50:8602–8604

Horton JC, Harding SE, Mitchell JR (1991a) Gel permeation chromatography – multi angle laser light scattering characterization of the molecular mass distribution of "Pronova" sodium alginate. Biochem Soc Trans 19:510–511

Horton JC, Harding SE, Mitchell JR, Morton-Holmes DF (1991b) Thermodynamic non-ideality of dilute solutions of sodium alginate studied by sedimentation equilibrium ultracentrifugation. Food Hydrocoll 5:125–127

Jumel K, Fiebrig I, Harding SE (1996) Rapid size distribution and purity analysis of gastric mucus glycoproteins by size exclusion chromotography/multi angle laser light scattering. Int J Biol Macromol 18:133–139

Laue TM, Shah BD, Ridgeway TM, Pelletier SL (1992) Computer-aided interpretation of analytical sedimentation data for proteins. In: Harding SE, Rowe AJ, Horton JC (eds) Analytical ultracentrifugation in biochemistry and polymer science. Royal Society of Chemistry, Cambridge, pp 90–125

Nikolajski M, Adams GG, Gillis RB, Besong DT, Rowe AJ, Heinze T, Harding SE (2014) Protein-like fully reversible tetramerisation and super-association of an aminocellulose. Sci Rep Nat 4:3861

Rowe AJ (1977) The concentration dependence of transport processes: a general description applicable to the sedimentation, translational diffusion, and viscosity coefficients of macromolecular solutes. Biopolymers 16:2595–2611

Rowe AJ (1992) The concentration dependence of sedimentation. In: Harding SE, Rowe AJ, Horton JC (eds) Analytical ultracentrifugation in biochemistry and polymer science. Royal Society of Chemistry, Cambridge, pp 394–406

Schuck P (2000) Size-distribution analysis of macromolecules by sedimentation velocity ultracentrifugation and Lamm equation modeling. Biophys J 78:1606–1619

Schuck P, Gillis RB, Besong D, Almutairi F, Adams GG, Rowe AJ, Harding SE (2014) SEDFIT-MSTAR: molar mass and molar mass distribution analysis of polymers by sedimentation equilibrium in the ultracentrifuge. Analyst 139:79–92

Van Holde KE (1985) Physical biochemistry. Prentice Hall, Indiana

Part VI
Applications of AUC: Biopharmaceuticals

Chapter 19
Use of Analytical Ultracentrifugation as an Orthogonal Method for Size Exclusion Chromatography: Assuring Quality for Therapeutic Protein Products and Meeting Regulatory Expectations

John F. Carpenter, David L. Bain, and Gibbes R. Johnson

Abstract Recombinant therapeutic proteins have revolutionized the treatment of numerous human diseases and disorders. From the advent of these products in the mid-1980s, there have been continual improvements in analytical methods for key degradation products and in our understanding of the mechanisms governing protein degradation. These advances have been especially important for analysis and control of protein aggregates. Not only are aggregate levels and types considered critical to product quality for therapeutic protein products, aggregates can lead to unwanted immunogenicity in patients. High-performance size exclusion chromatography (HP-SEC) has been the method of choice for decades and has served as the "workhorse" for aggregate analysis. However, often the HP-SEC results are not accurate measures of the monomer and aggregate levels (and sizes) that are present in the therapeutic protein drug product and can be misleading. To develop an accurate and reliable HP-SEC assay, results must be confirmed by an another method for quantifying and characterizing protein aggregates. To date, analytical ultracentrifugation (AUC) has proven to be the most powerful and useful method for this purpose. It took many years for AUC to be developed into a reliable method for quantitation and characterization of protein aggregates in therapeutic protein samples. But as a result of such efforts – an orthogonal method for HP-SEC – AUC is invaluable for the development and quality assurance of therapeutic protein products.

Keywords Therapeutic proteins • Aggregates • Size exclusion chromatography • Analytical ultracentrifugation • Critical quality attribute • Biosimilar

---

J.F. Carpenter (✉) • D.L. Bain
Department of Pharmaceutical Sciences, Center for Pharmaceutical Biotechnology, University of Colorado Anschutz Medical Campus, Aurora, CO 80045, USA
e-mail: John.Carpenter@ucdenver.edu

G.R. Johnson
Division of Therapeutic Proteins, Center for Drug Evaluation and Research, US Food and Drug Administration, Bethesda, MD 20892, USA

S. Uchiyama et al. (eds.), *Analytical Ultracentrifugation*,
DOI 10.1007/978-4-431-55985-6_19

Recombinant therapeutic proteins have revolutionized the treatment of numerous human diseases and disorders, such as cancers, multiple sclerosis, rheumatoid arthritis, inflammatory bowels disorders, and diabetes. From the advent of these products in the mid-1980s, there have been continual improvements in analytical methods for key degradation products and in our understanding of the mechanisms governing protein degradation. The scientific advances in these areas have led to vastly improved manufacturing processes and final formulations, as well as more stringent regulatory expectations for the characterization of therapeutic protein products. Over the last three decades, this iterative process of improved analytical methods and manufacturing, along with regulatory requirements for more in-depth and informative characterization, has led to great improvements in the quality of therapeutic proteins. And the improvements in product quality have directly benefitted patients, for example, through reduction in infusion and injection site reactions and fewer cases of loss of efficacy because of adverse immunogenicity.

These advances have been especially important for analysis and control of protein aggregates, and improved analytical methods have been critical in this expanded knowledge. With the concomitant increase in regulatory expectations (Cordoba-Rodriguez 2008), routine rigorous quantitation and characterization of aggregates are an essential part of successful development of a therapeutic protein product. Not only are aggregate levels and types considered critical to product quality for therapeutic protein products, aggregates can lead to unwanted immunogenicity in patients (Rosenberg 2006; Hermeling et al. 2004). An immune response induced by a therapeutic protein can have disastrous effects in patients, which include loss of efficacy, greater risk for severe infusion reactions, and neutralization of the endogenous counterpart of the therapeutic protein (Kessler et al. 2006). Thus, both for regulatory compliance and patient's health outcomes, it is absolutely essential that aggregates are controlled and that assays for aggregates are completely reliable.

For each therapeutic protein, it can be a daunting challenge to understand the causes of aggregates and how to control and limit their formation (Chi et al. 2003; Cromwell et al. 2006). Aggregation readily occurs at every step of a therapeutic protein's life cycle, from fermentation to administration to the patient. Even under optimal conditions that greatly favor the native state of a protein, aggregates can form at rates that are significant, considering processing times, expected shelf life, and delivery options (e.g., intravenous infusion after extensive dilution of stabilizing excipients) for most therapeutic proteins. Furthermore, therapeutic proteins are routinely subjected to aggregation-inducing stresses such as agitation, freeze-thawing, pH changes, liquid–solid interfaces, and exposure to particles of foreign materials. For each protein, it is important to have rigorous understanding of the effects of the stresses on protein degradation and how such damage can be minimized. These goals are accomplished by carefully studying and selecting solution and process conditions used in manufacturing, optimizing the final formulation, and choosing the least stressful route of administration(s). Such insights are critical for controlling protein aggregation during scale-up from production of nonclinical and clinical lots to commercial manufacturing and for appropriate control during routine manufacturing to assure consistent lot-to-lot product quality. Also, it is important to recognize that what may be considered relatively minor changes in processing conditions – such as altering the positioning of bottles of bulk drug substance in storage freezers – may have dramatic effects on protein aggregation.

An essential component of these efforts to control protein aggregation and to assure product quality is an assay that provides accurate and precise quantitation of protein aggregates. This method would aid in the characterization of these larger protein complexes and provide insight into their structural basis. A major goal for such assays is to assure that a commercial therapeutic product has the same quality and characteristics of the material used in pivotal clinical trials. Also, these assays confirm that each commercial batch of the therapeutic protein meets the specific quality attributes defined by the manufacturer. High performance size exclusion chromatography (HP-SEC) has been the method of choice for decades and has served as the "workhorse" for aggregate analysis. There are many positive attributes of the method. The required instruments and columns are readily available and their operation is widely familiar. HP-SEC can resolve various oligomeric species, even monomers from dimers, and quantitation is straightforward based on peak areas. Analysis times are relatively short, and the method can be run routinely by lab technicians without the need for advanced, specialized training. The results are highly reproducible, usually making it straightforward to validate the assays.

The problem is that often the HP-SEC results are not accurate measures of the monomer and aggregate levels (and sizes) that are present in the therapeutic protein drug product and can be misleading. And often these problems arise without any suspicions of the analytical scientists who have developed and employ the assay. In fact, there are unpublished examples of fully qualified and validated HP-SEC methods that grossly underreported the aggregate levels and/or mischaracterized aggregate sizes in therapeutic protein products. In one case, the method gave results of 98 % monomer and 2 % dimer for candidate formulation solutions, drug substance, and drug product. And even after drug product was placed in a boiling water bath, resulting in visible cloudiness due to massive aggregation, analysis by HP-SEC showed 98 % monomer and 2 % dimer. The method produced consistent results, but which were incorrect and misleading because aggregates were dissociating during the HP-SEC run. There are many other less dramatic unpublished and published examples of such failings of HP-SEC methods, and the published results have been discussed in recent reviews (Philo 2006, 2009; Arakawa et al. 2010; Carpenter et al. 2010).

Why does an HP-SEC method that provides such consistent results fail to detect and measure accurately what is actually in a protein sample? As also described in detail in the recent reviews (Philo 2006, 2009; Arakawa et al. 2010; Carpenter et al. 2010), there are a few main culprits:

1. When a sample is injected into the mobile phase, there is a substantial degree of dilution that might cause dissociation of some aggregates.
2. Protein monomers and aggregates may adsorb to the column matrix, but absorption is often greater for aggregates resulting in a depletion of non-monomer species.

3. Such adsorption is often due to charge–charge interaction between protein molecules and the matrix. To minimize such interactions, high salt concentrations are used in the mobile phase. These mobile phase conditions may cause alterations in the populations of aggregates. i.e., decrease in overall levels and/or redistribution among species.
4. High ionic strength may foster hydrophobic interactions between protein molecules and the matrix, which could result in preferential depletion of aggregates.
5. Subvisible particles may be filtered out of the sample by the in-line frit used to protect the HP-SEC columns.

The problem of protein adsorption to the matrix can be so problematic for some proteins that analytical scientists develop methods in which the column is "conditioned" by numerous injections of the protein sample to reduce this nonspecific binding capacity. With each subsequent injection, there is an increase mass recovery from the column, until the point at which adsorption of protein to the column is saturated. Then with each injection, there is presumably full recovery of the injected protein. Clearly, this approach has many potential problems, including subsequent sloughing of monomer and/or aggregates during subsequent analytical runs and no assurance that aggregate levels and distributions are not still being altered.

As has been learned through "on the job training" and the experience gained by analytical scientists over the past three decades, to develop an accurate and reliable HP-SEC assay, results must be confirmed by an another method for quantifying and characterizing protein aggregates. To date, analytical ultracentrifugation (AUC) has proven to be the most powerful and useful method for this purpose. But it took many years before AUC, which was firmly established in academic labs for rigorous study of protein assembly, to be developed into a reliable method for quantitation and characterization of the protein aggregates that are degradation products in therapeutic protein samples.

Interestingly, decades before the advent of the biopharmaceutical industry, AUC had been used in a study of protein aggregation during freeze-drying and the effects of various sugar additives. As part of this amazing formulation study in 1935, Brosteaux and Eriksson-Quensel (1935) used AUC to analyze samples of several different proteins before and after freeze-drying and rehydration. The results for post-rehydration samples were reported as homogeneous or inhomogeneous, with the latter terminology referring to samples in which the protein had aggregated. Not only did these innovative researchers provide important mechanistic insights into the stabilization of proteins by various sugars (many of which we use today as excipients in therapeutic protein products), they also demonstrated the utility and value of AUC in protein formulation studies.

In the modern biopharmaceutical industry analysis of protein aggregates with sedimentation velocity, AUC (SV-AUC) was pioneered by research scientists at a few companies, most of whom had been trained in the method during graduate and postdoctoral research in universities. Also some of the researchers who did the earliest work with AUC in industry initially focused characterizing reversible self-assembly of therapeutic proteins (e.g., recombinant insulin) but soon realized that the method could be used to analyze degradation aggregates (so-called irreversible aggregates). There were many challenges in these early efforts to develop SV-AUC methods for aggregates in therapeutic proteins including variability of results, difficulty in fitting and interpreting data, and effects of relatively high levels of stabilizing excipients in drug products (Berkowitz 2006; Liu et al. 2006; Pekar and Sukumar 2007; Gabrielson et al. 2007b, 2009; Brown et al. 2008; Arthur et al. 2009). Over time, a few industry research groups independently worked out solutions to many of these operational problems, and new data analysis routines allowed for fast and robust interpretation of results. Fortunately for the field, several industrial researchers working in these areas have published original research papers and insightful reviews on optimizing the use of SV-AUC for quantitation and characterization of aggregates and monomers in therapeutic protein samples (Berkowitz 2006; Liu et al. 2006; Pekar and Sukumar 2007; Gabrielson et al. 2007b, 2009, 2010; Brown et al. 2008; Arthur et al. 2009). As demonstrated in several studies, SV-AUC analysis can provide results that accurately reflect the monomer and aggregate composition in a protein sample because the sample is run in its original solution without dilution (e.g., Philo 2006, 2009; Gabrielson et al. 2007a; Hughes et al. 2009). Furthermore, unlike HP-SEC, in SV-AUC there is no solid phase to which protein molecules and aggregates can adsorb.

Therefore, today SV-AUC analysis is readily available (although it is considered expensive) for use as an orthogonal method to check and verify results during development of HP-SEC methods. Ideally, SV-AUC analysis would be able to confirm that a given HP-SEC method provides accurate and precise results for the levels of monomer, oligomers, and "insoluble aggregates" present in a sample, with full recovery of sample mass injected onto the column. These values would exactly match and reflect what was actually present in a sample whether it was a process intermediate, bulk drug substance, or final drug product. Furthermore, it would be clearly demonstrated that the HP-SEC method could quantify and characterize aggregates caused by a range of anticipated stresses such as agitation, freeze-thawing, and exposure to extremes of temperature and pH.

Achieving these goals can be extremely challenging because of common failures in the initial efforts to develop a proper HP-SEC method. As noted above, problems arise because the protein monomers and/or aggregates adsorb to the column resin and/or the mobile phase and dilution alters aggregate types and distribution. Approaches used to optimize an HP-SEC method focus on testing different mobile phase compositions and columns with different resin chemistry, including the newer chemistries available in ultrahigh-pressure SEC columns. At each step of the optimization process, the protein sample should be analyzed in its original formulation/solution directly by SV-AUC. In addition, the sample should be serially diluted into the SEC mobile phase under investigation and then analyzed by SV-AUC to determine protein concentration dependence and how the mobile phase itself may be altering the aggregate level and distribution. The testing of HP-SEC mobile phase composition and/or column type should continue until results from the HP-SEC match closely with data from SV-AUC analysis. There may be small differences between the percentages for each species detected. And in some cases, if there were any detectable species in the void volume of the HP-SEC column, SV-AUC may resolve these into multiple species, which would not occur in HP-SEC analysis (Philo 2006, 2009). However, in general, for each protein it should be feasible to develop a proper HP-SEC method and to confirm rigorously that it is suitable for accurately and precisely quantifying levels of monomer and aggregates.

It must be emphasized that this iterative approach using SV-AUC analysis to corroborate HP-SEC results is needed even for monoclonal antibodies for which scientists have a so-called platform HP-SEC method that they (and perhaps their managers) believe should work for antibodies with similar sequences. As has been shown in many studies (mostly unpublished), minor sequence changes can dramatically alter the pharmaceutical properties of a monoclonal antibody, such as solubility and propensity to aggregate. Similarly, such changes can also greatly affect how a given monoclonal antibody and its aggregates behave during HP-SEC runs. Though a "standard method" for similar monoclonal antibodies might be a good starting point for HP-SEC method development, it is essential that the requisite studies are done to assure that the method is optimized.

Given all of these issues, one might consider replacing the HP-SEC with SV-AUC as a critical quality analytical method. Unfortunately, the relatively long run times, low sample throughput, need for highly trained experts for sample running, and data analysis preclude the use of the SV-AUC analysis for routine quality assessment such as for lot release. On the other hand, the complementation of the SV-AUC with HP-SEC methods is not only useful to confirm the utility of the HP-SEC assay but is instrumental in product comparability exercises and in the analytical similarity assessments performed to demonstrate biosimilarity. In conclusion, as an orthogonal method for HP-SEC, SV-AUC is invaluable for the development and quality assurance of therapeutic protein products.

References

Arakawa T, Ejima D, Li T, Philo JS (2010) The critical role of mobile phase composition in size exclusion chromatography of protein pharmaceuticals. J Pharm Sci 99:1674–1692

Arthur KK, Gabrielson JP, Kendrick BS, Stoner MR (2009) Detection of protein aggregates by sedimentation velocity analytical ultracentrifugation (SV-AUC): sources of variability and their relative importance. J Pharm Sci 98:3522–3539

Berkowitz SA (2006) Role of analytical ultracentrifugation in assessing the aggregation of protein biopharmaceuticals. AAPS J 8:E590–E605

Brosteaux J, Eriksson-Quensel IB (1935) Study on the protein desiccation. Arch Biol Phys 12(4):209–226

Brown PH, Balbo A, Schuck P (2008) A Bayesian approach for quantifying trace amounts of antibody aggregates by sedimentation velocity analytical ultracentrifugation. AAPS J 10:481–493

Carpenter JF, Randolph TW, Jiskoot W, Crommelin DJ, Middaugh CR, Winter G (2010) Potential inaccurate quantitation and sizing of protein aggregates by size exclusion chromatography:

essential need to use orthogonal methods to assure the quality of therapeutic protein products. J Pharm Sci 99:2200–2208

Chi EY, Krishnan S, Randolph TW, Carpenter JF (2003) Physical stability of proteins in aqueous solution: mechanism and driving forces in nonnative protein aggregation. Pharm Res 20:1325–1336

Cordoba-Rodriguez R (2008) Aggregates in MAbs and recombinant therapeutic proteins: a regulatory perspective. BioPharm Int 21(11):44–53

Cromwell ME, Hilario E, Jacobson F (2006) Protein aggregation and bioprocessing. AAPS J 8:E572–E579

Gabrielson JP, Brader ML, Pekar AH, Mathis KB, Winter G, Carpenter JF, Randolph TW (2007a) Quantitation of aggregate levels in a recombinant humanized monoclonal antibody formulation by size-exclusion chromatography, asymmetrical flow field flow fractionation, and sedimentation velocity. J Pharm Sci 96:268–279

Gabrielson JP, Randolph TW, Kendrick BS, Stoner MR (2007b) Sedimentation velocity analytical ultracentrifugation and SEDFIT/c(s): limits of quantitation for a monoclonal antibody system. Anal Biochem 361:24–30

Gabrielson JP, Arthur KK, Kendrick BS, Randolph TW, Stoner MR (2009) Common excipients impair detection of protein aggregates during sedimentation velocity analytical centrifugation. J Pharm Sci 98:50–62

Gabrielson JP, Arthur KK, Stoner MR, Winn BC, Kendrick BS, Razinkov V, Svitel J, Jiang Y, Voelker PJ, Fernandes CA, Ridgeway R (2010) Precision of protein aggregation measurements by sedimentation velocity analytical ultracentrifugation in biopharmaceutical applications. Anal Biochem 296:231–241

Hermeling S, Crommelin DJ, Schellekens H, Jiskoot W (2004) Structure-immunogenicity relationships of therapeutic proteins. Pharm Res 21:897–903

Hughes H, Morgan C, Brunyak E, Barranco K, Cohen E, Edmunds T, Lee K (2009) A multi-tiered analytical approach for the analysis and quantitation of high-molecular-weight aggregates in a recombinant therapeutic glycoprotein. AAPS J 11:335–341

Kessler M, Goldsmith D, Schellekens H (2006) Immunogenicity of biopharmaceuticals. Nephrol Dial Transplant 21(Suppl 5):9–12

Liu J, Andya JD, Shire SJ (2006) A critical review of analytical ultracentrifugation and field flow fractionation methods for measuring protein aggregation. AAPS J 8:E580–E589

Pekar A, Sukumar M (2007) Quantitation of aggregates in therapeutic proteins using sedimentation velocity analytical ultracentrifugation: practical considerations that affect precision and accuracy. Anal Biochem 367:225–237

Philo JS (2006) Is any measurement method optimal for all aggregate sizes and types? AAPS J 8:E564–E571

Philo JS (2009) A critical review of methods for size characterization of non-particulate protein aggregates. Curr Pharm Biotech 10:358–372

Rosenberg AS (2006) Effects of protein aggregates: an immunologic perspective. AAPS J 8:E501–E507

Chapter 20
Biopharmaceuticals: Application of AUC-SV for Quantitative Analysis of Protein Size Distributions

Amanda A. Cordes, Kelly K. Arthur, and John P. Gabrielson

Abstract The major aim of any analytical purity method is to quantify the relative concentration of individual molecular species resolved by the method. During the development of biopharmaceuticals, products in which a heterogeneous distribution of protein molecules serves as the active ingredient, AUC-SV is often applied as a purity method to resolve and quantify size variants in the final product formulation. In this chapter, we briefly discuss the applications of AUC-SV in biopharmaceutical development. One such application is to confirm the accuracy of size exclusion chromatography (SEC) methods during SEC method development, ensuring the SEC methods are fit for purpose for product release and stability testing. In this context, we summarize the application of the AUC-SV method for the measurement of protein size distributions, with particular emphasis on how the method can be tailored to assist in SEC method development. Only when the limitations of AUC-SV are well understood, and mitigated, can the method be used effectively to meet a wide variety of protein product development challenges.

Keywords Analytical ultracentrifugation • Sedimentation velocity • Protein aggregation • Biopharmaceuticals • Therapeutic proteins • Method validation • Method development

---

A.A. Cordes
Department of Analytical Sciences, Amgen Inc.,, 1120 Veterans Boulevard, South San Francisco, CA 94080, USA

K.K. Arthur
Department of Analytical Sciences, Amgen Inc.,, 4000 Nelson Road, Longmont, CO 80503, USA

J.P. Gabrielson (✉)
Department of Analytical Sciences, Amgen Inc.,, 4000 Nelson Road, Longmont, CO 80503, USA

1450 Infinite Drive, Louisville, CO 80027, USA
e-mail: jgabriel@elionlabs.com

S. Uchiyama et al. (eds.), *Analytical Ultracentrifugation*,
DOI 10.1007/978-4-431-55985-6_20

20.1 Introduction

For more than 20 years, analytical ultracentrifugation (AUC) has served an important role in the development of biological products for pharmaceutical use (Laue and Stafford 1999; Lebowitz et al. 2002). Among its many applications, AUC has been applied in the biopharmaceutical industry for the characterization of proteins and protein complexes (Lebowitz et al. 2002). One prominent role for sedimentation velocity AUC (AUC-SV) in recent years has been to quantify aggregate levels in protein products (Berkowitz 2006; Gabrielson et al. 2006; Philo 2006). New therapeutic modalities, including viruses, antibody-drug conjugates, and bi-specific antibodies and antibody fragments, among others, represent a new challenge for the industry, and AUC-SV may be effectively applied to the characterization of size variant distributions of these products (Berkowitz and Philo 2007).

The amount of aggregated protein present in the final container of a formulated drug product is a critical quality attribute of many biopharmaceutical products because some types of aggregates, if present at sufficient levels, may pose a risk of immunogenicity for some patients receiving the medicine (Parenky et al. 2014; Rosenberg 2006). Therefore, protein aggregation must be adequately controlled during manufacturing, long-term storage, and administration. Although the acceptable level of aggregation is product specific, depending on factors such as the size and type of aggregate, product indication, target patient population, and dosing requirements, aggregate levels should be controlled to an acceptable level throughout the labeled storage duration of the product (Kozlowski and Swann 2006; Mahler et al. 2009).

Size exclusion chromatography (SEC) is the most widely used analytical method for measuring the amount of soluble aggregate present in biopharmaceutical products. SEC methods are typically rapid and precise, and the HPLC systems used to run SEC methods are easy to operate in a regulated quality control (QC) laboratory. For these reasons, SEC methods are often used for lot disposition and product stability testing. To ensure SEC methods are fit for purpose, potential error sources should be assessed during SEC method development, including potential matrix effects and product destabilization due to the dilution of the product into a mobile phase that differs from the formulation composition, nonspecific adsorption of the protein to the resin bead surface, and lack of resolution between different size variants near the exclusion limit of the column.

The AUC-SV method is a powerful tool for assessing potential risks during SEC development. The analyte (protein) can be measured directly in the formulation by AUC-SV, avoiding any risk related to the effects of dilution and buffer composition on the analysis results. However, AUC-SV also suffers from known limitations. The method throughput is low, and data analysis requires extensive analyst training and experience in data interpretation. Despite these limitations, when method capability is well understood and known error sources are controlled (Gabrielson and Arthur 2011; Gabrielson et al. 2010), AUC-SV serves as a valuable orthogonal method to SEC. Specific applications of AUC-SV will be discussed in Sect. 20.2, followed by a detailed evaluation of method limitations in Sect. 20.3. A balanced view of the AUC-SV method will enable useful application in biopharmaceutical development, as covered in Sect. 20.4.

20.2 Applications of AUC-SV for Biopharmaceutical Development

Biopharmaceutical products are naturally heterogeneous, with complex structures and size distributions. These products must be thoroughly characterized at the molecular level, including detailed evaluation of the presence and nature of the product-related variants. AUC-SV is an important tool to aid in the task of identifying and characterizing product-related size variants. It resolves aggregates in solution based on size, density, and shape characteristics of the size variant. Here, we discuss the use of AUC-SV during biopharmaceutical development: for quantification of aggregate levels in the final product.

20.2.1 The Utility of AUC-SV as an Orthogonal Method for Quantitation of Protein Aggregation

Separation of protein size variants in solution by AUC-SV is achieved by centrifuging a sample at a high rotational speed, such that molecules move at constant velocity in a radially outward direction despite the opposing force of diffusion. Sedimentation is dependent on both the size and shape of the species in solution. Larger size variants will sediment faster than smaller size variants, while more spherical size variants (i.e., those with aspect ratios closer to 1) will sediment faster than elongated species which experience a greater drag force. The underlying principle of the separation by AUC-SV makes it ideal for studying biopharmaceutical products with a distribution of size variants. Monomeric protein species can be separated from larger submicron aggregates in an AUC-SV experiment. The sedimentation profiles obtained during the AUC-SV experiment can then be fit using various computer software tools (e.g., SEDFIT, UltraScan, SEDANAL, DCDT (Demeler 2005; Schuck 2000; Stafford and Sherwood 2004)) to model the distribution of species best representing that sedimentation profile. The analysis of the resulting size distribution allows for the determination of the relative proportion of size variants in solution, similar to SEC analysis, but often under more native-like conditions. For example, for biopharmaceutical products in liquid formulation, AUC-SV enables the analysis of samples directly in the product formulation.

SEC is the primary method for determining aggregate levels and protein size distributions in a QC environment. Because product disposition decisions depend on the results of SEC, it is important to verify the accuracy of the SEC method during biopharmaceutical development. Both the AUC-SV and SEC analysis techniques result in a size distribution profile for the sample, yet the underlying separation principle is different. Therefore, AUC-SV is well suited for use as an orthogonal method to confirm SEC method accuracy (Philo 2006). Lower aggregate levels measured by SEC, compared to AUC-SV, can indicate SEC method-induced artifacts, such as dissociation of non-covalent aggregates, adsorption of aggregates to column material, insufficient column conditioning, the inability to accurately quantify or resolve aggregates eluting at the column exclusion volume, or the presence of aggregates so large they were removed by an in-line filter. When higher levels of aggregate are measured by SEC, it can indicate that the SEC method is generating aggregates during analysis or; in cases where the column is conditioned with aggregates, these aggregates may desorb during analysis. Application of the AUC-SV method can be a powerful way to either confirm the accuracy of the SEC method or to identify opportunities for SEC method improvements.

20.2.2 Use of AUC-SV to Guide SEC Method Development

There are inherent potential method risks associated with SEC analysis, including those related to sample dilution, nonspecific interactions, and lack of resolution between size variants. Thoughtful application of AUC-SV as an orthogonal technique during SEC method development can mitigate these risks and help ensure the final SEC method is fit for purpose.

One of the ways in which AUC-SV is a useful tool for SEC method development is that AUC-SV allows in situ analysis of soluble aggregates in final formulation conditions in the absence of a solid phase matrix that can perturb the solution behavior of the protein. The ionic strength of the SEC mobile phase and/or the presence of organic components in the buffer has the potential both to disrupt weakly associated aggregates and generate aggregates on the column during analysis. Thus the measurement itself has the potential to alter the aggregate profile during SEC analysis. In contrast to SEC, where samples are diluted into the column mobile phase upon injection, no change in solution conditions is required for AUC-SV analysis. The in situ analysis of aggregates in formulation conditions removes many analysis-based artifacts and thereby allows more accurate quantitation of aggregate levels. An example illustrating this advantage of AUC-SV is shown in Fig. 20.1. The analysis of an in-process sample of a small therapeutic protein (approximately 20 kDa molecular weight) by both SEC and AUC-SV revealed that non-covalent aggregates dissociated during SEC analysis, although they did not dissociate when analyzed by AUC-SV in the in-process sample buffer. Non-covalent reversible aggregates may not be critical to the quality of the drug, as they may dissociate upon delivery to patients; nevertheless AUC-SV serves as a powerful

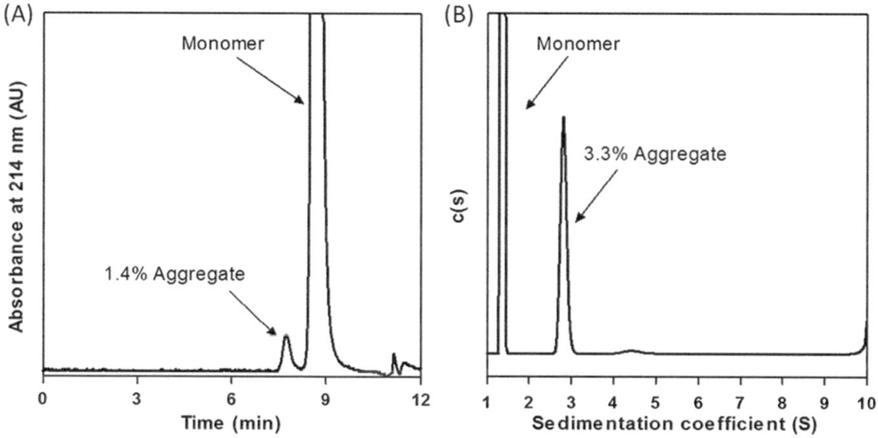

Fig. 20.1 SEC chromatogram (**a**) and AUC-SV c(s) distribution (**b**) for a 20 kDa therapeutic protein with low levels of aggregate. AUC-SV analysis was conducted in the buffer used for the process step from which the samples were pulled. The lower aggregate levels measured by SEC (approximately 50 % less) indicate the dissociation of non-covalent aggregates during analysis

method for identifying the presence of such species in the samples. In this example, the application of AUC-SV as an orthogonal technique provided insight into the nature of the aggregate species in the in-process samples, in addition to quantitation of aggregate levels.

Another characteristic of AUC-SV analysis is that it limits the potential for nonspecific protein adsorption to surfaces during analysis. During SEC analysis, protein aggregates can differentially adsorb to the bead surface of the SEC column resin. When this occurs, it leads to the underreporting of aggregates by SEC. Conditioning the column with a commercially available protein such as BSA or the protein of interest prior to analysis can minimize the issue by occupying surface area that would otherwise be available for nonspecific binding. However, the appropriate column conditioning material to use is not always obvious. In the case of analysis of an Fc-fusion protein by SEC, even though the column was well conditioned with a sample containing monomer and dimer, the larger aggregates present in a heavily degraded sample appeared to preferentially adsorb onto the resin. The analysis of the same sample by AUC-SV identified the low aggregate recovery by SEC and the need for additional column conditioning. After the SEC column was conditioned with a more appropriate sample type containing larger aggregates, improved recovery of the aggregate species was observed. The SEC chromatograms from multiple injections of the highly degraded sample onto an already conditioned column are shown in Fig. 20.2. Note the increasing recovery of large aggregates, which appears to reach a plateau after four injections, indicating the need for conditioning with aggregates for their accurate quantitation in the analytical runs. The application of AUC-SV in this case resulted in a successful improvement to the SEC method, enabling accurate recovery of samples containing large aggregates.

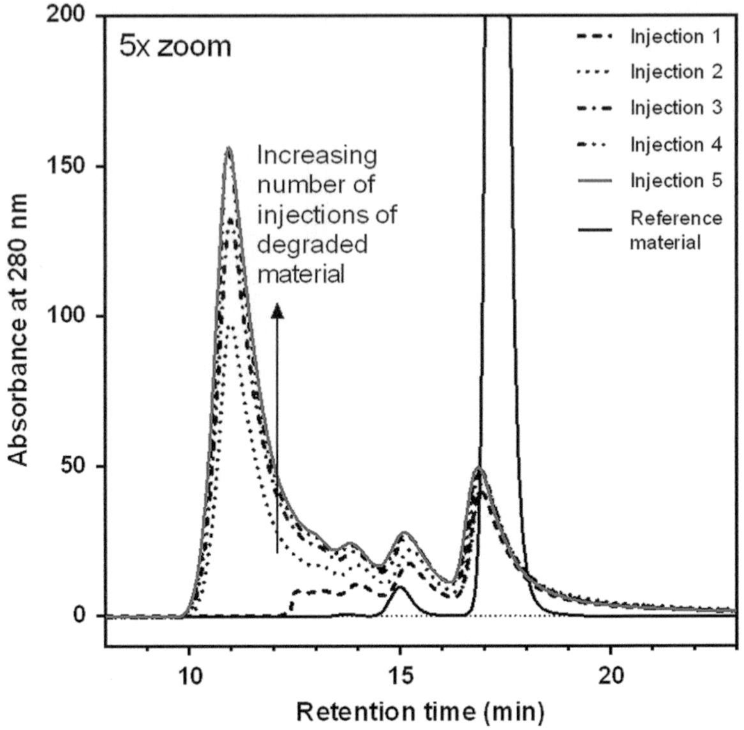

Fig. 20.2 Overlaid SEC chromatograms for reference material and repeated injections of a forcibly degraded protein sample. A large increase in the amount of aggregate detected is seen with increasing number of injections. This increase in aggregate level plateaus after the 4th injection, as the 4th and 5th injections overlay well. These data suggest the aggregate is binding to the column. After the column is well conditioned with aggregate, full recovery is achieved (Conclusion supported by AUC-SV results, not shown)

A third way in which AUC-SV can be beneficial during SEC method development is through the improved resolution of size variants it offers, especially for size variants which elute near the exclusion limit of the SEC column. This improved resolution provides important information, as co-elution of size variants during SEC analysis can lead to over- or underreporting of aggregate levels, depending on the size variant(s) of interest. Baseline resolution of the monomer from dimer and larger aggregates ensures that aggregate levels are not underreported due to co-elution with the monomer during SEC analysis. Improved resolution of specific size variants larger than monomer can provide insight into aggregation mechanisms, as it allows one to determine whether aggregate formation is due primarily to an increase in dimer, an increase in larger than dimer aggregates, or a mixture of both. The SEC chromatograms and AUC-SV c(s) distributions for heat degraded monoclonal antibody samples are shown in Fig. 20.3. Not only are the dimer and larger than dimer species resolved from each other by AUC-SV, but there is also resolution of

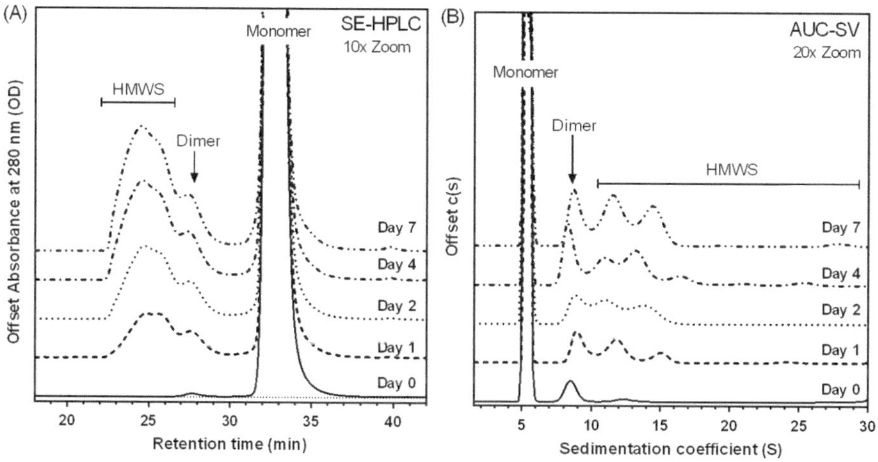

Fig. 20.3 Offset SEC chromatograms (**a**) and AUC-SV c(s) distributions (**b**) for samples of a heat degraded monoclonal antibody. AUC-SV yields better resolution between the dimer peak and peaks of species larger than dimer. Additionally, individual peaks for large aggregates are resolved by AUC-SV, while large aggregates elute as one broad peak by SEC, demonstrating the improved resolution of AUC-SV compared to SEC

the larger than dimer species into multiple separate peaks. The larger than dimer species elute as one broad peak during SEC analysis, and even the dimer co-elutes significantly with larger species. For samples such as these, AUC-SV can provide assurance the monomer and dimer peaks are adequately resolved by SEC and can also provide more detailed information about the size distribution profile.

The examples above illustrate the utility of AUC-SV to identify cases where SEC quantitation is impacted due to aggregate loss (either by dissociation or adsorption) or poor resolution of aggregate species. Therefore, AUC-SV analysis is a critical tool for SEC method development capable of identifying areas of SEC method improvement to ensure the accurate quantitation of aggregates.

20.3 Limitations of AUC-SV for Biopharmaceutical Development

Although there are characteristics of AUC-SV that make it well suited for the analysis of protein size distributions, AUC-SV suffers from its own set of limitations. The limitations of AUC-SV can be classified into two categories: (1) limitations that decrease the precision of the method and (2) limitations that decrease the robustness of the method. Because the detection limit (DL) of the method is a statistical quantity based on the precision of the method,[1] the factors listed below that impact the precision also contribute to the relatively high DL of AUC-SV as compared to SEC.

20.3.1 Precision of the AUC-SV Method

The lack of precision of AUC-SV measurements, compared to SEC, arises from many factors, only some of which can be controlled by the analyst. These factors include the cell centerpiece and housing, the orientation of the assembled cell in the rotor, the state of the instrument, and post-run data analysis (Gabrielson and Arthur 2011; Gabrielson et al. 2010; Pekar and Sukumar 2007). Damaged centerpieces can impact the sedimentation of protein molecules, leading to inaccurate measurements of aggregate levels (Gabrielson et al. 2010). Centerpieces represent one factor where analyst control is limited; although it may be easy to identify large defects, a slow decline in performance as the centerpiece ages may be harder to detect. However, one factor that can be controlled by the analyst is to treat cell centerpieces, windows, and housings as a single unit to minimize variability from substituting pieces across different cell assemblies and to more easily identify when a centerpiece or window requires replacement. Cell orientation in the rotor must also be carefully controlled as misalignment of even 1° can increase the reported aggregate level by 1.5 % (Gabrielson and Arthur 2011). Although custom-built tools for controlling alignment exist, these tools still rely on proper application by a trained analyst. Furthermore, the analyst has control over the alignment of the entire cell housing assembly, but not over the alignment of the centerpiece within the housing or of the channels in the centerpiece (Gabrielson et al. 2010).

The cell centerpieces and orientation are not the only factors affecting method precision. The operational state of the instrument itself has an effect on the precision of AUC-SV measurements. For example, cleanliness of the instrument optics system can impact the level of noise in the data and subsequently impact the level of precision. Post-run data analysis can also contribute to method variability and a decrease in precision. The size distribution profile is not directly measured during the AUC-SV experiment, but instead is the result of fitting a model to the raw sedimentation profiles collected during the experiment. The choice of the fitting parameters, including selection of which parameters to fit and initial conditions for those parameters, such as the area to fit, meniscus location, and frictional ratio, can impact both the quality of the fitted model and the final results.

An analyst conducting a well-controlled experiment at the current state of the art could expect a protein aggregate level measurement intermediate precision of 0.4 % aggregate (Gabrielson and Arthur 2011), although precision has been

[1]The DL is defined as 3.3 times the square root of the method variance (Eq. 20.2). A more detailed discussion of the DL can be found in Sect. 20.4.2.

shown to depend on the age of the centerpieces used in the experiment (Pekar and Sukumar 2007). While it is important for analysts to control factors affecting precision when possible, efforts of the analyst alone are not sufficient to drive large-scale improvement of AUC-SV precision. Therefore, continued development of the instrumentation and data analysis programs is necessary to improve the precision of the method and approach the level of precision offered by SEC analysis.

20.3.2 Robustness of the AUC-SV Method

Many of the factors that decrease the precision of the method also limit its robustness. The method is complicated and nonroutine; therefore execution of AUC-SV experiments requires extensive analyst training. There are numerous factors that must be controlled during the sample preparation and experimental run to ensure a successful AUC-SV experiment. When inadequately controlled, many of the factors that contribute to lack of precision, discussed in Sect. 20.3.1, can also negatively impact the robustness of the method. For example, cell leaks can increase when the centerpiece, windows, and housing are not treated as a unit but instead used as interchangeable parts. The method is also sensitive to sample handling, one example being the technique used to mix samples after placement into the assembled cells. More vigorous mixing (e.g., vortexing of samples) was shown in one case to increase variability of the amount of aggregate measured (unpublished results). Temperature variability can also have a large impact on the experimental results. Equilibration of the rotor to run temperature prior to the start of the experiment and external calibration of the rotor temperature can both be used to mitigate this source of error and increase the precision of the measurement (Gabrielson and Arthur 2011; Ghirlando et al. 2013; Zhao et al. 2014). Overall, because the sample preparation and data analysis require more analyst attention and judgment than in a typical SEC experiment, there is more opportunity for human error.

In addition to the increased opportunities for human error, the AUC-SV method also has a much lower throughput than SEC. The maximum number of samples that can be analyzed in a run is limited to 3 or 7, depending on the number of holes in the rotor used during the run. Because it requires a minimum of several hours to complete each run, in general it is difficult to complete more than two runs in a standard 8 h workday. The thermal equilibration of the rotor and sample required prior to the experimental analysis also limits the ease with which an AUC-SV run can be restarted if interrupted (e.g., in the case of power failure). Thus, the maximum throughput of the method is 14 samples per instrument per day, but in practice 5–10 samples per instrument per day represents the maximum sustainable run rate.

20.4 Tailoring the AUC-SV Method to Achieve Its Purpose

Both AUC-SV and SEC play a role in the analysis of protein aggregates in biopharmaceutical products. SEC is robust, high throughput, relatively easy to use, and amenable to implementation in a QC environment. However, given the potential of SEC analysis to perturb the aggregation state of the sample, the size distribution results obtained from this method may not always accurately reflect the size distribution of the original sample. AUC-SV gives one the ability to analyze the protein in its native solution conditions and provides a more accurate quantitation of aggregate levels. However, AUC-SV analysis is time consuming and requires extensive analyst training, limiting its utility for the purpose of routine analysis. In order to provide the best overall evaluation of the aggregate content throughout the product life cycle, the two methods can be used together, leveraging the advantages of each. Here we show an example of how AUC-SV can be used effectively when the intended purpose is well defined: to verify the accuracy of a SEC method.

20.4.1 Optimization of AUC Method Parameters

The first step in applying AUC-SV to verify the accuracy of the SEC method is to optimize the AUC-SV method parameters for the analysis of the protein of interest. This involves optimizing the parameters for both data collection and analysis. Parameters to take into consideration, along with suggested values and rationale for these choices, are listed in Table 20.1 (Gabrielson and Arthur 2011). Although there are many circumstances which would warrant changes to the values below, each exception is not exhaustively discussed here. Instead, these parameter values and settings are provided to serve as an initial starting point for method optimization. The rationale for the selection of these parameters is included to help guide the analyst in further refinement of the parameters for their particular AUC-SV method. The data analysis parameters are based on data analysis performed using SEDFIT software (NIH/NIBIB) and fitting to the c(s) distribution model (Schuck 2000).

20.4.2 Qualification of AUC-SV as a Characterization Method

Once the AUC-SV method parameters have been optimized, the method should be qualified for use as a characterization method. Although AUC-SV is not intended for use as a lot release method, principles from the ICH Q2 guideline for method validation can be used to design a qualification approach for AUC-SV. It is recommended at a minimum that the AUC-SV method be shown to be specific,

Table 20.1 AUC-SV method parameter summary

Experimental step	Parameter	Suggested value	Rationale
Data collection[a]	Detection	Absorbance detection	Correspond to the detection used for the SEC method (may be necessary to use a different wavelength based on differences in solution conditions between methods)
	Sample concentration	0.5–1.0 mg/mL	Target optical density of 0.5–1.0, to avoid nonideal sedimentation that occurs at high protein concentrations[b]
	Dilution buffer	Formulation buffer	Maintain as native-like solution environment as possible
	Reference sector buffer volume	440 uL	Fill reference and sample sectors such that meniscus is closer to the center of rotation in the reference cell than in the sample cell, while maintaining approximately equal volumes to prevent cell distortion[c]
	Sample volume	420 uL	
	Angular velocity	40,000 rpm (antibodies)	Centrifuge rapidly enough that sedimentation force overcomes opposing diffusion force
		50,000 rpm (smaller proteins)	
	Temperature	20 °C	Controlled temperature for analysis
	R_{min}	5.85	Ensure meniscus spike is included in data collection[d]
	R_{max}	7.05	Exclude area where back diffusion is significant (unless it is desired to model this effect during data analysis)[d]
	Radial scan increment	0.003	Provide good coverage of the concentration profile along the length of the cell
	Replicates	1	Maximize collection speed, increasing the likelihood of detecting large aggregates in each cell, if present
	Number of scans	Dependent on the number of samples analyzed, rotor speed, instrument differences	Collect sufficient scans to allow for complete sedimentation of the protein

(continued)

Table 20.1 (continued)

Experimental step	Parameter	Suggested value	Rationale
Data analysis[e]	Meniscus position	Align visually with positive meniscus absorbance spike maximum	Allow meniscus to vary during analysis, within the region of the sample meniscus, removing analyst bias
	Bottom	7.2	Fixed, based on geometry of centerpiece used
	Top fitting limit (from center of rotation)	+0.1 cm from meniscus	Fit area which does not contain any of the absorbance artifacts from the meniscus
	Bottom fitting limit (from center of rotation)	7.0 (no co-sedimenting solutes in buffer)	Varies. Include as much of the sedimentation profile as possible, without fitting areas where large concentration gradients are expected if co-sedimenting solutes (e.g., sugars) are present
		6.7 (co-sedimenting solutes in buffer)[f]	
	Frictional ratio	Allow parameter to vary from 1.2 to 1.6	A spherical particle has a frictional ratio of one; the less globular the protein, the higher the frictional ratio
	Baseline	Allow parameter to vary	Allow software to determine baseline, remove analyst bias
	Time invariant (TI) noise	Fit	Allows software to account for time invariant noise that may be present due to dust or scratches on the cell windows
	Resolution	200	Balance between sufficient resolution to obtain a meaningful c(s) profile and the time it takes to fit the model
	S_{min}	1	Sedimentation coefficient of species at lower end of expected size range. Value can be adjusted (increased or decreased) based on the sedimentation coefficient of the monomeric species

(continued)

Table 20.1 (continued)

Experimental step	Parameter	Suggested value	Rationale
	S_{max}	20	Sedimentation coefficient of species at upper end of expected size range. Value can be adjusted (increased or decreased) based on the sedimentation coefficients of the aggregate species present
	Scans included	Dependent on sedimentation profile	Include sufficient scans to capture sedimentation of the protein

[a]Parameter suggestions based on the use of a 12 mm two-sector centerpiece
[b]To fully understand the nonideal sedimentation behavior of a given protein, concentration dependence experiments may be necessary prior to selecting the final analysis concentration
[c]Assumes the use of absorbance optics for detection. The meniscus positions in the reference and sample sectors should be matched for interference optics
[d]R_{min} and R_{max} values should be set to optimize the time spent collecting real data about the sedimentation profile and reduce time spent collecting information in regions with no information about the sedimentation
[e]When parameters are allowed to vary during the data analysis, it is important to evaluate the final value for reasonableness. Unreasonable values (e.g., a meniscus position far removed from the visible meniscus artifact) can indicate a poor model fit by the software
[f]Assumes that co-sedimenting solutes are not included in the SEDFIT model applied. If using co-sedimenting solutes model, adjust bottom fitting limit to 7.0 cm

precise, and fit for its intended purpose (1996). This includes demonstration of the ability of AUC-SV to detect changes in the size distribution profile along with an assessment of the measurement precision, as well as determination of the detection and quantitation limits of the method.

The specificity of the AUC-SV method can be confirmed by demonstrating baseline resolution in the c(s) distribution between monomer and larger aggregates. This assessment of specificity takes advantage of the similarities between SEC chromatograms and AUC-SV c(s) distributions, as the ICH Q2 guideline states "For chromatographic procedures, representative chromatograms should be used to demonstrate specificity .... For critical separations, specificity can be demonstrated by the resolution of the two components which elute closest to each other" (1996). Resolution of the monomer and aggregate(s) is necessary both for the accurate determination of the monomer sedimentation coefficient (s*) and quantitation of aggregates in the sample. The example c(s) sedimentation profile for a monoclonal antibody shown in Fig. 20.4 demonstrates baseline resolution of the species and, therefore, specificity.

Assuming the method adequately resolves monomer and size variants larger than monomer, the precision of the AUC-SV method for the measurement of the aggregate levels can be evaluated. Sources of variability contributing to the lack of

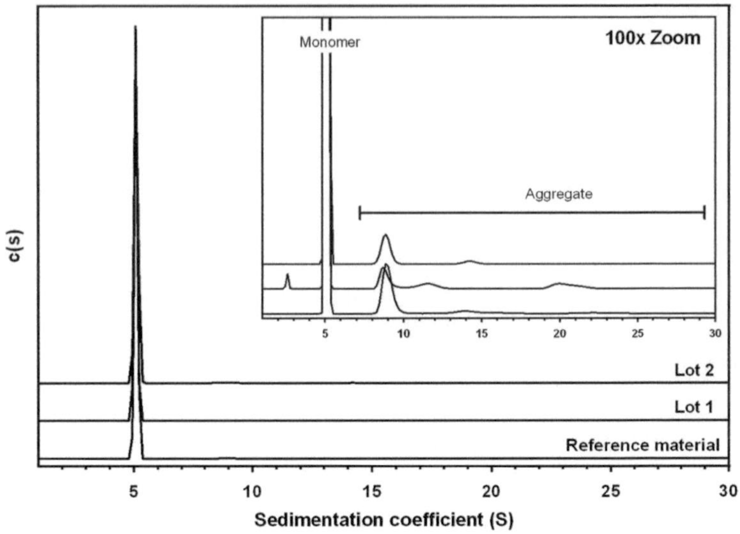

Fig. 20.4 Offset c(s) distributions at full scale and 100x enhanced (insert) for three separate lots of a monoclonal antibody. Specificity is satisfied by demonstrating baseline resolution in the c(s) distributions between components which have sedimentation coefficients close together

precision of the method include (Gabrielson and Arthur 2011):

- Run-to-run variance (σ_R^2)
- Residual variance[2] (σ_E^2)

According to ICH Q2, precision "may be considered at three levels: repeatability, intermediate precision, and reproducibility"(1996). Repeatability refers to the intra-assay precision of the method and reflects the precision of the method under the same operating conditions. For example, the variability of samples run on the same day using the same instrument would be represented by the repeatability of the method. The intermediate precision of the method reflects within-laboratory variability between samples run on different days, with different instruments, or by different analyses. Finally, the reproducibility of the method reflects the variability in analysis between different laboratories. Given the low throughput of AUC-SV, samples from a single experiment may need to be run on different days and/or on different instruments. These samples are typically analyzed in one laboratory. Thus, in many cases intermediate precision is the most relevant type of precision for AUC-SV. The intermediate precision (IP) of the AUC-SV method can be approximated

[2]Residual variance results from the intrinsic variability of the method. Refer to Sect. 20.3 for a more in-depth discussion of sources of method variability.

by the square root of the sum of the run and residual variances (Eq. 20.1):[3]

$$\text{Intermediate precision (IP)} = \sqrt{\sigma_R^2 + \sigma_E^2} \quad (20.1)$$

The intermediate precision can also be used to define the detection limit (DL) and quantitation limit (QL) of the method, using Eqs. 20.2 and 20.3 below (1996; Gabrielson and Arthur 2011):

$$\text{DL} = 3.3\sqrt{\sigma_R^2 + \sigma_E^2} \quad (20.2)$$

$$\text{QL} = 10\sqrt{\sigma_R^2 + \sigma_E^2} \quad (20.3)$$

The DL is the lowest concentration of a species that can be definitively known to be present. The QL represents the lowest concentration at which the analyte can not only be confidently detected but also quantified with suitable precision. Using the collective data from five protein products, DL and QL were calculated to be 1.2 % and 3.7 %, respectively (Gabrielson and Arthur 2011).

If levels of aggregate are above the QL, a more rigorous statistical comparison of aggregate levels between samples can be performed using an equivalence acceptance criterion (Burdick and Sidor 2013).

20.4.3 Concurrent Development and Qualification of a SEC Method

SEC is commonly employed as a lot release method in the biopharmaceutical industry to evaluate the purity of the product. Therefore, the development and qualification of the SEC method should be carried out in accordance with ICH Q2. It is best if the development and qualification of the SEC method is conducted concurrently (or at least not prior to) the development of the AUC-SV method in order to apply the knowledge gained from AUC-SV measurements during SEC method development. For products with an existing SEC method, it is still possible to verify the SEC method performance using AUC-SV. In such cases, assuming it is necessary to modify the SEC method after comparison to AUC-SV, an improved SEC method can be implemented after successful method revalidation.

[3] Initial statistical models (not shown) indicated dilution buffer effect was insignificant for measurements of aggregate levels. If there is concern about other potential effects, a broader statistical model can be used initially, eliminating terms as they are shown to be insignificant in order to arrive at a final model.

20.4.4 Validation of a SEC Method for Routine Use (Correlation of SEC to AUC-SV)

Once qualified AUC-SV and SEC methods exist, the AUC-SV method can be used as an important component of the SEC validation strategy through correlation of the AUC-SV and SEC methods for the determination of protein aggregation levels.

In order to determine the degree of correlation between the two methods, it is suggested an experiment be designed to measure the protein of interest after exposure to multiple relevant stress conditions (e.g., elevated temperature, exposure to UV light, and exposure to pH extremes) by both AUC-SV and SEC. Protein degradation mechanisms induced by multiple stress conditions often yield both covalent and non-covalent aggregates, as well as a wide range of aggregate sizes and structures (Paul et al. 2012). The chemical and physical nature of the aggregates (e.g., covalent bonds, unfolded structures) can be determined by isolation and characterization of specific peaks in the chromatogram.

The same degraded samples generated by multiple stress conditions should be measured by both AUC-SV and SEC. Given the lower precision of AUC-SV as compared to SEC, samples should be analyzed in triplicate by AUC-SV. A single replicate of each sample is usually sufficient for SEC analysis. Because AUC-SV suffers from fewer potential method risks than SEC in terms of measurement accuracy, AUC-SV should be considered the standard to which SEC results are compared. Thus by convention, when a correlation graph is constructed, the AUC-SV data are plotted on the abscissa, because AUC-SV is considered the independent variable, and the SEC data are plotted on the ordinate. In some cases it may be valuable to construct individual correlation plots for specific size variants, such as dimer and larger than dimer species, in addition to the correlation plot for total aggregate.

Example correlation plots for dimer, larger than dimer, and total aggregate for an IgG1 are shown in Fig. 20.5, with the data further differentiated by degradation pathway to provide additional information regarding how method performance depends on the nature of the aggregate. As shown in Fig. 20.5, the correlation between the two methods is linear for dimer, larger than dimer, and total aggregate measurements. The slopes of the linear regression for each data set also indicate that the methods are well correlated, with the slope and R^2 values close to 1. The slope of the larger than dimer aggregate correlation (1.3) suggests that SEC slightly overreports aggregate levels; this is likely due to the resolution of the method. SEC does not resolve dimer and larger aggregates as well as AUC-SV does, as shown in Fig. 20.3 for the thermally degraded samples of this product. The slope of the correlation for aggregates larger than dimer is greater than 1, which would be expected if dimer size aggregates are co-eluting with larger aggregates and, therefore, counted as larger aggregates. Resolution and integration differences

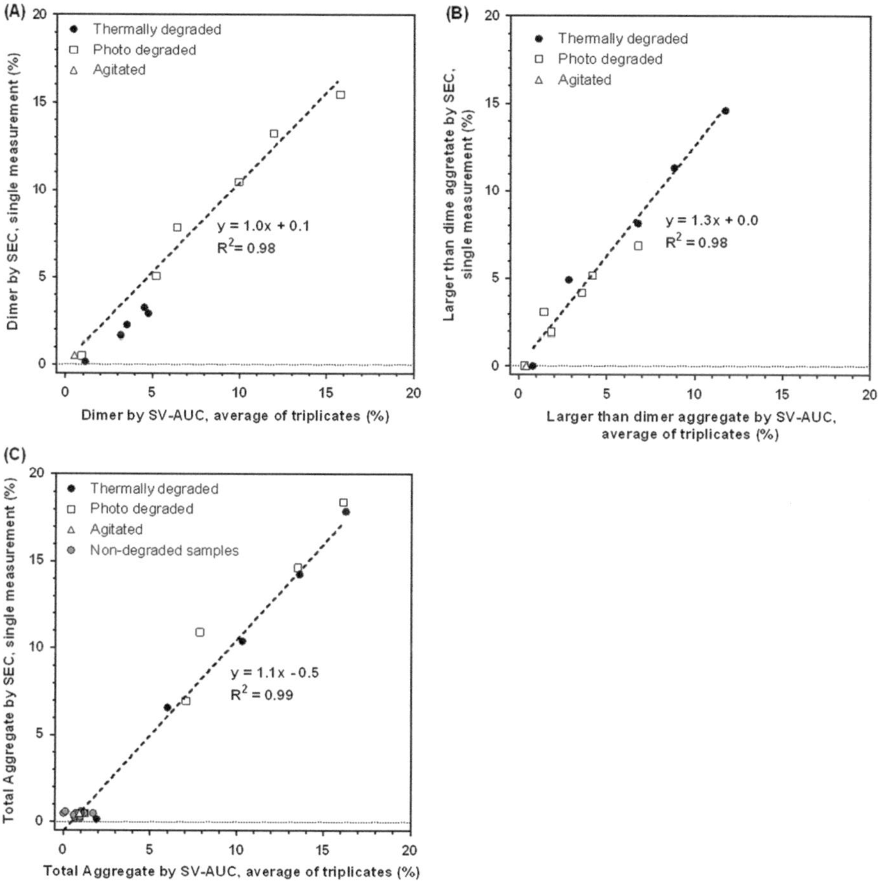

Fig. 20.5 Correlation plots for dimer (**a**), aggregates larger than dimer (**b**) and total aggregate (sum total of results from panels A and B) (**c**). There is a direct and linear correlation between the two methods, with correlation agreement between 90 and 120 %

between the two methods are likely the root cause for differences in quantitation between the two methods for the subpopulations of aggregates; overall the SEC and AUC-SV data are well correlated with the slope of the linear regression for the total aggregate correlation equal to 1.1.

The data shown in Fig. 20.5 also illustrate the different aggregate sizes present under different degradation conditions. For this particular protein, the thermal- and photodegraded samples had the highest levels of aggregates, while the agitation degraded samples had relatively low levels of aggregate. Nonreducing denaturing SEC analysis indicated that the aggregates in the photodegraded samples were primarily covalent (data not shown). Differences in the degree of correlation for samples generated by different degradation pathways can also be seen in Fig. 20.5. For example, in the dimer correlation plot (panel A of Fig. 20.5), the thermally degraded samples consistently fall below the data points from samples generated by other degradation mechanisms, indicating that SEC underreports dimer aggregates created by thermal stress in this case. Due to these differences in correlation related to the types of aggregates observed, it is recommended that a range of aggregate sizes and types be used to construct the correlation. In addition, sufficient data points above the QL should be included in order to accurately evaluate the slope of the response. The data in Fig. 20.5 show that both of these requirements can be accomplished by employing multiple degradation conditions, allowing for a true evaluation of the SEC and AUC-SV method correlation under relevant conditions that yield a suitable range of different types of aggregates.

Selection of degradation conditions to produce aggregates of varying conditions can also help identify method deficiencies during development. For example, if only degradation conditions that produce elevated levels of dimer are used, then it is not possible to identify preferential binding of larger aggregates to the SEC column material and the subsequent underreporting of those species by the SEC method (refer to example shown in Fig. 20.2). Similarly, if only covalent aggregates are produced by the selected degradation conditions, it will not be possible to identify cases where aggregates dissociate upon dilution in the SEC mobile phase (as shown in Fig. 20.1).

The example shown in Fig. 20.5 represents the correlation of the AUC-SV and SEC methods for one particular protein. By combining the correlation data for multiple products, it is possible to evaluate the correlation of the two methods more generally. The combined data for six proteins covering three protein structural classes are plotted in Fig. 20.6, with the associated correlation accuracies. Again, multiple degradation conditions were used to generate a range of aggregates of each protein, in order to evaluate the method correlation as generally as possible. The data in Fig. 20.6 demonstrate that overall AUC-SV and SEC correlate well for the detection of protein aggregates, and protein structural class is not a predictor of correlation.

Further useful information on the aggregate size distribution for the SEC method verification can be obtained by comparing the AUC-SV c(M) distribution to molar mass data obtained from in-line multi-angle static light scattering (MALS) coupled with the standard UV detection employed by the SEC method. The SEC-MALS chromatogram and AUC-SV c(M) distribution for the heat degraded sample from the correlation study in Fig. 20.5 are shown in Fig. 20.7. These data suggest good agreement between the methods, with the aggregate size distribution ranging from 300 to 900 kDa.

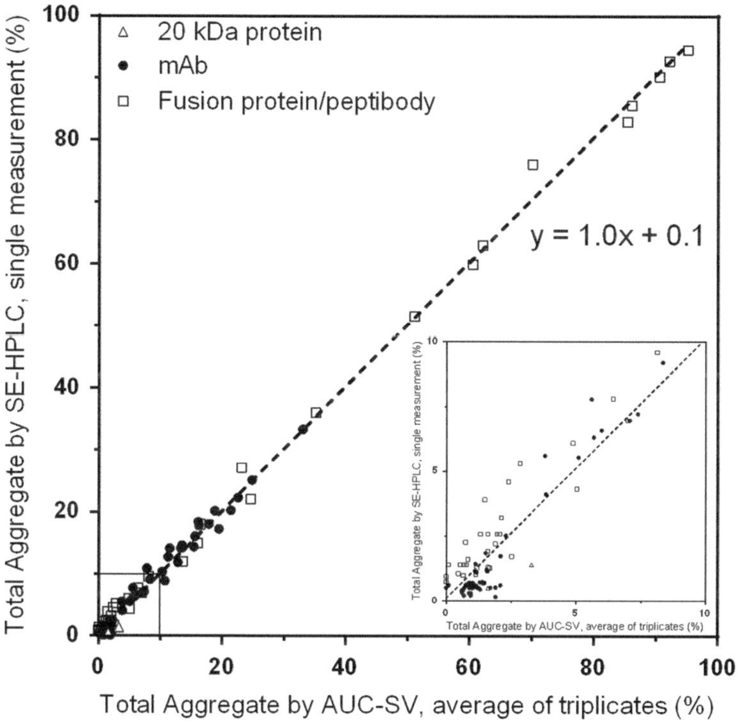

Fig. 20.6 Correlation plot for the total aggregate measured for six proteins covering three structural classes. Overall, AUC-SV and SEC correlate well for the detection and quantification of protein aggregates

When used as the basis for a method correlation study, the AUC-SV method is a powerful way to verify SEC method performance. A linear correlation between the methods with slope close to 1 indicates that the SEC method is appropriate for accurate detection and quantitation of protein aggregates. In addition, further information about the size range which can be detected by SEC analysis can be obtained from the comparison of AUC-SV c(M) distributions to SEC-MALS molar mass results. Poor correlation between the two methods can identify the need for SEC method improvements, for example, including additional conditioning samples if a particular aggregate species is preferentially adsorbing to the column resin or adjustments to the mobile phase composition to prevent aggregate dissociation upon injection. In summary, validation of the SEC method using AUC-SV ensures the SEC method is suitable for product release testing, stability testing, and other routine testing applications.

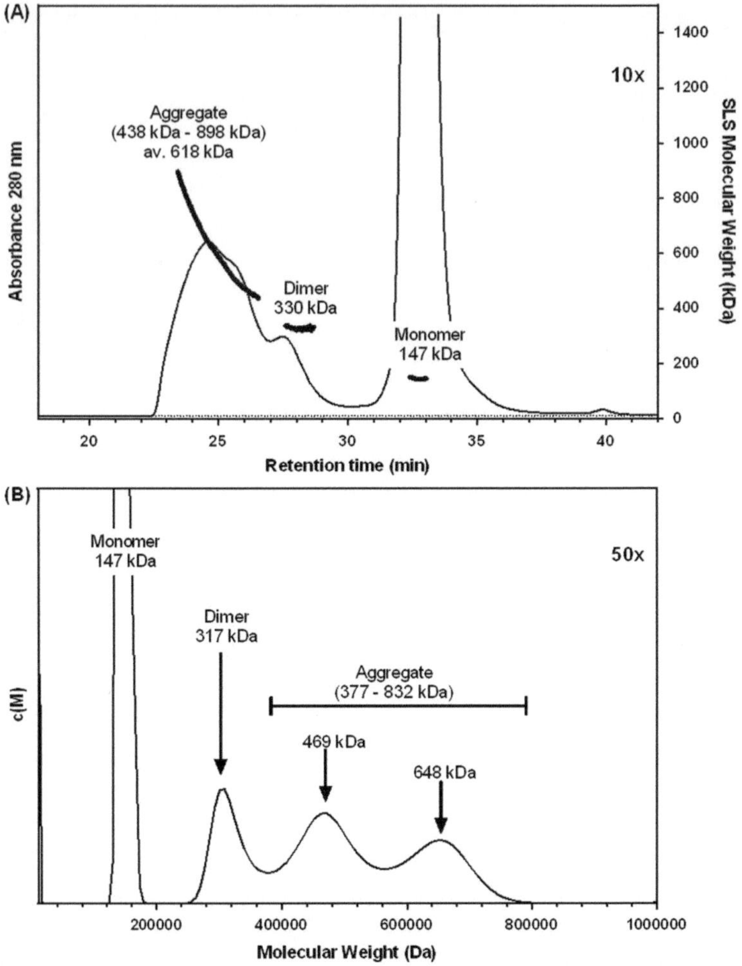

Fig. 20.7 SEC chromatogram coupled with static light scattering determined molar mass values (**a**) and AUC-SV c(M) distribution (**b**) demonstrating agreement between the two methods and indicating the presence of aggregates in a mass range from 130 to 1100 kDa

20.5 Conclusions

AUC-SV has served an important role in development of biopharmaceutical products for many years. Over time, the evaluation of the characteristics of the method has led to an improved understanding of how best to utilize AUC-SV during the development of biopharmaceutical products. Here we discussed the use of AUC-SV as an orthogonal method to SEC for quantitation of protein aggregates. AUC-SV offers three primary benefits as an orthogonal tool to SEC analysis, including the analysis of the sample under native-like solution conditions, minimization of nonspecific protein adsorption, and enhanced resolution of aggregate species. However, as a method for the analysis of protein size distributions, AUC-SV suffers from lower precision and throughput than SEC. Because the two methods are complementary, they are utilized best when run together for many applications during biopharmaceutical development. The development and validation of SEC methods is one such application where AUC-SV can serve to ensure the accurate quantitation of aggregates by SEC.

References

ICH Q2(R1) (1996) Validation of analytical procedures: methodology. In *International conference on harmonization of technical requirements for registration of pharmaceuticals for human use*. Available from: http://www.ich.org/fileadmin/Public_Web_Site/ICH_Products/Guidelines/Quality/Q2_R1/Step4/Q2_R1__Guideline.pdf

Berkowitz SA (2006) Role of analytical ultracentrifugation in assessing the aggregation of protein biopharmaceuticals. AAPS J 8:590–605

Berkowitz SA, Philo JS (2007) Monitoring the homogeneity of adenovirus preparations (a gene therapy delivery system) using analytical ultracentrifugation. Anal Biochem 362:16–37

Burdick RK, Sidor L (2013) Establishment of an equivalence acceptance criterion for accelerated stability studies. J Biopharm Stat 23:730–743

Demeler B (2005) UltraScan A comprehensive data analysis software package for analytical ultracentrifugation experiments. In: Scott DJ, Harding SE, Rowe AJ (eds) Modern analytical ultracentrifugation: techniques and methods. Royal Society of Chemistry, Cambridge, pp 210–229

Gabrielson JP, Arthur KK (2011) Measuring low levels of protein aggregation by sedimentation velocity. Methods 54:83–91

Gabrielson JP, Arthur KK, Stoner MR, Winn BC, Kendrick BS, Razinkov V, Svitel J, Jiang Y, Voelker PJ, Fernandes CA, Ridgeway R (2010) Precision of protein aggregation measurements by sedimentation velocity analytical ultracentrifugation in biopharmaceutical applications. Anal Biochem 396:231–241

Gabrielson JP, Brader ML, Pekar AH, Mathis KB, Winter G, Carpenter JF, Randolph TW (2006) Quantitation of aggregate levels in a recombinant humanized monoclonal antibody formulation by size-exclusion chromatography, asymmetrical flow field flow fractionation, and sedimentation velocity. J Pharm Sci 96:268–279

Ghirlando R, Balbo A, Piszczek G, Brown PH, Lewis MS, Brautigam CA, Schuck P, Zhao H (2013) Improving the thermal, radial, and temporal accuracy of the analytical ultracentrifuge through external references. Anal Biochem 440:81–95

Kozlowski S, Swann P (2006) Current and future issues in the manufacturing and development of monoclonal antibodies. Adv Drug Deliv Rev 58:707–722

Laue TM, Stafford WF (1999) Modern applications of analytical ultracentrifugation. Annu Rev Biophys Biomol Struct 28:75–100

Lebowitz J, Lewis MS, Schuck P (2002) Modern analytical ultracentrifugation in protein science: a tutorial review. Protein Sci 11:2067–2079

Mahler H-C, Friess W, Grauschopf U, Kiese S (2009) Protein aggregation: pathways, induction factors and analysis. J Pharm Sci 98:2909–2934

Parenky A, Myler H, Amaravadi L, Bechtold-Peters K, Rosenberg A, Kirshner S, Quarmby V (2014) New FDA draft guidance on immunogenicity. AAPS J 16:499–503

Paul R, Graff-Meyer A, Stahlber H, Lauer ME, Rufer AC, Beck H, Briguet A, Schnaible V, Buckel T, Boeckle S (2012) Structure and function of purified monoclonal antibody dimer induced by different stress conditions. Pharm Res 29:2047–2059

Pekar A, Sukumar M (2007) Quantitation of aggregates in therapeutic proteins using sedimentation velocity analytical ultracentrifugation: practical considerations that affect precision and accuracy. Anal Biochem 367:225–237

Philo JS (2006) Is any measurement method optimal for all aggregate sizes and types? AAPS J 8:564–571

Rosenberg AS (2006) Effects of protein aggregates: an immunologic perspective. AAPS J 8:E501–E507

Schuck P (2000) Size distribution analysis of macromolecules by sedimentation velocity ultracentrifugation and Lamm equation modeling. Biophys J 78:1606–1619

Stafford WF, Sherwood PJ (2004) Analysis of heterologous interacting systems by sedimentation velocity: curve fitting algorithms for estimation of sedimentation coefficients, equilibrium and rate constants. Biophys Chem 108:231–243

Zhao H, Balbo A, Metger H, Clary R, Ghirlando R, Schuck P (2014) Improved measurement of the rotor temperature in analytical ultracentrifugation. Anal Biochem 451:69–75

Chapter 21
Biopharmaceutical Evaluation of Intermolecular Interactions by AUC-SE

Shuntaro Saito and Susumu Uchiyama

Abstract Analytical ultracentrifugation sedimentation equilibrium (AUC-SE) is a useful technique to investigate the weak reversible intermolecular interactions among proteins in solution. It provides biophysical information such as the average apparent molecular weight, stoichiometry, and association constant of associating proteins. Several studies of intermolecular interaction in biopharmaceuticals by AUC-SE are introduced in this chapter. AUC-SE also provides the second virial coefficient (B_2), which represents the type, i.e., repulsive and attractive, and a magnitude of intermolecular interactions. The B_2 values obtained from the protein solution at low concentrations showed good correlation with aggregation and viscosity of MAb at high concentrations, indicating that B_2 can be an effective indicator of aggregation propensity and viscosity. These findings suggest that AUC-SE provides clues to understand the self-association in biopharmaceuticals and to establish effective manufacturing process, formulation, and administration of biopharmaceuticals.

Keywords Analytical ultracentrifugation sedimentation equilibrium • Self-association • Colloidal stability • Aggregation • Viscosity • Second virial coefficient • Antibody

21.1 Introduction

Development of therapeutic proteins such as monoclonal antibody (MAb) remains challenging because of their unique characteristics arising from their large molecular size and complex higher-order structures. Aggregation is one of the crucial

S. Saito
Biologics Technology Research Laboratories, Daiichi Sankyo Co., Ltd., 1-12-1, Shinomiya, Hiratsuka-shi, Kanagawa 254-0014, Japan
e-mail: saito.shuntaro.e5@daiichisankyo.co.jp

S. Uchiyama (✉)
Department of Biotechnology, Graduate School of Engineering, Osaka University, 2-1 Yamadaoka, Suita, Osaka 565-0871, Japan
e-mail: suchi@bio.eng.osaka-u.ac.jp

S. Uchiyama et al. (eds.), *Analytical Ultracentrifugation*,
DOI 10.1007/978-4-431-55985-6_21

issues in therapeutic proteins because they potentially impact on the protein activity, pharmacokinetics, and safety due to immunogenicity. High viscosity at a highly concentrated solution is also a crucial issue because it prevents effective manufacturing and limits the route and mode of administration to patients. Recently, it has become clear that reversible self-association of protein molecules could lead to aggregation and viscosity enhancement. Reversible self-association is attributed to the intermolecular interaction of protein molecules. This measure of molecular dispersity determined by the marginal balance of attractive and repulsive intermolecular interactions among protein molecules is known as colloidal stability. The Derjaguin–Landau–Verwey–Overbeck (DLVO) theory states that for a solution with high colloidal stability, the repulsive forces arising from electrostatic interactions among molecules are greater than the attractive forces attributed to van der Waals interactions (Verwey and Overbeck 1948).

Different types of intermolecular forces are involved in reversible intermolecular interactions, such as electrostatic interactions, van der Waals forces, and hydrophobic interactions. Environmental factors such as pH, ionic strength, and additives can alter the intermolecular interaction profile drastically. Therefore, colloidal stability has been assessed for optimizing formulations, stabilizing proteins, and determining the mechanisms of aggregation (Goldberg et al. 2011; Garidel et al. 2013). In these studies, several parameters were used as indicators of colloidal stability. The osmotic second virial coefficient (B_{22}) is one of the parameters that represents the degree of intermolecular interactions in relatively dilute solutions and can be used to assess colloidal stability. It is generally understood that a positive B_{22} value implies the presence of repulsive intermolecular interactions, whereas a negative B_{22} value indicates the presence of attractive intermolecular interactions (McMillan and Mayer 1945; Neal et al. 1998; van Holde et al. 2006). The second virial coefficient, B_2 ($B_2 = B_{22}/M_W{}^2$), can be determined experimentally using analytical ultracentrifugation sedimentation equilibrium (AUC-SE, Williams 1972). While B_2 can be determined by osmotic pressure measurement and static light scattering (SLS), AUC-SE is also a conventional but powerful method for the determination of B_2 from the concentration dependence of the apparent molecular weight ($M_{W,app}$) of a protein in solution. The B_2 value is a parameter that represents the degree of intermolecular interactions in dilute solutions.

In this chapter, we focus on intermolecular interaction analysis of biopharmaceuticals by AUC-SE. The role and usage of AUC-SE on the development of biopharmaceuticals are detailed in terms of the relationship of intermolecular interaction with the physical properties, i.e., aggregation propensity and viscosity.

21.2 Theory of AUC-SE Analysis to Determine the Intermolecular Interaction

AUC-SE is a conventional technique using analytical ultracentrifugation for determination of $M_{\mathrm{W,app}}$. On AUC-SE, the concentration gradient of solute formed by centrifugation is measured after the balance between centrifugal and when diffusional fluxes reach equilibrium. The experiments can be performed in free solution; therefore, there are no complications due to the interactions with matrices and surfaces. This method is applicable to samples over a wide range of concentrations and of various solvent compositions (Liu et al. 2005; Jiménez et al. 2007).

AUC-SE is carried out at relatively lower rotor speed (~ 20,000 rpm) than that for sedimentation velocity to create equilibrium concentration gradient. The optical system for acquisition of the concentration gradient in the cell is selected based on the dependence of the strength of intermolecular interaction. A higher concentration is required for weak intermolecular interactions. Rayleigh interference (IF) optics is suitable for a higher concentration sample compared to UV absorption optics. Preparative ultracentrifugation is also applicable for very highly concentrated samples (10 mg/mL ~). In this case, the concentrations of the solution fractionated from the cell are measured after completion of centrifugation.

$M_{\mathrm{W,app}}$ is estimated by nonlinear least-squares fitting of the concentration gradient according to Eq. 21.1:

$$c(r) = c_0 \exp\left[\frac{M_{\mathrm{W,app}}\,(1-\overline{\upsilon}\rho)\,\omega^2}{2RT}\left(r^2 - {r_o}^2\right)\right] + \mathrm{Baseline} \tag{21.1}$$

where $c(r)$ (mg/mL) is the protein concentration at the radial position r (cm), c_0 is the protein concentration at the reference radial position r_0, $\overline{\upsilon}$ (cm^3/g) is the partial specific volume of solute, ρ (g/cm^3) is the solvent density, ω (rad/s) is the angular velocity, T (K) is the absolute temperature, and R is the gas constant. The partial specific volume and solvent density can be calculated using the Sednterp software (Laue et al. 1992, http://bitcwiki.sr.unh.edu/index.php/Main_Page).

B_2 was obtained from the slope of the plot of the inverse of $M_{\mathrm{W,app}}$ against the concentration, according to Eq. 21.2:

$$\frac{1}{M_{\mathrm{W,app}}} = \frac{1}{W_{\mathrm{W}}} + 2B_2c \tag{21.2}$$

$$B_{22} = B_2 \times {M_{\mathrm{W}}}^2 \tag{21.3}$$

where M_W is the weight-average molecular weight at infinite dilution and c is the initial loading concentration (mg/mL). B_{22} was obtained using Eq. 21.3 (Winzor et al. 2007).

Several approaches have been used to evaluate the intermolecular interactions. B_2 values of protein solutions have been estimated using SLS (Narayanan and Liu 2003; Attri and Minton 2005; Alford et al. 2008) and self-interaction chromatography (SIC) (Tessier et al. 2002; Brun et al. 2009; Brun et al. 2010a; Brun et al. 2010b). The interaction parameter (k_d), which relates to B_2, can be derived from dynamic light scattering (DLS) (Zhang and Liu 2003).

The first advantage of AUC-SE as a tool for intermolecular interaction analysis is that it allows direct measurement of the molecular behaviors in the solution without any matrices and surfaces to interact with the solute unlike the case of SIC. There is a concern that B_2 values determined from SIC do not correspond to the interaction of proteins in solution because SIC measures the interaction of molecules in solution with molecules immobilized on the column matrix. In addition, degradation and conformational changes in proteins that occur during the preparation of columns and nonspecific interactions of proteins with the column matrix are also of concern.

The second advantage of AUC-SE is that B_2 from AUC-SE reflects the solute-solute interaction, whereas light scattering techniques such as DLS and SLS provide an indication of both solute-solute interactions and solute-solvent interactions (Deszczynski et al. 2006; Winzor et al. 2007).

The third advantage is that AUC-SE directly provides a B_2 value based on the $M_{W,app}$ which is independent on the shape and hydration state of molecules. Theoretically, k_D from DLS is composed of a thermodynamic term, B_2, and a hydrodynamic term, $\xi_1 + \overline{\nu}$, as shown in Eq. 21.4:

$$k_D = 2B_2M_W - (\xi_1 + \overline{\nu}) \tag{21.4}$$

where ξ_1 is the coefficient of the linear term in the virial expansion of the frictional coefficient as a function of solute concentration and increases with increasing protein solvation (Chari et al. 2009). Figure 21.1 shows the correlation of B_2 with k_D for two different MAbs in various formulation buffers (Saito et al. 2013). The values of ξ_1 and $\overline{\nu}$ are positive, creating a negative contribution to k_D as evidenced from the larger negative intercept as shown in Fig. 21.1, resulting in negative k_D values even in the case of repulsive interaction. Thus, the determination of the type of intermolecular interaction, i.e., repulsive or attractive, is difficult from k_D due to the contribution of the hydrodynamic term. In addition, the k_D value obtained by DLS may be misleading with respect to the magnitude of intermolecular interactions because k_D is related not only to B_2 but also to ξ_1, which depends upon the shape and hydration state of the proteins (Yamakawa 1962; Frost and Caroline 1976; Lehermayr et al. 2011). If the shape and size of the molecule change depending on the concentration, the k_D value does not reflect the only type of intermolecular

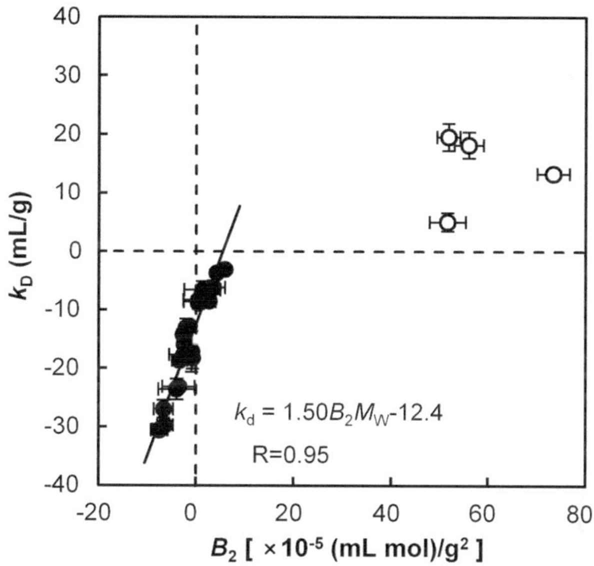

Fig. 21.1 Correlation of B_2 with k_D for MAbs: B_2 was obtained from AUC-SE and k_D was obtained from DLS. Regression lines were obtained from B_2 and k_D in different formulations at pH 5 and pH 7, except for those of MAb-B in the formulations without NaCl at pH 5 (○); these were eliminated for fitting to the regression line because of a deviation from the trend (Saito et al. 2013)

interaction. In fact, it should be noted that some exceptional relations between k_D and B_2 were observed. When B_2 exceeded 50×10^{-5} (mL mol)/g^2, the k_D values were smaller than those estimated from Eq. 21.4 as shown in Fig. 21.1 (open symbol). These results suggest that a higher degree of hydration and/or conformational change increases the frictional drag of the MAb.

These findings indicate that B_2 determined from AUC-SE is a better indicator of colloidal stability. Although AUC-SE is recognized as a low-throughput technique, the information obtained from this technique is highly valuable and can provide useful details about the type of dominant intermolecular interactions and the critical physical parameters of target biopharmaceuticals.

21.3 Analysis of Intermolecular Interaction in Biopharmaceuticals

MAb is one of the important classes of biopharmaceuticals, and several different therapeutic MAbs have recently entered into the market with many more under clinical trials, for treatment of diseases such as cancer, infectious diseases, allergies, autoimmune diseases, cardiovascular diseases, and inflammation (Reichert et al. 2005). MAbs in the same subclass have a similar primary structure composed of variable and constant regions. The variable regions, which occupy only less than

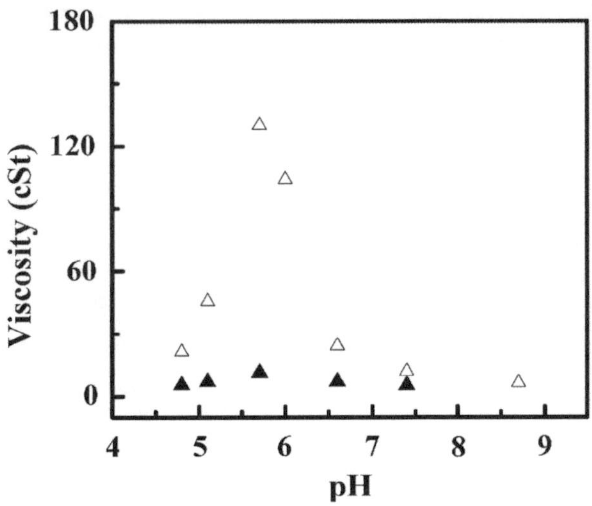

Fig. 21.2 The effects of pH on the kinematic viscosity of MAb1 solution containing 130 mg/mL protein, 2–17.5 mM of acetate or arginine with (▲) and without (△) 150 mM NaCl. The viscosity was measured using a Cannon-Fenske routine viscometer (Cannon Instrument) at 25 °C (Liu et al. 2005)

25 % of the primary structure, are composed of different amino acid sequences and play an essential role in the recognition and binding to antigens. It is known that individual MAb shows different physical properties, i.e., aggregation propensity and viscosity, even though most of their amino acid sequences are identical (Yadav et al. 2010; Sahin et al. 2010).

Liu et al. anticipated the investigation of the effects and contributions of weak reversible self-association on the viscosity of MAbs by AUC-SE (Liu et al. 2005). One specific MAb, MAb1, showed significantly higher viscosity at low salt concentrations than other MAbs (MAb2 and MAb3). The viscosity of MAb1 solution is highly dependent on the pH and salt concentration. The maximum value of viscosity was observed around pI of MAb1 at low salt concentrations, indicating that electrostatic attractive interactions are responsible for high viscosity of this antibody (Fig. 21.2). Singh et al. further investigated this attractive interaction in MAb1 and found that the viscosity correlated well with measured dipole moment. They, therefore, suggested that dipole-dipole interaction plays an important role in the electrostatic attractive interaction of MAb1 (Singh et al. 2014).

The contribution of weak reversible self-association on high viscosity of MAb1 was elucidated by AUC-SE analysis performed for highly concentrated solution (100 mg/mL) at low salt concentrations. Herein, AUC-SE analyses were conducted by using preparative centrifuge and a micro-fractionator. Obviously, MAb1 demonstrated a steeper concentration gradient (Fig. 21.3a) than MAb2 and MAb3. The average $M_{W,app}$ of MAb2 and MAb3 demonstrated similar curves and decreased as the concentration increased resulting in positive B_2 (Fig. 21.3b). In this study, the authors inferred that positive B_2 is attributed to the excluded volume effect.

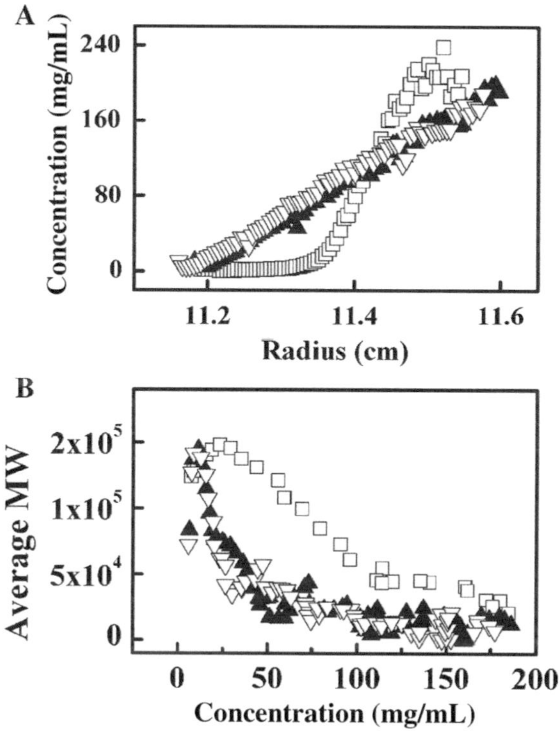

Fig. 21.3 The sedimentation equilibrium data of MAb1(□), MAb2 (▲), and MAb3 (∇) at 100 mg/mL, containing 240 mM trehalose, 40 mM histidine, and 0.04 % polysorbate 20 (**A**). The sedimentation equilibrium experiments were conducted at 15,000 rpm and 25 °C. The apparent weight-averaged molecular weights of MAbs were plotted as a function of protein concentration (**B**) (Liu et al. 2005)

Whereas, the average $M_{\mathrm{W,app}}$ of MAb1 increased, resulting in negative B_2, in the lower concentration range and then decreased as the concentration increased. This result suggests that attractive interaction and excluded volume effect may coexist at the same time. At the low protein concentration range, the attractive interaction is dominant, whereas repulsive interaction due to the excluded volume effect increased at higher concentration resulting in positive B_2. These AUC-SE results suggest that the reversible self-association of protein is mediated by electrostatic interaction and results in unusually high viscosity of MAb. It should be noted that this self-association is quite weak and is not detectable by sedimentation velocity and size-exclusion chromatography.

The weak self-association of MAb may be dominant in the concentrated solution. However, the detection of such weak self-association at high concentrations is challenging because of the contribution of nonspecific repulsive interaction such as excluded volume effect and electrostatic repulsion which may mask the weak attractive interaction. To analyze the weak self-association in highly concentrated solution, Jiménez et al. developed a model for fitting the AUC-SE data in which nonspecific repulsive interaction as a function of the protein concentration was taken into account (Jiménez et al. 2007). Herein, AUC-SE was conducted for IgG over a concentration up to 125 mg/mL. The UV absorption optics and preparative centrifugation were used to measure the concentration gradient depending on the

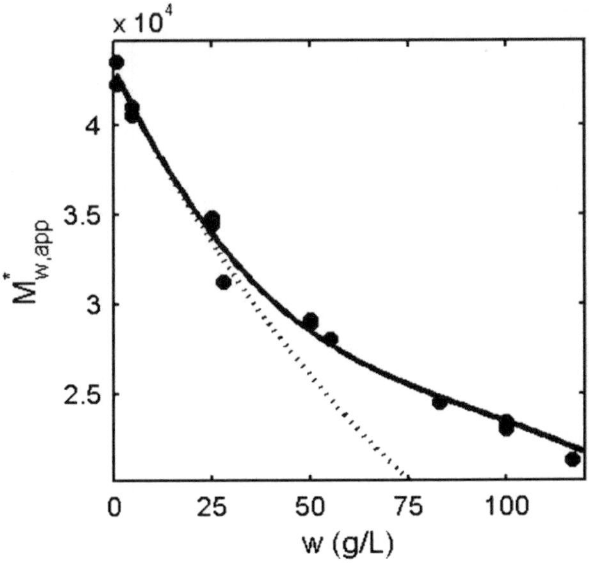

Fig. 21.4 $M_{W,app}$ of IgG plotted as a function of protein concentration. The *solid curve* is the best fit of model, assuming only trimer formation, with association constant log $K_3^0 = 5.0$ and $v_{eff} = 1.3$ cm^3/g. The *dashed curve* is calculated with the same value of v_{eff}, setting $K_3^0 = 0$. The magnitude of v_{eff} reflects both excluded volume effect and electrostatic interactions (Jiménez et al. 2007)

loaded concentration of the sample. The $M_{W,app}$ of IgG steeply decreased as a function of the concentration, which has characteristics of nonspecific repulsive interactions derived from excluded volume effect (Fig. 21.4). However, concentration dependence of $M_{W,app}$ did not follow the curve (dashed curve) which was calculated on the assumption that IgG does not self-associate, indicating the presence of attractive interactions. In this case, the simplest model capable of fitting the data to within experimental precision was a monomer-trimer self-association scheme. Thus, their technique using AUC-SE provides the information about stoichiometry and association constant of weak self-association of IgG. It should be noted that the concentration dependence of AUC-SE was quantitatively consistent with that obtained from osmotic pressure measurement. The values of $M_{W,app}$ and mass fractions of monomer and trimer as a function of concentration indicate that the fraction of trimer is approximately 1/3 at the concentration of 100 mg/mL and over 1/2 at a concentration of 200 mg/mL. The formation of oligomer at higher concentrations may have significant influence on the efficacy, physical and chemical stability, and physical properties that are compatible with manufacturing, storage, delivery, and administration of biopharmaceuticals. AUC-SE thus can provide the opportunity for the careful characterization of the weak self-association of biopharmaceuticals.

21.4 Evaluation of Intermolecular Interaction in Biopharmaceuticals

The solution pH, ionic strength, and addition of sugars have significant impact on the physical properties, i.e., aggregation propensity and viscosity, of proteins. Likewise, the concentration and kinds of surfactant and buffer component and concentration of protein are also parameters, which are known to influence the physical properties of proteins (Chi et al. 2003a; Bhambhani et al. 2012). Thus, tremendous efforts are necessary to evaluate all these parameters to optimize the formulation of biopharmaceuticals. The conventional strategies for formulation optimization are direct assessments of long-term stability and viscosity, which are time consuming and require a large amount of samples. In contrast, prediction of the aggregation propensity and viscosity based on the indicators of colloidal stability allows rational screening of various parameters with small sample consumption. In practice, several indicators of colloidal stability obtained from SLS, DLS, and SIC have been anticipated for the formulation optimization of biopharmaceuticals (Kumar et al. 2011; Sule et al. 2012; Garidel et al. 2013). The relationships between the colloidal stability quantified by these techniques and physical properties for various proteins are summarized in Table 21.1. The colloidal stabilities of various proteins such as MAb and lysozyme in the formulation varying in pH and the concentration and/or kind of buffer, ionic strength, sugar, and surfactant qualitatively correlated with aggregation and/or viscosity which indicated the protein association. Thus, an indicator of colloidal stability can be used for qualitative prediction of protein association, although careful data interpretation is necessary when scattering techniques and SIC are used for colloidal stability assessment as previously mentioned.

AUC-SE can provide an effective indicator, B_2, for the determination of the type and degree of colloidal stability. To evaluate AUC-SE as a tool for prediction of the aggregation propensity and viscosity, we investigated the relationship between B_2 and physical properties, where MAbs were used as model proteins. In the studies, the B_2 was measured by AUC-SE from three different concentrations less than 10 mg/mL in the different formulation buffers containing different concentrations of salt (Saito et al. 2013). The B_2 values were measured for two different humanized IgG_1 antibodies, MAb-A and MAb-B. Figure 21.5 shows NaCl concentration dependence of B_2 values at pH 5 and pH 7. At the lowest NaCl concentration, negative B_2 values of MAb-A at pH 5 and pH 7 showed the presence of attractive intermolecular interactions (Fig. 21.5a). The B_2 values increased as the NaCl concentration increased and inverted from negative to positive values. Such dependencies of B_2 on NaCl concentrations for MAb-A were similarly confirmed at pH 5 and pH 7; however, the B_2 values determined in each salt concentration at pH 7 were higher than those at pH 5, suggesting increased attractive intermolecular interactions at pH 5 than at pH 7. The dependence of B_2 on the NaCl concentration

Table 21.1 Relationship between colloidal stability and physical properties

Protein	Formulation factor	Colloidal stability				Physicochemical property			Correlation	Reference
		Parameter	Method	Conc. (mg/mL)	Value (k_D: mL/g) (B_2: $\times 10^{-4}$mol mL/g^2)	Property	Method	Conc. (mg/mL)		
rhGCSF	pH, ionic strength, excipient (sugar)	B_2	SLS	0.5–5	−23.4 ∼ 209	Aggregates (37 °C)	SEC	1.5	Partially qualitative correlation	Chi et al. (2003b)
IgG_1, BSA	pH, ionic strength, excipient (sugar, surfactant)	B_2	SEC-SLS	10	−1.14 ∼ −0.02	Aggregates (37 °C)	SEC	7.5	No correlation	Bajaji et al. (2006)
IgG_2	pH, ionic strength	k_D	DLS	≤12	−20 ∼ 70	Aggregates (25, 37 °C)	SEC	5, 50	Partially qualitative correlation	Saluja et al. (2007)
		B_2	SLS	≤10	1 ∼ 14				Partially qualitative correlation	
Lysozyme	pH, salt species and conc., excipient (sugar, PEG)	B_2	SIC	–	−6.5 ∼ 14	Aggregate (Agitation)	Turbidity	10	Partially qualitative correlation	Brun et al. (2009)

IgG_1	Ionic strength	k_D	DLS	≤12	−30.8 ~ 16.0	Viscosity	Viscometer	≤125	Partially qualitative correlation	Yadav et al. (2010)
Lysozyme	pH, ionic strength	B_2	SIC	–	−20 ~ 10	Aggregate (Agitation)	Turbidity	10	Partially qualitative correlation	Brun et al. (2010a)
IgG_1	pH, buffer species, ionic strength, excipient (amino acid)	B_2	SIC	–	−1.5 ~ 3.8	Aggregates (40 °C, agitation)	SEC, turbidity	–	Partially qualitative correlation	Brun et al. (2010b)
IgG_2 lysozyme	Anion species, buffer species	B_2	AUC-SV/DLS	–	−4 ~ 4	Aggregates (45 °C, agitation)	SEC	10	Qualitative correlation	Saluja et al. (2010)
		k_D	DLS		−6 ~ 55		Turbidity	70	Qualitative correlation	
IgG_1	Ionic strength	B_2	Membrane osmometry	≤100	−0.43 ~ 1.3	Viscosity	Viscometer	≤120	Qualitative correlation	Salinas et al. (2010)
			SLS	≤50						
			Zeta potential	2	−0.5 ~ 1.0	Aggregates (agitation)	Turbidity	≤120	Qualitative correlation	
					0.001 ~ 0.7				Partially qualitative correlation	

(continued)

Table 21.1 (continued)

Protein	Formulation factor	Colloidal stability				Physicochemical property			Correlation	Reference
		Parameter	Method	Conc. (mg/mL)	Value (k_D: mL/g) (B_2: $\times 10^{-4}$mol mL/g^2)	Property	Method	Conc. (mg/mL)		
IgG_1, DVD-Ig	Ionic strength	B_2	SLS	≤10	2.0 ~ 37.2	Aggregates (40 °C)	SEC	5, 150	Partially qualitative correlation	Kumar et al. (2011)
		k_D	DLS	≤10	−28.4 ~ 464				Partially qualitative correlation	
IgG_1	Ionic strength	B_2	SLS	2–1	−0.7 ~ 1.5	Viscosity	Viscometer	≤250	Partially qualitative correlation	Yadav et al. (2012)
		k_D	DLS	2–12	−19.8 ~ 11.1					
Albinterferon-α	pH, ionic strength, excipient (sugar, surfactant)	B_2	SLS	1–5	−3.6 ~ 4.3	Aggregates (40 °C, agitation)	SEC	–	Partially qualitative correlation	Chou et al. (2012)
IgG_1	pH	B_2	AUC-SE	1–10	−2.7 ~ 2.1	Aggregate (40 °C)	SEC	100	Qualitative correlation	Saito et al. (2012)
		k_D	DLS	2–10	−12.2 ~ –5.5	Viscosity	Viscometer	150	Partially qualitative correlation	

IgG_1	pH, ionic strength, excipient (sugar)	B_2	AUC-SE	1–10	−7.5 ~ 73.3	Aggregate (40 °C, freeze-thaw)	SEC	10	Qualitative correlation	Saito et al. (2013)
		k_D	DLS	2–10	−30.5 ~ 19.6				Partially qualitative correlation	
IgG_1, IgG_4	pH, ionic strength	k_D	DLS	≤12	−20.2 ~ 50	Viscosity	USR, DLS	200	Qualitative correlation	Neergaard et al. (2013)
IgG_1	pH, ionic strength, buffer species, excipient (sugar)	k_D	DLS	1–10	−28 ~ 15	Aggregate (agitation)	Turbidity	10	Qualitative correlation	Garidel et al. (2013)

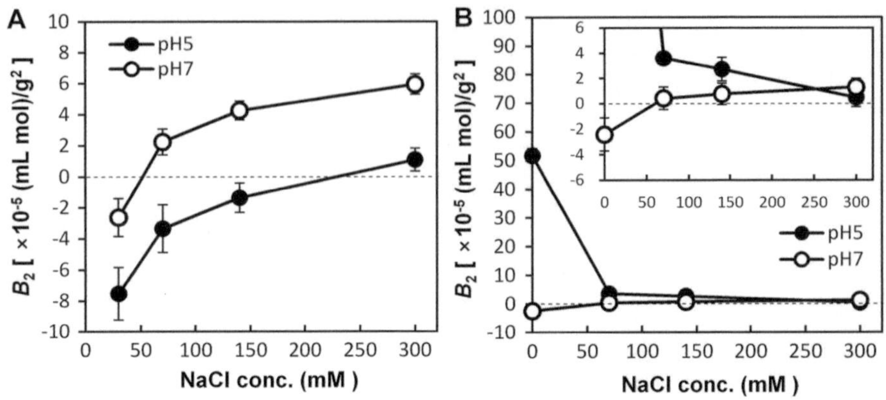

Fig. 21.5 B_2 for MAb-A (**a**) and MAb-B (**b**) in formulations containing different concentrations of NaCl were measured by AUC-SE. B_2 was obtained from the concentration dependence of $M_{\mathrm{W,app}}$ at three different concentrations, i.e., 1, 5, and 10 mg/mL, in 10 mM AcONa (pH 5) (•) and 10 mM sodium phosphate buffer (pH 7) (○) containing NaCl as indicated in the figures. The inset in B is an expanded figure. The lowest ionic strength buffer (initial buffer) contained 30 mM NaCl for MAb-A and no NaCl for MAb-B. The error bars were estimated from three independent experiments under the same conditions (Saito et al. 2013)

was similar to that for MAb-B at pH 7, at which the B_2 values were negative in solutions of the lowest ionic strength (Fig. 21.5b). The B_2 values increased as the NaCl concentration increased and inverted to positive values. In contrast, the B_2 value of MAb-B at pH 5 was high and positive, indicating the presence of strong repulsive intermolecular interactions. The B_2 values of MAb-B decreased strikingly with increasing the NaCl concentration, suggesting that NaCl greatly suppressed the repulsive intermolecular interactions. Thus, colloidal stability is highly sensitive to the buffer composition such as NaCl concentration and pH, and its behavior depends on the type of MAbs. These behaviors of MAbs can be explained by the proximity energies between two molecules (Laue 2012; Saito et al. 2013). Increase of ionic strength weakens the electrostatic interactions such as charge-charge, charge-dipole, and dipole-dipole interactions. In case charge-charge repulsive interaction is dominant like the case of MAb-B at pH 5, salts shield the charge-charge repulsive interaction resulting in the decrease of B_2 value. On the other hand, in case the attractive interactions by charge-dipole and dipole-dipole interactions are stronger than the charge-charge repulsive interaction, salts suppress the attractive interaction resulting in the increase of B_2 value like the case of MAb-A at pH 5 and pH 7 and MAb-B at pH 7.

The linear correlations were observed between B_2 values and aggregation rates in the formulations containing different concentrations of NaCl, indicating that B_2 representing the colloidal stability well reflects the aggregation propensities for all the MAbs under an accelerated storage condition at 40 °C, which are normally used in stability tests (Fig. 21.6). These results suggest that electrostatic interactions make major and quantitative contributions to the aggregation propensities of these MAbs

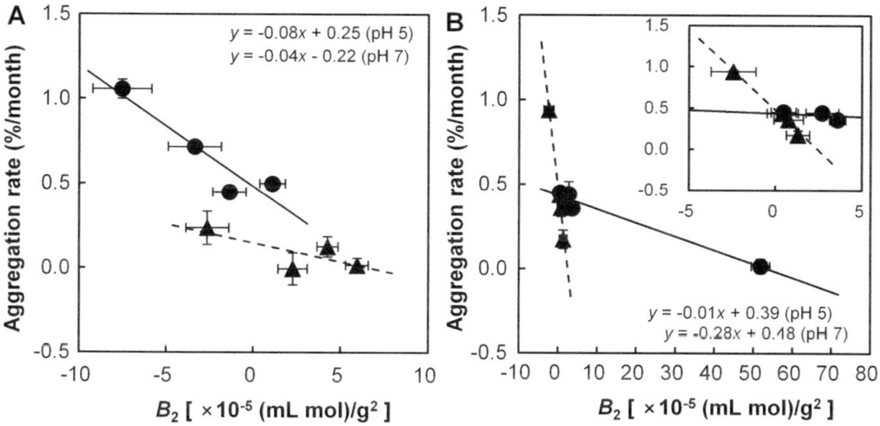

Fig. 21.6 Correlation of B_2 with aggregation propensity at 40 °C for MAb-A (**a**) and MAb-B (**b**) at pH 5 (*circles*) and pH 7 (*triangles*): the increase in the rate of aggregation was estimated from the increase in the population of aggregates after storage at 40 °C (%/month). The correlation coefficients were $R = -0.939$ (MAb-A: pH 5), $R = -0.719$ (MAb-A: pH 7), $R = -0.986$ (MAb-B: pH 5), and $R = -0.992$ (MAb-B: pH 7). The error bars were estimated from three independent experiments performed under the same conditions

and can therefore be predicted from the experimental estimates of B_2 prior to actual stability tests.

The effects on B_2 when the ionic strength changes were further investigated for four different MAbs (MAb-A, MAb-B, MAb-C, and MAb-D) at pH 5, pH 6, pH 7, and pH 8 at low ionic strength or high ionic strength. As shown in Fig. 21.7, the B_2 values of all the MAbs were affected to some extent as the ionic strength changed. The negative B_2 values for MAb-A at all pHs, MAb-B and MAb-C at pH 7 and pH 8, and MAb-D at pH 8 indicate the presence of weak attractive interactions at low ionic strength. Importantly, negative B_2 values were all inverted to positive values by the addition of 300 mM NaCl. Positive B_2 values were obtained at low ionic strength for MAb-B and MAb-C at pH 5 and pH 6 and MAb-D at pH 5, pH 6, and pH 7. These positive B_2 values were significantly reduced by the addition of 300 mM NaCl. These results indicate that salts both stabilize and destabilize the MAbs, depending on the type of electrostatic interaction, i.e., repulsion or attraction, at low ionic strength. Experimental evaluation of the colloidal stability is therefore of great importance for optimizing salts in formulations.

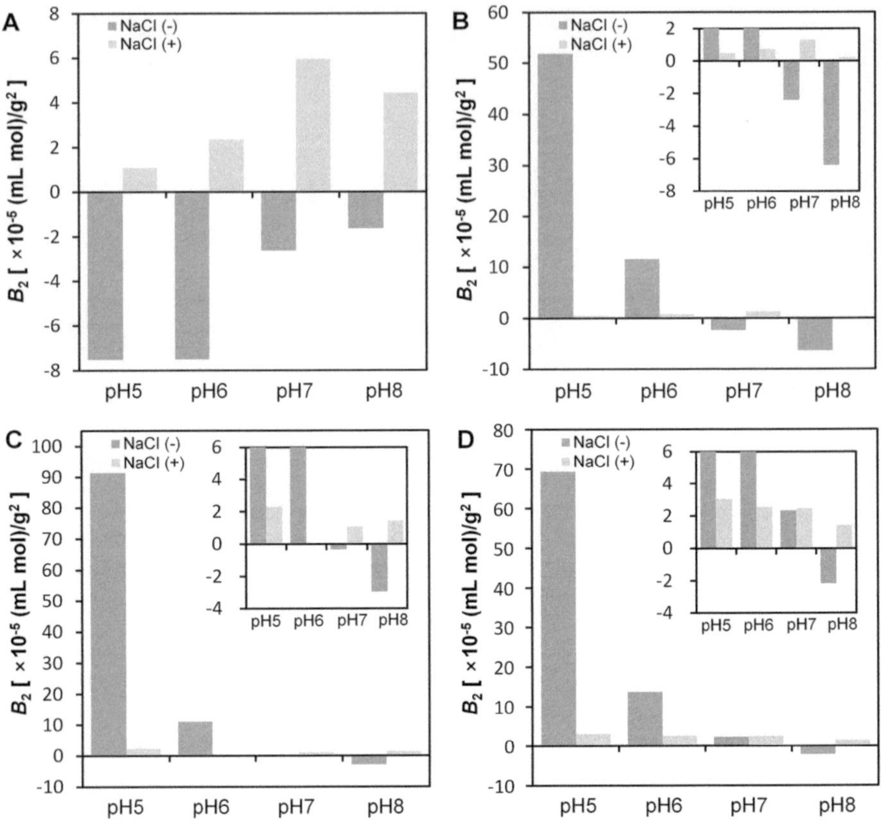

Fig. 21.7 Influence of ionic strength on B_2: B_2 for MAb-A (**a**), MAb-B (**b**), MAb-C (**c**), and MAb-D (**d**) were measured at four pHs. B_2 was obtained from the concentration dependence of $M_{W,app}$ obtained by AUC-SE at three different concentrations, i.e., 1, 5, and 10 mg/mL, in 10 mM AcONa (pH 5) and 10 mM sodium phosphate buffer (pH 6, pH 7, and pH 8) at the lowest ionic strength (30 mM NaCl for MAb-A and no NaCl for MAb-B, MAb-C, and MAb-D) or at the highest ionic strength (300 mM NaCl) (Saito et al. 2013)

21.5 Predictions of Aggregation Propensity and Viscosity at High Concentrations Based on B_2 at Low Concentrations

Development of high-concentration protein solutions is challenging because unfavorable phenomena such as increased viscosity and aggregates formation are frequently induced by increases in the concentration of protein solutions (Treuheit et al. 2002; Shire et al. 2004). These properties pose challenges in effective manufacturing, formulation development, analytical characterization, and administration of therapeutic proteins. Although direct measurements of the viscosity and aggregation propensity using high-concentration solutions provide useful information, the preparation of such solutions itself remains challenging, even on a small scale. In addition, it is technically difficult to measure the very weak intermolecular interactions that become dominant at concentrations above 100 mg/mL (Liu et al. 2005; Jiménez et al. 2007). Furthermore, even in cases where successful measurements have been made, interpretation of the data is extremely difficult since no concrete theory regarding the intermolecular interactions at high concentrations has been established yet.

In this section, we focus on the relationships between B_2 obtained at low concentrations and the viscosity and aggregation propensity of highly concentrated MAbs solutions. Herein, B_2 was determined from three concentrations less than 10 mg/mL in solutions of varying pH for three different MAbs. All MAbs, MAb-A, MAb-B, and MAb-C, are humanized IgG_1 which have the same constant region and different variable regions (Saito et al. 2012). B_2 determined by AUC-SE at different pH values is shown in Fig. 21.8. All three MAbs clearly show remarkable pH dependencies of B_2 values with only small standard deviations estimated from three repeated experiments.

The relationships between B_2 obtained at low concentrations and the viscosity and aggregation propensity of highly concentrated MAbs solutions are presented in Fig. 21.9. Aggregation propensity at 40 °C and viscosity were assessed at the concentration of 100 and 150 mg/mL, respectively. These concentrations are sufficiently high to select the subcutaneous and intramuscular administration routes, which are favored because they would reduce the burden on patients. Higher viscosities were observed at the pH levels where MAbs have lower B_2 values for all of the MAbs, as indicated in Fig. 21.9a. In addition, aggregation was accelerated under conditions that give a large negative B_2 attributable to the presence of attractive intermolecular interactions in all MAbs (Fig. 21.9b). Thus, colloidal

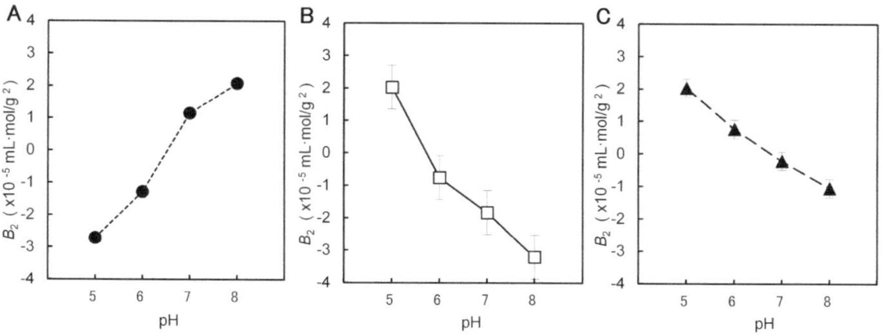

Fig. 21.8 pH dependence of B_2: B_2 for MAb-A (**a**), MAb-B (**b**), and MAb-C (**c**) were obtained from the concentration dependence of $M_{W,app}$ at three different concentrations in 10 mM sodium acetate buffer containing 140 mM NaCl (pH 5) or 10 mM phosphate buffer containing 140 mM NaCl (pH 6, 7, and 8). B_2 of MAb-A and MAb-C were obtained from concentrations 1, 5, and 10 mg/mL. B_2 of MAb-B was obtained from concentrations 5, 7.5, and 10 mg/mL (□)

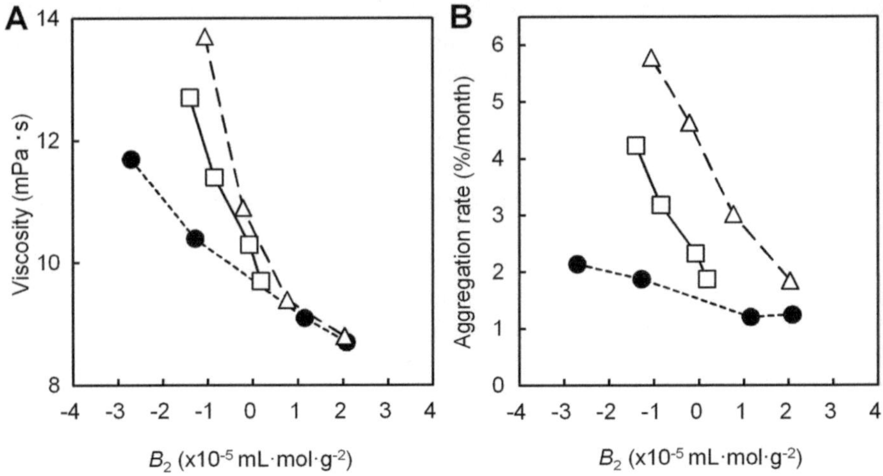

Fig. 21.9 Correlation of B_2 with viscosity (**a**) and aggregation propensity (**b**) for MAb-A (•), MAb-B, and MAb-C (△): B_2 values were obtained from AUC-SE. Viscosities were obtained at a concentration of 150 mg/mL, and aggregation rates at 40 °C were estimated based on SEC at a concentration of 100 mg/mL

stability is related to the enhancement of viscosity and formation of aggregates, and B_2, estimated from AUC-SE of MAbs solutions at low concentrations, is an effective indicator of viscosity and stability for highly concentrated solutions. It should be noted that the aggregation propensities and viscosities of three MAbs vary significantly with respect to conditions that provide larger negative B_2 values indicating the presence of stronger attractive interactions. On the other hand, the viscosities of the three MAb solutions were found to have similar values as B_2 increased and finally converged under conditions that provide positive B_2 values. These variations are attributed to the type of attractive interactions contributing to the viscosity enhancement and aggregation. The profile of attractive intermolecular interactions is largely dependent on the MAbs; therefore, careful interpretation is required for prediction of physical properties only from B_2 in particular which is negative.

21.6 Conclusions

AUC-SE is a useful analytical method not only for strong interactions such as antibody-antigen interaction but also for weak reversible interactions. These weak reversible interactions are thought to contribute to the physical properties of proteins, i.e., aggregation propensity and viscosity. In the presence of attractive interaction, viscosity is enhanced due to the transient cross-linked networks formed via weak reversible self-association as presented in Fig. 21.10a (Liu et al. 2005;

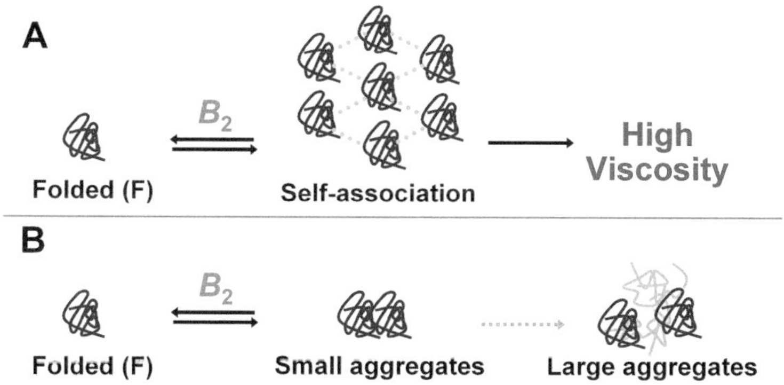

Fig. 21.10 Aggregation (**a**) and viscosity enhancement (**b**) pathways of proteins. B_2 is the second virial coefficient for proteins

Yadav et al. 2010; Chari et al. 2009). When protein molecules are closely located to each other, the short-range intermolecular interactions such as van der Waals force and hydrophobic interaction can become dominant and strongly assemble the molecules resulting in aggregation as presented in Fig. 21.10b (Chi et al. 2003a, b). In these pathways, B_2 can be an effective indicator of aggregation propensity and viscosity as long as the conformational stability is not impaired. In fact, good correlations of B_2 with aggregation propensity and viscosity have been reported as shown in this chapter. It should be noted that the effect of aggregates on viscosity requires careful investigation because the shape of aggregates can play an important role on viscosity. Elongated and flexible aggregates have an effect to enhance the viscosity, whereas spherical aggregates reduce the viscosity (Yadav et al. 2012; Schmit et al. 2014).

Generally, biopharmaceuticals show unique physical properties and are unstable when compared to low molecular weight compounds. Therefore, the manufacturing process, storage and shipping conditions, and administration routes have to be appropriately optimized with great effort. The analysis of intermolecular interaction by AUC-SE provides good opportunities to create an effective procedure for such optimization of biopharmaceuticals.

References

Alford JR, Kendrick BS, Carpenter JF, Randolph TW (2008) Measurement of the second osmotic virial coefficient for protein solutions exhibiting monomer-dimer equilibrium. Anal Biochem 377:128–133

Attri AK, Minton AP (2005) New methods for measuring macromolecular interactions in solution via static light scattering: basic methodology and application to nonassociating and self-associating proteins. Anal Biochem 337:103–110

Bajaji H, Sharma VK, Badkar A, Zeng D, Nema S, Kalonia DS (2006) Protein structural conformation and not second virial coefficient relates to long-term irreversible aggregation of a monoclonal antibody and ovalbumin in solution. Pharm Res 23:1382–1394

Bhambhani A, Kissmann JM, Joshi SB, Vokin DB, Kashi RS, Middaugh CR (2012) Formulation design and high-throughput excipient selection based on structural integrity and conformational stability of dilute and highly concentrated IgG_1 monoclonal antibody solutions. J Pharm Sci 101:1120–1135

Brun VL, Friess W, Bassarab S, Mühlau S, Garidel P (2010a) A critical evaluation of self-interaction chromatography as a predictive tool for the assessment of protein-protein interactions in protein formulation development: a case study of a therapeutic monoclonal antibody. Eur J Pharm Biopharm 75:16–25

Brun VL, Friess W, Schultz-Fademrecht T, Muehlau S, Garidel P (2009) Lysozyme-lysozyme self-interactions as assessed by the osmotic second virial coefficient: impact for physical protein stabilization. Biotechnol J 4:1305–1319

Brun VL, Friess W, Bassarab S, Garidel P (2010b) Correlation of protein-protein interactions as assessed by affinity chromatography with colloidal protein stability: a case study with lysozyme. Pharm Develop Technol 15:421–430

Chari R, Jerath K, Badkar AV, Kalonia DS (2009) Long- and short-range electrostatic interactions affect the rheology of highly concentrated antibody solutions. Pharm Res 26:2607–2618

Chi EY, Krishnan S, Randolph TW, Carpenter JF (2003a) Physical stability of proteins in aqueous solution: mechanism and driving forces in nonnative protein aggregation. Pharm Res 20:1325–1336

Chi EY, Krishnan S, Kendrick BS, Chang BS, Carpenter JF, Randolph TW (2003b) Roles of conformational stability and colloidal stability in the aggregation of recombinant human granulocyte colony-stimulating factor. Protein Sci 12:903–913

Chou DK, Krishnamurthy R, Manning MC, Randolph TW, Carpenter JF (2012) Physical stability of albinterferon-α2b in aqueous solution: effects of conformational stability and colloidal stability on aggregation. J Pharm Sci 101:2702–2719

Deszczynski M, Harding SE, Winzor DJ (2006) Negative second virial coefficients as predictors of protein crystal growth: evidence from sedimentation equilibrium studies that refutes the designation of those light scattering parameters as osmotic virial coefficients. Biophys Chem 120:106–113

Frost RA, Caroline D (1976) Diffusion of polystyrene in a theta mixed solvent (Benzene-2-Propanol) by Photon-correlation spectroscopy. Macromolecules 10:616–618

Garidel P, Blume A, Wagner M (2013) Prediction of colloidal stability of high protein formulations. Pharm Dev Technol 20(3):367–374

Goldberg DS, Bishop SM, Shah AU, Satish HA (2011) Formulation development of therapeutic monoclonal antibodies using high-throughput fluorescence and static light scattering techniques: role of conformational stability and colloidal stability. J Pharm Sci 100:1306–1315

Jiménez M, Rivas G, Minton AP (2007) Quantitative characterization of weak self-association in concentrated solutions of immunoglobulin G via the measurement of sedimentation equilibrium and osmotic pressure. Biochemistry 46:8373–8378

Kumar V, Dixit N, Zhou L, Fraunhofer W (2011) Impact of short range hydrophobic interactions and long range electrostatic forces on the aggregation kinetics of a monoclonal antibody and a dual-variable domain immunoglobulin at low and high concentrations. Int J Pharm 421:82–93

Laue T (2012) Proximity energies: a framework for understanding concentrated solutions. J Mol Recognit 25:165–173

Laue T, Shah BD, Ridgeway TM, Pelletier SL (1992) Analytical ultracentrifugation in biochemistry and polymer science. Royal Society of Chemistry, London, pp 90–125

Lehermayr C, Mahler HC, Mäder K, Fischer S (2011) Assessment of net charge and protein-protein interactions of different monoclonal antibodies. J Pharm Sci 100:2551–2562

Liu J, Nguyen MDH, Andya JD, Shire SJ (2005) Reversible self-association increases the viscosity of a concentrated monoclonal antibody in aqueous solution. J Pharm Sci 94:1928–1940

McMillan WG Jr, Mayer JE (1945) The statistical thermodynamics of multicomponent systems. J Chem Phys 13:276–305

Narayanan J, Liu XY (2003) Protein interactions in undersaturated and supersaturated solutions: a study using light and x-ray scattering. Biophys J 84:523–532

Neal BL, Asthagiri D, Lenhoff AM (1998) Molecular origins of osmotic second virial coefficients of proteins. Biophys J 75:2469–2477. Holde et al., 2006

Neergaard MS, Kalonia DS, Parshad H, Nielsen AD, Møller EH, van de Weert M (2013) Viscosity of high concentration protein formulations of monoclonal antibodies of the IgG1 and IgG4 subclass-prediction of viscosity through protein-protein interaction measurements. Eur J Pharm Biopharm 49:400–410

Reichert JM, Rosensweig CJ, Faden LB, Dewitz MC (2005) Monoclonal antibody successes in the clinic. Nat Biotechnol 23:1073–1078

Sahin E, Grillo AO, Perkins MD, Roberts CJ (2010) Comparative effects of pH and ionic strength on protein-protein interactions, unfolding, and aggregation for IgG_1 antibodies. J Pharm Sci 99:4830–4848

Saito S, Hasegawa J, Kobayashi N, Kishi N, Uchiyama S, Fukui K (2012) Behavior of monoclonal antibodies: relation between the second virial coefficient (B_2) at low concentrations and aggregation propensity and viscosity at high concentrations. Pharm Res 29:397–410

Saito S, Hasegawa J, Kobayashi N, Tomitsuka T, Uchiyama S, Fukui K (2013) Effects of ionic strength and sugars on the aggregation propensity of monoclonal antibodies: influence of colloidal and conformational stabilities. Pharm Res 30:1263–1280

Salinas BA, Sathish HA, Bishop SM, Harn N, Carpenter JF, Randolph TW (2010) Understanding and modulating opalescence and viscosity in a monoclonal antibody formulation. J Pham Sci 99:82–93

Saluja A, Badkar AV, Zeng DL, Kalonia DS (2007) Ultrasonic rheology of a monoclonal antibody (IgG2) solution: implications for physical stability of proteins in high concentration formulations. J Pharm Sci 96:3181–3195

Saluja A, Fesinmeyer M, Hogan S, Brems DN, Gokarn YR (2010) Diffusion and sedimentation interaction parameters for measuring the second virial coefficient and their utility as predictors of protein aggregation. Biophys J 99:2657–2665

Schmit JD, He F, Mishra S, Ketchem RR, Woods CE, Kerwin BA (2014) Entanglement model of antibody viscosity. J Phys Chem B 118:5044–5049

Shire SJ, Shahrokh Z, Liu J (2004) Challenges in the development of high protein concentration formulations. J Pharm Sci 93:1390–1402

Singh SN, Yadav S, Shire SJ, Kalonia DS (2014) Dipole-dipole interaction in antibody solutions: correlation with viscosity behavior at high concentration. Pharm Res 31:2549–2558

Sule SV, Cheung JK, Antochshuk V, Bhalla AS, Narasimhan C, Blaisdell S, Shameem M, Tessier PM (2012) Solution pH that minimizes self-association of three monoclonal antibodies is strongly dependent on ionic strength. Mol Pharm 9:744–751

Tessier PM, Lenhoff AM, Sandler SI (2002) Rapid measurement of protein osmotic second virial coefficients by self-interaction chromatography. Biophys J 82:1620–1631

Treuheit MJ, Kosky AA, Brems DN (2002) Inverse relationship of protein concentration and aggregation. Pharm Res 19:511–516

van Holde, KE, Johnson C, Ho PS (2006) Physical biochemistry, Pearson Education Inc., New York

Verwey EJW, Overbeck JTK (1948) Theory of stability of lyophobic colloids. Elsevier, Amsterdam

Winzor DJ, Deszczynski M, Harding SE, Wills PR (2007) Nonequivalence of second virial coefficients from sedimentation equilibrium and static light scattering studies of protein solutions. Biophys Chem 128:46–55

Williams (1972) Ultracentrifugation of macromelecules: modern topics, Academic Press, New York

Yadav S, Shire SJ, Kalonia DS (2010) Factors affecting the viscosity in high concentration solutions of different monoclonal antibodies. J Pharm Sci 99:4812–4829

Yadav S, Laue TM, Kalonia DS, Singh SN, Shire SJ (2012) The influence of charge distribution on self-association and viscosity behavior of monoclonal antibody solutions. Mol Pharm 9:791–802

Yamakawa H (1962) Concentration dependence of the frictional coefficient of polymers in solution. J Chem Phys 36:2995–3001

Zhang J, Liu XY (2003) Effect of protein-protein interactions on protein aggregation kinetics. J Chem Phys 119:10972–10976

Part VII
AUC of High-Concentration Systems and Non-ideal Solutions

Chapter 22
Johnston-Ogston Effects in Two Simulated Systems of Polystyrene Beads That Are Polydisperse with Respect to Density

Thomas P. Moody

Abstract This work qualitatively describes the results from analytical ultracentrifugation simulations of two model aqueous systems based on polystyrene beads that are polydisperse with respect to specific gravity. Each system includes 27 species of polystyrene beads, three of which account for 99 % of the total polystyrene-bead concentration of 0.10000 g/ml. These systems were contrived to exhibit Johnston-Ogston effects in the low-concentration polystyrene-bead species. The three high-concentration species are present as two two-species components in one system and are present as three single-species components in the other system. The method of simulation is an implementation of an integral, finite-element solution to the continuity equation for analytical ultracentrifugation. The solution differs from those previously described in several respects. To correctly implement their concentration dependence, the transport coefficients are defined as spatially independent parameters. To correctly evaluate the concentration-dependent transport coefficients at the time to be evaluated, the concentrations are calculated iteratively. By such an evaluation of the concentration-dependent transport coefficients at both the time already evaluated and the time being evaluated, the accuracy of each new set of concentrations is maximised. Computational artefacts are reduced by first calculating all concentrations in one order, then recalculating all concentrations in the opposite order and averaging the results. Additionally, simpler results of integration are obtained by using one-half the square of the radial position, rather than the radial position, as the spatial parameter of the continuity equation. Furthermore, a simple coupled-flow equation has been implemented.

Keywords Johnston-Ogston effect • Analytical ultracentrifugation • Simulation • Iterative • Integral • Finite element solution • Spatially independent transport coefficients • Irreversible thermodynamics • Density increment • Viscosity • Thermodynamic nonideality

T.P. Moody (✉)
Moody Biophysical Consulting, Vancouver, BC, Canada
e-mail: MoodyBiophysicalConsulting@gmail.com

S. Uchiyama et al. (eds.), *Analytical Ultracentrifugation*,
DOI 10.1007/978-4-431-55985-6_22

22.1 Introduction

This work qualitatively describes the results from analytical ultracentrifugation (AUC) simulations of two model aqueous systems based on polystyrene (PS) beads that are polydisperse with respect to density. These systems were contrived to exhibit Johnston-Ogston effects in those simulations (Moody 2012a). The method of simulation (Moody 2011) is an implementation of an integral, finite-element solution to the continuity equation. The method is built on that which Claverie et al. (1975) described in their solution to the Lamm equation but differs in several respects. To correctly implement their concentration dependence, the transport coefficients are defined as spatially independent parameters. To correctly evaluate the concentration-dependent transport coefficients at the time to be evaluated, the concentrations are calculated iteratively. By such an evaluation of the concentration-dependent transport coefficients at both the time already evaluated and the time being evaluated, the accuracy of each new set of concentrations is maximised. Computational artefacts are reduced by first calculating all concentrations in one order, then recalculating all concentrations in the opposite order and averaging the results. Additionally, simpler results of integration are obtained by using one-half the square of the radial position, rather than the radial position, as the spatial parameter of the continuity equation. Furthermore, a simple coupled-flow equation has been implemented. The simulation results shown here have been subjected to sedimentation-coefficient-distribution analysis in two other works (Moody 2012b, 2014), the earlier of which focuses on the development of the method of analysis, and the most recent of which applies that method of analysis to test the distinguishability of four mixtures of the PS-bead systems at three significantly different times.

22.2 Common Features of the Two Model Systems

The solute particles of the two model systems are based on the 30 nm diameter (±1 nm) product (part number 3030 A) of the Thermo Scientific 3000 Series Nanosphere Size Standards, the composition of which may include polystyrene (PS) and polystyrene divinylbenzene (PSDVB). (For simplicity, "PS bead" is used to describe such a particle.) In an early investigation by AUC (Sharp and Beard 1950), larger (approximately 253 nm diameter) particles of this composition were found to have a density of 1.053 g/ml, and that value is treated as a reference.

Isotope substitution is used to account for the polydispersity in density among PS-bead species. As PS beads are composed entirely of carbon and hydrogen, deuterium is the only stable heavy isotope available for substitution. With the expectation that the PS beads have a C-to-H ratio of 1, achievable amounts of substitutions of ^{1}H with D would suffice to produce the particles described here.

The absolute temperature pertaining to the AUC simulations is $T = 293.15$ K, which is treated as a reference. The time, t, at which AUC began was set to 0 s.

22.2.1 The Implicit Solvent Component of the Model Systems

The implicit solvent component of the model systems is an aqueous buffer consisting of 71.23 % D_2O (volume percent), 0.15 M NaCl, 20 mM NH_4HCO_3 and pH 7.0 at 293.15 K, at which temperature its density, ρ_0, and viscosity, η_0, are 1.08225 g/ml and 0.010196 poise, respectively, as calculated by Sednterp (Laue et al. 1992), version 1.09. The reference density of an unmodified PS bead, $\rho_{PS} = 1.05300$ g/ml, is padded with zeroes to match the significant digits of the values returned by Sednterp.

22.2.2 Common Characteristics of the PS Beads of the Model Systems

The model particles of each system consist of $n = 27$ species of PS beads. These species are indexed by k. The density of one species ($k = n$) is equal to ρ_{PS}. All other species ($1 \leq k < n$) have densities greater than ρ_{PS}. The density of species k is given by

$$\rho_k = \rho_{PS} + [(n-1)(3-\alpha_k) + 1 - k]\,\Delta\rho, \tag{22.1}$$

where $\Delta\rho = 0.00075$ g/ml is an increment of the density, $\alpha_{k<14} = 0$, $\alpha_{k=14} = 1$ and $\alpha_{k>14} = 2$. The implicit solvent is modelled as an aqueous buffer with sufficient D_2O to render its density, ρ_0, equal to $\rho_{k=14}$. Of the 26 model particles for which $\rho_k \neq \rho_0$, ρ_0 is less than ρ_k for half ($k < 14$), and ρ_0 is greater than ρ_k for the rest ($k > 14$). The solvent is treated as being incompressible, as is each solute species. In the limit as the total solute concentration approaches zero, the partial specific volume of species k approaches $1/\rho_k$. The latter relationship is exploited in Eqs. (22.5), (22.7), (22.11), (22.12), (22.13) and (22.15).

The sum of the concentrations of three solute species, H ($k = 1$), L ($k = 27$) and LH ($k = 14$), equals 99 % of the total solute concentration, $c = 0.10000$ g/ml. Each of the other 24 solute species is present at 1/24 of 1 % of c. Species LH is modelled as a heterodimer formed from one monomer of species H plus one monomer of species L. In the absence of concentration gradients, the molarity of species LH is half that of species L or H, while species L and H are equimolar. In subscripts, H, L and LH are interchangeable with $k = 1$, $k = 14$ and $k = 27$, respectively.

In the limit as c approaches zero, species H has the most positive sedimentation coefficient, species L has the most negative sedimentation coefficient, and species

LH has a sedimentation coefficient of zero. The transport parameters of the model particles Eqs. (22.4), (22.5), (22.6), (22.7), (22.8), (22.9), (22.10), and (22.14) and their concentration dependence Eqs. (22.11), (22.12), (22.13), and (22.15) are such that each of the low-concentration solutes ($1 < k < 14$, $14 < k < 27$) exhibits Johnston-Ogston effects in AUC simulations.

For every species except LH ($k = 14$), the solute particles are treated as spheres with an anhydrous radius of $R_{k \neq 14} = 1.5\text{E-}6$ cm, and thus a volume,

$$V_{k \neq 14} = \frac{4\pi R_{k \neq 14}^3}{3}, \tag{22.2}$$

of approximately 1.41372E-17 ml. The volume of species LH is $V_{LH} = V_L + V_H$. The molar mass of each species is given by

$$M_k = N_A \rho_k V_k, \tag{22.3}$$

where N_A is Avogadro's number. As LH is composed of L and H, $M_{LH} = M_L + M_H$. Species LH would not be spherical, but for the sake of making a subsequent equation (Eq. 22.7) applicable to all species, the anhydrous radius of its equivalent sphere is defined as $R_{LH} \equiv (3M_{LH}/4\pi N_A \rho_{LH})^{1/3} \cong 1.88988\text{E-}6$ cm. By virtue of this definition, $V_k = 4\pi R_k^3/3$ holds for each species, including LH ($k = 14$).

Using R_k (either known or obtainable from Eqs. (22.2) and (22.3)), the diffusion coefficient of species k in the limit as c approaches zero is calculated as

$$D_k^0 = \frac{RT}{6\pi N_A \eta_0 h_k R_k}, \tag{22.4}$$

where R is the ideal gas constant and h_k is the hydration factor of solute species k in the limit as c approaches zero. With $h_k = 1/0.65$ used for each solute species, $D_{k \neq 14}^0 = 9.12567\text{E-}8$ cm^2/s and $D_{k=14}^0 = D_{LH}^0 = 7.24305\text{E-}8$ cm^2/s.

The reduced molar mass of species k in the limit as c approaches zero is given by

$$\sigma_k^0 = \frac{\left(1 - \rho_0 \overline{v}_k^0\right) M_k \omega^2}{RT} = \frac{\left(1 - \rho_0/\rho_k\right) M_k \omega^2}{RT}, \tag{22.5}$$

where $\overline{v}_k^0 = 1/\rho_k$ is the partial specific volume of species k in the limit as c approaches zero, $\rho_0 = 1.08225$ g/ml is the density of the solution in the limit as c approaches zero, $(1 - \rho_0 \overline{v}_k^0)$ is the density increment of the system as it affects the transport of component k in the zero-concentration limit and $\omega = 2\pi(\text{RPM}/[60 \text{ s/min}])$ is the angular velocity of the rotor, the speed of which is reported in RPM.

The sedimentation coefficient of species k in the limit as c approaches zero is given by

$$s_k^0 = \frac{\sigma_k^0 D_k^0}{\omega^2}. \tag{22.6}$$

Using the right-hand sides of Eqs. (22.3), (22.4), and (22.5) to substitute for M_k, D_k^0 and σ_k^0, respectively, and using $4\pi R_k^3/3$ in place of V_k, Eq. (22.6) yields

$$s_k^0 = 2R_k^2\left(\rho_k - \rho_0\right)/9\eta_0 h_k. \tag{22.7}$$

Species 1 to 13, for which $\rho_k > \rho_0$, have positive s_k^0 values in the 6 Svedberg to 10 Svedberg range. Species 14, for which $\rho_k = \rho_0$, has a neutral s_k^0 value of 0 Svedberg. Species 15 to 27, for which $\rho_k < \rho_0$, have negative s_k^0 values in the -6 Svedberg to -10 Svedberg range.

22.2.3 *Concentration Dependence of Transport*

For each species, k, the concentration-dependent equations describing the reduced-molar-mass coefficient, the diffusion coefficient and the sedimentation coefficient, respectively, are

$$\sigma_k = \sigma_k^0\left(1 + \sum_{q=1}^{n} p_{k.q}c_q\right) \Big/ \left(1 + \sum_{q=1}^{n} y_{k.q}c_q\right), \tag{22.8}$$

$$D_k = D_k^0\left(1 + \sum_{q=1}^{n} y_{k.q}c_q\right) \Big/ \left(1 + \sum_{q=1}^{n} h_{k.q}c_q\right) \text{ and} \tag{22.9}$$

$$s_k = \frac{\sigma_k D_k}{\omega^2}, \tag{22.10}$$

where the parameters $p_{k,q}$, $y_{k,q}$ and $h_{k,q}$ are the second virial coefficients of the density increment, thermodynamic nonideality and viscosity, respectively (Moody 2011). The second virial coefficients are used to model the concentration-dependent effects of species q on the transport of species k.

The species-k-applicable transport coefficient that links the concentration of species q to the thermodynamic nonideality of the system was calculated as

$$y_{k,q} = 4\overline{\upsilon}_q^0 = 4/\rho_q, \tag{22.11}$$

which is based solely on the expected excluded-volume effect of a spherical solute with a partial specific volume of $\overline{\upsilon}_q^0$ in the limit as c approaches zero (van Holde 1985a – Solutions of Macromolecules).

The species-k-applicable transport coefficient that links the concentration of species q to the viscosity of the system was calculated as

$$h_{k,q} = 2.5\overline{\upsilon}_q^0 = 2.5/\rho_q, \tag{22.12}$$

where the coefficient 2.5 is the value of the intrinsic viscosity that, according to theory, applies to spherical particles (Eisenberg and Crothers 1979; van Holde 1985b – Viscosity). Parameters of species k have no effect on $h_{k,q}$ because the presence of one species is assumed to have no effect on the relationship between the volume fraction of any other species and η, the viscosity of the solution.

As D_k^0, $y_{k,q}$ and $h_{k,q}$ are independent of the density of any species, so too is D_k. Although $y_{k,q}$ and $h_{k,q}$ explicitly depend on parameters of species q only, the index k is retained because, in general, these coefficients may explicitly depend on parameters of species k as well. For example, the right-hand sides of Eqs. (22.11) and (22.12) could be multiplied by h_k to account for hydration. (As they do not account for hydration, Eqs. (22.11) and (22.12) may underestimate $y_{k,q}$ and $h_{k,q}$, respectively.)

The volume fraction of each species q is $\varphi_q = c_q/\rho_q \leq [\varphi_q]_{\max} \leq 1$, where $[\varphi_q]_{\max}$ is the maximum volume fraction of solute species q. With ρ_q assumed to be c-independent for all solutes, $[\varphi_q]_{\max} = [c_q]_{\max}/\rho_q$, where $[c_q]_{\max}$ is the maximum concentration of solute species q in the limit as every $c_{k \neq q}$ approaches zero. For each solute species, $[\varphi_q]_{\max}$ is assigned a value of 0.5.

The transport of species $k \neq 14$ is coupled to the effect of species q on the density of the system through $p_{k\neq 14,q}c_q$, where c_q is the concentration of species q, and

$$p_{k\neq 14,q} = \frac{\overline{\upsilon}_q^0}{[\varphi_q]_{max}}\left[\frac{\left(1-\rho_q\overline{\upsilon}_k^0\right)}{\left(1-\rho_0\overline{\upsilon}_k^0\right)} - 1\right] = -\frac{1}{\rho_q[\varphi_q]_{max}}\frac{(\rho_q-\rho_0)}{(\rho_k-\rho_0)}, \tag{22.13}$$

in which $(1-\rho_0\,\overline{\upsilon}_k^0)$ would be the density increment of the system with respect to the component to which species k pertains if there were no other component present besides a solvent with a density equal to ρ_0, and $(1-\rho_q\,\overline{\upsilon}_k^0)$ would be the density increment of the system with respect to the component to which species k pertains if there were no other component present besides a solvent with a density equal to ρ_q.

In the case of $\sigma_k^0 = 0$, a modification of Eq. (22.8),

$$\sigma_k = \frac{\left(\sigma_k^0 + \sum_{q=1}^{n} (\sigma p)_{k.q} c_q\right)}{\left(1 + \sum_{q=1}^{n} y_{k.q} c_q\right)}, \tag{22.14}$$

is used to calculate σ_k in terms of the set of $(\sigma p)_{k,q}$, each of which is the species-k-applicable transport coefficient that links the concentration of species q to the density increment of the system, and each of which is given by

$$(\sigma p)_{k.q} = \sigma_k^0 p_{k.q} = \frac{\overline{v}_q^0 \left[\rho_0 - \rho_q\right] \overline{v}_k^0 \omega^2 M_k}{\left[\varphi_q\right]_{max} RT} = \frac{\omega^2 M_k}{RT \rho_k} \left[\frac{\rho_0 - \rho_q}{\rho_q \left[\varphi_q\right]_{max}}\right]. \tag{22.15}$$

For $(\rho_0 - \rho_q)\omega^2 \neq 0$, Eq. (22.15) generally results in non-zero values of $(\sigma p)_{k,q}$, in which case, non-zero values of σ_k can occur even if $\sigma_k^0 = 0$.

22.3 Distinguishing Features of the Two Model Systems

The initial concentrations of species H ($k = 1$), LH ($k = 14$) and L ($k = 27$) were subject to the constraint that $c_H/M_H = 2(c_{LH}/M_{LH}) = c_L/M_L = m_{ref}$, where m_{ref} is a common reference concentration in mole/ml. The initial concentrations of species H, LH and L were also subject to the constraint that $c_{LH}/c_L c_H = K_a$ in the first system, where, treating all activity coefficients as equal to 1, K_a is the equilibrium constant of the mass-action association, $L + H \rightleftharpoons LH$, in which one monomer of species H and one monomer of species L form one heterodimer of species LH. Thus,

$$K_a = \frac{c_{LH}}{c_L c_H} = \frac{c_{H,LH,L} - c_L - c_H}{c_L c_H} = \frac{c_{H,LH,L} - m_{ref}\left(M_L + M_H\right)}{m_{ref}^2 M_L M_H} \quad \text{and} \tag{22.16}$$

$$K_a = \frac{c_{LH}}{c_L c_H} = \frac{\left(m_{ref}/2\right) M_{LH}}{m_{ref}^2 M_L M_H} = \frac{M_L + M_H}{2 m_{ref} M_L M_H}, \tag{22.17}$$

where $c_{H,LH,L} = c_H + c_{LH} + c_L$. Equating the right-hand sides of the preceding two equations and solving for m_{ref} yields

$$m_{ref} = \frac{2c_{H,LH,L}}{3\left(M_L + M_H\right)} = 3.582\text{E} - 9\text{mol/ml}, \qquad (22.18)$$

from which $K_a = 30.325$ ml/g is obtained via Eq. (22.16) or (22.17). The initial concentrations of species H, L and LH in either system are then determined, respectively, by $c_H = m_{\text{ref}} M_H$, $c_L = m_{\text{ref}} M_L$ and $c_{LH} = (m_{ref} M_{LH})/2$.

The model mass-action association linking species H, L and LH in the first system is given a forward rate constant of $k_{for} = 30{,}000$ [ml/g]/s. The reverse rate constant, k_{rev}, is equal to $k_{for}/K_a > 989$ s. Consequently, the equilibration of this reaction is effectively instantaneous on the scale of the maximum time increment (3 s) used in the AUC simulation of this system. For the second system, in which all species are single-species components, there is no K_a that applies to species H, L and LH because both k_{for} and k_{rev} are effectively equal to zero. Hence, the second system is described as that for which K_a is undefined.

22.4 Simulation Parameters

Each system was simulated (Moody 2011) in terms of the gravitational-potential-space parameter, $\xi = r^2/2$, where r is the radial position in the centrifuge. In terms of r, the meniscus position was set to 6 cm, and the base of the system was set to 7.2 cm. The number of spatial elements was set to 900. The spatial increment, $\Delta\xi$, was the same (8.8E-3 cm^2) for all adjacent pairs of spatial elements.

Data sets were saved at 0.5-min increments until $t = 1$ min. Data sets were then saved at 2.5-min increments until $t = 126$ min. Lastly, data sets were saved at 30-min increments until $t = 26.1$ h. For the purpose of data output, spatial data were saved in terms of radial position, $r = (2\xi)^{0.5}$.

Concentration data were saved in terms of g/ml for each individual species. Data were also saved in terms of g/ml for the sum of the concentrations of the three species, H ($k = 1$), LH ($k = 14$) and L ($k = 27$), that chemically equilibrate in the first system but exist as single-species components present at high concentrations in the second system. Additionally, the sum of the concentrations of species 2 to 13 plus 15 to 26 was multiplied by a signal factor of 1000 AU/[g/ml] and the data saved in terms of an arbitrary unit, AU, that is numerically equivalent to the mg/ml scale.

For concentration-dependent systems, it has been found that liberally realistic limits on D_k and σ_k reduce the likelihood of artefacts and instabilities in regions of high concentrations. On that basis, the following limits were imposed on D_k and σ_k:

for all D_k,

$$0.025D_k^0 \leq D_k \leq 160D_k^0; \tag{22.19}$$

for all $\sigma_k^0 > 0$,

$$-50\sigma_k^0 \leq \sigma_k \leq 50\sigma_k^0; \tag{22.20}$$

for all $\sigma_k^0 < 0$,

$$50\sigma_k^0 \leq \sigma_k \leq -50\sigma_k^0; \tag{22.21}$$

and for $\sigma_k^0 = 0$,

$$-5000/\text{cm}^2 \leq \sigma_k \leq 5000/\text{cm}^2. \tag{22.22}$$

The simulation programme (Moody 2011) includes a mass-conservation algorithm that can increase the time increment, Δt, when the system appears to be stable, or reduce Δt and return the system to the most recent stable time point in instances of computational instability. Without an upper limit on Δt, these opposing functions can lead to unproductive cycles of Δt changes. On the basis of preliminary test results, a maximum time increment, Δt_{max}, of 3 s was chosen as an upper limit in Δt. To permit the mass-conservation algorithm to reduce Δt in all instances of instability that would introduce major inaccuracies in the data, the minimum time increment, Δt_{min}, was set to 0 s.

For the mass-flow calculations, an upper limit of eight iterations per time increment was imposed. The acceptance criterion of convergence, which, when met, stops further mass-flow recalculations within a given Δt, was h/erg·s, where h is the cgs Planck constant.

For concentration-dependent systems, the inclusion of well-chosen third- and higher-order virial coefficients of the density increment, thermodynamic nonideality and viscosity might also reduce or eliminate the likelihood of artefacts and instabilities in regions of high concentrations and thus obviate the need for artificially imposed limits Eqs. (22.19), (22.20), (22.21), and (22.22) on D_k and σ_k.

For the reaction-flow calculations, for $t > 0$, a maximum of $N_{equil} = 2{,}500$ was imposed on the number of iterations allowed when recalculating a chemical equilibrium at a given spatial element. For $t = 0$, a limit of $N_{equil} = 5{,}000$ was used. For the mass-action association of the first system, the criterion for adequate equilibration was

$$\zeta K_a \leq c_{LH}/c_L c_H \leq K_a/\zeta. \tag{22.23}$$

At $t = 0$, $\zeta = 0.999$ was used. For all $t > 0$, $\zeta = 0.995$ was used. The concentration-change factor (Moody 2011) used in the reaction-flow algorithm, for which w is the iteration index, was

$$k_w = k_0 / \left(k_0 + [w - w_0]\right), \tag{22.24}$$

where k_0 was set to 3, w_0 was set to 1 for $t = 0$ and w_0 was set to 4 for $t > 0$.

Additional limits halted reaction-flow calculations when the computational effort was especially high but the information that could be gained was very low. When the concentration of one reactant species exceeded $\alpha = (10^{-21}\ \text{erg·s})/h$ times the sum of the concentrations of the product species and the other reactant species, no effort was made to recalculate the chemical equilibrium. When the concentration of each reactant species and the product species fell below $\beta = h(10^{21}\text{g/ml})/(\text{erg·s})$, any remaining product was converted to reactants.

22.5 Notation

For various parameters, the notation used here identifies each species by a subscript, such as k or a specific number. As such, the parameters in Equations or Inequalities 1 to 24 display a component-based notation. Though the correct notation would be the component-species sort (Moody 2011), that notation is more cumbersome and is not especially useful here, where only one system has any multispecies components, and those are outnumbered 12 to 1 by single-species components. Thus, where pertinent, such distinctions are noted in the text, rather than in the subscripts.

22.6 Results

For the first system, in which species H, L and HL participate in a mass-action association with $K_a = 30.325$ ml/g, the total concentration of LH was greater than zero during the first 66 min of simulated AUC. After the first 68.5 min of simulated AUC, the concentration of LH was equal to zero throughout the system. For the second system, in which all species are single-species components, the total concentration of LH remained constant throughout the entire course of simulated AUC. The observed Δt was equal to Δt_{max} at all times for both systems, and neither system exhibited instability at any time.

In both systems, the low-concentration species exhibit Johnston-Ogston effects (Figs. 22.1, 22.2, and 22.3 for the $K_a = 30.325$ ml/g system and Figs. 22.4, 22.5, and 22.6 for the system with K_a undefined) that result from each of those species

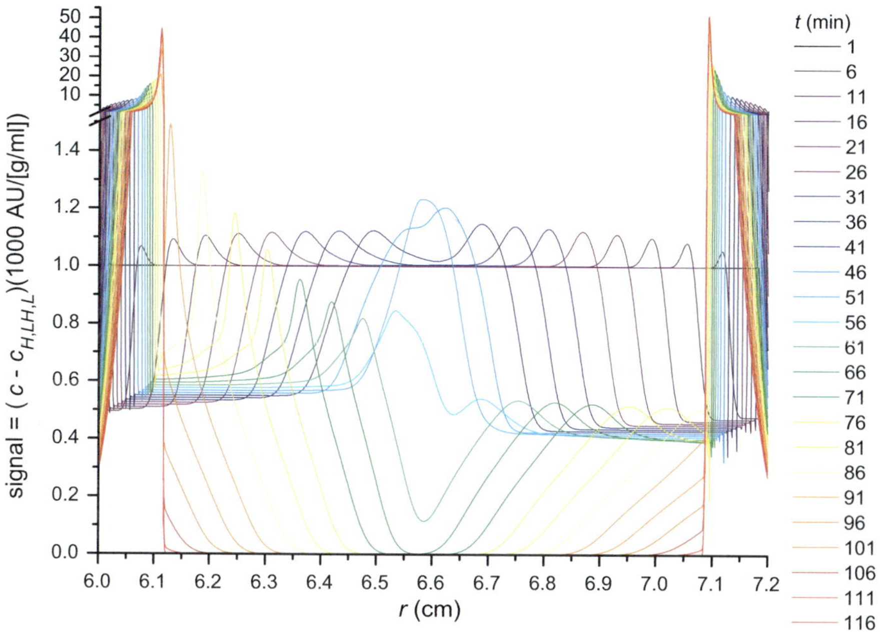

Fig. 22.1 $K_a = 30.325$ ml/g. (Compare with Fig. 22.4.) The sum of the signals from all species except species 1 (H), 14 (LH) and 27 (L), given by $(c - c_{H,LH,L})$(1000 AU/[g/ml]), versus r (cm) for 1 min $\leq t \leq$ 116 min

having a significantly higher-magnitude sedimentation coefficient ($|s_k|$) behind the boundary of the high-concentration solute with which it approximately comigrates (species 2 to 13 being comigratory with species H, and species 15 to 26 being comigratory with species L) than it does ahead of that boundary, relative to the direction in which the boundary travels. The Johnston-Ogston effects are primarily due to the concentration dependence of the density increment Eqs. (22.5), (22.8), (22.13), (22.14), and (22.15). The direction of the mass flow affects the intensity of Johnston-Ogston effects.

The volume of the sector-shaped system increases in proportion to ξ. Thus, the mass flow of a sedimenting species ($s_k > 0$) is in the direction of higher volume, and the mass flow of a floating species ($s_k < 0$) is in the direction of lower volume. Consequently, a sedimenting species is subject to a radial-dilution effect in which its concentration between its boundary and the pellet decreases with time such that $(\partial c_k/\partial t)_\xi < 0$ even in a region where $(\partial c_k/\partial \xi)_t = 0$. Likewise, a floating species is subject to a radial-concentration effect in which its concentration between its boundary and the supernatant increases with time such that $(\partial c_k/\partial t)_\xi > 0$ even in a region where $(\partial c_k/\partial \xi)_t = 0$. (For a neutrally buoyant species, in a region where both $(\partial c_k/\partial \xi)_t = 0$ and $s_k = 0$, $(\partial c_k/\partial t)_\xi = 0$.) As the Johnston-Ogston effects

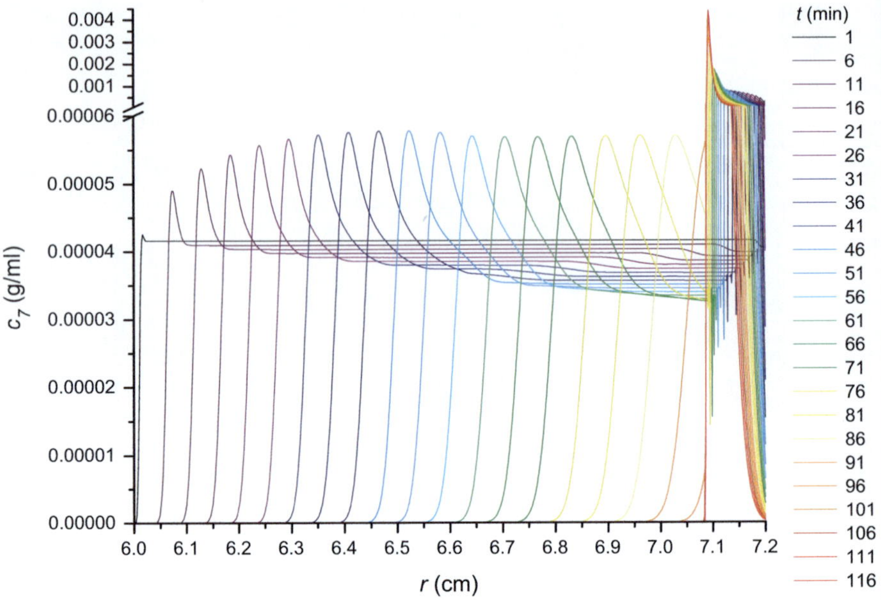

Fig. 22.2 $K_a = 30.325$ ml/g. (Compare with Fig. 22.5.) The concentration of species 7, c_7 (g/ml), versus r (cm) for 1 min $\leq t \leq 116$ min. These data typify those of species 2 through 13, for which $s_k^0 > 0$ and total $c_k =$ (total c)/2400

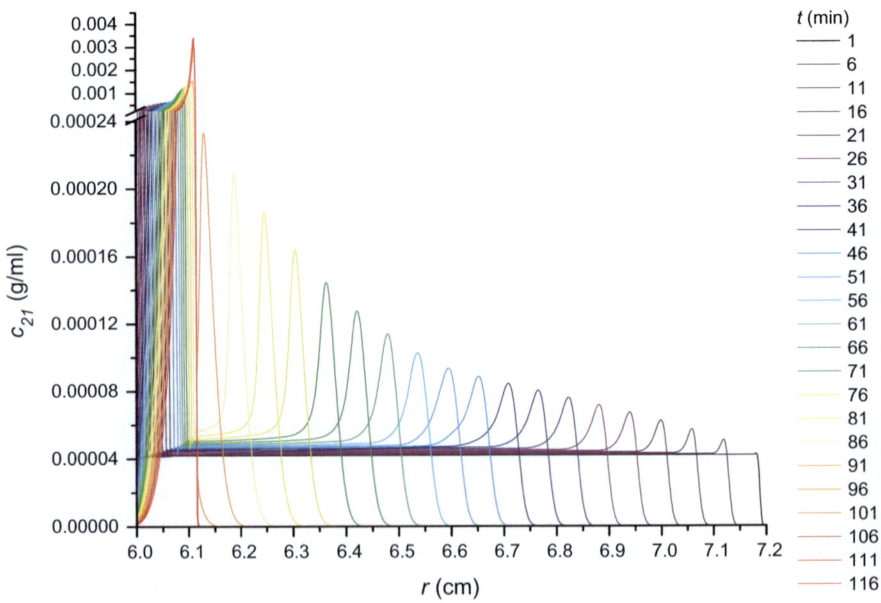

Fig. 22.3 $K_a = 30.325$ ml/g. (Compare with Fig. 22.6.) The concentration of species 21, c_{21} (g/ml), versus r (cm) for 1 min $\leq t \leq 116$ min. These data typify those of species 15 through 26, for which $s_k^0 < 0$ and total $c_k =$ (total c)/2400

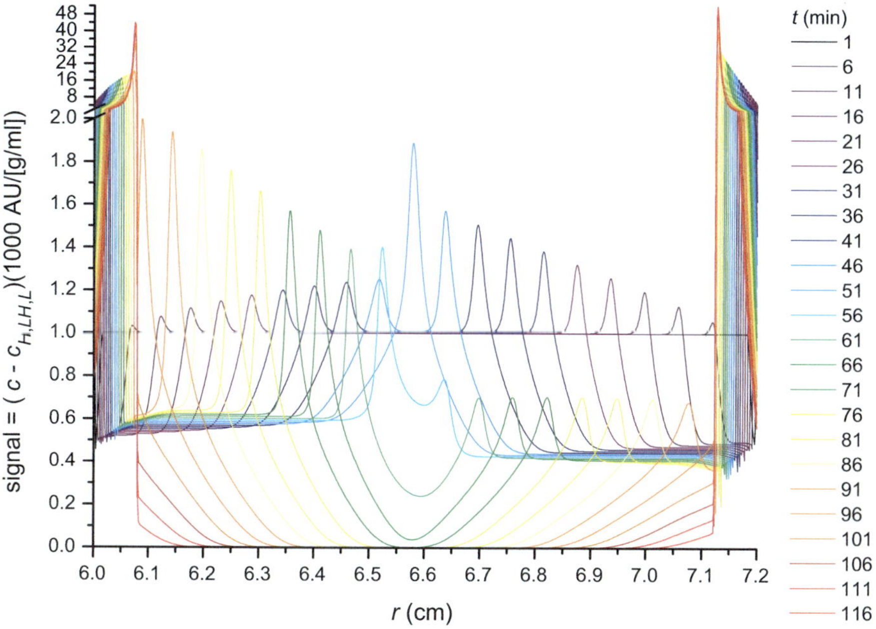

Fig. 22.4 K_a undefined. (Compare with Fig. 22.1.) The sum of the signals from all species except species 1 (H), 14 (LH) and 27 (L), given by $(c - c_{H,LH,L})$ (1000 AU/[g/ml]), versus r (cm) for 1 min $\leq t \leq$ 116 min

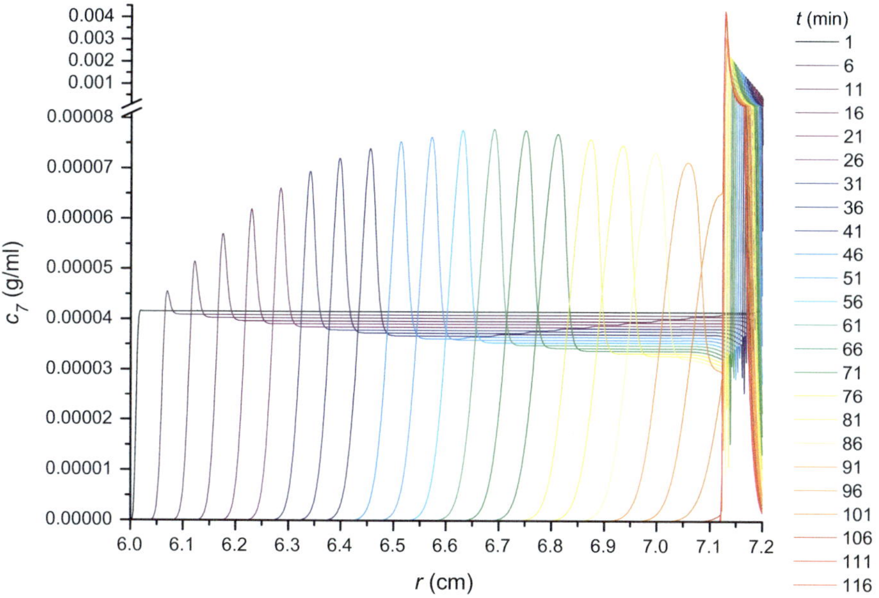

Fig. 22.5 K_a undefined. (Compare with Fig. 22.2.) The concentration of species 7, c_7 (g/ml), versus r (cm) for 1 min $\leq t \leq$ 116 min. These data typify those of species 2 through 13, for which $s_k^0 > 0$ and total $c_k = (\text{total } c)/2400$

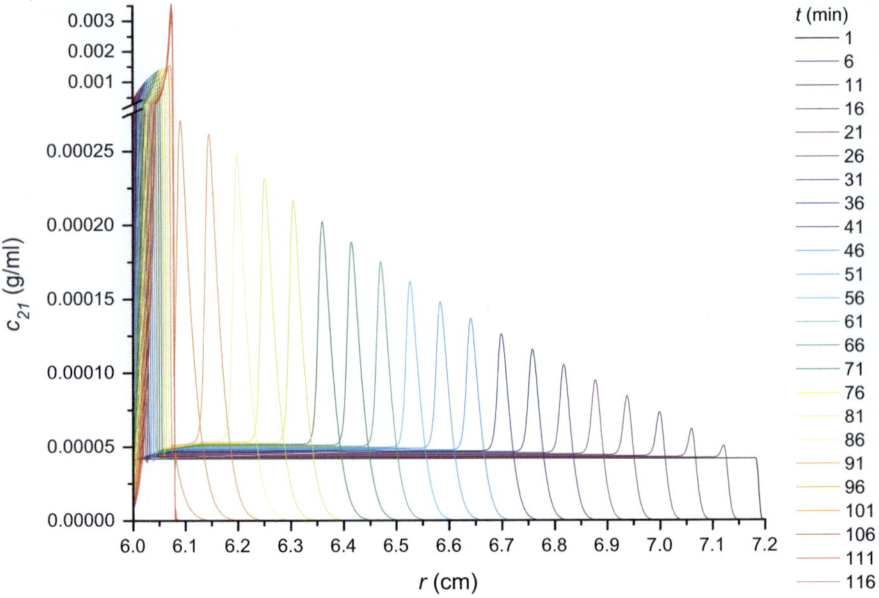

Fig. 22.6 K_a undefined. (Compare with Fig. 22.3.) The concentration of species 21, c_{21} (g/ml), versus r (cm) for 1 min $\leq t \leq$ 116 min. These data typify those of species 15 through 26, for which $s_k^0 < 0$ and total c_k = (total c)/2400

are concentration dependent, and as the geometry of the system produces radial-dilution or radial-concentration effects depending on the direction of mass flow, the Johnston-Ogston effects tend to continuously intensify with time for floating species but eventually start to weaken with time for sedimenting species.

As the system with K_a undefined approaches equilibrium (Fig. 22.7), the concentration of single-species component LH rises to a peak at either the pellet formed mainly by H or the supernatant formed mainly by L. This effect may be an artefact of using severely truncated virial expansions to describe the density increment, thermodynamic nonideality and viscosity of the system in regions where the concentrations of LH, L or H are extremely high. (See Eqs. (22.8), (22.9), (22.10), (22.11), (22.12), (22.13), (22.14), and (22.15)).

22.6.1 $K_a = 30.325$ ml/g: Two Components Account for Three of the Species Present

The depletion of species H from the meniscus outward and the depletion of species L from the base inward give rise, respectively, to the left-hand and right-hand boundaries of species LH. As the left-hand and right-hand boundaries of species

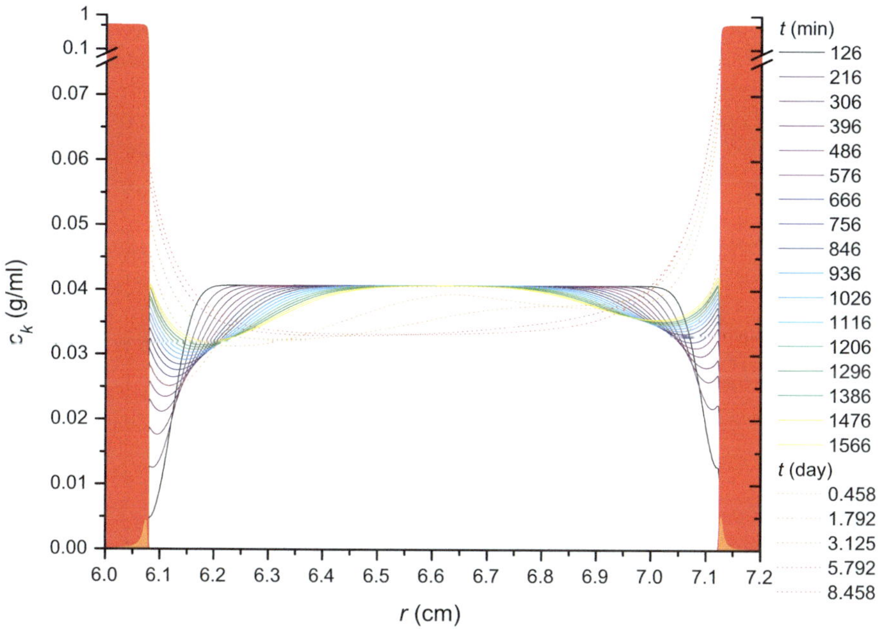

Fig. 22.7 K_a undefined. Concentration, c_k (g/ml), versus radial position, r (cm), during the approach to equilibrium. The concentration axis is scaled logarithmically above the break. The solid lines show the concentration of species 14 (*LH*) versus r from $t = 126$ min to $t = 1566$ min, in 90-min increments, for the 27-component system. The concentrations of species 1 (*H*, which forms most of the pellet) and species 27 (*L*, which forms most of the supernatant) at $t = 116$ min are shown in solid red. The sum of the concentrations of species 2 to 13 (which form a small part of the pellet) and species 15 to 26 (which form a small part of the supernatant) at $t = 116$ min are shown in solid orange. The dotted lines show the concentration of species 14 (*LH*) versus r from $t = 0.458$ days to $t = 8.458$ days for a system that differs from the 27-component one by including only the high-concentration species, *H*, *L* and *LH*, for which K_a is undefined

LH move toward the centre of the system (Fig. 22.8: $K_a = 30.325$ ml/g), *LH* dissociates to form the oppositely migrating species, *H* and *L*, causing the dips in the concentrations of those species seen in Figs. 22.9 and 22.10 ($K_a = 30.325$ ml/g), respectively. The effect occurs even with all $(\sigma p)_{k,q}$ $p_{k,q}$, $y_{k,q}$ and $h_{k,q}$ set to zero, and all $c_k = 0$ for $1 < k < 14$ and $14 < k < 27$.

22.6.2 K_a Undefined: All Species Are Single-Species Components

Changes in c_{LH} (Fig. 22.11: K_a undefined) are entirely driven by changes in the concentrations of the system's other solute species, particularly *H* (Fig. 22.12: K_a undefined) and *L* (Fig. 22.13: K_a undefined). As c_H increases, s_{LH} becomes more

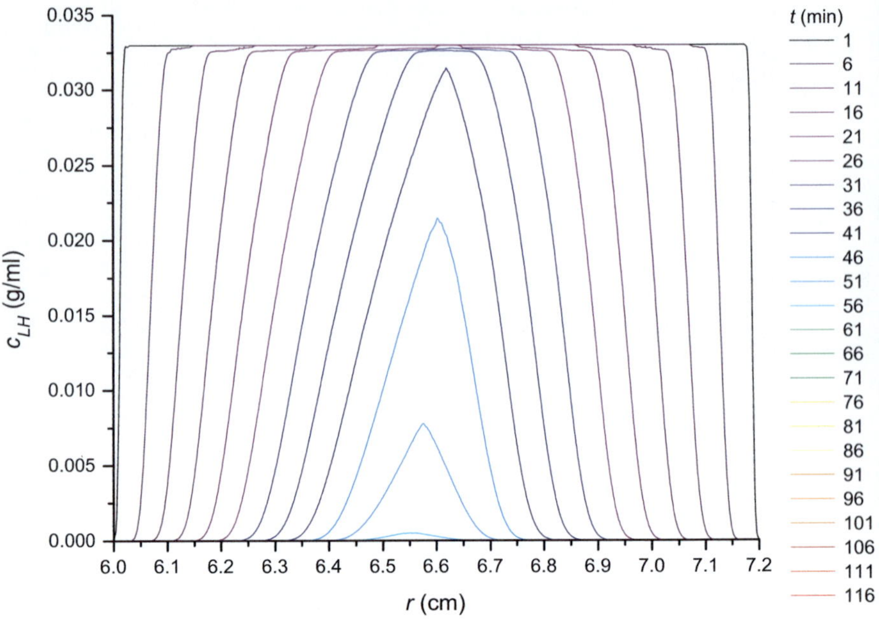

Fig. 22.8 $K_a = 30.325$ ml/g. (Compare with Fig. 22.11.) The concentration of species 14 (LH), c_{LH} (g/ml), versus r (cm) for 1 min $\leq t \leq$ 116 min

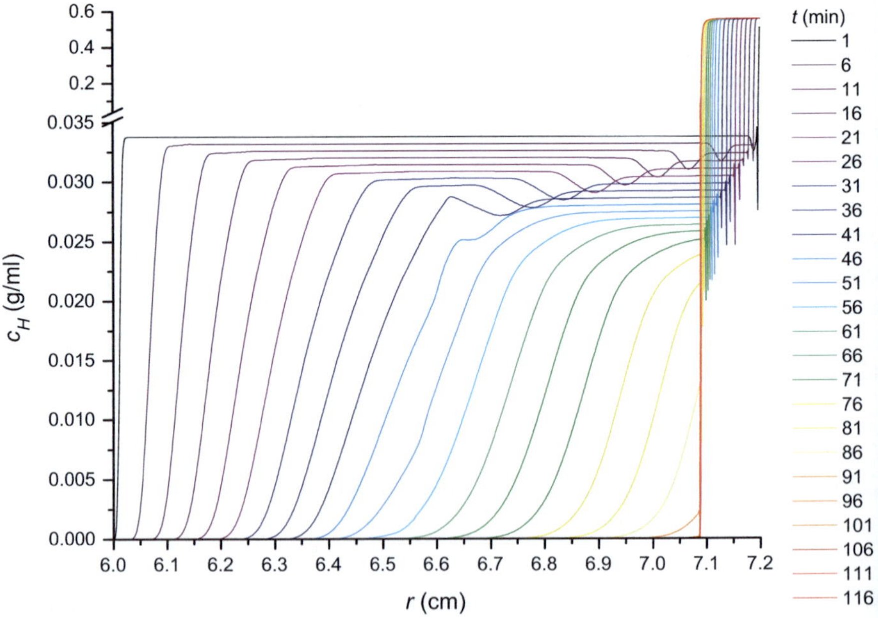

Fig. 22.9 $K_a = 30.325$ ml/g. (Compare with Fig. 22.12.) The concentration of species 1 (H), c_H (g/ml), versus r (cm) for 1 min $\leq t \leq$ 116 min

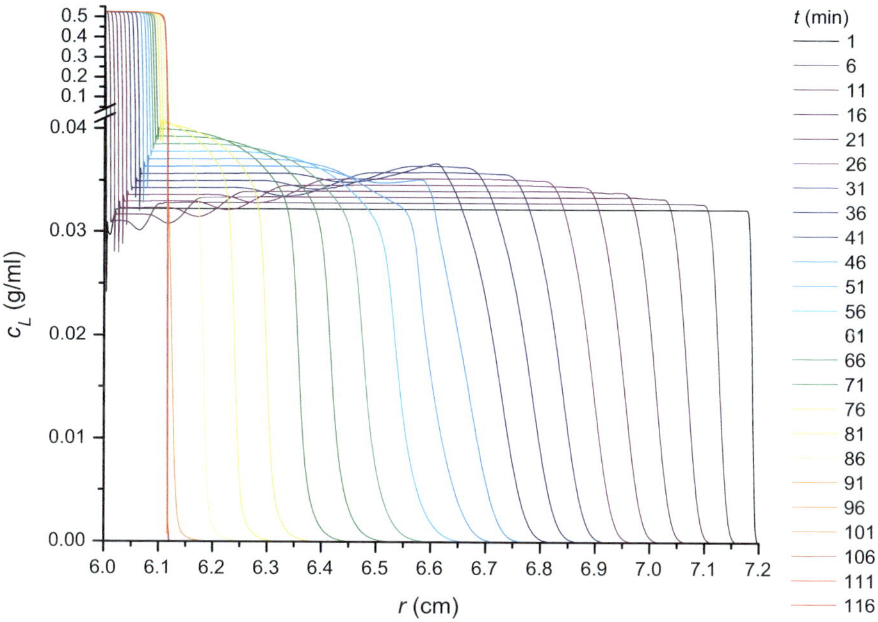

Fig. 22.10 $K_a = 30.325$ ml/g. (Compare with Fig. 22.13.) The concentration of species 27 (L), c_L (g/ml), versus r (cm) for 1 min $\leq t \leq$ 116 min

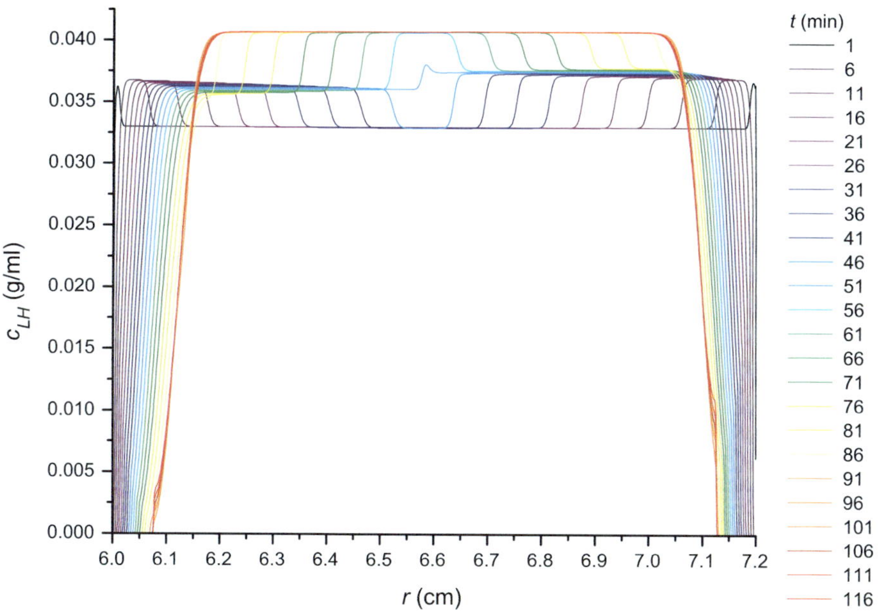

Fig. 22.11 K_a undefined. (Compare with Fig. 22.8.) The concentration of species 14 (LH), c_{LH} (g/ml), versus r (cm) for 1 min $\leq t \leq$ 116 min

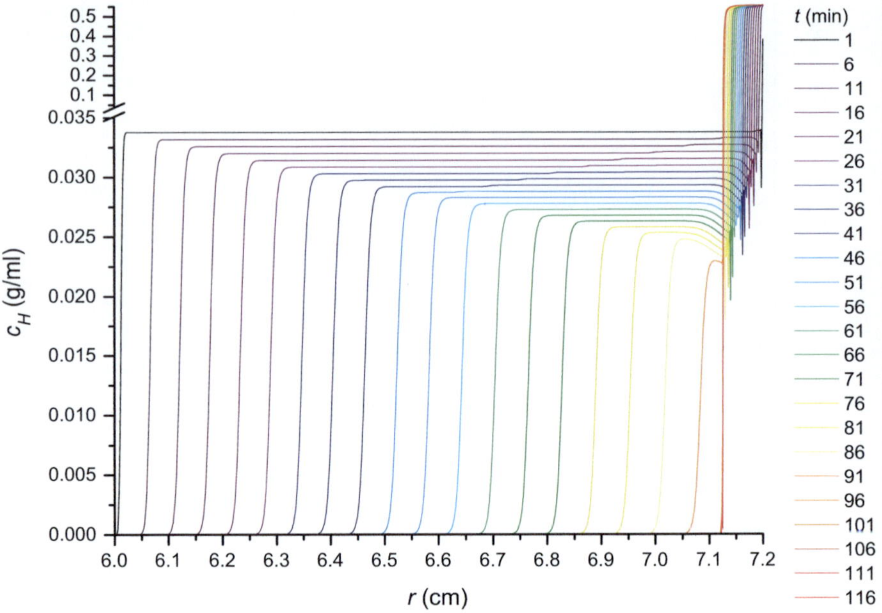

Fig. 22.12 K_a undefined. (Compare with Fig. 22.9.) The concentration of species 1 (H), c_H (g/ml), versus r (cm) for 1 min $\leq t \leq$ 116 min

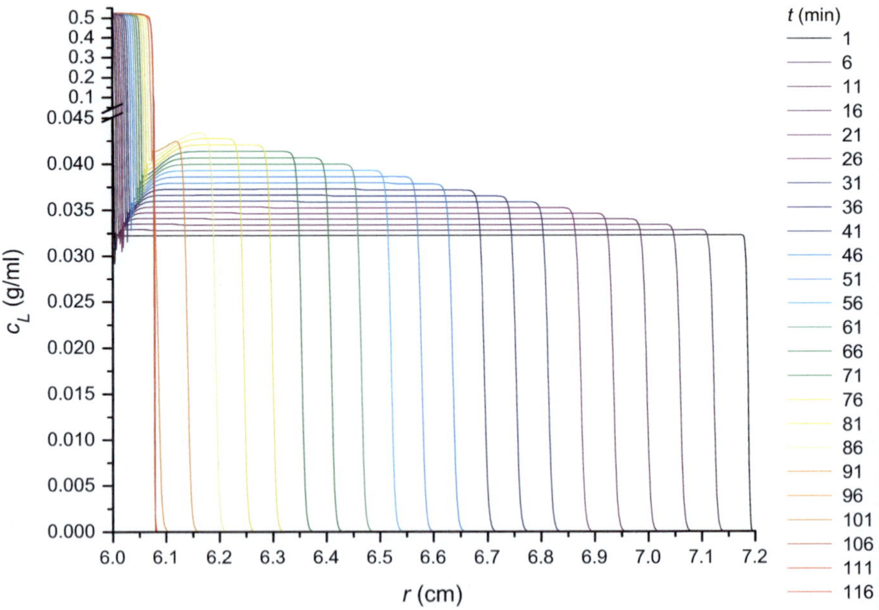

Fig. 22.13 K_a undefined. (Compare with Fig. 22.10.) The concentration of species 27 (L), c_L (g/ml), versus r (cm) for 1 min $\leq t \leq$ 116 min

negative. As c_L increases, s_{LH} becomes more positive. Thus, where c_H accumulates in the pellet or c_L accumulates in the supernatant, c_{LH} decreases. (With respect to species *LH*, the effects of species 2 to 13 are similar to the effects of species *H*, and the effects of species 15 to 26 are similar to the effects of species *L*.)

Species *LH* has no effect on the density of the system because $p_{k,LH} = 0$ for all *k*. However, concentration gradients in other species produce density gradients, and those density gradients redistribute species *LH*. The resulting concentration gradients in species *LH* give rise to a radial dependence in the viscosity and thermodynamic nonideality of the system. Of these, only the radial dependence in the viscosity imparts a radial dependence to s_H and s_L. Where c_{LH} plummets toward the pellet and supernatant, the decrease in viscosity causes $|s_H|$ and $|s_L|$ to increase enough to cause the small decreases seen in c_H toward the pellet and c_L toward the supernatant (Figs. 22.12 and 22.13, respectively). (With respect to species *LH*, the effects on species 2 to 13 are similar to the effects on species *H*, and the effects on species 15 to 26 are similar to the effects on species *L*.) A radial dependence in thermodynamic nonideality is also implicated, however, as the effect is weakened if all $y_{k,q}$ Eq. (22.11) are set to zero, even with all $(\sigma p)_{k,q}$, all $p_{k,q}$ and all $h_{k,q}$ governed by Eqs. (22.15), (22.13), and (22.12), respectively. (All $y_{k,14}$ and all $h_{k,q \neq 14}$ can be set to zero without weakening the effect, however.) Setting all $(\sigma p)_{LH,q}$ equal to zero renders $s_{LH} = 0$ everywhere at all times, with the result that c_{LH} does not redistribute with time. Setting all $h_{k,LH}$ equal to zero renders $s_{k \neq 14}$, including s_H and s_L, insensitive to c_{LH}. Thus, setting all $h_{k,LH}$ equal to zero, or setting all $(\sigma p)_{LH,q}$ equal to zero, eliminates the effect.

The density of species *LH* is such that it should neither sediment nor float in a solvent-*LH* system. Perturbations in the solution density by the mass transport of the other species from $t = 0$ to $t = 116$ min (Figs. 22.4, 22.12 and 22.13: K_a undefined) caused species *LH* to redistribute in that period (Fig. 22.11: K_a undefined), however. Figure 22.7 (K_a undefined) shows the further redistribution of species *LH* after species 1 to 13 have formed a nearly time-invariant pellet, and species 15 to 27 have formed a nearly time-invariant supernatant. In the pellet and the supernatant, the concentration of species *LH* approaches zero. As the system approaches equilibrium, the concentration of species *LH* does not approach a radially independent value in the space between the supernatant and the pellet, even after $t = 8.458$ days. Instead, at the boundary between the bulk of the solution occupied by species *LH* and either the pellet or the supernatant, the concentration of species *LH* rises to a peak. At either boundary, the large and steep gradient in the concentration of species *LH* drives mass transport of *LH* into the pellet or supernatant, while the solution density in either the pellet or the supernatant drives mass transport of *LH* out of those regions. A model that included third- and higher-order virial coefficients of the density increment, thermodynamic nonideality and viscosity might reduce or eliminate this effect, or even produce the opposite effect, wherein the concentration of *LH* would be lower at the supernatant and pellet boundaries than it would be toward the centre of the system.

References

Claverie J-M, Dreux H, Cohen R (1975) Sedimentation of generalized systems of interacting particles. I. Solution of systems of complete Lamm equations. Biopolymers 14:1685–1700

Eisenberg DS, Crothers DM (1979) Transport processes. In Physical chemistry with applications to the life sciences. Benjamin/Cummings, Menlo Park, pp 700–745

Laue TM, Shah BD, Ridgeway TM, Pelletier SL (1992) Computer aided interpretation of analytical sedimentation data for proteins. In: Harding SE, Rowe AJ, Horton JC (eds) Analytical ultracentrifugation in biochemistry and polymer science. Royal Society of Chemistry, Cambridge, pp 90–125

Moody TP (2011) An irreversible thermodynamic description of analytical ultracentrifugation (AUC) applied to a solution of the time- and gravitational-potential-space-dependent Lamm equation. http://moodybiophysicalconsulting.blogspot.com/

Moody TP (2012a) Johnston-Ogston effects in AUC simulations of two model systems based on polystyrene beads that are polydisperse with respect to specific gravity. http://moodybiophysicalconsulting.blogspot.com/

Moody TP (2012b) The apparent sedimentation coefficient, s^*, and its distribution function, $g(s^*)$, within $-\infty < s^* < \infty$. http://moodybiophysicalconsulting.blogspot.com/

Moody TP (2014) Distinguishing hypothetical systems of PS beads by $g(s^*)$ analysis of simulated AUC data to which noise has been added, and quantifying the statistical significance of any observed distinguishability. http://moodybiophysicalconsulting.blogspot.com/

Sharp DG, Beard JW (1950) Size and density of polystyrene particles measured by ultracentrifugation. J Biol Chem 185:247–253

van Holde KE (1985a) Solutions of Macromolecules. In: Physical Biochemistry, 2nd ed. Prentice-Hall, Englewood Cliffs, pp 24–50

van Holde KE (1985b) Viscosity. In: Physical Biochemistry, 2nd ed. Prentice-Hall, Englewood Cliffs, pp 164–181

Chapter 23
Analysis of Nonideal, Interacting, and Noninteracting Systems by Sedimentation Velocity Analytical Ultracentrifugation

Walter F. Stafford

Abstract Analysis of ideal associating systems has been described in detail previously (Rivas et al., Methods 19:194–212, 1999): the reader is referred to that article for the basic theory. We will extend that analysis to nonideal and self- and hetero-associating systems by adding terms for deviations from both thermodynamic ideality and hydrodynamic ideality. In this chapter we will consider several effects of non-ideality on the sedimentation process.

Keywords Analytical ultracentrifugation • Thermodynamic non-ideality • Hydrodynamic non-ideality • Self-association • Hetero-association • Diffusion • Sedimentation velocity • Nonlinear curve fitting

23.1 Background Theory

To put things in perspective, let's start with a quote from Williams et al. (1958):

> It is a well-known fact that the sedimentation methods have enjoyed a spectacular success in protein chemistry. It is now apparent that because of his enthusiasm for the transport method the protein chemist has on occasion allowed himself to be carried to some excesses. For instance, ideal equations descriptive of behavior in two-component systems with no volume change on mixing have been used to describe the experimental observations in multicomponent and not entirely ideal systems. Apparent single translational friction coefficients have been combined with other data and assumptions to provide information about the shape and volume of protein and polysaccharide molecules, when several such coefficients must have been involved.

This problem still exists some 50 odd years later. It is especially prevalent among the users of easy-to-use, "black-box" software packages that have become so popular. It is hoped that this chapter will provide a rigorous approach to the problems of treating non-ideality in associating systems. Some approximations are

W.F. Stafford (✉)
Boston Biomedical Research Institute, c/o 12 Francis Avenue, Cambridge, MA 02138, USA
e-mail: wstafford3@walterstafford.com

S. Uchiyama et al. (eds.), *Analytical Ultracentrifugation*,
DOI 10.1007/978-4-431-55985-6_23

inevitably necessary to provide tractable relationships. These assumptions and their consequences must be kept in mind when analyzing data. Mike Johnson has outlined the "rules" that must be heeded when performing least squares analyses of any type of data (Johnson 1992), which are as follows:

1. The model must be correct.
2. The noise must be normally distributed with mean of zero.
3. The data must be free of systematic error.
4. The data set must have a large number of data points.
5. The observations must be independent.
6. The independent variable must be free of experimental uncertainty

It should be obvious that if the model is incorrect, the parameter values obtained from it will be meaningless. This will happen, for instance, if one were to try to fit to an interacting system with a noninteracting model or vice versa or a nonideal system with an ideal model.

23.1.1 *Self-Associations*

23.1.1.1 Self-Association: Two Species

A simple monomer to N-mer self-association can be described by two equations, a relation for mass action and a statement of conservation of mass for this one-component system:

$$\mathrm{nA} \rightleftharpoons \mathrm{A_n} \tag{23.1}$$

$$K_{1,n} = \frac{a_{A_n}}{a_A^n} = \left(\frac{\gamma_{A_n}}{\gamma_A^n}\right)\frac{C_{A_n}}{C_A^n} \tag{23.2}$$

$$C_A^o = C_A + nC_{A_n} \tag{23.3}$$

where a denotes the thermodynamic activity; γ_i, the molar activity coefficient; C, the molar concentration; K, the molar equilibrium constant; and C_A^o, the total concentration of component A.

23.1.1.2 Self-Associations: Multispecies

Higher-order self-associations or sequential associations can, in general, be represented by a series of reactions and one conservation relation since this is a one-component system:

$$2\mathrm{A_1} \rightleftharpoons \mathrm{A_2} \tag{23.4}$$

$$\mathrm{A_2+A_1} \rightleftharpoons \mathrm{A_3} \tag{23.5}$$

$$A_3 + A_1 \rightleftharpoons A_4 \tag{23.6}$$

$$\ldots \tag{23.7}$$

$$A_{n-1} + A_1 \rightleftharpoons A_n \tag{23.8}$$

$$K_{i-1,n} = \left(\frac{\gamma_{A_n}}{\gamma_{A_{i-1}}\gamma_A}\right)\frac{C_{A_n}}{C_{A_{i-1}}C_A}; \qquad i = 2, 3, \ldots, n \tag{23.9}$$

$$C_A^o = C_{A_1} + 2C_{A_2} + 3C_{A_3} + 4C_{A_4} + \ldots + nC_{A_n} \tag{23.10}$$

23.1.2 Hetero-Associations

23.1.2.1 Hetero-Association: Bimolecular, Single Step

A simple two-component hetero-associating system can be represented by a mass action relation and two conservation of mass relations, one for each of the two components, A and B:

$$A + B \rightleftharpoons AB \tag{23.11}$$

$$K_{AB} = \left(\frac{\gamma_{AB}}{\gamma_A\gamma_B}\right)\frac{C_{AB}}{C_AC_B} \tag{23.12}$$

$$C_A^o = C_A + \frac{M_A}{M_{AB}}C_{AB} \tag{23.13}$$

$$C_B^o = C_B + \frac{M_B}{M_{AB}}C_{AB} \tag{23.14}$$

23.1.2.2 Hetero-Association: Bimolecular, Two Step

A somewhat more complicated, but frequently encountered, system is a two-step (e.g., antigen-antibody system) hetero-association of the following form:

$$A + B \rightleftharpoons AB \tag{23.15}$$

$$AB + B \rightleftharpoons AB_2 \tag{23.16}$$

$$K_{AB} = \left(\frac{\gamma_{AB}}{\gamma_A\gamma_B}\right)\frac{C_{AB}}{C_AC_B} \tag{23.17}$$

$$K_{AB_2} = \left(\frac{\gamma_{AB_2}}{\gamma_{AB}\gamma_B}\right)\frac{C_{AB_2}}{C_{AB}C_B} \tag{23.18}$$

$$C_A^o = C_A + \frac{M_A}{M_{AB}} C_{AB} + \frac{M_A}{M_{AB_2}} C_{AB_2} \tag{23.19}$$

$$C_B^o = C_B + \frac{M_B}{M_{AB}} C_{AB} + \frac{2M_B}{M_{AB_2}} C_{AB_2} \tag{23.20}$$

Other more complicated single- and multicomponent systems can be derived by simple extension of these relationships.

23.2 General Discussion of Non-ideality

Sedimentation velocity analysis of nonideal systems requires taking into account both hydrodynamic and thermodynamic non-ideality (Stafford and Sherwood 2004). Hydrodynamic non-ideality arises due to the displacement of solvent by the sedimenting macromolecules. Because the centrifuge cell is a closed system, the result is a "backflow" (a countercurrent, if you will) of solvent displaced by the macromolecule, which impedes its transport relative to the cell's coordinate system. The backflow is affected by the shape of the macromolecule and its charge and the ionic strength, which in turn determine its effective Stokes radius (Fuoss and Onsager 1961). This backflow leads to a decrease in both sedimentation coefficient and diffusion coefficient with increasing concentration. The backflow contribution is proportional to the concentration of macromolecule. It affects both sedimentation and diffusional transport equally (see Appendix 1). The reader is also referred to the monograph by Katchalsky and Curran, Chapter 9, especially Equations 9–19 (Katchalsky and Curran 1967) for further enlightenment.

23.2.1 Hydrodynamic Non-ideality

23.2.1.1 Single Macromolecular Component, Single-Species Systems

We can express the hydrodynamic non-ideality through the frictional coefficient. The frictional coefficient, f^o, at infinite dilution is given by

$$f^o = N_A 6\pi \eta_o R_s \tag{23.21}$$

where R_s is the Stokes radius, N_A is Avogadro's number, and η_o is the viscosity of water at 20 °C.

The concentration dependence of f is given by

$$f = f^o(1 + k_s c) \tag{23.22}$$

as a first-order function of concentration. From the definition of the sedimentation coefficient as

$$s = \frac{M(1 - v\rho)}{f} \tag{23.23}$$

we can express the concentration dependence of the sedimentation coefficient as

$$s(c) = \frac{s^o}{(1 + k_s c)} \tag{23.24}$$

where s^o is the value of s at infinite dilution, and in the absence of thermodynamic concentration dependence, we can likewise express the hydrodynamic concentration dependence of the diffusion coefficient as

$$D(c) = \frac{D^o}{(1 + k_s c)} \tag{23.25}$$

23.2.2 *Thermodynamic Non-ideality*

The chemical potential, μ_i, for species i, is given by

$$\mu_i = \mu_i^o + RTln(y_i c_i) \tag{23.26}$$

Taking the total differential of both sides and dividing through by RT, we have

$$\frac{d\mu_i}{RT} = dln(c_i) + dln(y_i) \tag{23.27}$$

Expanding in terms of the corresponding partial derivatives, we have an expression for the concentration dependence of the activity coefficient as a function of the concentration of all species present.

$$dln(y_i) = \sum_{j=1}^{n} \frac{\partial ln(y_i)}{\partial ln(c_j)} dln(c_j) \tag{23.28}$$

This can be rewritten as

$$dln(y_i) = \frac{\partial ln(y_i)}{\partial ln(c_i)} dln(c_i) + \sum_{\substack{j=1 \\ j \neq i}}^{n} \frac{\partial ln(y_i)}{\partial ln(c_j)} dln(c_j) \tag{23.29}$$

Now, substituting Eq. 23.29 into Eq. 23.27 , we have

$$d\mu_i/RT = dln(c_i) + \frac{\partial ln(y_i)}{\partial ln(c_i)} dln(c_i) + \sum_{\substack{j=1 \\ j\neq i}}^{n} \frac{\partial ln(y_i)}{\partial ln(c_j)} dln(c_j) \tag{23.30}$$

and rearranging

$$d\mu_i/RT = dln(c_i)\left(1 + \frac{\partial ln(y_i)}{\partial ln(c_i)}\right) + \sum_{\substack{j=1 \\ j\neq i}}^{n} \frac{\partial ln(y_i)}{\partial ln(c_j)} dln(c_j) \tag{23.31}$$

This equation is essentially Equation 9 of Goldberg (1953).

The driving force for diffusion is the gradient of the chemical potential, and so the diffusional flux, J_i^d, is proportional to the gradient of the chemical potential. From Fick's first law in terms of the chemical potential gradient, we can write:

$$J_i^d = -\frac{D_i c_i}{RT}\left(\frac{\partial \mu_i}{\partial r}\right)_t \tag{23.32}$$

Thermodynamic non-ideality, as we mentioned above, is manifest through the concentration dependence of the activity coefficient according to the following equations relating the gradient of the chemical potential to the concentration gradients. In cylindrical coordinates, in the case of diffusion in the absence of sedimentation, the continuity equation can be written as

$$\left(\frac{\partial c}{\partial t}\right)_\xi = -\nabla J_i^d \tag{23.33}$$

$$\nabla J_i^d = \frac{1}{r}\frac{\partial}{\partial r}\left[rJ_i^d\right] = \frac{\partial}{\partial \xi}\left[-r\frac{D_i c_i}{RT}\left(\frac{\partial \mu_i}{\partial r}\right)_t\right] \tag{23.34}$$

Expressing Eq. 23.33 in cylindrical coordinates and multiplying top and bottom of the right hand side by r, we have

$$\nabla J_i^d = \frac{\partial}{\partial \xi}\left[-2\xi\frac{D_i c_i}{RT}\frac{d\mu_i}{d\xi}\right] \tag{23.35}$$

where $\xi = r^2/2$.

Now, expanding the gradient of μ_i in terms of concentrations and activity coefficients, we have

$$\frac{1}{RT}\frac{d\mu_i}{d\xi} = \frac{dln(c_i)}{d\xi} + \sum_{j=1}^{n} \frac{\partial ln(y_i)}{\partial ln(c_j)}\frac{dln(c_j)}{d\xi} \tag{23.36}$$

This can be factored:

$$\frac{1}{RT}\frac{d\mu_i}{d\xi} = \frac{dln(c_i)}{d\xi} + \frac{\partial ln(y_i)}{\partial ln(c_i)}\frac{dln(c_i)}{d\xi} + \sum_{\substack{j=1\\ j\neq i}}^{n} \frac{\partial ln(y_i)}{\partial ln(c_j)}\frac{dln(c_j)}{d\xi} \tag{23.37}$$

$$\frac{1}{RT}\frac{d\mu_i}{d\xi} = \frac{dln(c_i)}{d\xi}\left(1 + \frac{\partial ln(y_i)}{\partial ln(c_i)}\right) + \sum_{\substack{j=1\\ j\neq i}}^{n} \frac{\partial ln(y_i)}{\partial ln(c_j)}\frac{dln(c_j)}{d\xi} \tag{23.38}$$

For a single species the concentration dependence of the activity coefficient has been treated historically by representing the concentration dependence with a polynomial in powers of concentration. This polynomial is often truncated after the first-order term, so that:

$$ln(y_i) = 2B_iM_ic_i + 0c^2 + \ldots \tag{23.39}$$

$$\frac{\partial ln(y_i)}{\partial c_i} = 2B_iM_i \tag{23.40}$$

substituting

$$\left(1 + c_i\frac{\partial ln(y_i)}{\partial c_i}\right) = (1 + 2B_iM_ic_i) \tag{23.41}$$

Both hydrodynamic non-ideality and thermodynamic non-ideality affect the diffusion coefficient, and we can write

$$D_i(c_i) = D_i^o\left[\frac{1 + c_i\frac{\partial ln(y_i)}{\partial c_i}}{1 + k_{s,i}c}\right] = D_i^o\left[\frac{1 + 2BMc_i}{1 + k_{s,i}c_i}\right] \tag{23.42}$$

However, in general, in a solution containing multiple species, we must include cross-terms to reflect the influence of all other species on each other. Expanding the non-ideality term in a first-order power series, we can write

$$ln(y_i) = 2\sum_{j=1}^{n} B_{i,j}M_jc_j \tag{23.43}$$

For example, for $n = 3, j = 1, 2, 3$, we have

$$ln(y_1) = 2B_{1,1}M_1c_1 + 2B_{1,2}M_2c_2 + 2B_{1,3}M_3c_3 \tag{23.44}$$

$$ln(y_2) = 2B_{2,1}M_1c_1 + 2B_{2,2}M_2c_2 + 2B_{2,3}M_3c_3 \tag{23.45}$$

$$ln(y_3) = 2B_{3,1}M_1c_1 + 2B_{3,2}M_2c_2 + 2B_{3,3}M_3c_3 \tag{23.46}$$

taking the total differential of each side:

$$dln(y_1) = 2B_{1,1}M_1dc_1 + 2B_{1,2}M_2dc_2 + 2B_{1,3}M_3dc_3 \tag{23.47}$$

$$dln(y_2) = 2B_{2,1}M_1dc_1 + 2B_{2,2}M_2dc_2 + 2B_{2,3}M_3dc_3 \tag{23.48}$$

$$dln(y_3) = 2B_{3,1}M_1dc_1 + 2B_{3,2}M_2dc_2 + 2B_{3,3}M_3dc_3 \tag{23.49}$$

So now we can write that

$$\frac{\partial ln(y_i)}{\partial ln(c_j)} = 2B_{i,j}M_jc_j; \tag{23.50}$$

giving nine, n^2, partial derivative terms relating the second virial self and cross coefficients, so that the so-called thermodynamic factor, let's call it N_i,

$$N_i = \left(1 + \sum_{j=1}^{n} \frac{\partial ln(y_i)}{\partial ln(c_j)}\right)$$

after substituting, becomes

$$N_i = \left(1 + \sum_{j=1}^{n} 2B_{i,j}M_jc_j\right) \tag{23.51}$$

Now, for three species:

$$N_1 = (1 + \{2B_{1,1}M_1c_1 \ + 2B_{1,2}M_2c_2 + 2B_{1,3}M_3c_3\}) \tag{23.52a}$$

$$N_2 = (1 + \{2B_{2,1}M_1c_1 \ + 2B_{2,2}M_2c_2 + 2B_{2,3}M_3c_3\}) \tag{23.52b}$$

$$N_3 = (1 + \{2B_{3,1}M_1c_1 \ + 2B_{3,2}M_2c_2 + 2B_{3,3}M_3c_3\}) \tag{23.52c}$$

in matrix form

$$\mathbf{N} = \left(1 + \mathbf{B}'\mathbf{c}\right) \tag{23.53}$$

and for 3 species, we have

$$\mathbf{N} = \begin{bmatrix} N_1 \\ N_2 \\ N_3 \end{bmatrix} \tag{23.54}$$

$$\mathbf{B}' = \begin{vmatrix} 1 \ B'_{11} \ B'_{12} \ B'_{13} \\ 1 \ B'_{21} \ B'_{22} \ B'_{23} \\ 1 \ B'_{31} \ B'_{32} \ B'_{33} \end{vmatrix} \tag{23.55}$$

where $B'_{i,j} = 2B_{i,j}M_j$
and

$$\mathbf{c} = \begin{vmatrix} 1 \\ c_1 \\ c_2 \\ c_3 \end{vmatrix} \tag{23.56}$$

The full expressions for the diffusion coefficients become

$$D_1 = D_1^0 \frac{(1 + B'_{1,1}c_1 + B'_{1,2}c_2 + B'_{1,3}c_3)}{(1 + k_{1,1}c_1 + k_{1,2}c_2 + k_{1,3}c_3)} \tag{23.57a}$$

$$D_2 = D_2^0 \frac{(1 + B'_{2,1}c_1 + B'_{2,2}c_2 + B'_{2,3}c_3)]}{(1 + k_{2,1}c_1 + k_{2,2}c_2 + k_{2,3}c_3)} \tag{23.57b}$$

$$D_3 = D_3^0 \frac{(1 + B'_{3,1}c_1 + B'_{3,2}c_2 + B'_{3,3}c_3)}{(1 + k_{3,1}c_1 + k_{3,2}c_2 + k_{3,3}c_3)} \tag{23.57c}$$

It should be noted that in this treatment we have assumed that the cross diffusion coefficients, $D_{i,j}$, are sufficiently small that they can be ignored (see below).

Equations 23.57a, 23.57b, and 23.57c can be written in matrix notation as

$$\mathbf{D} = \mathbf{D^0} \frac{(1 + \mathbf{B'c})}{(1 + \mathbf{k_s c})} \tag{23.58}$$

where $\mathbf{B'}$ is called the "BM matrix" and $\mathbf{k_s}$ the "k_s" matrix

$$f_1 = f_1^o\,(1 + k_{1,1}c_1 + k_{1,2}c_2 + k_{1,3}c_3) \tag{23.59}$$

$$f_2 = f_2^o\,(1 + k_{2,1}c_1 + k_{2,2}c_2 + k_{2,3}c_3) \tag{23.60}$$

$$f_3 = f_3^o\,(1 + k_{3,1}c_1 + k_{3,2}c_2 + k_{3,3}c_3) \tag{23.61}$$

For the k_s matrix for hydrodynamic non-ideality, we have

$$\mathbf{f} = \mathbf{f^o}\,(1 + \mathbf{k_s c}) \tag{23.62}$$

Let

$$\mathbf{F} = (1 + \mathbf{k_s c}) \tag{23.63}$$

where (e.g., for 3 species)

$$\mathbf{F} = \begin{vmatrix} f_1/f_1^o \\ f_2/f_2^o \\ f_3/f_3^o \end{vmatrix} \tag{23.64}$$

$$\mathbf{k_s} = \begin{vmatrix} 1 & k_{s,11} & k_{s,12} & k_{s,13} \\ 1 & k_{s,21} & k_{s,22} & k_{s,23} \\ 1 & k_{s,31} & k_{s,32} & k_{s,33} \end{vmatrix} \tag{23.65}$$

$$\mathbf{c} = \begin{vmatrix} 1 \\ c_1 \\ c_2 \\ c_3 \end{vmatrix} \tag{23.66}$$

The $\mathbf{k_s}$ matrix and the $\mathbf{B}'$ matrices are implemented in SEDANAL as known, fixed parameters that must be obtained by independent measurements of the concentration interdependence of binary mixtures of the components in separate experiments. Experiments are underway to use these terms for the studies of proteins at high concentration (Correia 2015). These experiments are being carried out on model systems comprising binary mixtures of purified proteins in pairs initially to measure both the B' matrix and the Ks matrix with plans to apply them to proteins in serum.

23.2.3 Cross Diffusion Coefficients

Here we attempt to accommodate the cross diffusion coefficients $D_{i,j}$. Generally, the cross diffusion coefficients have been assumed to be insignificant and, therefore, mostly ignored. However, they may become important at higher concentrations found in biological fluids like serum. The cross diffusion coefficients might need to be included because the gradient of any one component in a mixture will become a driving force for diffusion of the other components in the mixture. First we consider the ideal case and then add in the thermodynamic non-ideality terms:

$$J_d^1 = -D_{11}\frac{c_1}{RT}\frac{\partial \mu_1}{\partial r} - D_{12}\frac{c_2}{RT}\frac{\partial \mu_2}{\partial r} - D_{13}\frac{c_3}{RT}\frac{\partial \mu_3}{\partial r} \tag{23.67a}$$

$$J_d^2 = -D_{21}\frac{c_1}{RT}\frac{\partial \mu_1}{\partial r} - D_{22}\frac{c_2}{RT}\frac{\partial \mu_2}{\partial r} - D_{23}\frac{c_3}{RT}\frac{\partial \mu_3}{\partial r} \tag{23.67b}$$

$$J_d^3 = -D_{31}\frac{c_1}{RT}\frac{\partial \mu_1}{\partial r} - D_{32}\frac{c_2}{RT}\frac{\partial \mu_2}{\partial r} - D_{33}\frac{c_3}{RT}\frac{\partial \mu_3}{\partial r} \tag{23.67c}$$

Ordinarily, we assume that

$$D_{i,j} = \delta_{i,j} D_{i,j} \tag{23.68}$$

where $\delta_{i,j}$ is the Kronecker delta.

Under nonideal conditions, taking into account the activity coefficients by expanding the chemical potential gradient in terms of concentrations and activity coefficients, we have

$$J_d^1 = -D_{11} \frac{c_1}{RT} \frac{\partial ln(y_1 c_1)}{\partial r} - D_{12} \frac{c_2}{RT} \frac{\partial ln(y_2 c_2)}{\partial r} - D_{13} \frac{c_3}{RT} \frac{\partial ln(y_3 c_3)}{\partial r} \tag{23.69a}$$

$$J_d^2 = -D_{21} \frac{c_1}{RT} \frac{\partial ln(y_1 c_1)}{\partial r} - D_{22} \frac{c_2}{RT} \frac{\partial ln(y_2 c_2)}{\partial r} - D_{23} \frac{c_3}{RT} \frac{\partial ln(y_3 c_3)}{\partial r} \tag{23.69b}$$

$$J_d^3 = -D_{31} \frac{c_1}{RT} \frac{\partial ln(y_1 c_1)}{\partial r} - D_{32} \frac{c_2}{RT} \frac{\partial ln(y_2 c_2)}{\partial r} - D_{33} \frac{c_3}{RT} \frac{\partial ln(y_3 c_3)}{\partial r} \tag{23.69c}$$

expanding

$$\begin{aligned} J_d^1 = & -D_{11} \frac{c_1}{RT} \frac{\partial ln(c_1)}{\partial r} \left(1 + \sum_{j=1}^{n} \frac{\partial ln(y_1)}{\partial ln(c_j)} \right) \\ & -D_{12} \frac{c_2}{RT} \frac{\partial ln(c_2)}{\partial r} \left(1 + \sum_{j=1}^{n} \frac{\partial ln(y_2)}{\partial ln(c_j)} \right) \\ & -D_{13} \frac{c_3}{RT} \frac{\partial ln(c_3)}{\partial r} \left(1 + \sum_{j=1}^{n} \frac{\partial ln(y_3)}{\partial ln(c_j)} \right) \end{aligned} \tag{23.70a}$$

$$\begin{aligned} J_d^2 = & -D_{21} \frac{c_1}{RT} \frac{\partial ln(c_1)}{\partial r} \left(1 + \sum_{j=1}^{n} \frac{\partial ln(y_1)}{\partial ln(c_j)} \right) \\ & -D_{22} \frac{c_2}{RT} \frac{\partial ln(c_2)}{\partial r} \left(1 + \sum_{j=1}^{n} \frac{\partial ln(y_2)}{\partial ln(c_j)} \right) \\ & -D_{23} \frac{c_3}{RT} \frac{\partial ln(c_3)}{\partial r} \left(1 + \sum_{j=1}^{n} \frac{\partial ln(y_3)}{\partial ln(c_j)} \right) \end{aligned} \tag{23.70b}$$

$$J_d^3 = -D_{31}\frac{c_1}{RT}\frac{\partial ln(c_1)}{\partial r}\left(1+\sum_{j=1}^{n}\frac{\partial ln(y_1)}{\partial ln(c_j)}\right)$$
$$-D_{32}\frac{c_2}{RT}\frac{\partial ln(c_2)}{\partial r}\left(1+\sum_{j=1}^{n}\frac{\partial ln(y_2)}{\partial ln(c_j)}\right) \quad (23.70c)$$
$$-D_{33}\frac{c_3}{RT}\frac{\partial ln(c_3)}{\partial r}\left(1+\sum_{j=1}^{n}\frac{\partial ln(y_3)}{\partial ln(c_j)}\right)$$

Now we can substitute the first-order approximate virial expansions:

$$\sum_{j=1}^{n}\frac{\partial ln(y_i)}{\partial ln(c_j)} = \sum_{j=1}^{n} 2B_{i,j}M_jc_j$$

Expanding as above (Eqs. 23.52a, 23.52b, and 23.52c) and substituting, we can write out explicitly for the flux of each component including both cross diffusion coefficients and cross second virial coefficients:

$$J_d^1 = -D_{11}\frac{c_1}{RT}\frac{\partial ln(c_1)}{\partial r}(1+2B_{1,1}M_1c_1+2B_{1,2}M_2c_2+2B_{1,3}M_3c_3)$$
$$-D_{12}\frac{c_2}{RT}\frac{\partial ln(c_2)}{\partial r}(1+2B_{2,1}M_1c_1+2B_{2,2}M_2c_2+2B_{2,3}M_3c_3) \quad (23.71a)$$
$$-D_{13}\frac{c_3}{RT}\frac{\partial ln(c_3)}{\partial r}(1+2B_{3,1}M_1c_1+2B_{3,2}M_2c_2+2B_{3,3}M_3c_3)$$

$$J_d^2 = -D_{21}\frac{c_1}{RT}\frac{\partial ln(c_1)}{\partial r}(1+2B_{1,1}M_1c_1+2B_{1,2}M_2c_2+2B_{1,3}M_3c_3)$$
$$-D_{22}\frac{c_2}{RT}\frac{\partial ln(c_2)}{\partial r}(1+2B_{2,1}M_1c_1+2B_{2,2}M_2c_2+2B_{2,3}M_3c_3) \quad (23.71b)$$
$$-D_{23}\frac{c_3}{RT}\frac{\partial ln(c_3)}{\partial r}(1+2B_{3,1}M_1c_1+2B_{3,2}M_2c_2+2B_{3,3}M_3c_3)$$

$$J_d^3 = -D_{31}\frac{c_1}{RT}\frac{\partial ln(c_1)}{\partial r}(1+2B_{1,1}M_1c_1+2B_{1,2}M_2c_2+2B_{1,3}M_3c_3)$$
$$-D_{32}\frac{c_2}{RT}\frac{\partial ln(c_2)}{\partial r}(1+2B_{2,1}M_1c_1+2B_{2,2}M_2c_2+2B_{2,3}M_3c_3) \quad (23.71c)$$
$$-D_{33}\frac{c_3}{RT}\frac{\partial ln(c_3)}{\partial r}(1+2B_{3,1}M_1c_1+2B_{3,2}M_2c_2+2B_{3,3}M_3c_3)$$

In summary, in general, for n components, we have a total of n^2 diffusion coefficients and n^2 first-order virial coefficients giving a total of $2n^2$ parameters to characterize diffusion in a nonideal n component system to first order in concentration. The total number of diffusion coefficients cannot be reduced by invoking the Onsager reciprocal relations (Onsager 1931a,b) which do not allow us to say that $D_{i,j} = D_{j,i}$.

Since, in general,

$$\begin{vmatrix} D_{11} & D_{12} & D_{13} \\ D_{21} & D_{22} & D_{23} \\ D_{31} & D_{32} & D_{33} \end{vmatrix} = \begin{vmatrix} L_{11} & L_{12} & L_{13} \\ L_{21} & L_{22} & L_{23} \\ L_{31} & L_{32} & L_{33} \end{vmatrix} \begin{vmatrix} \mu_{11} & \mu_{12} & \mu_{13} \\ \mu_{21} & \mu_{22} & \mu_{23} \\ \mu_{31} & \mu_{32} & \mu_{33} \end{vmatrix}$$

where the $L_{i,j}$ are the Onsager phenomenological coefficients, and

$$\mu_{i,j} \equiv \frac{\partial \mu_i}{\partial c_j}$$

For sedimentation velocity, we have Katchalsky and Curran (1967) (Eqns 9–29 through 9–50) the case that the observed sedimentation coefficients are functions of the chemical potential gradients of the other components leading to the following set of relations between the phenomenological coefficients, $L_{i,j}$, and the sedimentation coefficients. These equations result from a consideration of the gradient of the total potential gradient including both diffusion and sedimentation since for each component

$$\mu_i = \mu_i^o + \mu_i^c - M_i(1 - \rho\bar{v}_i)\omega^2 r^2/2 \tag{23.72}$$

where μ_i^c is the chemical potential, and the last term is the centrifugal potential. The standard reference potential term, μ_i^o, drops out upon differentiation, and we have

$$J_1 = s_1 c_1 \omega^2 r - D_{11}\frac{\partial c_1}{\partial r} - D_{12}\frac{\partial c_2}{\partial r} + D_{13}\frac{\partial c_3}{\partial r} \tag{23.73}$$

$$J_2 = s_2 c_2 \omega^2 r - D_{21}\frac{\partial c_1}{\partial r} - D_{22}\frac{\partial c_2}{\partial r} + D_{23}\frac{\partial c_3}{\partial r} \tag{23.74}$$

$$J_3 = s_3 c_3 \omega^2 r - D_{31}\frac{\partial c_1}{\partial r} - D_{32}\frac{\partial c_2}{\partial r} + D_{33}\frac{\partial c_3}{\partial r} \tag{23.75}$$

Leading to:

$$s_1 = \frac{1}{c_1}\left[L_{11}M_1(1 - \bar{v}_1\rho) + L_{12}M_2(1 - \bar{v}_2\rho) + L_{12}M_3(1 - \bar{v}_3\rho)\right]$$

$$s_2 = \frac{1}{c_2}\left[L_{21}M_1(1 - \bar{v}_1\rho) + L_{22}M_2(1 - \bar{v}_2\rho) + L_{22}M_3(1 - \bar{v}_3\rho)\right]$$

$$s_3 = \frac{1}{c_3}\left[L_{31}M_1(1 - \bar{v}_1\rho) + L_{32}M_2(1 - \bar{v}_2\rho) + L_{32}M_3(1 - \bar{v}_3\rho)\right]$$

where $L_{i,j}/c_i = D_{i,j}/RT$, and we have

$$s_1 = \left[\frac{D_{11}}{RT}M_1(1-\bar{v}_1\rho) + \frac{D_{12}}{RT}M_2(1-\bar{v}_2\rho) + \frac{D_{13}}{RT}M_3(1-\bar{v}_3\rho)\right]$$

$$s_2 = \left[\frac{D_{21}}{RT}M_1(1-\bar{v}_1\rho) + \frac{D_{22}}{RT}M_2(1-\bar{v}_2\rho) + \frac{D_{23}}{RT}M_3(1-\bar{v}_3\rho)\right]$$

$$s_3 = \left[\frac{D_{31}}{RT}M_1(1-\bar{v}_1\rho) + \frac{D_{32}}{RT}M_2(1-\bar{v}_2\rho) + \frac{D_{33}}{RT}M_3(1-\bar{v}_3\rho)\right]$$

Under conditions (e.g., high concentrations) in which the $L_{i,j}, i \neq j$, are not negligible, it will be necessary to include the $D_{i,j}$ in simulations using Lamm equation modeling. For cases in which the cross-terms are negligible, these equations reduce to the familiar Svedberg equations for each component.

23.2.4 Nonideal, Interacting Systems: Effects of Non-ideality on the Equilibrium Constant

We can write the equilibrium expression in terms of either molar or mass concentrations: in terms of activities and molar concentrations and then converting to weight concentrations we have, for a monomer-dimer system:
Molar:

$$K_{1,2} = \frac{a_2}{a_1^2} = \frac{\gamma_2 C_2}{(\gamma_1 C_1)^2} \tag{23.76}$$

Weight:

$$k_{1,2} = K_{1,2}\frac{M_2}{M_1^2} = \frac{y_2 c_2}{(y_1 C_1)^2} \tag{23.77}$$

$$k_{1,2,\text{obs}} = \frac{c_2}{c_2^2} = k_{1,2}\left(\frac{y_2}{y_1^2}\right)^{-1} \tag{23.78}$$

It is general practice, without much justification, to assume that

$$y_2 = y_1^2 \tag{23.79}$$

And in general for a multispecies self-associating system, it is mathematically convenient to assume

$$y_n = y_1^n \tag{23.80}$$

This approximation assumes that all species contribute the same non-ideality per unit mass (Tanford 1961) and obviously will not be valid if there is a significant change of overall charge or change in excluded volume upon association-disassociation. However, this approximation allows us to write, for the monomer-dimer case, that

$$k_{1,2,\mathrm{obs}} = k_{1,2} \tag{23.81}$$

and for the general monomer to n-mer case:

$$k_{1,n,\mathrm{obs}} = k_{1,n} \tag{23.82}$$

The reader is referred to the papers by Dennis Roark and David Yphantis (Roark and Yphantis 1969, 1971; Yphantis and Roark 1972) for a more thorough treatment of non-ideality in associating systems.

The situation is somewhat more complicated for hetero-associating systems since such an approximation generally cannot be easily made if the two proteins have significantly different charges and/or excluded volumes or if charges are canceled upon association. In the simplest bimolecular association, the three species will in general have sufficiently different properties (charge and/or excluded volume) that a simple relationship between activity coefficients usually cannot be written. One would need specific information concerning each species in the reaction. The necessary information could, in principle, be obtained by confined membrane electrophoresis (CME) (Filoti et al. 2015) for each of the species in the reaction for the effective charge on the macromolecule. Excluded volume differences would be more difficult to describe.

Nevertheless, some reasonable assumptions can be made in many cases. For example, in the two-component system

$$\mathrm{A} + \mathrm{B} \rightleftharpoons \mathrm{AB}$$

assuming there are no significant changes in charge or excluded volume upon association, the non-ideality of species AB can be assumed to be the mass weighted average of the contributions from species A and B.

For mixtures of noninteracting components, the species (i.e., components) can be separated and studied independently to measure their individual properties and in pairs to study their mutual effects on each other to obtain the cross-terms. Measuring the cross-terms for an interacting system would be extremely difficult.

Sedimentation transport is described by the Lamm equation (Lamm 1929). For a single macromolecular species, i, we have for $c_i = c_i(r, t)$ that

$$\left(\frac{\partial c_i}{\partial t}\right)_r = -\frac{\partial}{r\partial r}\left[\omega^2 s_i(c_i)c_i - rD_i(c_i)\left(\frac{\partial c_i}{\partial r}\right)_t\right] \tag{23.83}$$

where c_i is the mass concentration of species i, r is the radius, t is the sedimentation time in seconds, ω is the angular velocity of the rotor in radians-sec^{-1}, s is the sedimentation coefficient in seconds, and D is the diffusion coefficient in centimeters^2-sec^{-1}

At this point it is useful to define some terms and point out the differences between a species and a component. A component, in the thermodynamic sense, is a chemical entity that can, in principle, be added to or removed from a solution independently of other species. A species is an individual chemical entity that may be a component or be a constituent of a component.

The Gibbs phase rule applied to the systems ordinarily encountered in the context of analytical ultracentrifugation tells us that, at constant temperature and pressure or in an incompressible solution, the number of macromolecular components is equal to the number of species minus the number of chemical reactions between them. For example, a monomer-dimer, self-associating system is a single-component system comprising two species. It's a single component because if one could remove the dimers, the remaining monomers would self-associate to form a mixture of monomer and dimers, and conversely, the dimers would dissociate to form a mixture of monomers and dimers. That is, monomers and dimers cannot be separated. At constant temperature and pressure, this system has one degree of freedom, namely, the total macromolecular concentration, which alone determines the composition (i.e., the fraction of monomers and dimers at equilibrium) given a particular value of the equilibrium constant.

For a two-component system, the number of degrees of freedom is 2. Again, at constant temperature and pressure, the composition of the solution is determined by the total concentration of each component given the equilibrium constants.

23.3 Curve Fitting: Numerical Solutions to the Lamm Equation

Curve fitting techniques allow us to combine data from several optical systems to fit to data from solutions that contain components that have different extinction properties. Model-independent methods, i.e., those not involving curve fitting, give us only the sum total of all the contributions from all species added together in one signal. For example, absorbance optics give us only the total absorbance for any particular solution of several species that may have quite different optical properties. Model-dependent methods, such as least squares curve fitting, allow us to extract the contributions from several species having different optical properties as long as we know the extinction coefficients of each species in a mixture.

To allow us to combine data sets from several optical systems, each of which may have different units and noise levels, we compute the sum of the weighted squares of the residuals, *WSSR*, over all the radial points, scans, and cells in a combined global fit (Stafford and Sherwood 2004), where the weighting factors are the inverse of the variance of the data which, in general, is a function of radius. For example, the absorption optical system of the Beckman XL-A records the standard deviation of the absorbance or intensity along with its standard deviation at each radial position. The inverse of the square of the standard deviation can be used by SEDANAL to weight the squared residuals. With the interference optical system, the standard variation of the scan is essentially independent of position, and so a single weighting factor equal to the inverse of the standard deviation of the fringe displacement can be used. For fluorescence optics, both the magnitude and the standard deviation are much larger numerically than data from the other systems. However, when those data are normalized by weighting the squared residuals by the inverse of their variance, the signal-to-noise becomes comparable to weighted residuals from the other optical systems and they can be compared. Because the dimensionless weighted squares of the residuals are non-denominate numbers, they can be added together without violating any laws of mathematics.

$$WSSR = \frac{1}{LMN}\sum_{i=1}^{L}\sum_{j=1}^{M}\sum_{k=1}^{N}\left(\frac{dev_{i,j,k}^{2}}{\sigma_{i,j,k}^{2}}\right) \tag{23.84}$$

where N is the number of points over the range being fitted in each cell; M is the number of scans in each cell; L is the number of cells; $dev_{i,j,k}$ is the residual; and $\sigma_{i,j,k}$ is the standard deviation of the data. Thus, $(1/dev_{i,j,k})^2$, i.e., the inverse of the variance, becomes the weighting factor $w_{i,j,k}$.

$$WSSR = \frac{1}{LMN}\sum_{i=1}^{L}\sum_{j=1}^{M}\sum_{k=1}^{N}\left(w_{i,j,k}dev_{i,j,k}^{2}\right) \tag{23.85}$$

For example, in SEDANAL we curve fit the Lamm equation to time difference curves, and so the residual is computed as

$$dev_{i,j,k} = \Delta S_{i,j,k}(r_i, t_j, t_{j+M/2}) - \sum_{l=1}^{l=n}\alpha_l \Delta C_{i,j,k,l}(r_i, t_j, t_{j+M/2}) \tag{23.86}$$

where $\Delta S_{i,j,k}(r_i, t_j, t_{j+M/2})$ is the time difference curve computed from the signal; n is the number of species in the model being fitted; α_l is the extinction coefficient for species l; and $\Delta C_{i,j,k,l}(r_i, t_j, t_{j+M/2})$ is the corresponding time difference curve computed from the current guesses in the solution of the Lamm equation. The reader is referred to the original paper for the details (Stafford and Sherwood 2004).

The time difference curves are computed as follows:

For the signal, S:

$$\Delta S_{i,j,k}(r_i, t_j, t_{j+M/2}) = S_{i,j,k}(r_i, t_{j+M/2}) - S_{i,j,k}(r_i, t_j) \tag{23.87}$$

And likewise, for the solutions, C, to the Lamm equation:

$$\sum_{l=1}^{l=n} \alpha_l \Delta C_{i,j,k,l}(r_i, t_j, t_{j+M/2}) = \sum_{l=1}^{l=n} \alpha_l C_{i,j,k,l}(r_i, t_{j+M/2}) - \sum_{l=1}^{l=n} \alpha_l C_{i,j,k,l}(r_i, t_j) \tag{23.88}$$

Weighting the residuals from different optical systems normalizes the residuals so that residuals from different optical systems can be combined in a global fit. Effectively, the result is that we minimize the reduced chi-square values over all the data sets. Since there are a large number of data points (usually several tens of thousands) and the noise on the data is normally distributed, this fitting procedure is the method of maximum likelihood (Bevington and Robinson 2003). This procedure maximizes the probability that the guesses for the parameters are correct.

Appendix 1: Hydrodynamic Non-ideality

It has been well established that at infinite dilution, the frictional coefficients for sedimentation and diffusion are equal (Schachman 1959). Above we claim that their dependence on concentration is also the same. We can show this to be true with a proof by contradiction, namely, that the hydrodynamic concentration dependencies of sedimentation and diffusion, k_{sed} and k_{diff}, are equal. We start by assuming they are not equal (i.e., $k_{\text{sed}} \neq k_{\text{diff}}$) and proceed to show that this assumption leads to a contradiction. This hydrodynamic concentration dependence arises purely because of the frame of reference we are using, i.e., the cell is of constant volume, and so displaced solvent has to move against the macromolecule and into the volume that was occupied by the macromolecule while the macromolecule is translating in response to the forces acting upon it, whether or not the force arises from the gradient of the centrifugal potential (the driving force for sedimentation) or the gradient of the chemical potential (the driving force for diffusional transport) (Katchalsky and Curran 1967). This relationship has been postulated in the past (Harding and Johnson 1985) but never explicitly proven.

Consider the fluxes arising from sedimentation and diffusion:

$$J_{\text{sed}} = \omega^2 rc \left[\frac{s_o}{1 + k_{\text{sed}} c} \right] \tag{23.89}$$

$$J_{\text{diff}} = -D_o \left[\frac{\left(1 + \frac{\partial ln(y)}{\partial ln(c)}\right)}{1 + k_{1\text{diff}} c} \right] \left(\frac{\partial c}{\partial r} \right)_t \tag{23.90}$$

At sedimentation equilibrium, the flux due to sedimentation is equal to the flux due to diffusion throughout the cell. And we have that

$$J_{\text{sed}} = J_{\text{diff}} \tag{23.91}$$

and we have

$$\omega^2 rc \left[\frac{s_o}{1 + k_{\text{sed}} c} \right] = -D_o \left[\frac{\left(1 + \frac{\partial ln(y)}{\partial ln(c)}\right)}{1 + k_{\text{diff}} c} \right] \left(\frac{\partial c}{\partial r} \right)_t \tag{23.92}$$

This can be seen from the Lamm equation (Eq. 23.83) by setting $(\partial c / \partial t)_r = 0$ and rearranging:

$$\omega^2 \frac{s_o}{D_o} = \left[\frac{1 + k_{\text{sed}} c}{1 + k_{\text{diff}} c} \right] \frac{1}{rc} \left(\frac{\partial c}{\partial r} \right)_t \left(1 + \frac{\partial ln(y)}{\partial ln(c)} \right) \tag{23.93}$$

This equation agrees with the standard thermodynamic derivation if and only if k_{sed} is equal to k_{diff}.

$$\omega^2 \frac{s_o}{D_o} = \frac{1}{rc} \left(\frac{\partial c}{\partial r} \right)_t \left(1 + \frac{\partial ln(y)}{\partial ln(c)} \right) \tag{23.94}$$

Invoking the Svedberg equation, we have

$$\frac{M(1 - \bar{v}\rho)\omega^2}{RT} = \frac{1}{rc} \left(\frac{\partial c}{\partial r} \right)_t \left(1 + \frac{\partial ln(y)}{\partial ln(c)} \right) \tag{23.95}$$

which agrees with the thermodynamic derivation given by Williams et al. (1958). And so, it must be the case that

$$\left[\frac{1 + k_{\text{sed}} c}{1 + k_{\text{diff}} c} \right] = 1$$

Therefore, our initial assumption that $k_{\text{sed}} \neq k_{\text{diff}}$ must have been false, and we have proven that k_{sed} is equal to k_{diff} and that any difference in the concentration dependence of sedimentation and diffusion arises only through the thermodynamic non-ideality term which in turn arises from the concentration dependence of the activity coefficients (cf. Eq. 23.41) of each of the species present in the solution (Goldberg 1953; Sherwood and Stafford 2016). The equality of k_{sed} and k_{diff} has been demonstrated experimentally for sucrose (LaBar and Baldwin 1963). It should be pointed out that $k_{\text{sed}} c$ and $k_{\text{diff}} c$ (i.e., linear dependence on concentration) can be replaced by any function of concentration, $g(c)$, that describes the concentration dependence of the observed frictional coefficient.

References

Bevington PR, Robinson DK (2003) Data reduction and error analysis for the physical sciences, 3rd edn. McGraw-Hill, New York

Correia JJ (2015) An examination of the Johnston-Ogston effect using the AVIV-FDS and SEDANAL global direct boundary fitting. In: Uchiyama S et al. (eds) Analytical ultracentrifugation: instrumentation, software, and applications. Springer, Tokyo

Filoti DI, Shire SJ, Yadav S, Laue TM (2015) Comparative study of analytical techniques for determining protein charge. J Pharm Sci 104:2123–2131

Fuoss RM, Onsager L (1961) Thermodynamic potentials of symmetrical electrolytes. PNAS 47:818–825

Goldberg RJ (1953) Sedimentation in the ultracentrifuge. J Phys Chem 57:194–197

Harding SE, Johnson P (1985) The concentration-dependence of macromolecular parameters. Biochem J 231:543–547

Johnson ML (1992) Why, when, and how biochemists should use least squares. Anal Biochem 206(2):215–225. Review

Katchalsky A, Curran PF (1967) Nonequilibrium thermodynamics in biophysics. Harvard University Press, Cambridge

LaBar FE, Baldwin RL (1963) The sedimentation coefficient of sucrose. J Am Chem Soc 85(20):3105–3108

Lamm O (1929) Die differentialgleichung der ultrazentrifugierung. Ark Mat Astron Fys 21B(2): 1–4

Onsager L (1931a) Reciprocal relations in irreversible processes i. Phys Rev 37:405–426

Onsager L (1931b) Reciprocal relations in irreversible processes. ii. Phys Rev 38:2265–2279. doi:10.1103/PhysRev.38.2265. http://link.aps.org/doi/10.1103/PhysRev.38.2265

Rivas G, Stafford WF, Minton A (1999) Characterization of heterologous protein-protein interactions using analytical ultracentrifugation. Methods 19:194–212

Roark DE, Yphantis DA (1969) Studies of self-associating systems by equilibrium ultracentrifugation. Ann NY Acad Sci 164(1):245–278

Roark DE, Yphantis DA (1971) Equilibrium centrifugation of nonideal systems. The Donnan effect in self-associating systems. Biochemistry 10(17):3241–3249

Schachman HK (1959) Ultracentrifugation in biochemistry. Academic, New York

Sherwood PJ, Stafford WF (2016) Sedanal: model-dependent and model independent analysis of sedimentation data. In: Uchiyama S, Arisaka F, Stafford WF, Laue T (eds) Analytical ultracentrifugation: instrumentation, software, and applications. Springer, Tokyo, pp xx–xx

Stafford WF, Sherwood PJ (2004) Analysis of heterologous interacting systems by sedimentation velocity: curve fitting algorithms for estimation of sedimentation coefficients, equilibrium and kinetic constants. Bioph Chem 108:231–243

Tanford C (1961) Physical chemistry of macromolecules. Wiley, New York

Williams JW, Van Holde KE, Baldwin RL, Fujita H (1958) Theory of sedimentation analysis. Chem Rev 58:715–744

Yphantis DA, Roark DE (1972) Equilibrium centrifugation of nonideal systems. Molecular weight moments for removing the effects of nonideality. Biochemistry 11(16):2925–2934

Chapter 24
Techniques for Dissecting the Johnston-Ogston Effect

John J. Correia, Daniel F. Lyons, Peter Sherwood, and Walter F. Stafford

Abstract The development of the fluorescence detection system (Aviv-FDS) for the AUC allows a single fluorescently labeled species to be quantitatively characterized against a highly concentrated and heterogeneous background. During our use of the FDS to characterize ELP, a novel drug delivery vector (see Lyons et al., Biophys J 104:2009–2021, 2013), in serum, we encountered the Johnston-Ogston (J-O) effect. The J-O effect is a classical anomaly in sedimentation velocity theory and practice describing the nonideal sedimentation properties of a component as a function of high concentrations of other components. We examined the J-O effect using recent advances in AUC hardware, the AU-FDS (AVIV Biomedical), and data analysis methods, DCDT+ and SEDANAL global direct boundary fitting. We empirically quantified the self- and cross-sedimentation nonideality properties of ELP and the two most ubiquitous serum proteins, albumin (~35–40 mg/ml) and γ-globulins (~10–15 mg/ml). We have verified and measured the presence of cross-term hydrodynamic nonideality by running SV studies on a fluorescently labeled component (~100 nM) in a titration experiment with high concentrations of unlabeled components. This has been accounted for through the introduction of a 3×3 nonideality matrix of Ks values into SEDANAL. ELP experiments with mixtures of albumin and γ-globulins were also performed in an attempt to recapitulate the J-O behavior of a serum solution. Clearly, other components or effects contribute to the serum J-O effect. Additional experiments with lipids, lipidated serum albumin, and PEG solutions are planned. These studies lay the groundwork for bringing quantitative hydrodynamic analyses into crowded environments and will allow

Supported by the UMC AUC Facility and NSF MRI grant 1040372

J.J. Correia (✉) • D.F. Lyons
Department of Biochemistry, University of Mississippi Medical Center, 2500 N State St, Jackson, MS 39216, USA
e-mail: jcorreia@umc.edu

P. Sherwood
Interactive Technology, Inc., 20 Woodside Glen Ct., Oakland, CA 94602, USA

W.F. Stafford
Boston Biomedical Research Institute, c/o 12 Francis Ave, Cambridge, MA 02138, USA

S. Uchiyama et al. (eds.), *Analytical Ultracentrifugation*,
DOI 10.1007/978-4-431-55985-6_24

measurement of hydrodynamic and equilibrium macromolecular properties in a physiological state.

Keywords Johnston-Ogston effect • Nonideality • Hydrodynamic nonideality • Thermodynamic nonideality • AU-FDS • Serum hydrodynamics • Cross-term nonideality • Analytical ultracentrifugation • Sedimentation velocity

24.1 Introduction

Macromolecular therapeutics (antibodies, drug delivery vectors, etc.) are often delivered by direct injection into the blood stream. Biophysical characterization of macromolecular therapeutics may thus require analysis of their behavior in serum and plasma to understand their stability, state of aggregation, and interaction with serum macromolecules. A rigorous method of analysis is to perform sedimentation velocity experiments with the Aviv AU-FDS recently developed in the Laue lab (MacGregor et al. 2004; Kroe and Laue 2009; Kingsbury and Laue 2011) and being further developed by Aviv (Zhao et al. 2014). This method allows in principle a determination of size, shape, state of aggregation, and interaction with other macromolecules. Given the high concentration of proteins in serum (70–100 g/l), one must also take into account both hydrodynamic and thermodynamic nonideality. The early ultracentrifuge studies on milk and serum proteins date back to the 1930s and 1940s (McFarlane 1935; Ogston 1937; Johnston and Ogston 1946) and focused on mixtures of lactoglobulin, bovine serum albumin, and bovine serum γ-globulin. These studies reported a boundary anomaly that became known as the Johnston-Ogston effect. While many theories involving disaggregation and cement substances were proposed, the ultimate explanation of the boundary anomaly consistently observed in these studies involved differences in the rate of sedimentation of a slow component in the presence and absence of faster components (Johnston and Ogston 1946). The concentration of the slow component appeared to increase, while the concentration of the fast component appeared to decrease. It was subsequently proposed that the magnitude of the anomaly can be predicted from the properties of the fast component (Harrington and Schachman 1953). In the Harrington study, TMV (tobacco mosaic virus) was used as the fast component, and the observation was its impact on the concentration of BSV (bushy stunt virus) was greater than that observed by Johnston and Ogston who used a more symmetric molecule BSA as the fast component. The original analysis treated the data as a total concentration effect, while Harrington and Schachman established that the individual components must be taken into account separately (Harrington and Schachman 1953; Trautman et al. 1954). This conclusion derives from the empirical dependency of the s value of the slow component on the concentrations of all components s(c) and phenomenologically reflects both excluded volume and charge effects. (While there are theoretical predictions for excluded volume effects, see Chap. 25, by Rowe (2015) in this volume for a summary, there is no general theory for hydrodynamic s(c) where charge is also a major component.) If the form and magnitude of this dependency are similar for different components, then a total concentration was applied in these prior studies. If the *s* vs *c* dependencies are different (Soda et al. 1967), then each component much be accounted for separately. This dependency may also give rise to hidden nonideality and self-association if the two phenomena cancel out (Muramatsu and Minton 1989; Lyons et al. 2013a, 2014). As we will show below, the magnitude of the s(c) dependences is pair-wise specific and may vary with the dominant component.

Here we present studies of ELP, a novel drug delivery vector (Lyons et al. 2013a, 2014) in mixtures of bovine serum albumin and γ-globulins. The approach involves performing sedimentation velocity (SV) experiments as a function of the concentration of each component. The data are analyzed by DCDT+ and initially plotted as $1/s$ vs c to extract hydrodynamic nonideality Ks terms that reflect interactions between components. Methods to numerically simulate J-O effects have been described previously (Correia et al. 1976). However, experimental data analysis is best accomplished within the framework of rigorous application of the modern AUC data analysis program SEDANAL (Stafford and Sherwood 2004). This chapter outlines the experimental and computational approach required to use SEDANAL for the biophysical analysis of macromolecular therapeutics (antibodies, drug delivery vectors, etc.) in heterogeneous solution of high-concentration mixtures. The ultimate goal is to understand the requirements for rigorous studies in human serum (Kroe and Laue 2009; Demeule et al. 2009).

24.2 Methods

To study the sedimentation behavior of ELP in serum, one must account for the interaction of the therapeutic with the major components of serum, albumin (Sigma G7906) and γ-globulins (the total IgG fraction from serum; Sigma G5009). A cys-ELP construct was expressed and purified as described in Lyons et al. (2014). Cys-ELP was chosen since it contains a single-Cys residue as an attachment site for fluorescent probes (Lyons et al. 2014). Serum typically contains 30–40 mg/ml albumin and 10–15 mg/ml γ-globulins or serum IgG. These values can dramatically change during disease states (Anderson and Anderson 2002). Sedimentation velocity of ELP must be performed as a function of the concentration of all the major components, albumin, γ-globulins, and ELP itself. This is based upon the empirical expression for a sedimentation coefficient:

$$s_1 = \frac{s_1^o}{(1 + K_{11}c_1 + K_{12}c_2 + K_{13}c_3 + \cdots)}$$

where s_i is the sedimentation coefficient of a macromolecule, s^o is the sedimentation coefficient extrapolated to zero concentration of all components, K_{ij} values are hydrodynamic nonideality coefficients reflecting the concentration dependent of s_i

and the c_i's are concentrations of each component in mg/ml. Thus, to study and understand the behavior of ELP (component 1) in a mixture of serum albumin (component 2) and γ-globulin (component 3, referred to as IgG here), sedimentation velocity experiments must be done as a function of the concentration of albumin, IgG, and ELP themselves. SV experiments also need to be done for albumin as a function of its own concentration, as well as the concentrations of IgG and ELP, as well as for IgG as a function of its concentration, and as a function of the concentrations of albumin and ELP. This means performing nine sets of experiments over a wide range of component concentrations that allows the determination of the nine K_{ij} values as outlined in the equations below. (Note this is the nomenclature of Williams et al. (1958) where solvent is component 0.)

$$s_1 = \frac{s_1^o}{(1 + K_{11}c_1 + K_{12}c_2 + K_{13}c_3)}$$

$$s_2 = \frac{s_2^o}{(1 + K_{21}c_1 + K_{22}c_2 + K_{23}c_3)}$$

$$s_3 = \frac{s_3^o}{(1 + K_{31}c_1 + K_{32}c_2 + K_{33}c_3)}$$

The diagonal K_{ii} terms are the self-nonideality terms, while the $K_{ij\ (i \neq j)}$ terms are referred to as the cross-nonideality terms. Three sets of experiments involve the components alone to determine the self-nonideality terms and can usually be performed over a range of concentrations with absorbance optics and a combination of 3 mm and 12 mm centerpieces. Alternatively, one can run a small amount of fluorescently labeled material (100–150 nM) as a function of unlabeled material (Lyons et al. 2013a, 2014), especially if you want to achieve total concentrations approaching 100 mg/ml. Experiments are performed by a mixture of both methods here. For the cross-term experiments, fluorescently labeled material (100–250 nM) was run as a function of the concentration of unlabeled material. (Note Lyons et al. (2013b) have shown the FDS optics is linear up to 500–1000 nM.)

All experiments are performed in an AUC equipped with an Aviv AU-FDS system on samples equilibrated into phosphate-buffered saline (PBS) at pH 7.4 by spun columns, at 20 C (19.69 C by calibration; Liu and Stafford 1995), and 50,000 rpm in Spin Analytical double-sector centerpieces. Buffer density was measured in an Anton Paar DM-5000 density meter. Partial specific volumes were also measured in an Anton Paar DM-5000 density meter (ELP, 0.7234 ml/mg; BSA 0.7533 ml/mg; IgG, 0.7263 ml/mg). Buffer viscosity was estimated with Sednterp. The meniscus position for all fluorescence data sets discussed in this study was determined from a pseudo-absorbance scan at 50,000 rpm taken immediately after the completion of a fluorescence run. Analysis of SV data was performed by DCDT+ to generate g(s*) distributions (Stafford 1992; Philo 2006). Global fitting was performed by direct boundary analysis with SEDANAL (Stafford and Sherwood 2004).

24.3 Self-Nonidealty Measurements

An example of SV analysis of ELP as a function of concentration is shown in Fig. 24.1 where concentrations are in weight units and s values are normalized and corrected to $s_{20,w}$ values. Note since BSA concentrations are typically 35 mg/ml, the range of concentrations investigated should be up to 40 or 50 mg/ml to validate a linear dependence of s on c. If there is curvature, SEDANAL is equipped to include second-order terms in concentration. The decrease in the weight-average sedimentation coefficient observed with increasing concentration in the g(s*) plots is caused by hydrodynamic nonideality (Fig. 24.1a). The linear fit (Fig. 24.1b) determines a K_{11} value of 0.0224 ml/mg for ELP acting on itself.

The hydrodynamic properties of BSA were measured using the same procedure. First, the hydrodynamic properties of BSA were measured at low concentrations (0.41–1.26 mg/mL) using the absorbance optical system (Fig. 24.2a). The

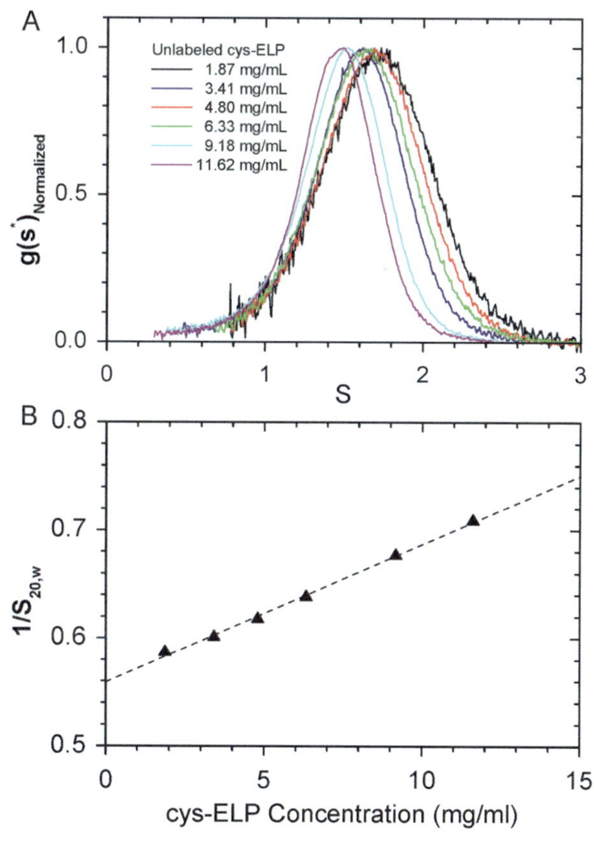

Fig. 24.1 Hydrodynamic characterization of cys-ELP1. All measurements were performed in PBS + 1 mM TCEP. (**a**) Normalized g(s*) distributions. (**b**) Plotting the inverse weight-average sedimentation coefficient determined from the g(s*) distribution as a function of concentration. The slope is equal to the magnitude of the hydrodynamic nonideality (K_{11}) divided by the sedimentation coefficient extrapolated to zero concentration ($s^o = 1.79$ s and $K_{11} = 0.0224$ ml/mg)

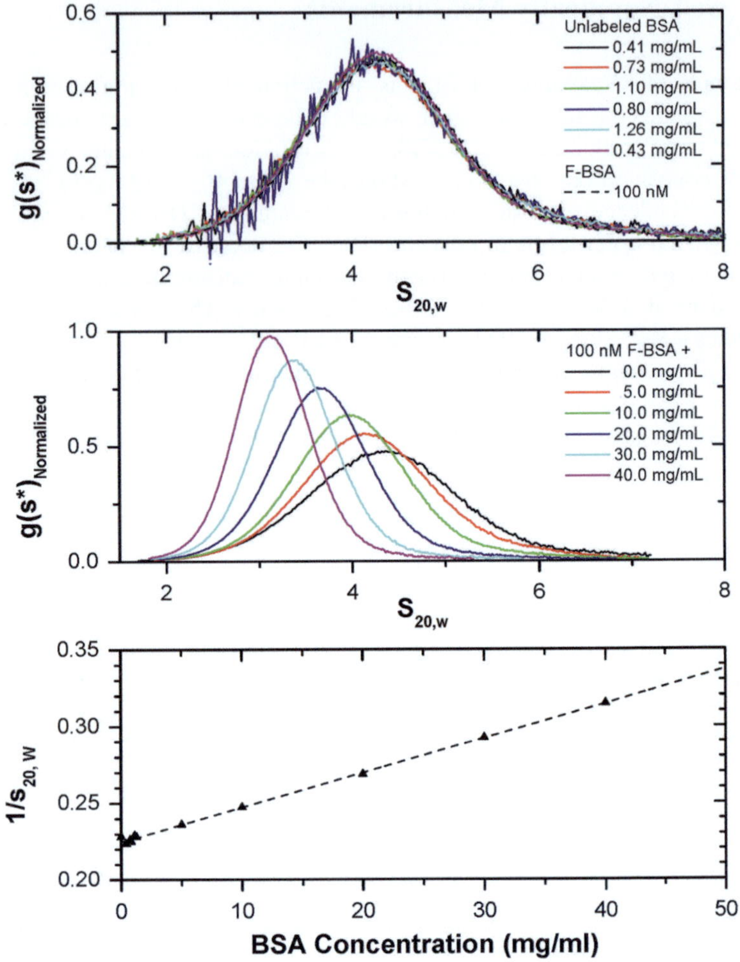

Fig. 24.2 Hydrodynamic characterization of BSA. (**a**) g(s*) analysis of a titration series of unlabeled BSA (0.4–1.1 mg/mL) analyzed using absorbance optics. One g(s*) trace of F-BSA (100 nM) is included for ease of comparison. (**b**) g(s*) analysis of a titration series of F-BSA (0.01–40 mg/mL) using fluorescence optics. (**c**) Plotting the weight-average sedimentation coefficient determined from the g(s*) analysis as a function of BSA concentration. A linear fit revealed $s^o = 4.45$ s and $K_{22} = 0.0093$ ml/mg

physiological concentration of BSA extends beyond the range measureable using absorbance. Therefore, BSA was covalently labeled with fluorescein (F-BSA), and the fluorescence optical system (AVIV AU-FDS) was used. F-BSA at 100 nM was sedimented in the presence of up to 40 mg/mL unlabeled BSA (Fig. 24.2b). This allowed the concentration-dependent behavior to be examined over a wide range while keeping the concentration of fluorescent material low to avoid inner-filter effects (MacGregor et al. 2004; Lyons et al. 2013b). A g(s*) analysis of all SV experiments (absorbance and fluorescence) suggests that BSA behaves as a nonideal monomer with approximately 4 % irreversible dimer. The K_{22} was determined by plotting the inverse weight-average sedimentation coefficient determined from the g(s*) analysis as a function of concentration (Fig. 24.2c; $K_{22} = 0.0093$ mg/mL).

The hydrodynamic properties of IgG were determined by the same method as cys-ELP and BSA (Fig. 24.3). Six samples spanning a low-concentration range

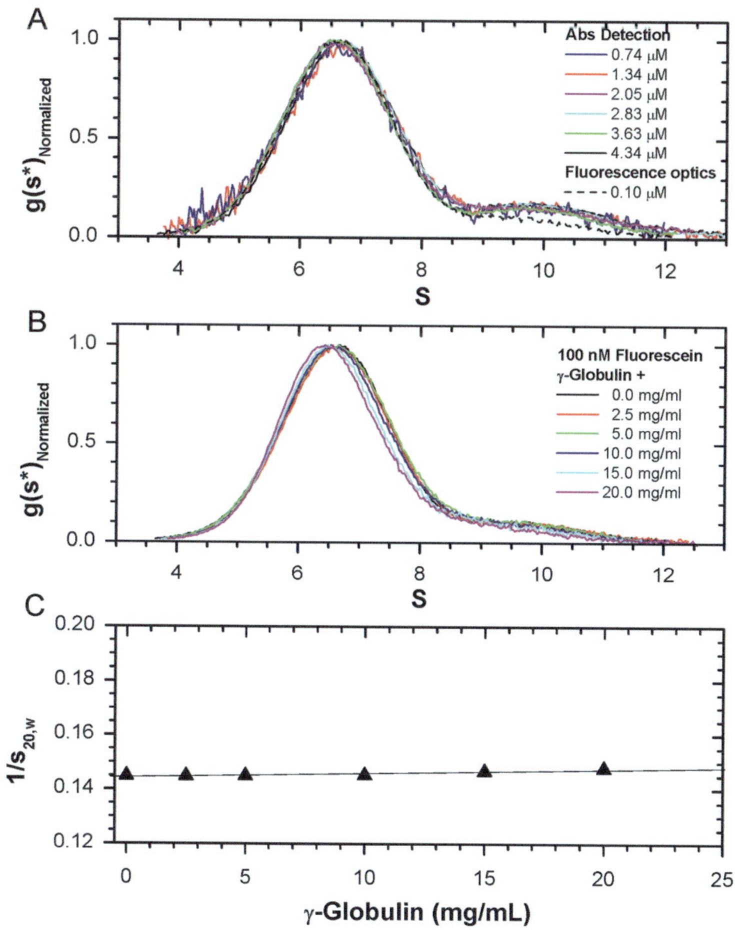

Fig. 24.3 Analyzing the concentration-dependent behavior of IgG. (**a**) g(s*) analysis of a titration series of unlabeled IgG (0.1–0.68 mg/mL) analyzed using absorbance optics. One g(s*) trace of F-IgG is included for ease of comparison. (**b**) g(s*) analysis of a titration series of 100 nM F-IgG titrated with unlabeled IgG (0–20 mg/mL) using fluorescence optics. (**c**) Plotting the weight-average sedimentation coefficient (measured from g(s*) analysis panels (**a**) and (**b**)) as a function of concentration. A linear fit revealed $s^o = 6.93$ s and $K_{22} = 0.0015$ ml/mg

(0.74–4.34 uM) were analyzed using absorbance SV (Fig. 24.3a). The g(s*) distributions exhibited no hydrodynamic nonideality over this low-concentration range. IgG was then labeled with fluorescein (F-IgG), in order to examine a physiologically relevant concentration range, similar to the analysis of BSA. A linear fit of $1/\underline{s}_{20,w}$ vs c returned a value for $K_{33} = 0.0015$ mg/mL. This is a surprisingly small value for K_{33} given that IgG is typically larger and more extended in shape than BSA. The flatness of the curve probably reflects weak self-association masking nonideality in this heterogeneous mixture of antibodies (Lilyestrom et al. 2013). As described above K_{ii} is a phenomenological constant and will be used as such in subsequent analysis without an explicit model for self-association.

24.4 Cross-Term Nonideality Measurements

A mixture of two components was used to measure empirically the concentration-dependent effect that one sedimenting component exerts on another sedimenting component. A constant concentration of labeled protein was sedimented in a concentration series of unlabeled component (e.g., 0–40 mg/mL component 2). The sedimentation coefficient of the labeled component can then be plotted as a function of the concentration of unlabeled component through the following relationship, first published by Johnston and Ogston (1946):

$$s_1 = \frac{s_1^o}{(1 + K_{11}c_1 + K_{12}c_2)}$$

By sedimenting at a low concentration of species 1 (100–250 nM), the term $K_{11}c_1$ becomes negligible, and the relationship can be simplified and rearranged to:

$$\frac{1}{s_1} = \frac{1}{s_1^o} + \frac{K_{12}}{s_1^o}c_2$$

where the y-intercept is $1/s^o{}_1$ and the slope, $K_{12}/s^o{}_1$, equals the cross-term hydrodynamic nonideality that the unlabeled species exerts on the labeled component divided by $s^o{}_1$. The magnitude of the hydrodynamic nonideality each component exerts on another species is determined empirically.

The first cross-term interaction parameters measured were between cys-ELP1 and BSA (Fig. 24.4). F-cys-ELP1 at 150 nM was sedimented at a series of concentrations of unlabeled BSA (0–53 mg/mL) spanning a physiologically relevant concentration range. An increasingly negative concentration gradient is observed in the g(s*) traces (Fig. 24.4a) reflecting the J-O effect. This is caused by the buildup of cys-ELP1 behind the sedimenting BSA boundary. Qualitatively cys-ELP sediments

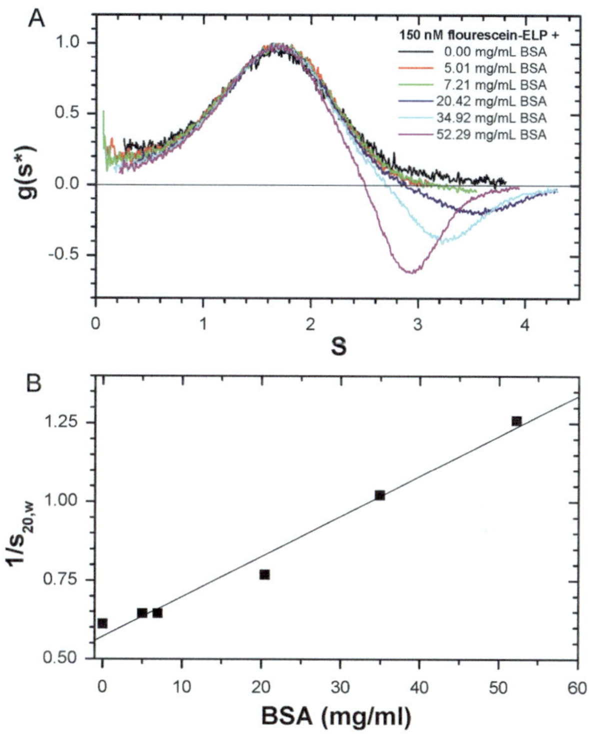

Fig. 24.4 Measuring the K_{12} that BSA exerts upon cys-ELP1. (**a**) g(s*) distributions of 150 nM F-cys-ELP1 sedimenting in 0–53 mg/mL unlabeled BSA. (**b**) Plotting the weight-average sedimentation coefficient determined from the g(s*) distributions as a function of BSA concentration. A linear fit revealed $K_{12} = 0.0223$ ml/mg. The measured s^o (1.75 s) is very close to the s^o measured for cys-ELP1 alone (1.79 s)

slower in the presence of BSA (γ phase as defined by Trautman et al. (1954); see there in Fig. 24.1) but faster in the region depleted of BSA (β phase as defined by Trautman et al. (1954)), thus causing a buildup of ELP concentration at the βγ interface. The effect that the presence of BSA exerts upon cys-ELP1 was measured by plotting the inverse weight-average sedimentation coefficient (determined from the g(s*) analysis) as a function of BSA concentration (Fig. 24.4c) and determined to be $K_{12} = 0.0223$ ml/mg. This is significantly greater than the nonideality that BSA exerts upon itself ($K_{22} = 0.0093$ ml/mg) and suggests cross-term hydrodynamic nonideality is determined by the pair of fast-slow components and not just the fast component.

The cross-term hydrodynamic nonideality between cys-ELP1 and IgG (Fig. 24.5) was measured using the same procedure that was used to examine the interactions between cys-ELP and BSA (Fig. 24.4). 100 nM F-cys-ELP was sedimented at a series of concentrations of unlabeled IgG (0–15 mg/mL) to determine how IgG would affect the sedimentation properties of cys-ELP1. Interestingly, the g(s*) distributions (Fig. 24.5a) overlay perfectly. Since IgG exhibits extremely weak

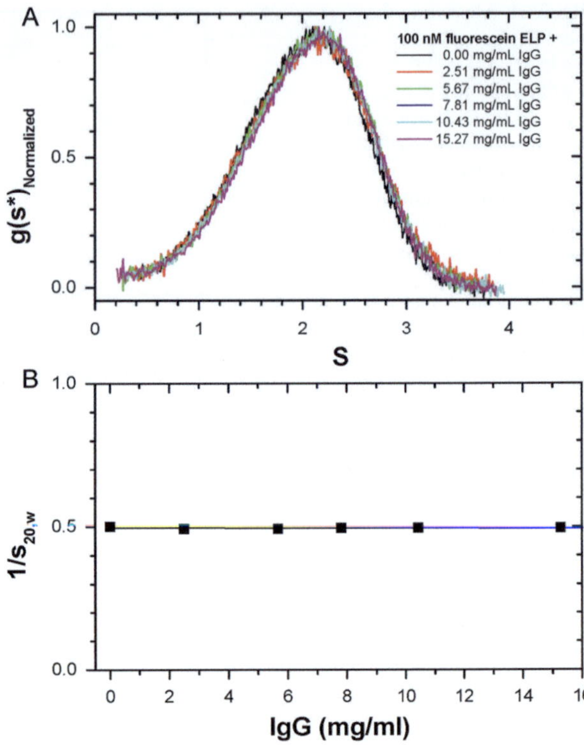

Fig. 24.5 Measuring the K_{13} that IgG exerts upon cys-ELP1. (**a**) g(s*) distributions of 100 nM F-cys-ELP1 sedimenting in 0–15 mg/mL unlabeled IgG. (**b**) Plotting the inverse weight-average sedimentation coefficient determined from the g(s*) distributions as a function of IgG concentration. A linear fit revealed $s^o = 1.88$ s and $K_{13} = -0.0064$ ml/mg

self-nonideality ($K_{33} = 0.0015$ g/mL), this suggests that weak association is also occurring between ELP and some fraction of IgG, in effect, masking the expected cross-term nonideality. The hydrodynamic nonideality was quantified by plotting the inverse weight-average sedimentation coefficient of cys-ELP as a function of unlabeled IgG concentration ($K_{13} = -0.0064$). A negative hydrodynamic concentration dependence is consistent with weak association masking nonideality. The absence of a negative concentration gradient is consistent with a small K_{ij} term.

The cross-term interactions between BSA and IgG were also examined (Figs. 24.6 and 24.7). 250 nM F-BSA was sedimented in a concentration series of unlabeled IgG (0–15 mg/mL IgG) to determine K_{23}, the effect that IgG has on BSA (Fig. 24.6a, b). The g(s*) distributions clearly exhibit the negative concentration gradient also seen in the data of cys-ELP sedimenting in BSA (Fig. 24.4), which is characteristic of the J-O effect. The hydrodynamic nonideality that IgG exerts upon BSA was determined by plotting the inverse weight-average sedimentation coefficient of BSA with increasing IgG concentration (Fig. 24.6b). The K_{23} was measured to be 0.0080 mg/mL. This is the similar to the K_{22} that BSA exerts on itself ($K_{22} = 0.0093$ mg/mL) but smaller than what might be expected for a large asymmetric molecule.

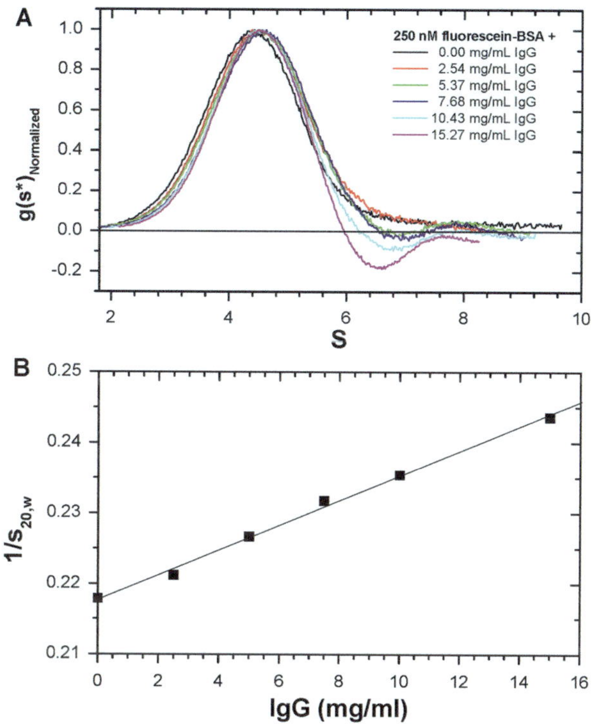

Fig. 24.6 Measuring the K_{23} that IgG exerts upon BSA. (**a**) g(s*) distributions of 250 nM F-BSA sedimenting in 0–15 mg/mL unlabeled IgG. (**b**) Plotting the inverse weight-average sedimentation coefficient determined from the g(s*) distributions as a function of IgG concentration. A linear fit revealed $s^o = 4.59$ s and $K_{23} = 0.0080$ ml/mg

250 nM F-BSA was then sedimented at a series of concentrations of unlabeled IgG (0–40 mg/mL) to determine K_{32}, the effect of BSA on IgG. As expected, BSA exerts hydrodynamic nonideality on IgG, as seen as a decrease in the weight-average sedimentation coefficient obtained from the g(s*) analysis (Fig. 24.7a). The hydrodynamic nonideality was quantified (Fig. 24.7b) and determined to be $K_{32} = 0.0085$ g/mL. Note that since IgG sediments faster than BSA, there is no boundary anomaly or J-O effect.

Experiments were also done on ELP effects on BSA ($K_{21} = 0.0234$) and IgG ($K_{31} = 0.0206$), and both data sets exhibited linear plots of $1/s$ vs c. These g(s*) patterns were strongly nonideal on average but also revealed heterogeneity in boundary shape that may reflect IgG diversity, heterogeneity in ELP excluded volume, or weak binding interactions between both pairs of components. These normalized data are summarized below with the other data sets in Fig. 24.8.

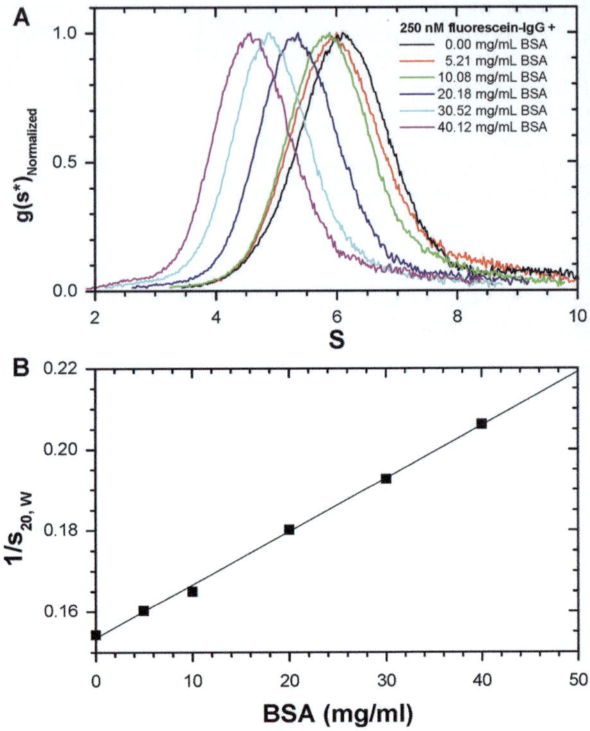

Fig. 24.7 Measuring the K_{32} that BSA exerts upon IgG. (**a**) g(s*) distributions of 250 nM IgG sedimenting in 0–40 mg/mL unlabeled BSA. (**b**) Plotting the inverse weight-average sedimentation coefficient determined from the g(s*) distributions as a function of BSA concentration. A linear fit revealed $s^o = 6.52$ s and $K_{32} = 0.0085$ g/mL

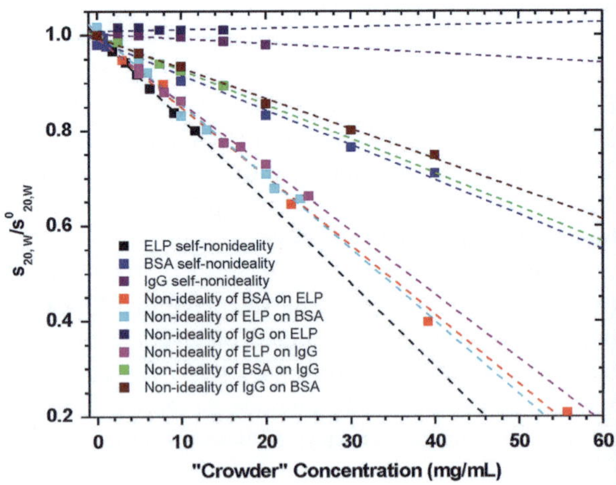

Fig. 24.8 Normalized plot of s/s^o vs c for each component and interaction. The weight-average sedimentation coefficient at each concentration was divided by the sedimentation coefficient extrapolated to zero concentration. This normalized all s vs concentration curves so that they extrapolate to 1. The slope of the line is equal to $-K_{ij}$ (summarized in Table 24.1)

24.5 Summary of the Nonideality Data and Application to SEDANAL

The full set of SV nonideality data are summarized in a normalized plot of s/s_o vs concentration of crowding agent (Fig. 24.8). The data cluster into three main regions corresponding to the self-nonideality of each component (cys-ELP1, BSA, or IgG). The hydrodynamic nonideality cross terms derived from this analysis (summarized in Table 24.1) appear to be symmetric for ELP and albumin ($K_{12} = K_{21}$) and albumin and IgG ($K_{23} = K_{32}$) and are similar in magnitude to the K_{ii} of the component with the greatest endogenous self-nonideality. The ELP and IgG cross terms (K_{13} and K_{31}) are not equal, apparently because some fraction of total IgG self-associates with ELP when IgG is in excess, thus masking the nonideality. (There are thought to be 10 million different sequences of antibodies in typical serum (Anderson and Anderson 2002) making this suggestion likely.) When ELP is in excess, no association masking nonideality is evident (indicated by a large K_{31} value = 0.0206) probably due to the small concentration of the IgG fraction that recognizes ELP. There is some excess boundary spreading in both experiments where ELP is the crowding agent that may reflect a broad distribution of ELP shapes (Lyons et al. 2013a, 2014).

To test the analytical utility of this approach, labeled Cys-ELP was run in the presence of 32 mg/ml BSA and 20 mg/ml γ-globulin and the data analyzed with SEDANAL (Fig. 24.9). All sedimentation coefficients and K_{ij} values were fixed. Only the ELP signal intensity was allowed to float by varying the apparent fluorescence extinction coefficient. The fit captures the general features of the J-O effect including a negative concentration gradient while returning random residuals. It is worth noting that with nanomolar ELP, the only significant K_{ij} terms will be for BSA on ELP (K_{21}) and γ-globulin on ELP (K_{31}). The other terms are not important in an FDS experiment where only the signal for Cys-ELP is monitored. This demonstrates that the approach works in principle as a phenomenological model that uses a 3×3 matrix of empirical K_s values on a system composed of three macromolecular components. Data collected in either plasma or serum (data not shown) exhibit a strong J-O effect and cannot be fit well with this 3×3

Table 24.1 Summary of the matrix of hydrodynamic K_{ij} values determined by sedimentation velocity experiments for a three-component system involving Cys-ELP, BSA, and γ-globulin

	Cys-ELP	Albumin	IgG
Cys-ELP	0.0224	0.0223	−0.0064
Albumin	0.0234	0.0093	0.0080
IgG	0.0206	0.0085	0.0015

Note the matrix is labeled K_{11}, K_{12}, and K_{13} across the top row. This matrix is entered into SEDANAL through a developer option that allows nonideality terms to be specified for concentration dependence in a three-species model. Any weak association reactions implied by in the data are treated as phenomenological through the $1/(1 + K_{ij}c)$ terms

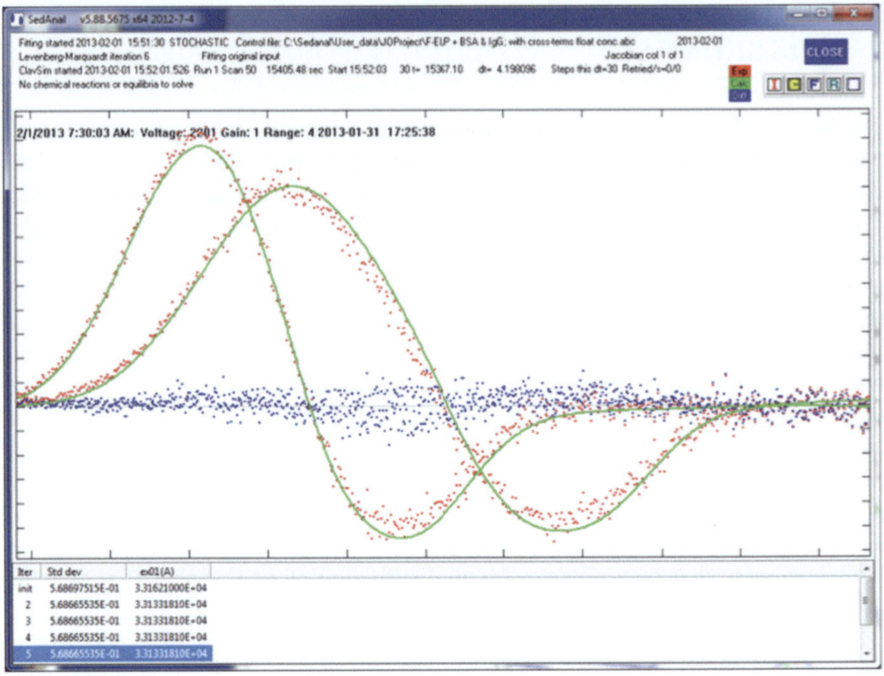

Fig. 24.9 SEDANAL fitting of labeled Cys-ELP in BSA and IgG. Fluorescence labeled Cys-ELP (100 nM) was run in 32 mg/ml BSA and 20 mg/ml IgG and analyzed with SEDANAL using a three-species model with self- and cross-terms hydrodynamic nonideality (Table 24.1)

matrix approach in SEDANAL. The peaks are sharper than expected reflecting the influence of other serum factors, possibly lipids and lipoproteins.

We have not yet mentioned thermodynamic nonideality through the second virial coefficient term, BM_1. Fitting the data for self-nonideality (Figs. 24.1, 24.2 and 24.3) globally with SEDANAL to a model including both K_s and BM_1 (where self-nonideality K_{ii} values were fixed) returned BM_1 values for ELP, BSA, and IgG of 0.0480, 0.0055, and 0.0316, respectively. Thermodynamic nonideality influences D in SV experiments and thus boundary width, according to the following equation (Stafford and Sherwood 2004; Stafford, Chap. 23, this volume, 2015).

$$D = D^o \frac{(1 + BM_1 c)}{(1 + K_s c)}$$

Since K_{ij} also influences D, the thermodynamic BM_1 and hydrodynamic K_{ij} couple and partially cancel one another in fitting; however, concentration dependence of s helps to constrain K_{ij}. This is further complicated by the heterogeneity, especially in the IgG samples. The best fit in Fig. 24.9 does not include BM_1 values, and thus BM_1 is not significant in this three-component system (Scott et al. 2014; Stafford Chap. 23, this volume, 2015). Previous work had suggested both K_{ij}

and BM1 could be extracted from velocity experiments and other orthogonal techniques (Saluja et al. 2010), although these studies did not consider heterogeneous samples like bovine γ-globulins. Recently John Philo and colleagues at APL (Philo and Maluf 2015; Berkowitz and Philo 2014) have described a short-column sedimentation equilibrium method for extracting second virial coefficients B_{22} at high therapeutic protein concentrations of pure macromolecules, using an approach first described by Yphantis (1960) and Correia and Yphantis (1992). Future sedimentation equilibrium experiments for this serum project will attempt to extract BM_1 values independent of hydrodynamic effects for appropriate non-heterogeneous systems. The heterogeneity of IgG samples will be avoided by working with high concentrations of pure monoclonal antibodies. Thermodynamic nonideality cross terms may be more challenging to measure, but sedimentation equilibrium with the AU-FDS will be explored. The nonideality of serum lipid can in part be handled by removing lipid by precipitation or phase separation and studying what remains behind. The goal is to develop software tools and experimental protocols that allow therapeutic proteins to be studied biophysically in serum and formulation mixtures. This chapter represents an initial effort toward that goal.

References

Anderson NL, Anderson NG (2002) The human plasma proteome. Mol Cell Proteomics 1(11):845–867

Berkowitz SA, Philo JS (2014) Characterizing biopharmaceuticals using analytical ultracentrifugation. In: Houde DJ, Berkowitz SA (eds) Biophysical characterization of proteins in developing pharmaceuticals. Elsevier, Amsterdam, pp 211–260

Correia JJ, Yphantis DA (1992) Equilibrium sedimentation in short solution columns. In: Harding SE, Rowe AJ, Horton JC (eds) Analytical ultracentrifugation in biochemistry and polymer sciences. Royal Society of Chemistry, Cambridge, pp 231–252

Correia JJ, Johnson ML, Weiss GH, Yphantis DA (1976) Numerical study of the Johnston-Ogston effect in two-component systems. Biophys Chem 5:255–264

Demeule B, Steven J, Shire SJ, Liu J (2009) A therapeutic antibody and its antigen form different complexes in serum than in Phosphate Buffered Saline (PBS): a study by analytical ultracentrifugation. Anal Biochem 388(2):279–287

Harrington WF, Schachman HK (1953) Analysis of a concentration anomaly in the ultracentrifugation of mixtures. J Am Chem Soc 75:3533–3539

Johnston JP, Ogston AG (1946) A boundary anomaly found in the ultracentrifugal sedimentation of mixtures. Trans Faraday Soc 42:789–799

Kingsbury JS, Laue TM (2011) Fluorescence-detected sedimentation in dilute and highly concentrated solutions. Methods Enzymol 492:283–304

Kroe RR, Laue TM (2009) NUTS and BOLTS: applications of fluorescence-detected sedimentation. Anal Biochem 390:1–13

Lilyestrom WG, Yadev S, Shire SJ, Scherer TM (2013) Monoclonal antibody self-association, cluster formation and rheology at high concentrations. J Phys Chem B 117:6373–6384

Liu S, Stafford WF (1995) An optical thermometer for direct measurement of cell temperature in the Beckman instruments XL-A analytical ultracentrifuge. Anal Biochem 224:199–202

Lyons DF, Le V, Bidwell GL III, Kramer WH, Lewis EA, Raucher D, Correia JJ (2013a) Structural and hydrodynamic analysis of a novel drug delivery vector: ELP[V5G3A2-150]. Biophys J 104:2009–2021

Lyons DF, Lary JW, Husain B, Correia JJ, Cole JL (2013b) Are fluorescence – detected sedimentation velocity data reliable? Anal Biochem 437(2):133–137

Lyons DF, Le V, Kramer WH, Lewis EA, Raucher D, Correia JJ (2014) The effect of basic cell penetrating peptides on the structural, thermodynamic and hydrodynamic properties of a novel drug delivery vector: ELP[$V_5G_3A_2$-150]. Biochemistry 53(6):1081–1091

MacGregor IK, Anderson AL, Laue TM (2004) Fluorescence detection for the XLI analytical ultracentrifuge. Biophys Chem 108:165–185

McFarlane AS (1935) The ultracentrifugal analysis of normal and pathological serum fractions. Biochem J 29:1209–1226

Muramatsu N, Minton AP (1989) Hidden self-association of proteins. J Mol Recognit 4:166–171

Ogston AG (1937) Some observations on mixtures of serum albumin and globulin. Biochem J 31:1952–1957

Philo JS (2006) Improved methods for fitting sedimentation coefficient distributions derived by time-derivative techniques. Anal Biochem 354:238–246

Philo J, Maluf NK (2015) New approaches to investigating the self-association and colloidal stability of protein pharmaceuticals at high concentration, HOS 2015. http://www.ap-lab.com/sedimentation_equilibrium.htm#example

Saluja A, Fesinmeyer M, Hogan S, Brems DN, Gokarn YR (2010) Diffusion and sedimentation interaction parameters for measuring the second virial coefficient and their utility as predictors of protein aggregation. Biophys J 99:2657–2665

Scott DJ, Harding SE, Winzor DJ (2014) Concentration dependence of translational diffusion coefficients for globular proteins. Analyst 139:6242–6248

Soda A, Fujimoto T, Nagasawa M (1967) The Johnston-Ogston effect in sedimentation. J Phys Chem 71:4274–4281

Stafford WF (1992) Boundary analysis in sedimentation transport experiments: a procedure for obtaining sedimentation coefficient distributions using the time derivative of the concentration profile. Anal Biochem 203:295–301

Stafford WF, Sherwood PJ (2004) Analysis of heterologous interacting systems by sedimentation velocity: curve fitting algorithms for estimation of sedimentation coefficients, equilibrium, and kinetic constants. Biophys Chem 108:231–243

Trautman R, Schumaker VN, Harrington WF, Schachman HK (1954) The determination of concentrations in the ultracentrifugation of two-component systems. J Chem Phys 22:555–559

Williams JW, vanHolde KE, Baldwin RL, Fujita H (1958) The theory of sedimentation analysis. Chem Rev 58:715–806

Yphantis DA (1960) Rapid determination of molecular weights of peptides and proteins. Ann N Y Acad Sci 88:586–601

Zhao H, Ma J, Ingaramo M, Andrade E, MacDonald J, Ramsey G, Piszczek G, Patterson G, Schusk P (2014) Accounting for photophysical processes and specific signal intensity changes in fluorescence-detected sedimentation velocity. Anal Chem 86:9286–9292

Chapter 25
Acquisition and Analysis of Data from High Concentration Solutions

Tabot M.D. Besong and Arthur J. Rowe

Abstract The problems associated with ultracentrifugal analysis of macromolecular solutions at high (>10 mg/ml) arc rcvicwcd. Espccially for thc casc of solutes which are non-monodisperse, meaningful results are not readily achievable using sedimentation velocity approaches. It is shown however by both simulation and analysis of practical data that using a modified form of an algorithm (INVEQ) published in other contexts, sedimentation equilibrium (SE) profiles can be analysed successfully, enabling topics such as oligomer presence or formation to be defined.

To achieve this, it is necessary to employ an approach in which the solution density, which in an SE profile is radius-dependent, is taken into consideration. Simulation suggests that any reasonable level of solute concentration can be analysed.

Keywords Analytical ultracentrifuge • Concentrated solutions • Sedimentation velocity • Oligomers

25.1 Introduction

There is considerable current interest in high concentration solutions of biologicals. Such solutions in general, and of proteins in particular, can in many cases exhibit high levels of solubility in aqueous solvents. In particular, these solvents will often be of an ionic strength level and pH value closely similar to that found in native biological tissue – as in the ubiquitous 'Dulbecco's Physiological Saline'. Clearly, in these tissues, the environment is normally a 'high concentration' or 'crowded' one (Chapter in this volume by Minton). Analysis of macromolecular solutes, such

T.M.D. Besong
Functional Nanomaterials Laboratory, Physical Sciences and Engineering, King Abdullah University of Science and Technology, Thuwal 23955-6900, Saudi Arabia

A.J. Rowe (✉)
NCMH School of Biosciences, University of Nottingham, Sutton Bonington LE12 5RD, UK
e-mail: arthur.rowe@nottingham.ac.uk

S. Uchiyama et al. (eds.), *Analytical Ultracentrifugation*,
DOI 10.1007/978-4-431-55985-6_25

as enzymes, transport or other proteins, would thus optimally be conducted at the level of concentration found within a living cell.

Classically biochemists have conducted their experiments under high dilution conditions – often at solute levels less than 1 mg/ml, and frequently (as in gel column based analyses) at concentrations lower still. 'Extrapolation to infinite dilution' still tends to be regarded as normative for securing solute parameters free of 'non-ideality' effects. There is nothing incorrect about this, as such. If the aim is to secure an unequivocal value for the molecular weight of (let us say) a multi-enzyme complex, then working at the highest dilution compatible with good signal/noise ratio must be optimal. Yet the acquisition of knowledge of the behaviour of genuinely concentrated/crowded solutions is of real and growing importance.

25.1.1 Reasons for Interest

In bio/pharma practice, there is an increasing use of drug formulation for storage or delivery at concentrations of 20–200 mg/ml, as described at a recent Bioprocessing Summit (http://www.bioprocessingsummit.com/Protein-Formulations/). There are very few methods capable of yielding data on the dispersive state of macromolecular systems at concentrations of up to 100+ mg/ml. Static light scattering has been shown to be applicable in this range, and a 'hard-sphere' approximation found to be valid for elucidation of the concentration dependence terms (Fernández and Minton 2009). This method, although highly informative with respect to the behaviour of single species, is not readily applicable to the analysis of solutes showing degrees of presence of oligomeric species, as is often the case in bio/pharma systems. More exotic approaches can certainly be of value, as evidenced by a recent study of antibody self-association via the analysis of the interparticle distance-dependent plasmon wavelength (Sule et al. 2011): but our aim has been to develop a methodology capable of being implemented on widely available hardware.

25.1.2 Why an AUC Approach Could be of Value

For solutions of proteins at low (<2 mg/ml) concentration, the AUC used in the sedimentation velocity mode is currently accepted as a method of choice for the characterization of the oligomeric state of biomacromolecules: orthogonal to gel-column based methods which are operative at lower concentrations, but which present to the solute a surface with which it may interact. Good practice is probably to use both approaches. At higher concentrations, the use of gel-columns is not technically possible, and the sedimentation velocity analysis presents major difficulties. We discuss the extent to which these difficulties can be over come, and then describe developing, novel methods by means of which the alternative mode of AUC analysis (Sedimentation Equilbrium Analysis – SEA) can be employed with success.

25.2 Sedimentation Velocity Analysis (SVA)

25.2.1 What is SVA and What Can It Tell Us?

A historic strength of SVA has from the start been that the sedimentation at high centrifugal field of a macromolecular solution yields via established software, a diagram (a 'sedimentation velocity profile') graphing out the sedimentation rates of the components ('s-values') present and their relative concentrations. The default assumption is that each 'peak' seen in (e.g) a c(s) vs s profile is an individual solute component.

However, the possibility that even the presence of single peak may not unequivocally denote the presence of a single species needs to be explored: as a chemical equilibration could be involved. A well-known historical presence as an early procedure for defining levels of self-association in a monomer-dimer system is the work of Gilbert and colleagues (Gilbert and Gilbert 1980). These authors showed that for a monomer-dimer system in relatively rapid equilibration, only a single, slightly asymmetrical sedimenting boundary is seen, reflecting the weight-averaged sedimentation rate of the monomer and dimer species. This s value increases with total solute concentration, and this is often taken as diagnostic of a self-associating system: since for a single, monomeric species the s value will always *decrease* in value with increasing concentration (Rowe 1977). Analysis of monomer-dimer equilibrium 'isotherms' via a set of $s_{w,app}$ value over a range of concentrations will always require a knowledge of the c-dependence of the s values of both monomer and dimer, in addition to their extrapolated (to zero concentration) s values. Software solutions have been published facilitating this, involving in every case the assumption that a single, linear coefficient of regression (k_s) suffices (Stafford and Sherwood 2004; Dam and Schuck 2005). For dilute solutions – less than a few mg/ml for globular proteins – the assumption is very adequate.

However, if we wish to explore the presence and interaction of species at high concentration, the s-c dependence is known to be very definitely non-linear (Rowe 1992) and a more sophisticated approach is needed if we are to be able to study interaction at high solute concentration. In this area, only a very limited range of work has been reported. The c-dependence of sedimentation has been considered by both fluid dynamicists and biophysical chemists: although in neither case has either group paid any full attention to work in the 'other area'. We present a very brief summary of the most important studies carried out and show that a coherent, common analysis is possible.

25.2.2 *Fluid Dynamics of Sedimentation at High Solute Concentration*

The work of Burgers and colleagues in Amsterdam was and is the prime source of theory in this field (Burgers 1941, 1942). Burgers demonstrated the need to distinguish between the c-dependence defined on a 'dynamic' basis from that defined on a 'kinematic' basis, the terms having the same meaning as familiar in viscometry (i.e. the former 'correcting' for solution density, the latter not so doing). The extrapolated values, at infinite dilution, do not differ – unlike an intrinsic viscosity value. It was shown in the first part of the treatment that the limiting slope of the regression of s upon c for spherical particles had a value of 4 (v/v) for the dynamic case and 5 (v/v) for the kinematic value. These values related to the effect of 'return flow' of the fluid forward convected by the particles in linear 'Stokes flow'. Importantly, however, the presence of an additional effect was postulated, arising from an essentially colligative effect of the bulk viscosity of the suspension/solution of particles. This raised the predicted value to 5.875/6.875 for the two cases respectively. Many experimental studies have been performed over later decades, using particles such as pollen grains, glass spheres or latex particles. The estimates yielded for k_s therefore have generally been much closer to the former prediction than to the latter. A later analysis by Batchelor (1972) has however been widely quoted, predicting values for the limiting k_{s} just slightly lower than those of Burgers.

A later and more detailed analysis of the problem has however been given by Brady and Durlovsky (1988). Unlike the earlier analyses of Burgers and Batchelor, the Brady-Durlovsky treatment covers the entire range of possible solute concentrations, up to limiting volume packing fraction (ϕ_{p}). Batchelor's treatment of the problem is critically analysed and found to need correction. Brady and Durlovsky show that whilst a simple Rotne-Prager approximation can yield an accurate description of the sedimentation velocity of neutral spheres all the way from dilute suspension to maximal packing fraction, the earlier pairwise additive approximation used by Batchelor cannot: and indeed leads to results which are aphysical above a ~23 % volume fraction. Both analytically and by Stokesian dynamics it is shown that while a suspension of neutrally buoyant particles may be modeled as a homogenous fluid with an effective viscosity, a sedimenting suspension cannot be modeled in this way. Thus the Rotne-Prager approximation, used in this study, captures with accuracy the features of many-body interaction in sedimentation of spherical particles. Finally, Brady and Durlovksy present, using the Percus-Yevick hard-sphere approximation, an analytical expression which they show to agree well with the whole range of experimental evidence. The limiting regression coefficient is found to be 5.0 v/v (assumed to be kinematic).

This treatment of Brady and Durlovsky appears to be definitive. It has been unchallenged since its publication, and for the system to which it relates (suspension of spheres) it is as accurate a definition as is likely to be found.

25.2.3 *Biophysical Studies of Sedimentation Over the Range of Solute Concentration*

The most important contribution under this heading was a definition by Cheng and Schachman (1955) of both the validity of the description of hydrodynamic parameters (sedimentation rate, intrinsic viscosity) in terms of Stokes/Einstein theory for small spherical particles and of a numerical value for the coefficient of regression of sedimentation rate (k_s) for polystyrene spheres. The latter was found to be 4.06 v/v (dynamic) or 5.06 v/v (kinematic). This astonishingly 'modern' paper, containing experimental work of a precision unequalled with any other solute system, has never, so far as can be discovered, been referenced in the world of fluid dynamics. Although a variety of semi-empirical relationships have been proposed, no general treatment was published: until in a study by Rowe (1977) a simple, general equation was derived, applicable not only to spheres but to all conformations of particles for which a charge effect was either absent or susceptible to being suppressed by neutral electrolyte

$$k_s = 2\overline{v}\left\{\frac{V_s}{\overline{v}} + \left(\frac{f}{f_0}\right)^2\right\} \tag{25.1}$$

where $\overline{v}$ is the partial specific volume, V_s the 'swollen specific volume' and f/f_0 the frictional ratio of the solute. This relationship was shown to give consistent results for a wide range of (mostly protein) solutes. The basis for the derivation of this equation is in the insight that (extended) Stokes' Law the frictional ratio is a term which must define a 'swept volume' of solvent, which in a closed cell must return. Unsurprisingly, the predicted limiting regression coefficient for sedimentation of spheres is 5 v/v (kinematic) or 4 v/v (dynamic) in exact agreement with Burger's values (for 'backflow'), Cheng and Schachman's empirical values for PSL spheres (4.06), and with the values from the Brady and Durlovsky treatment, as given above. We note than in virtually all practical work it is the *kinematic* sedimentation coefficient which is measured.

An extension of the Rowe approach (Rowe 1979), using an approach based upon the mathemati,cs of finite probability space, has given an expression which within plausible experimental error is numerically the same over the whole range of possible solute faction to that given by Brady & Durlovsky (Fig. 25.1):

$$s^c = s^0\left\{1 - \left(2c\left(V_s + \overline{v}F^2\right) - \left(2\phi_p - 1\right)/\phi_p{}^2\right)/\left(2c\overline{v}F^3 + 1\right)\right\} \tag{25.2}$$

where s^c and s^0 are the sedimentation rates at volume fraction c and 0 respectively, F is the solute frictional ratio, and ϕ_p is the limiting packing fraction (=0.64 for perfect spheres, range of 0.4–0.5 for proteins). This equation describes well the behaviour of a range of proteins (Rowe 1992).

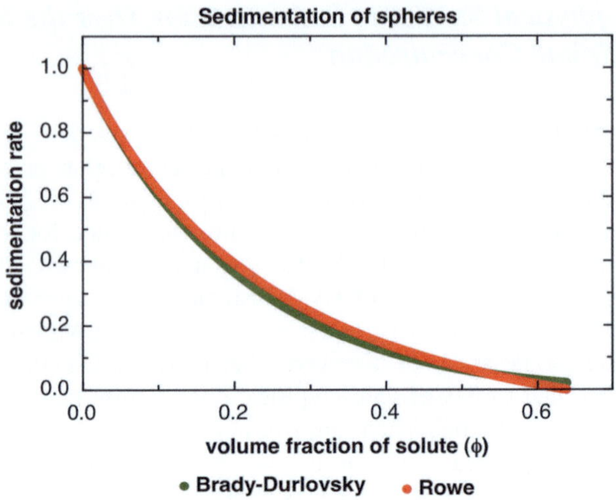

Fig. 25.1 Plots of the sedimentation coefficient relative to the coefficient s^0 at infinite dilution for spherical particles against volume fraction (ϕ) occupied by the particles: computed by the equation of Brady and Durlovsky (1988) and by the extended equation for s-c dependence of Rowe (1979). The values computed are kinematic values, as normally estimated in experimental work

25.2.4 Fluid Dynamics and Biophysical Analysis – A Common Approach?

There is seen to be no real difference in outcomes between the two approaches. Fluid dynamics provides a broad-ranging insight into mechanisms, whilst biophysical analysis of systems such as proteins yields results much better defined than pollen grains (for example) can offer.

25.2.5 Practical Aspects of SV Analysis of High Concentration Solutions

Modification of an Isotherm approach for determination of K_a values for interacting (monomer-dimer) systems to incorporate Eq. (25.2) is simple: albeit hardly necessary for globular proteins at modest concentration, for which a simple linear k_s term suffices (Stafford and Sherwood 2004; Dam and Schuck 2005). However when systems such as carbohydrates (of narrow fraction nature) are concerned, then using a fitting program incorporating Eq. (25.1) may be advantageous. An example is the definition of a weak reversible interaction in a carbohydrate system (Patel et al. 2007).

In more general terms, what difficulties arise in attempting to use SV analysis on high concentration systems? Immediately it is obvious that expecting to simply log (say) 100 radial scans and then run them through SEDFIT or other analytical software is bound to fail. High absorption levels will rule out the use of absorption optics, even with short optical path length cells, so interference optics will be mandatory. But with any optical system the very high gradients of refraction associated with a sedimenting boundary will usually take the signal out of aperture of the optics, and a 'black band' in the region of a boundary will be all that is visible. On top of that, the shape of the boundary will be massively distorted by self-sharpening, and the Johnston-Ogston effect (Soda et al. 1967) causing distortion of the apparent ratios of boundary concentration values will be seriously accentuated by the 'moving together' of adjacent boundaries, cause by c-dependence. Finally, for interference optics, unless the optics are focused on the *exact* two-thirds plane of focus, the linear relationship between fringe displacement and concentration increment breaks down. Re-focussing the XL-I optics is non-trivial, and if short optical path length cells are used then a satisfactory outcome is almost impossible to achieve.

Most of the above problems can be alleviated, to some degree at least, by lowering the rotor speed. For analysis it is more or less necessary to use older 'boundary movement' methods, as any sort of Lamm equation fitting must fail. All these problems being recognized, work has been done and results obtained, with patience, and acceptance of long duration experiments. Multi-component systems are very hard to interpret, but single component systems can to some extent yield information of interest. For example, in a study (Errington et al. 2002) in which s vs c data sets for apoferritin and ovalbumin were analysed via Eq. (25.2), an increase in the V_S value found from the whole data set over the 'infinite dilution' value was seen, most clearly for ovalbumin. Interpretation of this result is complex: pressure effects were also found to be present, but even so these findings could not have been obtained by other methods.

However, practical interest in the behaviour of proteins at high concentration does for the most part focus on topics such as stability, interactions and oligomer formation or dissolution. The SV approach is clearly limited in the contribution which it makes in this area. Our most recent focus has therefore been on the development of new methods using sedimentation equilibrium.

25.3 Sedimentation Equilibrium Analysis (SE)

As described above, lowering the rotor speed circumvents many of the difficulties associated with carrying out AUC analyses of solutions of high solute concentration. Steep gradients of refraction can be avoided, and issues such as boundary location or Johnston-Ogston effects are irrelevant. Pressure effects are essentially non-existent. With high concentrations, an excellent signal level (refraction increment over length of solution column used) can be obtained at low rotor speeds – expressed in terms of the reduced floatational molecular weight $\boldsymbol{\sigma}$, defined by

$$\sigma = M\,(1 - \overline{v}\rho)\,\omega^2/RT \tag{25.3}$$

where M is the solute molecular weight, $\overline{v}$ its partial specific volume, R the gas constant and T the temperature (K) means that an experiment at $\boldsymbol{\sigma} = <0.5$ is perfectly feasible. All this is obvious: so why is it that – so far as we know – there is little or no reported work of this type?

The answer to this latter question is obvious. No analytical software is available which is directed at answering the question raised above. SE analysis traditionally is used to determine 'average molecular weights', or (even worse) 'apparent molecular weights': not to seek to disentangle the relative contributions made (for example) by oligomeric species to the final recorded signal under various high concentration conditions. Our wish would be to know the *true* molecular weights of identified species, not some sort of 'apparent' value.

We have recently made significant progress in this area. The algorithms and functions/programs which we employ are in essence already in the public domain and have been so for several years: we have simply – perhaps belatedly – realized that extensions can be devised which very much address the above requirements. The power of SE methods has recently been extended to give distributions of molecular weights, c(M) vs M profiles, via computation of c($\boldsymbol{\sigma}$) vs $\boldsymbol{\sigma}$ profiles from a routine called MultiSig (Gillis et al. 2013). But securing these profiles calls for low solute concentrations to be employed, to minimize c-dependence terms. Our ultimate aim is to produce from high concentration data c(M) vs M profiles which are continuous distributions of ideal molecular weights. There is still some way to go, but already we can define ways in which profiles of oligomeric species can be computed to give profiles resembling those obtained in c(s)/c(M) analysis via SV – with similar precision, but with 'ideal' molecular weights assigned to the individual species present. As always with new algorithms we test our new routines against simulated data. Initially, we perform checks using 'perfect' (i.e. machine precision data), but this is merely to test that no algebraic or programming errors are present. Then prior to testing with 'real' experimental data, we test with 'perfect' data to which simulated normal random error has been added. Universally we use ±0.005 fringe as the standard deviation of the error (Ang and Rowe 2010).

25.3.1 Method of Simulation for Analysis of SE Data

The particular form of the equation for non-ideal sedimentation equilibrium analysis which we employ (the INVEQ form – below) has many unique virtues, but care needs to be taken in its implementation. We use existing routines in the analysis/plotting package pro Fit™ (Uetikon am See, Switzerland) on a MacBook Pro i7 Quadcore platform. The routines employed are widely available for multiple hardware platforms, but are conveniently clustered together in pro Fit™ for sequential use and subsequent plotting. There are four stages in the fit process:

1. A manual fit is performed, using the user-interactive parameter 'drag' facility, until the fit achieved appears visually close to the plot of the data set. For this manual fit to be successful, initial parameter values must be 'sensible', and a close watch must be kept to ensure that algebraically possible but practically absurd parameter values are not selected.
2. A 'search' – type fit follows. We use a Robust fit, but a Simplex fit would be an equally good alternative e.g. Schuck et al. (2014).
3. After 'passing' the parameters from the Robust fit, a Levenberg-Marquardt fit is performed.
4. After a successful outcome, a Levenberg-Marquardt fit incorporating a full Error Analysis (500 fits, confidence interval 0.67) is performed. A tabulation of the 500 estimates of each floated parameter is returned. These are plotted against parameter value in each case, and if the distribution seen is plausibly close to normal, then a Gaussian fit is performed, with the SD deviation returned (by Levenberg-Marquardt) being taken as a standard error estimate for the parameter in question. An example is given below (Fig. 25.2)

There is a very particular reason for carrying out this last 'bootstrapping' error analysis. With the system we are working with, we find that a significant proportion of our fits can return up to 5 % or more parameter values which are serious 'outliers': numerical values differing by up to 2–3 orders of magnitude from the main corpus of values returned. In other words, if reliance is placed on the outcome of a single fit, then there is a ~5 % chance of this outcome being in serious error. It is of course simple to select just the central, main distribution values for the final, Step 4 gaussian fit.

25.3.1.1 M_INVEQ – A Revised Formulation for SE Analysis of High Concentration Data

In addressing the problem, we start an approach – the INVEQ algorithm – which we have already demonstrated (Rowe 2011) as enabling ultra-weak K_a values to be retrieved from data gathered at a suitably elevated concentration. The normal equation for non-ideal sedimentation equilibrium which represents $c = f(r)$ is simply inverted to give the form shown below, which represents $r = f(c)$:

$$r = \Big(\big(\ln\left((c_r + E) / (c_{\mathrm{ref}} + E)\right) + 0.5^* \left(\sigma_{w,r} / \left(1 + 2\mathrm{BM}\left(c_r + E\right)\right)\right)^* r_{ref}^{\ 2}\big) / \left(0.5\left(\sigma_{w,r} / \left(1 + 2\mathrm{BM}\left(c_r + E\right)\right)\right)\right) \Big)^{0.5} \tag{25.4}$$

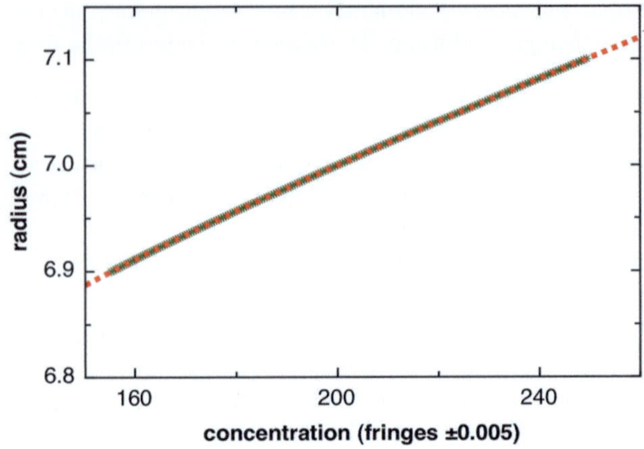


```
Iterations:  462                                   Robust Fit
-------------------------------------------
Chi squared        =       0.5276
Goodness of fit           =       1.0000

Parameters:
c_referenc =      200.0003
sigma        =        0.5029
r_referenc  =        7.0000
c_dimer      =        1.0000e-10
c_trimer     =        1.0000e-10
BM             =        1.2097e-3
E                =       -0.9166
```

```
----------------------------------------------------
Chi squared        =       0.5261
Goodness of fit           =       1.0000        Levenberg-Marquardt Fit

Parameters:                                  Standard deviations:
c_referenc =      200.0003           Δc_referenc       =       1.0074e-2
sigma        =        0.5029          Δsigma              =       4.0531e-2
r_referenc  =        7.0000
c_dimer      =        1.0000e-10
c_trimer     =        1.0000e-10
BM             =        1.2097e-3         ΔBM                  =       1.4066e-4
E                =       -0.9166          ΔE                    =      12.4809

-------------------------------------------
Confidence intervals (       68.300% ):
c_referenc 199.9895          ...        200.0101
sigma          0.4611            ...          0.5411
BM               1.0699e-3        ...          1.3400e-3
E               -11.7181           ...         12.8564
```

Fig. 25.2 Plot of simulated data points for a system at sedimentation equilibrium distribution of a solution at 200 fringe cell loading concentration; $\sigma = 0.5000$; BM = 1.2000 (fringe units) and baseline offset $E = 0$ fringes. The line fitted by the M_INVEQ routine is plotted.
The process of fitting via M_INVEQ is shown for the two fitting stages employed, Robust Fit followed by Levenberg-Marquardt fit

where r is the radial value at which the observed concentration c_r (after correction for baseline offset E) is located: $\sigma_{w,r}$ is the weight-averaged σ value at that radial position; c_{ref} is the observed concentration at a defined radial reference position r_{ref} and 2BM is the second virial term. It has been shown by Ang and Rowe (2010) that the incorporation of a third virial term (3 CM) is meaningful, and realistic values for the same can be obtained: but such values are found in practice, for proteins in neutral solution, to be very small in magnitude, and whilst noting the possibilities, we do not at the moment pursue them further.

The great advantage of the use of the above equation giving $r = f(c)$ over the use of the original formulations (of $c = f(r)$) is that the latter are *transcendental* in nature, and hence not readily solved or fitted: whereas the inverted form is a simple relationship, amenable to normal fitting procedures. Both by simulation and by application to a range of practical problems, this INVEQ algorithm has been well validated (Rowe 2011). There are no approximations involved in the (trivial) derivation of Eq. (25.4).

It seemed sensible, in our present enquiry, to seek to find a way in which an equation already validated for estimating Ka values in interacting systems at relatively high solute concentration might be employed, in modified form, to study systems where σ was *not* necessarily known, and/or where there could be a number of components greater than two (monomer, dimer) present. To this end we have carried out a broad range of simulations and present some of the most pertinent.

Initially, we simply checked that it was possible to estimate values for the ideal reduced molecular weight σ^0 for a monodisperse, non-ideal system. Fitting simulated data with ± 0.005 fringe noise; $\sigma = 0.5000$; 200 fringe concentration at reference radial position (7.000 cm); BM $= 0.0012$ (fringe concentration; $E = 0$) we find that these values can be retrieved via the use of M_INVEQ to a very good level of precision and accuracy (Figs. 25.2 and 25.3). The $\pm$ % errors are estimated to be <0.02 % in reference concentration; 0.2 % in σ; 0.27 % in BM and 0.25 % in (c_ref + E). These are clearly very acceptable levels of uncertainty.

We now address the issue of the analysis of more than one solute component. Given the widespread interest in the formation of small oligomers in solutions of 'purified' proteins, we have explored the level to which M_INVEQ analysis can assess the level of presence of dimer and trimer species. To achieve this, M_INVEQ incorporates an amended $\sigma_{\mathrm{w,r}}$ term in Eq. (25.4). Our original INVEQ routine, used for estimation of ultra-weak K_{a} values, had the $\sigma_{\mathrm{w,r}}$ term defined in the program by the current value for the monomer and dimer concentration yielded by the current value for the total concentration. That this is possible arises from the fact that the sedimentation potential at any given radial position is the simple algebraic sum of all contributing species: there being *no* contribution of 'non-ideality' effects to the individual sedimentation potential values. Hence if a fraction of the total solute

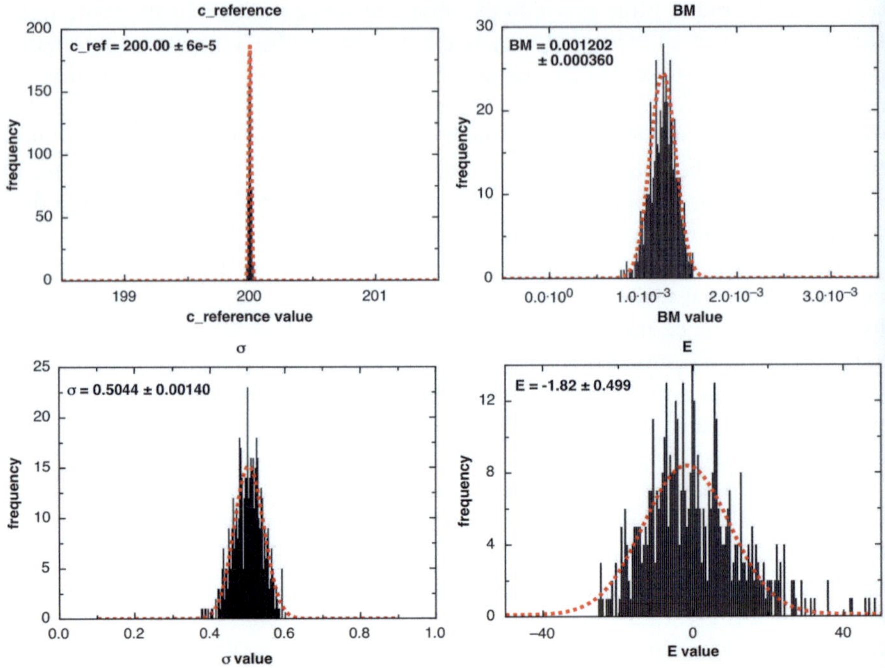

Fig. 25.3 Plots of the distribution of the 500 estimates of the parameters computed via the error analysis employed in the final Levenberg-Marquardt fit (Fig. 25.2). The standard errors displayed in each panel are derived from fitting a single Gaussian function to the values in the distribution plotted

*c*_dimer is in dimer form, and *c*_trimer the fraction in trimer form, then since all fractions must sum to 1 we can write

$$\sigma_{w,r} = (1 - c\_\text{dimer}\_c\_\text{trimer}) + c\_\text{dimer} + c\_\text{trimer}\Big) \tag{25.5}$$

where *c*_dimer and *c*_trimer can be floated parameters.

Our test data for simulation of fitting of such a system employed a cell load of 200 fringes, allowed for 13 % dimer and 7 % trimer with a (single) second virial term (BM = 0.0018 fringe units). As can be seen (Fig. 25.4) precise estimates are returned for all parameters. Notably the estimate for the weight of the monomer species is unaffected by the presence of 13 % dimer and 7 % trimer species. It is therefore considered that the mode of analysis offers a feasible approach to the analysis of the oligomeric profile, at a monomer-dimer-trimer level at high concentration. This is the primary aim of our work, but it is also necessary to demonstrate that results can be yielded at more usual protein concentrations (a few mg/ml at most) which are consistent with the results obtained by conventional sedimentation velocity analysis.

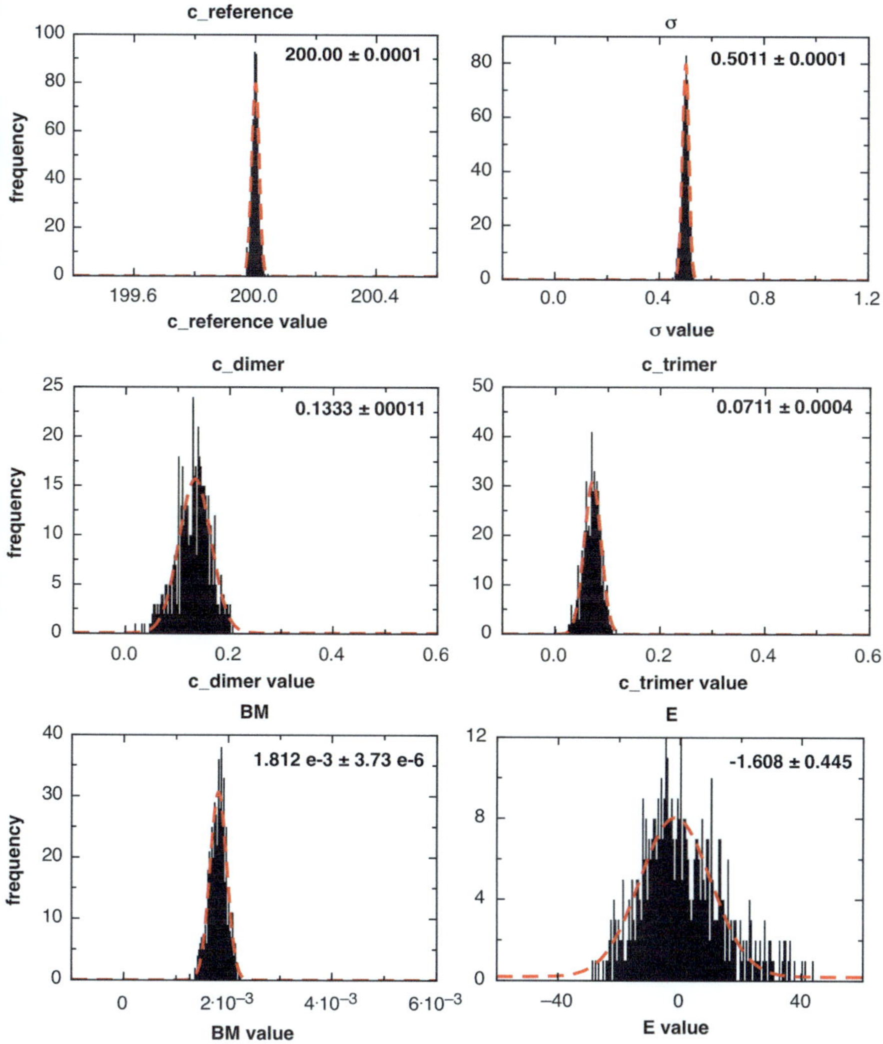

Fig. 25.4 Estimates for the parameters computed for a simulated data set with parameters identical to those used for fitting to a single species data set (Fig. 25.2), except for the composition of the solute comprising 80 % monomer, 13 % dimer and 7 % trimer. The fit (not shown) was of a quality equivalent to that shown in Fig. 25.2. In each individual panel, the value of the estimate for the named parameter and the corresponding estimate for the precision of the estimate

25.3.2 *Application of M_INVEQ Analysis to Real Systems*

An immediate problem arose with the application of this method. It has long been considered and has very recently been demonstrated from a sound theoretical standpoint (Spruijt and Beisheuvel 2014) that the density term (ρ) in the definition of the effective floatation weight (σ) of a solute (Eq. (25.3)) must be the density of the *solution*, not of the *solvent*. Although this matter is trivial at (for example) 2 mg/ml solute, it has a serious effect in the region of 10+ mg/ml. Furthermore, at sedimentation equilibrium, as the solute concentration is a function of radial position, so the solution density term will vary, and hence σ becomes a function of radial distance. It is thus necessary to incorporate a simple additional term into our definition (Eq. (25.4)) of the radial distance at which a given solute concentration is located, and this has been effected.

25.3.2.1 Application of M_INVEQ to BSA Solutions Containing Dimer and Trimer

A solution of Bovine Serum Albumin (BSA) at ~30 mg/ml in Dulbecco's PBS has been employed. Sedimentation velocity analysis of this solution, diluted to ~2 mg/ml, revealed (Fig. 25.5) the presence of 13 % dimer and 7 % trimer (with a very small < <1 % level of tetramer, which we have ignored). M_INVEQ analysis of this diluted sample gave results consistent with an SV analysis (Fig. 25.5). At 30 mg/ml, M_INVEQ analysis incorporating a variable solution density term yields a similarly good fit (Figs. 25.6 and 25.7) but with a slightly higher level of trimer (9.9 %) found at this elevated solute concentration. This is discussed below (Sect. 25.4). The value returned for σ yields an estimate of 65,533 Da for the monomer weight of BSA, which is acceptable.

We are in the course of exploring the application of this M_INVEQ routine to a range of systems. Our approach is to explore initially how M_INVEQ tackles the distribution of oligomers at solute concentrations < = 10 mg/ml, where comparison can be made with existing knowledge, based upon SV or existing SE methods. The expectation has to be that there will be consistency found. For example, we have further validated a recent approach which we have made to characterizing the lightest component in a polydisperse system (Gillis et al. 2013; Nikolajski et al. 2013), where a rotor speed was selected which would leave a system for analysis which was predominantly just the lightest component. We now have, in M_INVEQ, a simple routine, which can check quantitatively for the presence of higher oligomers. Figure 25.8 illustrates the analysis of an IgG sample under 'close to meniscus depletion' speed conditions: it is clear that dimers/trimers are only retained in the solution column analysed to the level of <1% in each case – but they are detectable and quantifiable.

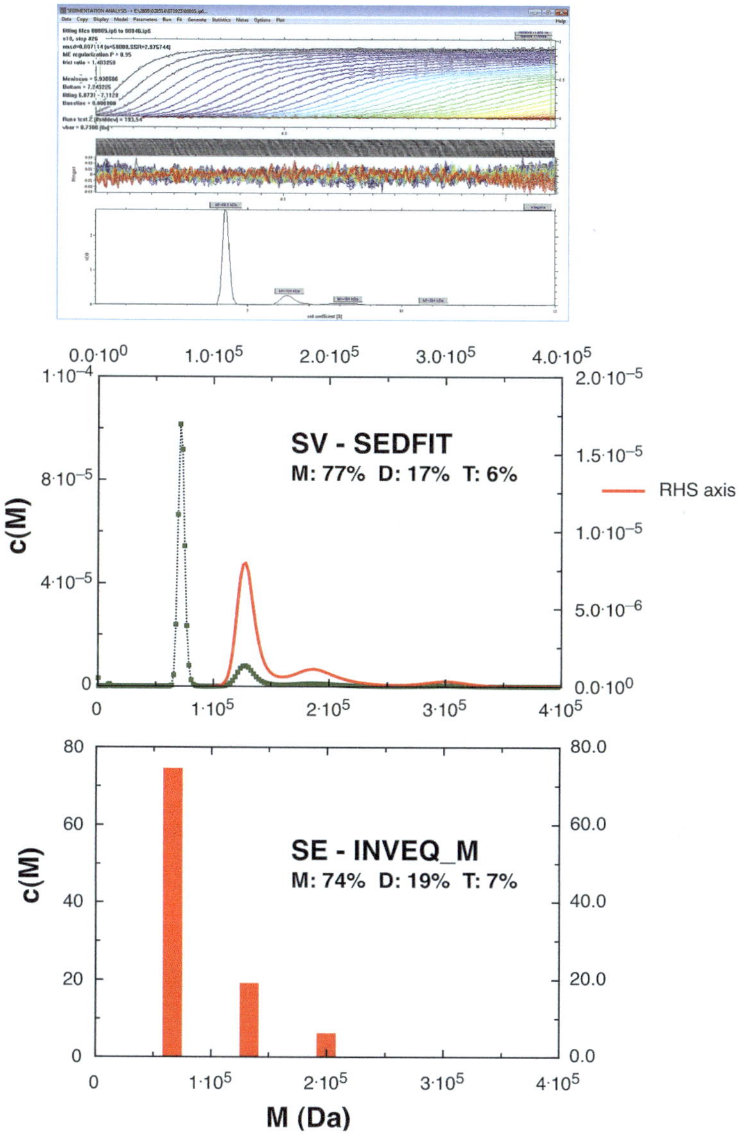

Fig. 25.5 A comparison between the results for the estimation of monomer, dimer and trimer levels in a solution of BSA at ~2 mg/ml: upper panel – sedimentation diagram with SEDFIT fit' central panel – c(M) profile from SEDFIT analysis of this sedimentation velocity run, with % estimates for the level of the oligomeric species present; bottom panel – histogram of the estimates of the level of the three species present obtained by analysis of a sedimentation equilibrium run via the M_INVEQ routine

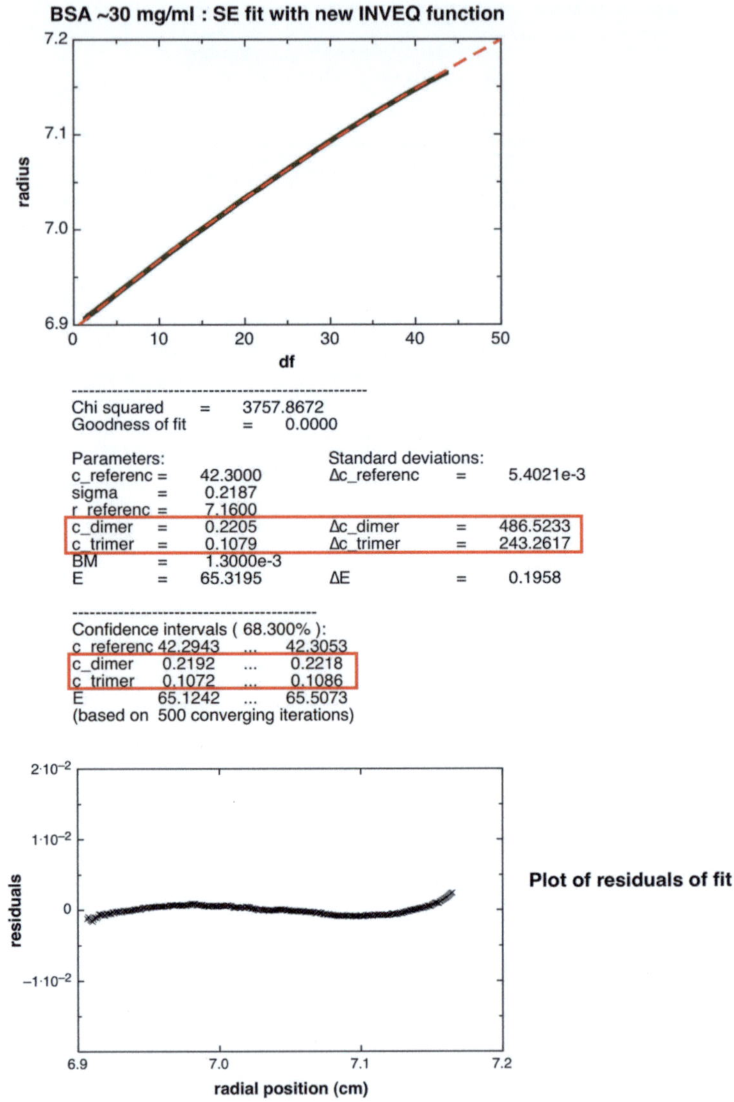

Fig. 25.6 Plot of the sedimentation equilibrium data of BSA ~30 mg/ml and of the fit made using M_INVEQ. The residuals of the fit, shown above the main plot, can be interpreted to show that over all the points within the data set the SEM (= 0.006 fringe in this case) of the residuals in fringe number accords well with results given for other systems by Ang and Rowe (2010)

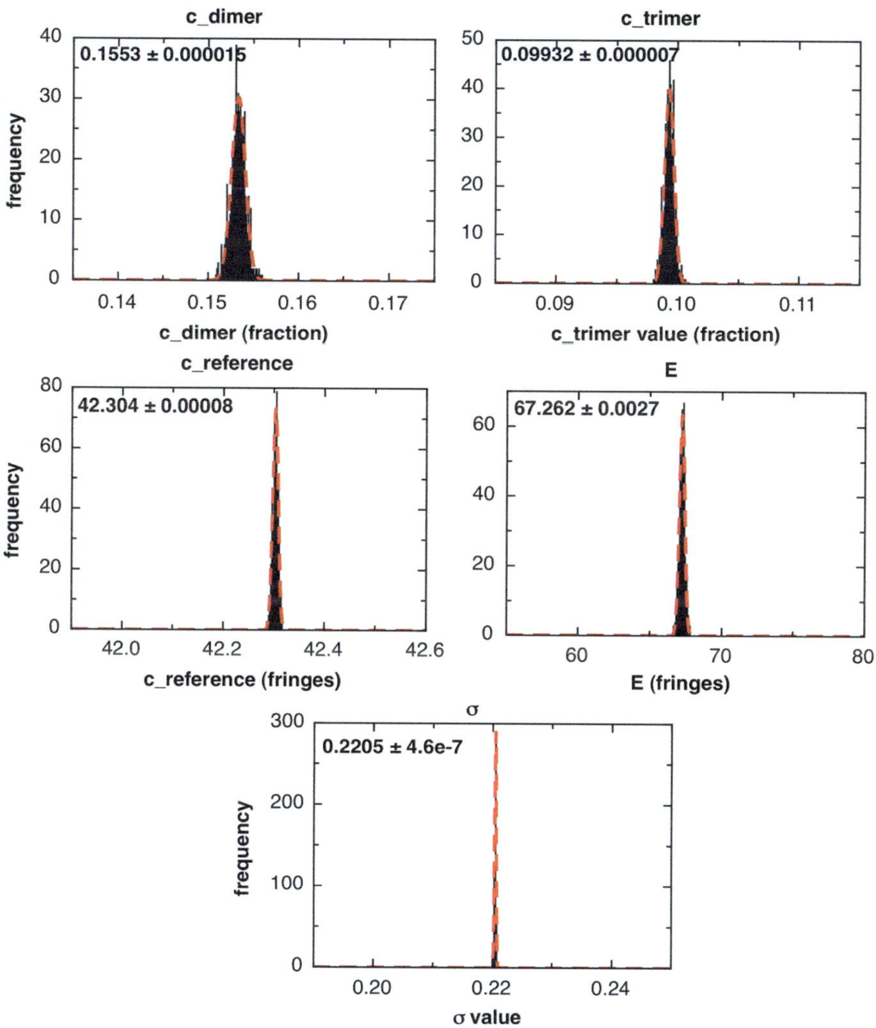

Fig. 25.7 Plots of the distribution of the 500 estimates of the parameters computed in the error analysis employed in the final Levenberg-Marquardt fit (Fig. 25.6). The standard errors displayed in each panel are derived from fitting a single Gaussian function to the values in the distribution plotted

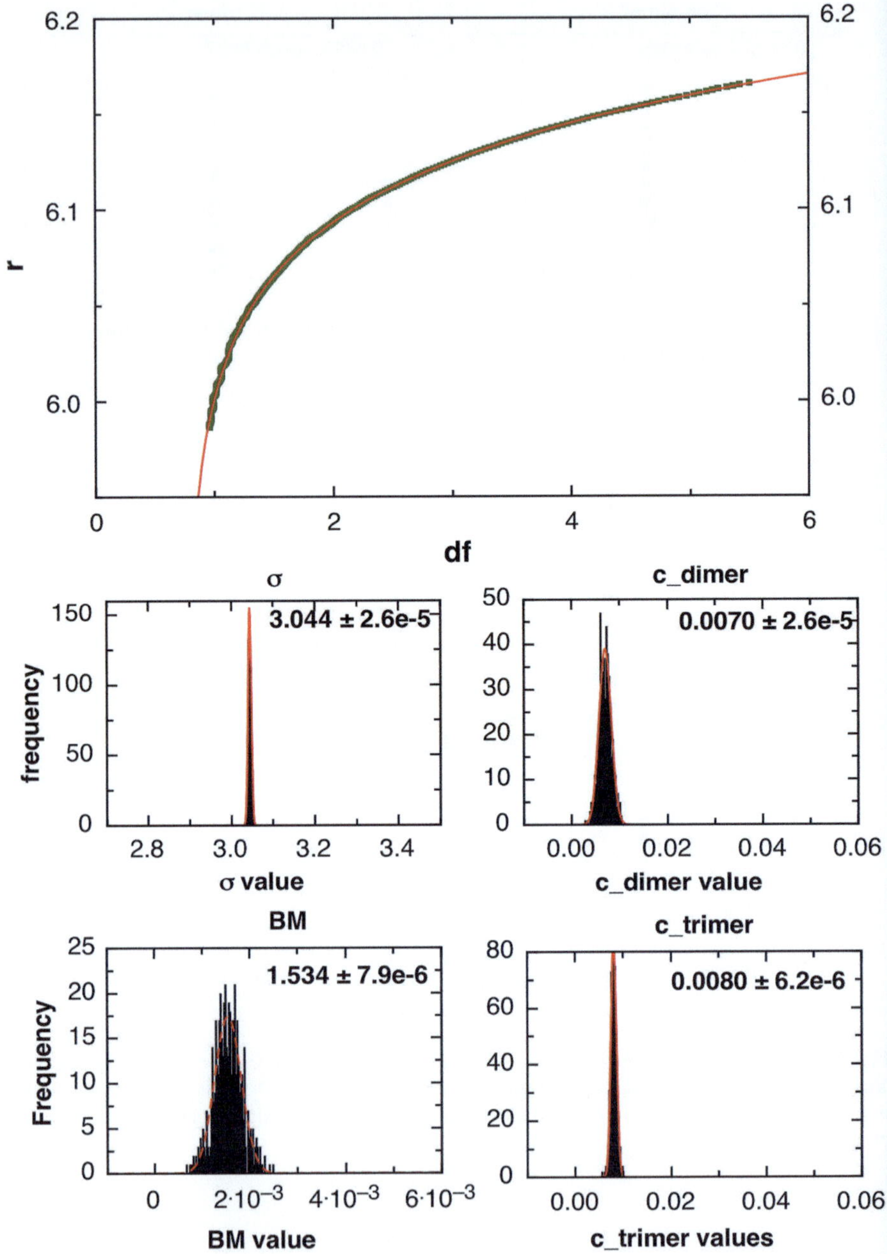

Fig. 25.8 Plot, fit and M_INVEQ analysis of SE data for a glycosylated IgG, SMYL (Besong 2013). The data set is taken from the inner channel of a multi-channel run. Whether the σ value is fixed (based upon the known molecular weight and computed partial specific volume) or floated the outcome is essentially the same: as can be seen – in this case from the confidence intervals – that there is less than 1 % of dimer or trimer present, although the actual presence of a small amount of these oligomers is highly probable

25.4 Discussion

The use of AUC analysis for the characterization of high concentration solutions of biopolymers, at least via basic SV/SE methods in an analytical rotor, has over the years been extremely limited. The difficulties, mostly optical, encountered when working in SV mode has long been appreciated: and although there is now sound theory to underpin this mode of analyses, at least in the absence of un-suppressed charge effects (Sect. 25.2.2 above), the one less usual parameter which can be returned (V_s – the ‘swollen specific volume’) is not very simple to interpret.

We are now starting to see however that the use of SE could be a more productive and alternative approach. The recently published MultiSig algorithm (Gillis et al. 2013) makes it possible to retrieve molecular weight profiles for polydisperse systems over an up to nine-fold size range, but its use is limited to dilute solutions, and it is not, in our experience, easily applicable even under those conditions to the study of a small range of well defined species, as is the case with oligomeric systems. Embarking on a totally new approach, building upon an algorithm already defined and broadly applied for estimation of ultra-weak K_a values, we have developed M_INVEQ, which we show by both detailed simulation and application to real systems can return

1. An ideal molecular weight value from a single SE scan
2. An ideal weight-averaged molecular weight value for a narrow-distribution polymer (not shown)
3. An ideal molecular weight value for the base species in an oligomerizing system, and an estimate of good precision for the total solute mass and the relative proportion of the various oligomers at any cell radial position: for the latter one would normally use the ‘hinge point’ value, which is referenced to the sample as cell loaded

The excellent level of precision demonstrated by simulation and practice should not be surprising: the total signal being analysed is by definition high, so providing that the equation being fitted is algebraically correct, then *assuming the fit to be stable*, the parameter values returned will be closely defined. It may be, of course, that the use of a single (BM) term to cover non-ideality is insufficient. In this case, it is trivial to insert an additional term (CM) from the virial expansion: it is already known that this term is very small, but the INVEQ algorithm which is the basis of M_INVEQ has been shown to be capable of detecting same (Ang and Rowe 2010). Returning to the issue of stability, we do find that in some cases the fit can be unstable, and this seems to be correlated with fully floating the virial term. However, simple analysis shows that a $\pm 10\ \%$ uncertainty in BM results in only a $\pm \sim 1\ \%$ change in σ value returned, the matter is readily tackled by close ‘bracketing’ of parameter estimates. A more surprising fact which has emerged, however, is that the precision of our analyses performed on real data can clearly be superior to that which we find in simulation (cf the results shown in Figs. 25.4 and 25.5). How can this be possible? The suggestion that we tentatively offer is that what we see is a consequence of a difference in structure of the noise in the two cases. For simulation we follow general custom and add pure ‘shot noise’, where there is no correlation of any sort between the sign/amplitude of noise added to the $(i+1)$-th term, as compared to the i-th term. But it has been established (Ang and Rowe 2010) that real SE data has an error structure capable of being resolved, using Fourier analysis, into a major term (LFAN – low frequency anharmonic noise) and a minor term (true ‘shot noise’). The latter has an r.m.s. amplitude of ± 0.002 fringe. There is thus an overall strong correlation between the total noise found in successive data values, the set of which may perhaps be more susceptible to fitting to a continuous function than is the case with ‘shot noise only’, as in our simulations. We do not, unfortunately, have an obvious way of simulating LFAN.

A concern might be that we have not considered the possibility of a small change in partial specific volume being associated with high solute concentration. This we consider to be unlikely – it is long-established practice to assume that no such change occurs. We do find that the σ value returned for BSA is greater at 30 mg/ml solute than in dilute solution – but only to an extent that would be expected on the basis of an increased solution density, as this term is incorporated into a definition of σ (Eq. 25.3).

We note above (Sect. 25.3.1) that our fitting routines can give rise to a range of extreme ‘outliers’ in parameter estimates returned. This necessitates the use of a full, 500-fold iteration error analysis. On occasion however even this may fail, as the fit becomes (as we note above) unstable, exploring ranges of parameter values which make no sense. To circumvent this type of outcome we now constrain certain critical parameter values (i.e. we specify for the fit a range within which values can be explored). This is trivial within the software package used, but may not be possible across a range of fitting/graphing software.

An interesting fact, which becomes apparent in practical work, is that surprisingly low levels of σ (i.e. of rotor speed) can be employed. This limits the problems associated with excessively steep gradients developing. Our practical results with BSA were obtained with a σ value as low as 0.22. This very positive outcome relates to the fact that the signal values, cell meniscus to cell base, are numerically very large, even at such low σ values. Finally, we must address the simple question: “Is it of importance to be able to characterize the oligomeric state of concentrated solutions of proteins?” There is no way of knowing, until a range of systems have been analysed. But given the increasing use of high concentration formulations in bio/pharma practice, it is clear that even ‘negative’ outcomes – i.e. study of a dilution suffices – would be of value.

We believe that we have validated the potential use of the AUC in SE mode for the evaluation of oligomeric content and potentially more. For the widely used ‘standard protein’ BSA, the results given above (Sect. 25.3.2.1) at ~30 mg/ml concentration show a small but significant increase in level of trimer (from 7 to 9.9 %), which could be interpreted in terms of a very weak self-association of dimer and monomer. It will be interesting to see this topic pursued further, and especially interesting to discover whether effects related to high concentration can be observed in solutions of Mabs/IgG samples.

References

Ang S, Rowe AJ (2010) Evaluation of the information content of sedimentation equilibrium data in self-interacting systems. Macromol Biosci 10:798–807

Batchelor GK (1972) Sedimentation in a dilute dispersion of spheres. J Fluid Mech 52:245–268

Besong (2013) On the hydrodynamics of antibody glycoforms, pure protein and polysaccharide systems in response to bioprocessing. PhD dissertation University of Nottingham, UK

Brady JF, Durlovsky LJ (1988) The sedimentation rate of disordered suspensions. Phys Fluids 31:717–727

Burgers JM (1941) On the influence of the concentration of a suspension upon the sedimentation velocity (in particular for a suspension of spherical particles). Proc Acad Sci Amsterdam 44:1045–1051 & 1177–1184

Burgers JM (1942) On the influence of the concentration of a suspension upon the sedimentation velocity (in particular for a suspension of spherical particles). Proc Acad Sci Amsterdam 45:9–16 & 126–128

Cheng PY, Schachman HK (1955) The validity of the Einstein viscosity law and Stokes law of sedimentation. J Polym Sci 16:19–30

Dam J, Schuck P (2005) Sedimentation velocity analysis of heterogeneous protein-protein interactions: sedimentation coefficient distributions $c(s)$ and asymptotic boundary profiles from Gilbert-Jenkins theory. Biophys J 89:651–666

Errington N, Mistry P, Rowe AJ (2002) Protein hydration varies with protein crowding and with applied pressure: a sedimentation velocity study. Progr Colloid Polym Sci 119:58–63

Fernández C, Minton AP (2009) Static light scattering from concentrated protein solutions II: experimental test of theory fro protein mixtures and weakly self-associating proteins. Biophys J 96:1992–1998

Gilbert GA, Gilbert LM (1980) Ultracentrifuge studies of interactions and equilibria: impact of interactive computer modeling. Biochem Soc Trans 8:520–522

Gillis RB, Adams GG, Heinze T, Nikolajski M, Harding SE, Rowe AJ (2013) MultiSig: a new high-precision approach to the analysis of complex biomolecular systems. Eur Biophys J 42:777–786

Nikolajski M, Adams GG, Gillis RB, Besong DTB, Rowe AJ, Heinze T, Harding SE (2013) Protein-like fully reversible tetramerisation and super-association of an aminocellulose. Sci Rep 4:3861

Patel TR, Harding SE, Ebringerova A, Deszczynski M, Hromadkova Z, Togola A, Paulsen BS, Morris GA, Rowe AJ (2007) Weak self-association in a carbohydrate system. Biophys J 93:741–749

Rowe AJ (1977) The concentration dependence of transport processes: a general description applicable to the sedimentation, translational diffusion, and viscosity coefficients of macromolecular solutes. Biopolymers 16:2595–2611

Rowe AJ (1979) An extended theory of the concentration dependence of sedimentation. Euromech 120:5.2

Rowe AJ (1992) The concentration dependence of sedimentation. In: Harding SE, Horton JC, Rowe AJ (eds) Analytical ultracentrifugation in biochemistry & polymer science. The Royal Society of Chemistry, London

Rowe AJ (2011) Ultra-weak reversible protein-protein interactions. Methods 54:157–166

Schuck P, Gillis RB, Besong TMD, Almutairi F, Adams GG, Rowe AJ, Harding SE (2014) SEDFIT-MSTAR: molecular weight and molecular weight distribution analysis of polymers by sedimentation equilibrium in the ultracentrifuge. Analyst 139:79–92

Soda A, Fujimoto T, Nagasawa M (1967) Johnston-Ogston effect in sedimentation. J Phys Chem 71:4274–4281

Spruijt E, Beisheuvel PM (2014) Sedimentation dynamics and equilibrium profiles in multicomponent mixtures of colloidal particles. J Phys Condens Matter 26:1–16

Stafford WF, Sherwood PJ (2004) Analysis of heterologous interacting systems by sedimentation velocity: curve fitting algorithms for estimation of sedimentation coefficients, equilibrium and kinetic constants. Biophys Chem 108:231–243

Sule SV, Sukumar M, Weiss WF, Marcelino-Cruz AM, Sampie T, Tessler PM (2011) High-throughput analysis of concentration-dependent antibody self-association. Biophys J 10:1749–1757

Part VIII
New Applications of AUC

Chapter 26
Detection and Quantitative Characterization of Macromolecular Heteroassociations via Composition Gradient Sedimentation Equilibrium

Allen P. Minton

Abstract A novel approach to the detection and characterization of macromolecular heteroassociations is described, based upon a combination of short-column sedimentation equilibrium measurements and the utilization of a composition gradient spanning the entire range of mole fractions of interacting components. Simulations are presented to demonstrate that the method described provides a high degree of discrimination between alternative models for heteroassociation equilibria. Comparison with experimental protocols utilized previously shows that the new approach provides equivalent information with less sample and a considerably shorter duration of centrifugation.

Keywords Protein-protein interactions • Protein-nucleic acid interactions • Reversible associations • Analytical ultracentrifugation • Short-column centrifugation

26.1 Introduction

The measurement and analysis of sedimentation-diffusion equilibrium (sedimentation equilibrium or SE for short) is one of the earliest biophysical techniques applied to the study of biological macromolecules. Initially employed to determine molecular weights in solution (Svedberg and Pedersen 1940; Yphantis 1963), SE has subsequently been exploited as a powerful tool for the study of macromolecular

A.P. Minton (✉)
Laboratory of Biochemistry and Genetics, National Institute of Diabetes and Digestive and Kidney Diseases, National Institutes of Health, U.S. Department of Health and Human Services, Building 8, Room 226, Bethesda, MD, USA
e-mail: minton@helix.nih.gov

S. Uchiyama et al. (eds.), *Analytical Ultracentrifugation*,
DOI 10.1007/978-4-431-55985-6_26

interactions in dilute, semi-dilute, and concentrated solutions (see, e.g., Adams and Williams 1964; Teller 1973; Zorrilla et al. 2004).

The characterization of macromolecular heteroassociations via measurement and analysis of SE in dilute solution has been extensively analyzed (Hsu and Minton 1991; Minton 1997; Rivas et al. 1999) and employed to good effect in a number of published studies (Bailey et al. 1996; Ghirlando 2011; Philo et al. 1996; Vistica et al. 2004). The experimental protocol generally employed is as follows: (1) Multiple solutions containing two chemical components (e.g., two different proteins or a protein and a nucleic acid), called A and B, in different molar ratios are prepared. (2) Each of these solutions is loaded into a sector-shaped centrifuge cell in an amount, typically on the order of 100 ul, sufficient to create a solution column with a height of at least 3 mm. (3) The rotor containing these solutions is centrifuged at a rotor speed sufficient to attain a readily measurable equilibrium gradient of signal (absorbance at one or more wavelengths or refractive index) for a time sufficient to attain sedimentation equilibrium. After the signal gradient is recorded, the rotor speed is increased and the procedure repeated until equilibrium data for all samples are acquired at two or three rotor speeds. (4) An equilibrium model is proposed specifying the concentrations of each of a number of macromolecular complexes as a function of the total concentrations of A and B. (5) This model is combined with equations specifying the concentration and signal gradients of each sedimenting species at SE, together with conditions specifying conservation of the known total amounts of A and B in each solution sample. The global model thus obtained provides a quantitative specification of the equilibrium dependence of each measured signal upon radial distance and rotor speed in each solution, as a function of the number of complexes proposed in each model and the equilibrium relations between them (Teller 1973). (6) The global model is then fit to all of the equilibrium gradients simultaneously by allowing the equilibrium constants in the model to systematically vary to obtain a minimum value of the sum of squared residuals.

In the present report we introduce an alternative implementation of SE for the characterization of macromolecular heteroassociations that reduces the requirement for sample, substantially reduces the time required to obtain all necessary data, increases the experimenter's ability to discriminate between alternative models, and greatly simplifies analysis and global modeling of the data. In the following section, we provide the quantitative relationships underlying the new approach and present simulations illustrating the sensitivity of the analysis to variation in reaction schemes and model parameters. The analysis described is then applied to data obtained in a recent study of the interaction between two proteins (in preparation).

26.2 Experimental Procedure

Stock solutions of macromolecular solutes A and B in buffer are prepared containing w/v concentrations w_A and w_B, respectively. A series of solution mixtures are then prepared by mixing volume fraction f_A of solution A and volume fraction $(1-f_A)$ of solution B, where $0 \leq f_A \leq 1$. In principle, the more mixtures of different composition, the better, but a minimum of six mixtures is recommended. The mixtures so prepared will contain A at a concentration of $f_A \times w_A$ and B at a concentration of $(1-f_A) \times w_B$. A quantity of each stock solution and solution mixture sufficient to create a sample column of 1–1.5 mm is loaded into one sample well of a six-channel analytical centerpiece (Yphantis 1963), while a slightly larger volume of buffer is loaded into the corresponding reference well. As many as 24 such samples, loaded into an eight-hole analytical rotor, may then be simultaneously centrifuged in the analytical ultracentrifuge at a speed estimated to provide a readily measurable gradient of signal (absorbance or refractive index) until equilibrium is attained (Teller 1973). The duration of centrifugation required to reach SE is roughly proportional to the square of the length of the sample column, so a 1.5 mm sample column is expected to equilibrate approximately four times more rapidly than a 3 mm column (Laue 1992; Yphantis 1960). Following recording of the equilibrium gradient, the rotor speed may be increased and centrifugation continued until a second equilibrium gradient is attained and recorded. If desired, this procedure may be repeated at an additional higher rotor speed. The acquisition of data at two or three rotor speeds provides a check on attainment of sedimentation equilibrium and an improvement in the reliability of data analysis, as discussed below.

Following data acquisition, the gradient of signal $S(r)$ is transformed into the linearized gradient ln $S(r^2)$. An example of a linearized scan of one six-hole centerpiece is shown in Fig. 26.1a. A straight line is then fit to the data obtained from each individual solution sample, as shown in Fig. 26.1b. The slope of the best-fit straight line is proportional to the sample-average signal-average buoyant molar mass, defined by

$$M^*_{S,av} = \frac{2RT}{\omega^2}\frac{\mathrm{d}\ln S}{\mathrm{d}r^2} \tag{26.1}$$

where R denotes the molar gas constant, T the absolute temperature, and ω the angular velocity of the rotor. Provided that ln S is sufficiently well described as a linear function of r^2, as is usually the case for a short sample column at sedimentation equilibrium, the value of $M^*_{S,av}$ so calculated may be regarded as a property characteristic of the loading concentrations of A and B, denoted by w_A and w_B, respectively, in *w/v* units, or, alternatively, f_A. The results of the experiment are then recorded as a table of $\{w_A, w_B, M^*_{S,av}\}$.

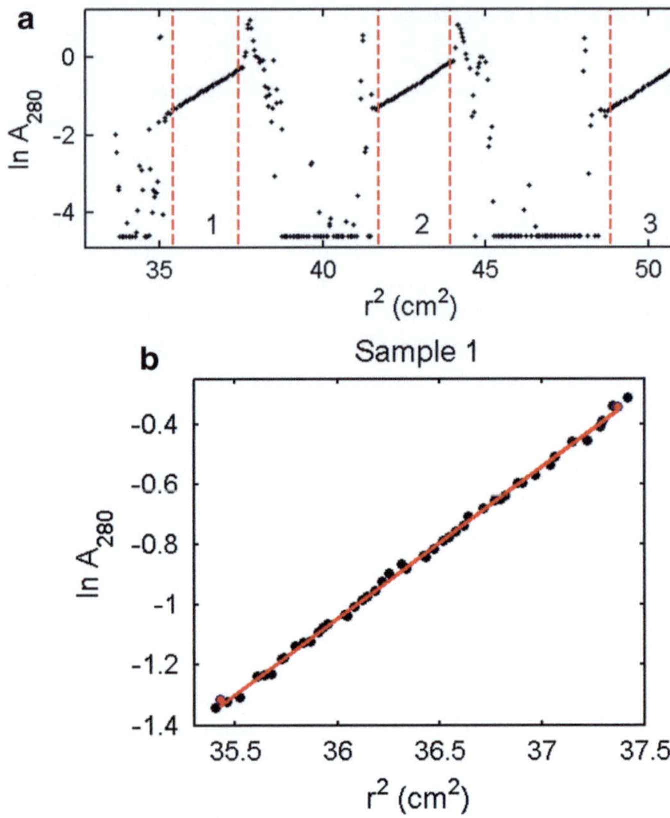

Fig. 26.1 (**a**) Linearized scan of a sample cell containing a three-hole centerpiece with three sample solutions. Unpublished data of Arai et al. (in preparation). (**b**) Best fit of a straight line to the data obtained from the leftmost of the three samples in 1A

26.3 Analysis of the Dependence of $M^*_{S,av}$ upon w_A and w_B

Consider a solution containing multiple sedimentable species having a composition A_iB_j.

The signal-average buoyant molar mass is expressed as a function of solution composition by (Hsu and Minton 1991; Minton 1997)

$$M^*_{s,av} = \frac{\sum_{i,j} M^*_{ij} s_{ij}}{\sum_{i,j} s_{ij}} \tag{26.2}$$

where M^*_{ij} and s_{ij} respectively denote the buoyant mass and specific signal (signal per unit *w/v* concentration) of species A_iB_j. The buoyant mass of A_iB_j is given by

$$M^*_{ij} = iM^*_{\mathrm{A}} + jM^*_{\mathrm{B}} = iM_{\mathrm{A}}\left(\frac{\mathrm{d}\rho}{\mathrm{d}w_{\mathrm{A}}}\right)_{\mu} + jM_{\mathrm{B}}\left(\frac{\mathrm{d}\rho}{\mathrm{d}w_{\mathrm{B}}}\right)_{\mu} \tag{26.3}$$

where M_X and $(\mathrm{d}\rho/\mathrm{d}w_X)_\mu$ respectively denote the molar mass of component *X* (*A* or *B*) and the density increment of component *X* measured at dialysis equilibrium (Eisenberg and Casassa 1964). The specific signal of A_iB_j is given by

$$s_{ij} - (is_{\mathrm{A}}M_{\mathrm{A}} + js_{\mathrm{B}}M_{\mathrm{B}})\, c_{ij} \tag{26.4}$$

where s_X denotes the specific signal of pure component *X*. For example, if the signal is absorbance at a particular wavelength, then s_X is the specific extinction coefficient of *X* at that wavelength. It is noted that Eq. (26.4) is predicated on the assumption that the signals of A and B are independent of the states of association of both components. When the specific signal is refractive increment, it is safe to assume that this assumption is true. When the signal is absorbance or fluorescence intensity – especially in the case of the latter signal – the assumption should be tested experimentally prior to undertaking the present analysis. A simple test for absorbance is described in the appendix.

With the exception of species concentrations, c_{ij}, every quantity on the right-hand sides of Eqs. (26.2), (26.3), and (26.4) may be measured independently in solutions containing either pure A or pure B. Thus modeling the dependence of $M^*_{S,av}$ upon w_{A} and w_{B} requires only a model specifying each of the c_{ij} as functions of w_{A} and w_{B}. Such models have been described previously in numerous publications (see, e.g., Minton 1997), but for convenience and uniformity of notation, we recapitulate briefly here.

The concentrations of each sedimenting species are related to the concentrations of pure monomeric A and B (c_{10} and c_{01}, respectively) by an equilibrium constant:

$$K_{ij} = \frac{c_{ij}}{c^i_{10}c^j_{01}} \tag{26.5}$$

The total concentrations of A and B are then given by

$$c_{\mathrm{A,tot}} = \frac{w_{\mathrm{A}}}{M_{\mathrm{A}}} = \sum_{i,j} ic_{ij} = \sum_{i,j} iK_{ij}c^i_{10}c^j_{01} \tag{26.6a}$$

and

$$c_{\mathrm{B,tot}} = \frac{w_{\mathrm{B}}}{M_{\mathrm{B}}} = \sum_{i,j} jc_{ij} = \sum_{i,j} jK_{ij}c^i_{10}c^j_{01} \tag{26.6b}$$

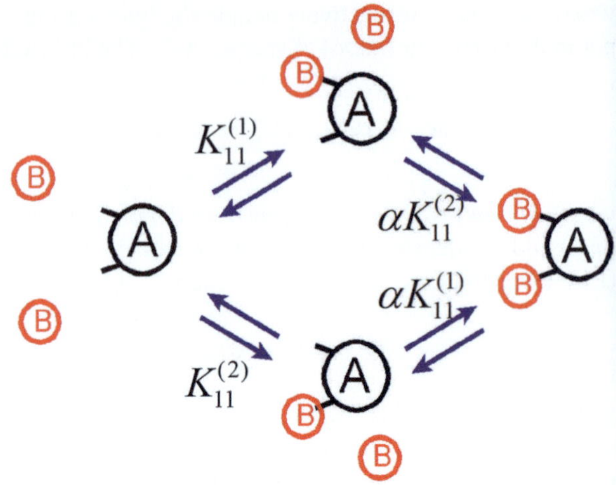

Fig. 26.2 Schematic depiction of the equilibrium association between univalent and bivalent macromolecules, as described in the text

Given values of w_A, w_B, M_A, M_B, and all of the K_{ij}, Eqs. (26.6a) and (26.6b) may be solved analytically or numerically for the values of the two remaining unknowns, c_{10} and c_{01}, and given these values, Eq. (26.5) may be used to solve for the values of the remaining c_{ij}. Then Eqs. (26.2), (26.3), and (26.4), together with independently measured values of the specific signals and density increments of pure A and B, may be used to calculate the value of $M^*_{S,av}$. We illustrate this approach using the specific model example described below.

Let A contain two separate sites which may bind B with distinct affinities. The binding of B to one site may either facilitate or inhibit binding of B to the second site. This model, illustrated schematically in Fig. 26.2, may be expressed quantitatively as follows:

$$K_{11} = K_{11}^{(1)} + K_{11}^{(2)} = c_{11}/c_{10}c_{01} \tag{26.7a}$$

$$K_{12} = \alpha K_{11}^{(1)} K_{11}^{(2)} = c_{12}/c_{10}c_{01}^2 \tag{26.7b}$$

where $K_{11}^{(1)}$ and $K_{11}^{(2)}$ denote site-specific binding constants as indicated in Fig. 26.2, and the parameter α quantifies the extent to which binding of B to one site on A facilitates ($\alpha > 1$) or inhibits ($\alpha < 1$) binding of B to the second site. It follows that

$$c_{A,tot} = \frac{w_A}{M_A} = c_{10} + K_{11}c_{10}c_{01} + K_{12}c_{10}c_{01}^2 \tag{26.8a}$$

and

$$c_{B,tot} = \frac{w_B}{M_B} = c_{01} + K_{11}c_{10}c_{01} + 2K_{12}c_{10}c_{01}^2 \tag{26.8b}$$

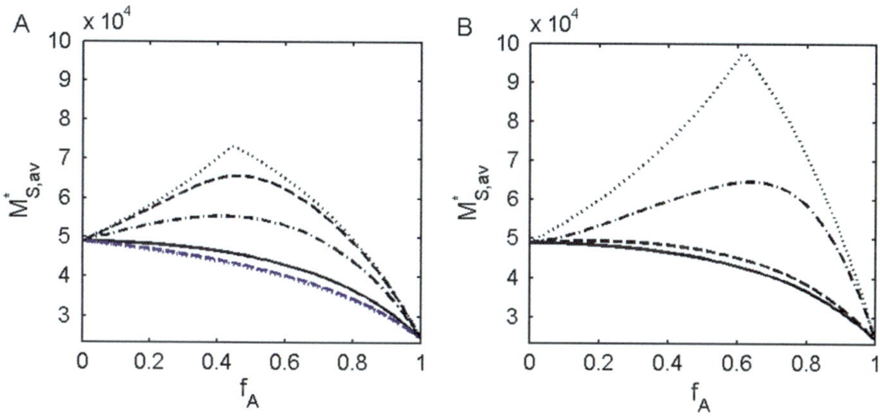

Fig. 26.3 Dependence of $M^*_{S,av}$ upon f_A calculated using Eqs. (26.2), (26.3), (26.4), (26.7), and (26.8) with the following values of experimentally measurable constants: $M_A = 100$ kg, $M_B = 200$ kg, $\varepsilon_A = 1.9$ L/g, $\varepsilon_B = 1.2$ L/g, $d\rho/dw_A = d\rho/dw_B = .245\ cm^3/g$, $w_{A,stock} = 0.25g/L$, $w_{B,stock} = 0.4g/L$, and the following equilibrium constants. (**a**) $K^{(2)}_{11} = \alpha = 0$ for all curves, and $K^{(1)}_{11} = 0$ (*blue dotted*), $10^4\ M^{-1}$ (*blue dash*), $10^5\ M^{-1}$ (*black solid*), $10^6\ M^{-1}$ (*black dot-dash*), $10^7\ M^{-1}$ (*black dash*), and $10^{10}\ M^{-1}$ (*black dotted*). (**b**) $K^{(1)}_{11} = 10^5M^{-1}$ and $K^{(2)}_{11} = \alpha = 0$ (*solid*), $K^{(2)}_{11} = K^{(1)}_{11} = 10^5M^{-1},\ \alpha = 0$ (*dash*), $K^{(2)}_{11} = K^{(1)}_{11} = 10^5M^{-1},\ \alpha = 100$ (*dot-dash*), and $K^{(2)}_{11} = K^{(1)}_{11} = 10^6M^{-1},\ \alpha = 100$ (*dotted*)

We shall use Eqs. (26.2), (26.3), and (26.4) and Eqs. (26.7a and b) and (26.8a and b) to calculate the dependence of $M^*_{S,av}$ upon f_A for several sets of experimentally measurable constants and model parameters, given in the caption to Fig. 26.3. The calculated dependence of $M^*_{S,av}$ upon f_A for binding of B to a single site on A is plotted for a range of binding constants in Fig. 26.3a. The calculated dependence of $M^*_{S,av}$ upon f_A for binding of B to both one and two sites on A is plotted for a range of relative amounts of binding to one and two sites in Fig. 26.3b. It may be seen that both the shape and maximum value of the plot are sensitive to variation in model parameters over a wide range, indicating that least-squares modeling of different equilibrium models to a sufficiently comprehensive and precise experimental measurement of the dependence of $M^*_{S,av}$ upon f_A should enable both determination of the simplest model capable of providing a satisfactory fit and a reasonably accurate determination of the values of the model parameters providing the best fit of that model.

26.4 Experimental Test

The technique of composition gradient sedimentation equilibrium has been utilized, together with other techniques, to characterize the interaction between two proteins, peroxiredoxin (Prx), and NADH oxidase (Nox) from *Amphibacillus xylanus* (Kitano

Fig. 26.4 Experimentally measured dependence of $M^{*}_{S,av}$ upon f_{Prx}. *Data points* indicate the mean of six independent measurements made at three rotor speeds in each of two replicate samples. *Error bars* correspond to ±2 standard deviations of measurement. The *plotted curve* represents the best fit of a model for 1-1 association, i.e., $K_{11}^{(2)} = \alpha = 0$. Best-fit parameter values and standard errors of estimate are $M_{Prx} = 105$ (+5, −1) kg, $M_{Nox} = 183$ (+7, −5) kg, and $\log_{10}K_{11}^{(1)}$ $(M^{-1}) = 5.0$ (+0.2, −0.4)

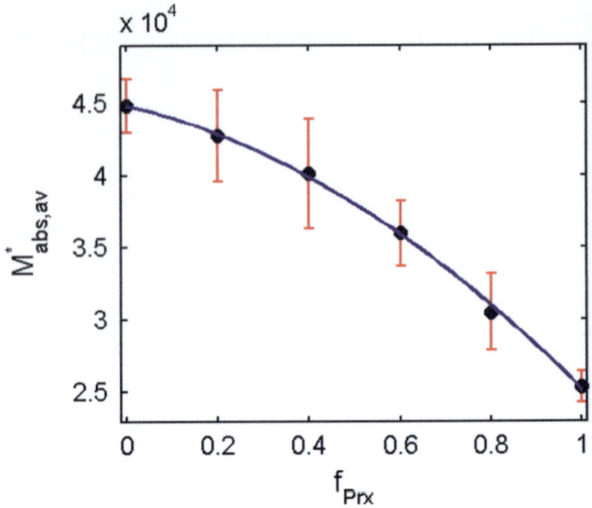

et al. 1999). A full account of this study will be presented elsewhere (in preparation), but some results relevant to the present purpose are presented here with the permission of coauthors of the full report. Prior to the sedimentation equilibrium experiments, the density increments and specific extinction coefficients at 280 nm of solutions of pure Nox and pure Prx were measured via conventional methods. It was ascertained that the extinction coefficients of both proteins did not change significantly when the proteins were mixed at the highest concentrations attained in the study. Solutions containing Nox, Prx, and mixtures containing volume fractions of Prx solution equaling 0.2, 0.4, 0.6, and 0.8 were then prepared. Replicates of all six solutions were centrifuged as described above, and the absorbance gradient of each solution at sedimentation equilibrium was recorded at three rotor speeds. The signal-average buoyant molar mass of each solution was calculated as described above, and a summary of the results is shown in Fig. 26.4, together with a curve calculated using the model described by Eqs. (26.7a and b) and (26.8a and b) together with the best-fit parameter values given in the figure caption. These values are in excellent agreement with results obtained from analysis of independent measurements of the composition dependence of static light scattering, to be described in the full report.

26.5 Discussion

The strategy of cross-gradient variation of the concentrations of interacting components was first applied successfully to the analysis of macromolecular heteroassociations via measurement of static light scattering (Attri and Minton 2005). The present report describes the application of this strategy to the measurement and analysis of macromolecular heteroassociations via short-column SE. While the conventional approach described in the Introduction section can yield the information desired – stoichiometry of significant complexes and the equilibrium constants for their formation – the use of concentration cross-gradients in conjunction with short sample columns yields the same information with considerably less sample, a two- to fourfold reduction in the duration of centrifugation, and a much simpler analysis of data. We are planning to test the technique using eight-channel short-column centerpieces (Yphantis 1960), and if satisfactory data can be obtained using these centerpieces, sample requirements and the duration of experiments will be reduced even further.

Acknowledgments The author thanks Fumio Arisaka (Tokyo Institute of Technology; present address: Nihon University) for providing experimental data in advance of publication. This research is supported by the Intramural Research Program of the National Institute of Diabetes and Digestive and Kidney Diseases.

Appendix: Verification of the Absence of Hyper- or Hypo-chromicity

The stock solutions of A and B are placed in the separated compartments of a tandem absorbance cell (515A, NSG Precision Cells, Farmingdale, NY), and a UV-visible spectrum is recorded, which is the sum of the spectra of the two separate solutions. Then the contents of the two solutions are mixed via inversion of the tandem cell, and the spectrum is rerecorded. Identity of the spectra before and after mixing is sufficient to indicate the absence of either significant heteroassociation or significant hyper- or hypo-chromism upon heteroassociation. In either case, the analysis presented in the text is valid over the range of compositions established by mixing the two stock solutions in any proportion.

References

Adams ET Jr, Williams JW (1964) Sedimentation equilibrium in reacting systems. II. Extensions of the theory to several types of association phenomena. J Am Chem Soc 86:3454–3461

Attri AK, Minton AP (2005) Composition gradient static light scattering: a new technique for rapid detection and quantitative characterization of reversible macromolecular heteroassociations in solution. Anal Biochem 346:132–138

Bailey MF, Davidson BE, Minton AP, Sawyer WH, Howlett GJ (1996) The effect of self-association on the interaction of the *Escherichia coli* regulatory protein TyrR with DNA. J Mol Biol 263:671–684

Eisenberg H, Casassa EF (1964) Thermodynamic analysis of multicomponent solutions. Adv Protein Chem 19:287–395

Ghirlando R (2011) The analysis of macromolecular interactions by sedimentation equilibrium. Methods 54:145–156

Hsu C, Minton AP (1991) A strategy for efficient characterization of macromolecular heteroassociations via measurement of sedimentation equilibrium. J Molec Recognition 4:93–104

Kitano K, Niimura Y, Nishiyama Y, Miki K (1999) Stimulation of peroxidase activity by decamerization related to ionic strength: AhpC protein from *Amphibacillus xylanus*. J Biochem 126:313–319

Laue TM (1992) Short column sedimentation equilibrium analysis for rapid characterization of macromolecules in solution. Beckman Instruments Inc., Spinco Business Unit, Application Note DS-835

Minton AP (1997) Alternative strategies for the characterization of interactions in multicomponent solutions via measurement of sedimentation equilibrium. Prog Polym Sci 107:11–19

Philo JS, Aoki KH, Arakawa T, Narhi LO, Wen J (1996) Dimerization of the extracellular domain of the erythropoietin receptor by EPO: one high-affinity and one low-affinity interaction. Biochemistry 35:1681–1691

Rivas G, Stafford W, Minton AP (1999) Characterization of heterologous protein-protein interactions using analytical ultracentrifugation. Methods 19:194–212

Svedberg T, Pedersen KO (1940) The ultracentrifuge. Oxford Press, Oxford

Teller DC (1973) Characterization of proteins by sedimentation equilibrium in the analytical ultracentrifuge. Methods Enzymol 27:346–441

Vistica J, Dam J, Balbo A, Yikilmaz E, Mariuzza RA, Rouault TA, Schuck P (2004) Sedimentation equilibrium analysis of protein interactions with global implicit mass conservation constraints and systematic noise decomposition. Anal Biochem 326:234–256

Yphantis DA (1960) Rapid determination of molecular weights of peptides and proteins. Ann N Y Acad Sci 88:586–601

Yphantis DA (1963) Equilibrium ultracentrifugation of dilute solutions. Biochemistry 3:297–317

Zorrilla S, Jiménez M, Lillo P, Rivas G, Minton AP (2004) Sedimentation equilibrium in a solution containing an arbitrary number of solute species at arbitrary concentrations: theory and application to concentrated solutions of ribonuclease. Biophys Chem 108:89–100

MIX
Papier aus verantwortungsvollen Quellen
Paper from responsible sources
FSC® C105338
FSC
www.fsc.org

If you have any concerns about our products,
you can contact us on
ProductSafety@springernature.com

In case Publisher is established outside the EU,
the EU authorized representative is:
Springer Nature Customer Service Center GmbH
Europaplatz 3, 69115 Heidelberg, Germany

Printed by Libri Plureos GmbH
in Hamburg, Germany